ELECTRONIC COMMUNICATIONS
Modulation and Transmission

ELECTRONIC COMMUNICATIONS
Modulation and Transmission

Robert J. Schoenbeck

Merrill Publishing Company
A Bell & Howell Information Company
Columbus   Toronto   London   Melbourne

Cover Photo: Marilyn Martin

Published by Merrill Publishing Company
A Bell & Howell Information Company
Columbus, Ohio 43216

This book was set in Century Schoolbook

Administrative Editor: Steve Helba
Developmental Editor: Don Thompson
Production Coordinator: Rex Davidson
Art Coordinator: Mark D. Garrett
Cover Designer: Cathy Watterson
Text Designer: Cynthia Brunk

Library of Congress Catalog Card Number: 87–61601
International Standard Book Number: 0–675–20473–9
Printed in the United States of America
1 2 3 4 5 6 7 8 9—92 91 90 89 88

Only a crazy man would fail to dedicate his first book to his wife. My wife's name is Kathryn Delores Schoenbeck. I call her Kate. She has sacrificed much in order that I may complete this work. This giving effort increases my already enormous indebtedness to her.

Thanks, Kate.

MERRILL'S INTERNATIONAL SERIES IN ELECTRICAL AND ELECTRONICS TECHNOLOGY

PREFACE

The following objectives for this text arise from my personal philosophy of teaching. I feel that a text should serve not only for the benefit of the reader, but also as a pedagogically sound outline for a course of instruction. A text should be sufficiently clear to enable the reader to understand the material well by its reading, with a realism that approaches hands-on experience. I also feel that a text should be more comprehensive than the course for which it is used; thus, some material can be used both as a reference source and as a source of further examples and illustrations by the student and technician. Finally, I feel that the text should be able to stand alone, with minimal need for supplemental documentation and references. I hope that each student and instructor finds that all of these objectives have been met in this text.

Of all the fields in electronics, electronic communications is perhaps the most rapidly expanding, particularly regarding new job opportunities. At one time this area of study was limited primarily to amplitude modulation (AM), frequency modulation (FM), and single sideband (SSB). Electronic communications has since grown to include radar, satellite, and microwave communications. More recently, the field has been expanded by the new technologies of lasers, fiber optics, optoelectronics, and cellular communications. Each of these communications fields has its own advantages, limitations, and peculiar problems. To address each of these special needs, the technician must be familiar and comfortable both with the basics and with each new branch of the communications field. For this reason, a solid foundation is laid in this text with three chapters on amplitude modulation, since this is essential for understanding those topics that follow.

The text has been written in an informal, conversational tone. In the schematics, every effort has been made to lead the reader step-by-step through each concept.

Mathematics have been kept to a minimum, so that the student will learn the central concept being presented rather than be intimidated by a torrent of math. Finally, review questions have been included at the end of each section to crystallize and reinforce the key concepts of that section.

The early chapters (Chapters 1, 2, and 4) introduce the principles, transmitters, and receivers of AM signals, including balanced coverage of sideband transmission (Chapter 3). A similar introduction to FM principles, transmitters, and receivers is then presented in Chapters 5, 6, and 7, followed by a practical treatment of transmission lines in Chapter 8, Smith charts in Chapter 9, and antennas in Chapter 10.

Once these basics have been covered, the text concentrates on the quickly expanding areas of microwave systems and devices (Chapter 11), wave propagation (Chapter 12), and digital communications with coverage of LAN, local area networking, and cellular communications in Chapter 13. Since many telecommunications systems are being converted to fiber optic systems, this important area is given a practical, balanced coverage in Chapter 14. Technicians must understand the principles of television to work effectively on computer monitors, as well as TVs. Thus, Chapter 15 not only explains television principles, but also includes stereo TV and HDTV (high density television). Closely related to television is the topic of satellites, which is explained in Chapter 16.

I have benefitted from the assistance of a number of people in reviewing, writing, and producing this text. I would like to take this opportunity to thank the following reviewers, who provided many helpful, constructive suggestions: Paul Bierbauer, DeVry Institute of Technology, Chicago, IL; Robert Blodgett, DeVry Institute of Technology, Lombard, IL; John Clark, ITT Technical Institute, Aurora, CO; Tony DeSilva, DeVry Institute of Technology, Columbus, OH; Anthony Iula, Sylvania Technical Institute, Waltham, MA; John Meese, DeVry Institute of Technology, Columbus, OH; Mike Reilly, Minnesota School of Business, Minneapolis, MN; Charles Schenkenberger, DeVry Institute of Technology, Decatur, GA; Terry Stivers, ITT Technical Institute, St. Louis, MO; Thomas Stultz, Sylvania Technical Institute, Waltham, MA; and Mohammad Tasooji, DeVry Institute of Technology, Los Angeles, CA.

Robert J. Schoenbeck

CONTENTS

CHAPTER NINE
SMITH CHARTS

CHAPTER FOURTEEN
FIBER OPTICS   517

ELECTRONIC COMMUNICATIONS
Modulation and Transmission

CHAPTER ONE

AMPLITUDE MODULATION PRINCIPLES

1.1   INTRODUCTION

It should come as no surprise that the voice messages we send to each other are at the low end of the frequency spectrum. Frequency, of course, refers to the cyclical changes, or number of vibrations per second, of the sound waves. These frequencies may be as low as 25 Hz, and they range as high as 16 kHz for some musical instruments. The audio frequency range is approximately from 20 Hz to 20 kHz and is based on the limits of human hearing.

Message forms other than audio can extend to frequencies much higher than 20 kHz. Commercial video falls within the frequency region between 30 Hz and 4.2 MHz, while closed-loop video and data information may use frequencies up to 10 MHz. Regardless of the nature of the information to be transferred or the frequency limits involved, this text will refer to *all* such signals as **message signals.** The magnitude of the message voltage will be denoted E_s, and its frequency will be termed the **message frequency** (f_s). It is understood that the message signal is intentional and will have the same meaning to the sender and the receiver. This definition of the message signal sets it apart from random signals, such as noise.

1.2   MODULATION

A grossly understated definition of the process called **modulation** is an "up-shifting" of the message frequencies (intact) to a range more useful for transmission. A message that contains all of the frequencies between 1 Hz and 5000 Hz, a bandwidth of 5 kHz, could be shifted upward in frequency to the range of 100,001 Hz to 105,000 Hz—and retain the 5000 Hz bandwidth. If this is done correctly, the message will not lose its meaning; the receiver will simply down-shift the signal to its original span of 1 Hz to 5 kHz. This, of course, is an oversimplification, because the conversion requires several intermediate steps.

1.2.1 The Need to Modulate

If two musical programs were played at the same time within hearing distance, it would be difficult for anyone to listen to one source and not hear the second source. Since all musical sounds have approximately the same frequency range, from about 50 Hz to 10 kHz, it would strain the human ear to separate the two messages for any length of time. If the desired program can be shifted up to a band of frequencies between 100 kHz and 110 kHz, and the second program shifted up to the band between 120 kHz to 130 kHz, then both programs will still have a 10 kHz bandwidth, and the listener can (by band selection) retrieve the program of his or her (their) choice. The receiver would downshift only the selected band of frequencies to the audible range of 50 Hz to 10 kHz.

A second, more technical reason to shift the message signal to a higher frequency is related to antenna size. In Chapter 10, you will see that antenna size is inversely proportional to the frequency to be radiated. For a frequency of 1 kHz, the required antenna size is about 88 miles in length, whereas a signal at 100 mHz requires an antenna about 4.5 feet in length.

The third reason for modulating a high-frequency carrier is that RF (radio frequency) energy will travel a greater distance than the same amount of energy transmitted as acoustical (sound) power.

Two signals are necessary to implement modulation: one low-frequency signal and one high-frequency signal. The message is in the low-frequency band. The high-frequency signal is called the "carrier" because it, in effect, carries the low-frequency message signal. In addition, the low-frequency signal changes one or more characteristics of the high-frequency carrier signal to induce modulation.

1.2.2 Forms of Modulation

The carrier signal is a sine wave at the carrier frequency. Equation 1.1 shows that the sine wave has three characteristics that can be altered.

$$\text{Instantaneous voltage } (e) = E_{c(\max)} \sin(2\pi f_c t + \theta) \tag{1.1}$$

The terms that may be varied are the carrier voltage E_c, the carrier frequency f_c, and the carrier phase angle θ.

Amplitude modulation is an increase or decrease of the carrier voltage (E_c), with all other factors remaining constant. **Frequency modulation** is a change in the carrier frequency (f_c), with all other factors remaining constant. **Phase modulation** is a change in the carrier phase angle (θ). The phase angle cannot change without also effecting a change in frequency. Therefore, phase modulation is in reality a second form of frequency modulation (see Chapter 5).

There are variations to these basic forms of modulation, and they will be studied in turn. In this chapter, the major topic is amplitude modulation (AM). The message signal is the *modulating* signal, and the carrier is the *modulated* signal.

1.2.3 Visual Concepts

A carrier signal with its peak amplitude (size) changed to take on the shape of the message signal is shown in Figure 1.1. Notice that during time (a), the message signal has zero amplitude and the carrier wave has a peak amplitude of E_c. At time (b) the message signal is positive, with a value of $+E_s$, and the carrier peak amplitude has increased to $E_c + E_s$. At time (c) the message signal is passing through zero, and the carrier peak amplitude returns to a level equal to E_c. At time (d) the message signal is at its most negative peak, $-E_s$, and the carrier signal size has decreased to the value of $E_c - E_s$. During time (e) the message signal returns to zero, and the carrier peak voltage is again steady at E_c. There are several important observations to be made here.

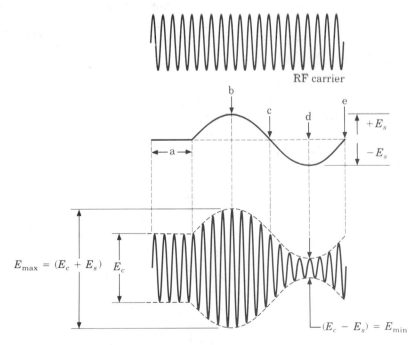

FIGURE 1.1
Time domain plot of the AM wave.

1. The change in the amplitude of the carrier is caused by, and is directly proportional to, the polarity and amplitude of the message signal.
2. The zero level of the message signal corresponds to the peak level of the unmodulated carrier signal. When the message signal goes positive, the carrier signal is made greater in amplitude than the unmodulated carrier by an amount equal to the peak message signal. When the modulating message signal goes negative, the carrier peak level is made less than the peak level of the unmodulated carrier wave by an amount equal to the peak message signal.
3. The positive and negative peaks of the modulated carrier wave are mirror images of each other for conditions of E_{max} and E_{min}.
4. When the peak level of the message signal is equal to the peak level of the carrier signal, the carrier is said to be fully modulated. This is called *100% modulation*. In Figure 1.1, at (b) the total amplitude is $E_c + E_s = 2E_c$, and at point (d) the total amplitude is $E_c - E_s = 0$, the conditions for 100% modulation.

The display in Figure 1.1 is called the *time domain presentation* of the amplitude-modulated wave since it is a plot of the wave amplitude with respect to time. It shows that the carrier frequency is constant while its amplitude changes in accord with the shape of the message signal size and frequency, and that the carrier frequency is several times higher than the highest message signal frequency.

1.2.4 Side Frequencies

Two signal frequencies can be mixed linearly, and the resulting signal is the algebraic sum of the two voltages at any point in time, as in an adder. The output waveform for the linear addition of two waves is shown in Figure 1.2a. When the same two signal frequencies are plotted on a graph of signal amplitude compared to frequency, it would appear as in Figure 1.2b, called a *frequency domain* graph.

When a nonlinear device is used to mix the same two signal frequencies, the amplitude of one signal will change the amplitude of the other signal through time. Figure 1.3 shows a circuit, with waveforms, that illustrates this modulation process (nonlinear mixing). With *no* message signal applied to the transistor's emitter, the carrier signal at the base will be amplified and will appear at the output, unchanged in shape through time T_1.

During time T_2, when the message signal at the emitter goes positive, the effect is not to amplify the signal, but to *lower the gain* of the transistor for the base input carrier signal. (Remember that raising the emitter voltage has the effect of lowering the base voltage.) Because the gain decreases, the output signal strength decreases during T_2. For the time period T_3, the opposite effect holds true. The emitter voltage is lowered (which is the same as raising the

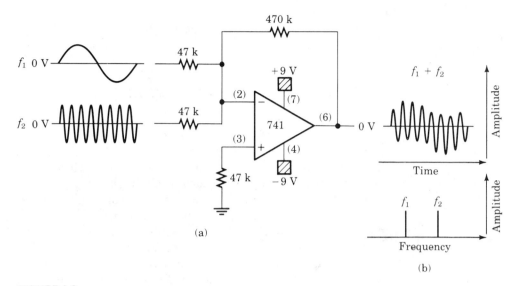

FIGURE 1.2
Linear mixing (summing).

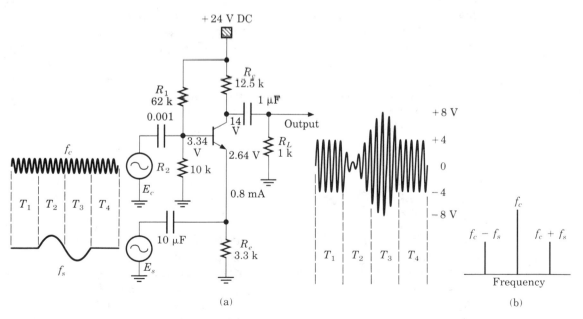

FIGURE 1.3
Amplitude modulation.

FIGURE 1.4

Upper and lower sidebands, each 3 kHz.

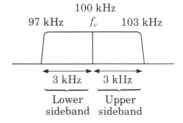

base), the gain increases since the base is held constant by the voltage-divider bias circuit of R_1 and R_2, and the output signal strength increases. Increasing and decreasing the bias of the transistor changes the gain.

Modifying the amplitude of one signal to conform to the amplitude of a second signal is called amplitude modulation. When the signals are blended in this manner, the frequencies as well as the voltages are blended. That is, the voltage will be modified as shown in Figure 1.3, and the frequencies of the output wave will contain the two original frequencies plus the *sum* of the frequencies and the *difference* between the frequencies. These sum and difference frequencies are called the **side frequencies** or sideband frequencies. These are shown in Figure 1.3b. Compare Figures 1.2 and 1.3, and you will recognize the difference between "mixing" (the simple addition of two voltages as in Figure 1.2), and "modulation" (altering the size or shape of one voltage by another voltage as in Figure 1.3). When mixing two signals, the top and bottom peaks of the wave move in the *same* direction, changing position (but not size) along the graph. When modulating two signals, the top and bottom peaks of the wave move in opposite directions, making the wave larger or smaller along the graph.

These opening statements have considered only one modulating signal frequency. When the message signal contains a whole group of frequencies, say, all frequencies from 1 Hz to 3000 Hz, then the modulated carrier would have an **upper band** of frequencies and a **lower band** of frequencies, as shown in Figure 1.4. For a carrier frequency of 100 kHz, the transmitted signal of Figure 1.4 would be a band of all frequencies from 97 kHz to 103 kHz.

1.2.5 Modulation Factor and the Percentage of Modulation

Figures 1.1 and 1.3 have waveshapes that are expressed as being 100% modulated. What does this mean, and is 100% the only extent of modulation? A carrier signal by itself is unmodulated, or is at zero modulation; its amplitude is constant. When the amplitude of the carrier signal is changed, the degree of change is a measure of the percentage of modulation. For example, a change of one-half the amplitude of the carrier (peak to peak) is 50% modulation. In Figure 1.5, notice that a decrease of 50% at one part of the wave results in an increase of 50% at another part of the wave. Thus, 100% modulation can be

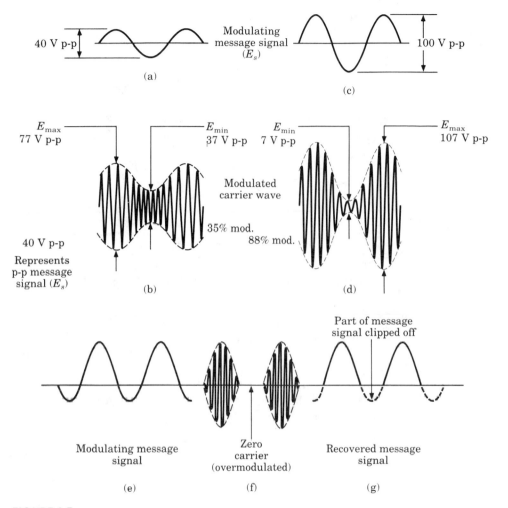

FIGURE 1.5
The amplitude-modulated wave.

described as an amplitude change that causes the carrier voltage to go to 0 V at one point of the wave and to two times the unmodulated carrier voltage at another point of the wave. Figure 1.5 shows several cases for different percentages of modulation.

The **modulation factor** is obtained by viewing the modulated wave on an oscilloscope and recording the peak-to-peak values of the wave at its crest and at its trough:

$$\text{Modulation factor } (m) = \frac{E_{\max} - E_{\min}}{E_{\max} + E_{\min}} \qquad \textbf{(1.2)}$$

(when the modulated wave is viewed) or

$$m = \frac{E_{\text{message peak}}}{E_{\text{carrier peak}}}$$

(when the two unmodulated voltages are known).

EXAMPLE 1.1

In Figure 1.5b, E_{max} (p-p) = 77 V, E_{min} (p-p) = 37 V. Thus,

$$m = \frac{77\ \text{V} - 37\ \text{V}}{77\ \text{V} + 37\ \text{V}} = \frac{40\ \text{V}}{114\ \text{V}} = 0.35$$

The percentage of modulation is the modulation factor (m) times 100, or

$$\% \text{ modulation} = m(100) = \frac{E_{\text{max}} - E_{\text{min}}}{E_{\text{max}} + E_{\text{min}}} \times 100 \tag{1.3}$$

In this example, the percentage of modulation is 35%.

The reader is encouraged to calculate the percentage of modulation in Figure 1.5d to verify that it is 87.72%.

One hundred percent modulation is the most desirable and 0% the least desirable. Overmodulation is shown in Figure 1.5e, f, and g and should be avoided at all costs.

The unmodulated carrier, not shown in Figure 1.5, will have a voltage midway between E_{max} and E_{min}.

$$\text{Carrier voltage } (E_c) = \frac{E_{\text{max}} + E_{\text{min}}}{2} \tag{1.4}$$

In Figure 1.5, the voltage of the unmodulated carrier is

$$E_c = \frac{77\ \text{V} + 37\ \text{V}}{2} = 57\ \text{V p-p}$$

and the applied message signal voltage is simply

$$E_{\text{message p-p}}\ (E_s) = \frac{E_{\text{max}} - E_{\text{min}}}{2} \tag{1.5}$$

$$= \frac{77\ \text{V} - 37\ \text{V}}{2} = 20\ \text{V p-p}$$

EXAMPLE 1.2

A modulated carrier is viewed on an oscilloscope and has a crest voltage of 44 V p-p. The bottom point (or trough) of the wave measures 6 V p-p.
a. What is the modulation factor? b. What is the percentage of modulation?
c. What is the peak-to-peak unmodulated carrier voltage?

SOLUTION

a. modulation factor = (44 V − 6 V)/(44 V + 6 V) = 0.76

b. % modulation = (0.76)(100) = 76%

c. Carrier voltage = (44 V + 6 V)/2 = 25 V p-p

Most commonly, the carrier voltage is known and a desired percentage of modulation is to be incorporated. As the message signal is applied, the oscilloscope display will reveal the values of both E_{max} and E_{min} of the modulated carrier. Rather than use several trial-and-error attempts, the crest and trough of the wave can be predetermined by

$$E_{max} = E_c(1 + m) \quad \text{and} \quad E_{min} = E_c(1 - m) \qquad (1.6)$$

Reversing Example 1.2, we can determine the crest and trough voltages of a 25 V p-p carrier with 76% modulation:

$$E_{max} = 25 \text{ V p-p } (1 + 0.76) = 44 \text{ V p-p}$$
$$E_{min} = 25 \text{ V p-p } (1 - 0.76) = 6 \text{ V p-p}$$

Overmodulation The FCC (Federal Communication Commission of the U.S.A.) recommends that AM broadcast stations maintain a carrier modulation between 85% and 95%, not to exceed 100% modulation. The overmodulated carrier signal in Figure 1.5f distorts the message signal and reduces the quality of the transmission. This is the broadcasters' concern.

Clipping the modulated carrier sine wave as in Figure 1.5g makes it look like the beginning of a square wave, which is rich in harmonic frequencies. These frequencies can fall outside of a station's assigned frequency band and interfere with another station's transmissions. This interference is the major concern of the FCC, and there are laws to protect stations from harmful interference from other broadcasters.

1.2.6 The Mathematics of the AM Wave

Equation 1.1 is the instantaneous voltage expression for a single sine wave. The term E_c identifies the carrier voltage and f_c the carrier frequency. For the sake of simplicity, assume time = 0 and phase = 0. The message signal under these conditions will also be a sine wave with the characteristics:

$$E_{message} = e = E_{s(max)}\sin(2\pi f_s)t \qquad (1.7)$$

where

$$E_s = \text{the message signal peak voltage}$$
$$f_s = \text{the message signal frequency}$$

To see what happens when one sine wave alters the amplitude of another sine wave, it is common to use the trigonometric expression $(\sin X)(\sin Y) =$

½ cos(*X* + *Y*) + ½ cos(*X* − *Y*). The modulation factor *m*, which shows the relative amplitude of one wave compared to the other, is included to complete the expression for the amplitude-modulated wave:

$$e_{tot} = E_c \sin(2\pi f_c)t$$
$$- \frac{mE_c}{2} \cos 2\pi(f_c + f_s)t$$
$$+ \frac{mE_c}{2} \cos 2\pi(f_c - f_s)t$$

(1.8)

No in-depth math skills are required to follow the important observations set forth in this equation:

1. The first line of Equation 1.8 represents the carrier signal and includes the terms for the carrier voltage (E_c) and the carrier frequency (f_c).
2. The remainder of the equation describes the side frequency signals. When the factor of modulation (m) is zero, then both remaining terms go to zero.
3. When the carrier is modulated, the second line of the equation describes the upper sideband frequency, which is the carrier frequency plus the message frequency ($f_c + f_s$). It also tells us the voltage of the upper sideband compared with the unmodulated carrier voltage, related to the factor of modulation (m) as $mE_c/2$.
4. The third line of Equation 1.8 describes the lower sideband frequency. It is the carrier frequency minus the message frequency ($f_c - f_s$). The voltage in the lower sideband is the same as the voltage in the upper sideband relative to the carrier voltage and the factor of modulation, $mE_c/2$.
5. The minimum voltage to be transmitted will be the unmodulated carrier voltage E_c (for $m = 0$). The maximum voltage to be transmitted will be $2E_c$ (for 100% modulation), with a voltage of $E_c/2$ in the lower sideband and a voltage of $E_c/2$ in the upper sideband.

1.2.7 Recap

At this point, the basic concepts of amplitude modulation should be clear. A mental picture should begin to take shape that will allow you to recognize what changes to the modulated wave will occur as any of the variable terms are modified. All of the concepts discussed so far in this chapter are brought together in the three-dimensional overview of Figure 1.6. This diagram should be given a rigorous examination to ensure a full understanding of the components of the AM wave.

Viewing the modulated wave in the frequency domain shows the location and relative amplitude of the three signal frequencies. Viewing the modulated wave in the time domain of Figure 1.6 shows the three distinct sine waves and

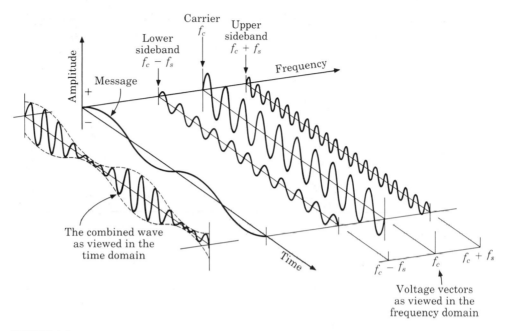

FIGURE 1.6
Three-dimensional layout of the modulated carrier signal with its side frequencies.

their amplitudes. When the three waves are compressed into a two-dimensional plane, as an oscilloscope would do, the result is the pattern of the modulated wave of Figure 1.4.

1.2.8 Modulating Voltage Ratios

As stated earlier, the peak message signal voltage is equal to the peak carrier voltage at 100% modulation. Then the total voltage in the wave at 100% modulation must be equal to two times the unmodulated carrier voltage. The total sideband voltage is equal to the carrier voltage; the voltage of each sideband is equal to one-half the carrier voltage. Equation 1.8 is a simplified expression that addresses only the voltage and percentage of modulation of the wave. It can be expanded to

$$\text{Total voltage } (E_{\text{tot}}) = E_c + \frac{mE_c}{2} + \frac{mE_c}{2} \qquad (1.9)$$

Notice that the two sideband voltages are equal, and that neither may exceed one-half of the carrier voltage. Also notice that the sideband voltages are a function of the percentage of modulation, which means that each sideband voltage will always be some value between zero and one-half of the carrier voltage.

EXAMPLE 1.3 ══

A transmitter has a 75 V_{rms} carrier at 1.25 MHz applied to a 56.25 Ω antenna. The carrier is then modulated to 40% with a 5 kHz message signal.
a. What are the side frequencies? b. How much voltage is in each sideband signal?

SOLUTION

    a.   The upper side frequency is

$$f_c + f_s = 1.25 \text{ MHz} + 0.005 \text{ MHz} = 1.255 \text{ MHz}$$

        The lower side frequency is

$$f_c - f_s = 1.25 \text{ MHz} - 0.005 \text{ MHz} = 1.245 \text{ MHz}$$

    b.   The voltage in each sideband frequency is

$$mE_c/2 = (0.4)(75 \text{ V})/2 = 15 \text{ V}_{rms}$$

There will be 30 V_{rms} total in the two sidebands, which is added to the 75 V_{rms} of the carrier, making the total transmitted signal voltage 105 V_{rms}. (See Figure 1.7.)

FIGURE 1.7
The voltage and frequency from Example 1.3.

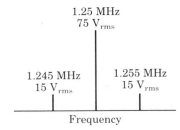

1.25 MHz
75 V_{rms}

1.245 MHz
15 V_{rms}

1.255 MHz
15 V_{rms}

Frequency

1.3   **POWER RELATIONSHIPS**

The power in AC circuits is determined in the same way as for DC circuits. That is, Power $= E^2/R$. From Example 1.3, the carrier power is $(75 \text{ V}_{rms})^2/56.25 \ \Omega = 100$ W. Power is always stated in true rms watts; therefore, the voltage must be stated in rms volts in the calculation of transmitted power.

    The sideband power is also the square of the voltage divided by the resistance, and for Example 1.3 is $(15 \text{ V}_{rms})^2/56.25 \ \Omega = 4$ W in one sideband. The total power to be transmitted is the *sum* of all of the individual powers:

$$P_c + P_{usb} + P_{lsb} = 100 + 4 + 4 = 108 \text{ W}$$

FIGURE 1.8
The power and frequency from Example 1.3.

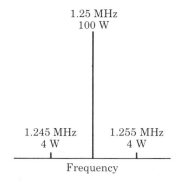

The process is simplified in the following equation:

$$\text{Total power} = P_t = P_c + \frac{m^2 P_c}{4} + \frac{m^2 P_c}{4}$$

$$= P_c + \frac{m^2 P_c}{2} = P_c\left(1 + \frac{m^2}{2}\right)$$

(1.10)

As the factor of modulation (m) changes, so does the sideband power. At 100% modulation, $m = 1$, and since $1^2 = 1$, the power in each sideband will be one-quarter of the power in the carrier (from Equation 1.10). The combined sideband power is one-half of the carrier power, and the total transmitted power is the sum of all powers. At 100% modulation, the total power is 1.5 times the unmodulated carrier power.

In Example 1.3, there is 100 W in the carrier and 4 W in each sideband, for a total power of 108 W. The message power for this transmission is in the sidebands, so only 8 W out of 108 W are for message. This arrangement is only 7.5% efficient and is shown in Figure 1.8.

When the same carrier is modulated to 100%, the sideband power, from Equation 1.10, is 100 W divided by 2, or 50 W, which is added to the carrier power for a total power of 150 W. The efficiency is 50 W of sideband power divided by the total power of 150 W, or 33%. Herein lies the major disadvantage of amplitude modulation: the system has a maximum efficiency of only 33%. The carrier power is helpful in getting the signal across many miles to the receiver, but it contains no message power and is therefore considered to have no value in the message recovery process.

EXAMPLE 1.4

A carrier signal of 600 V_{rms} is applied to a 75 Ω antenna.

    a.   What is the carrier power?
    b.   What is the total sideband power at 75% modulation?
    c.   What is the total radiated power at 75% modulation?

d.   What is the total sideband power at 100% modulation?
e.   What is the total radiated power at 100% modulation?
f.   What is the change in carrier power from 75% to 100% modulation?

SOLUTION

a.   The carrier power is $E^2/R = (600)^2/75 = 4800$ W.
b.   The sideband power at 75% modulation $= (0.75^2)(4800)/2 = 1350$ W.
c.   The total power is $4800 + 1350 = 6150$ W.
d.   At 100% modulation, the sideband power is $(1.0^2)(4800)/2 = 2400$ W.
e.   The total radiated power at 100% is $4800 + 2400 = 7200$ W (1.5 times the unmodulated carrier power).
f.   The carrier power does *not* change with modulation and therefore remains at 4800 W.

1.4   COMPONENT PHASORS OF THE AM WAVE

We have seen three ways to describe the modulated AM wave: (1) the time domain diagram of Figure 1.1, (2) the voltage and power diagrams in the frequency domain of Figures 1.7 and 1.8, and (3) the mathematical equivalent of the AM wave in Equations 1.9 and 1.10. One more description of the AM wave may seem redundant, but it will fortify what has already been stated, as well as make later concepts easier to understand.

Figure 1.9 shows one cycle of the signal (wave 1) used to modulate a 10-cycle carrier signal (wave 3). Waves 2 and 4 are the resulting side frequencies. Wave 2 is the 10 Hz carrier minus the 1 Hz message signal (a lower side frequency of 9 Hz). Wave 4 is the 10 Hz carrier plus the 1 Hz message (an 11 Hz upper side frequency). When waves 2, 3, and 4 are vectorally added at all instances of time across Figure 1.9, their sum voltage is that of amplitude-modulated wave 6.

Close examination shows that the phase of the upper side frequency changes at a faster rate than the phase of the carrier signal (11 Hz compared to 10 Hz). The phase of the lower side frequency changes at a slower rate than the phase of the carrier signal (9 Hz compared to 10 Hz).

We know that the phase of the carrier signal constantly changes as the wave progresses through each cycle, but if we use this phase as a reference, we can see that the phase of the lower side frequency (changing at a slower rate) constitutes a vector moving slowly in a counterclockwise direction from the phase vector of the carrier signal. Moreover, the phase of the upper side frequency (changing at a faster rate) forms a vector moving slowly in a clockwise direction from the phase vector of the carrier signal.

Wave 5, at (a), shows the two sideband vectors each 90° out of phase from the carrier vector. The sideband phasors are equal in amplitude and opposite in

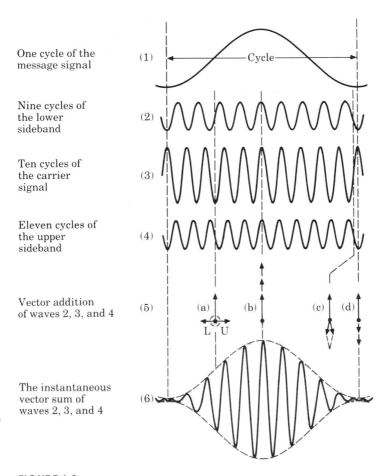

One cycle of the
message signal

Nine cycles of
the lower
sideband

Ten cycles of
the carrier
signal

Eleven cycles of
the upper
sideband

Vector addition
of waves 2, 3, and 4

The instantaneous
vector sum of
waves 2, 3, and 4

FIGURE 1.9
A vector comparison of the component parts of the amplitude-modulated wave.

phase and will therefore cancel each other. The accumulated voltage at this point is the voltage of the carrier only.

In wave 5, at (b), the two sideband vectors are in phase with the carrier and the three voltages sum to equal two times the carrier voltage, or E_{max} of the modulated wave.

At (d) in wave 5, the two sideband vectors are in phase with each other, but 180° out of phase from the carrier signal. Thus, they will completely cancel the voltage in the carrier wave to become the trough (E_{min}) of the modulated wave.

At (c) in wave 5, the vector sum of the two sideband voltages is not quite equal to the carrier voltage (E_c) and so do not completely cancel the carrier voltage.

The purpose of this discussion is to emphasize the concept that the AM wave is the vector sum of the two sidebands, as counter-rotating vector voltages, added to the reference carrier vector voltage. It should be understood that when the carrier voltage is used as a reference, the lower sideband voltage is a *clockwise*-rotating vector voltage with an angular position equal to the *counterclockwise*-rotating vector voltage of the upper sideband, and that the trigonometric sum of the sideband voltages will directly *add to* or *subtract from* the carrier vector voltage to form the AM wave.

1.5   ASSIGNABLE FREQUENCY SPECTRUM

The total range of frequencies between 200 kHz and several hundred GHz has been allocated for shared or exclusive use by various transmitting services. The vast majority of the assigned bands are for specific services, which may be further subdivided by territory. For example, the AM radio broadcast band in the North American continent is assigned the frequency range between 535 kHz and 1605 kHz. A station at 670 kHz in Phoenix, Arizona, for instance, is limited to a radius of 150 miles by its maximum power rating. This station would not (under normal circumstances) interfere with a station in Detroit, Michigan, that is also assigned the frequency of 670 kHz and a radiating radius of 150 miles. Distance is the determining factor. Both stations have a specific band assignment, a specific carrier frequency assignment, and, because of the power restrictions, a territorial limit.

Mobile AM communication services, such as class D citizens band radio, are assigned 40 specific station carrier frequencies within the band from 26.965 MHz to 27.405 MHz everywhere within the North American continent, that is, with *no* territorial limits. Channel 19 of the citizens band radio is the same frequency of 27.085 MHz in *every* city within its legal area. Here the frequency is assigned to a channel number, not to a station. A total listing of all frequency assignments would be too lengthy. It is sufficient to recognize that the entire spectrum from several hundred kHz to several hundred GHz are committed to some service. Once these crowded conditions are understood, we can appreciate the need for rigid control on the use of the airwaves and the limits imposed on signal transmissions.

The frequency designations used in communications are listed in Table 1.1 and the broadband frequency classifications in Table 1.2.

TABLE 1.1
Frequency groups

Exponent	Group	Abbreviation
10^0	Hertz	Hz
10^3	Kilohertz	kHz
10^6	Megahertz	MHz
10^9	Gigahertz	GHz
10^{12}	Terahertz	THz
10^{15}	Petahertz	PHz
10^{18}	Exahertz	EHz

TABLE 1.2
Frequency bands

30 Hz–300 Hz	ELF	Extremely low frequencies
300 Hz–3 kHz	VF	Voice frequencies
3 kHz–30 kHz	VLF	Very low frequencies
30 kHz–300 kHz	LF	Low frequencies
300 kHz–3 Mhz	MF	Medium frequencies
3 Mhz–30 Mhz	HF	High frequencies
30 Mhz–300 Mhz	VHF	Very high frequencies
300 MHz–3 GHz	UHF	Ultrahigh frequencies
3 GHz–30 GHz	SHF	Super high frequencies
30 GHz–300 GHz	EHF	Extremely high frequencies
300 GHz–3 THz		Infrared
3 THz–30 THz		
30 THz–300 THz		Visible light spectrum
300 THz–3 PHz		Ultraviolet light
3 PHz–30 PHz		X rays
30 PHz–300 PHz		
300 PHz–3 EHz		Gamma rays
3 EHz–30 EHz		Cosmic rays

1.6   BAND SELECTION

The commercial AM radio broadcast band will be used here as the vehicle for discussion of the limits and ratings used to regulate the transmission of amplitude modulated carrier signals. This is because there are many published universal rulings for this band that are relevant to other forms of modulation treated in later chapters.

1.6.1 Commercial AM Broadcast Band

The commercial AM band of frequencies was opened for the intended use of voice communications only. With the equipment available at that time, the audio frequency range was limited to 5 kHz. Modulating the carrier with a 5 kHz message signal required a transmission channel of 10 kHz. Therefore, 10 kHz was allocated to each of the channels in the AM broadcast band, and the band of frequencies from 535 kHz to 1605 kHz was set aside for this use only. The lowest channel in the band is identified as channel 1, having a carrier frequency of 540 kHz. Its lower side frequency extends down to 535 kHz, and its upper side frequency up to 545 kHz. The highest channel in the band is numbered channel 107 and has a carrier frequency of 1600 kHz. The upper sideband frequency for this channel reaches 1605 kHz, and the lower side frequency goes down to 1595 kHz.

Table 1.3 lists all 107 channels in the AM broadcast band, shows the frequency of each channel, and identifies its original status ("C" for clear channel, "R" for regional channel and "L" for local channel, designations that will be explained later).

TABLE 1.3

Standard AM radio broadcast channels in the United States

Carrier Frequency (kHz)	Channel Class	Number	Carrier Frequency (kHz)	Channel Class	Number	Carrier Frequency (KhZ)	Channel Class	Number
540	C	1	900	C	37	1260	R	73
550	R	2	910	R	38	1270	R	74
560	R	3	920	R	39	1280	R	75
570	R	4	930	R	40	1290	R	76
580	R	5	940	C	41	1300	R	77
590	R	6	950	R	42	1310	R	78
600	R	7	960	R	43	1320	R	79
610	R	8	970	R	44	1330	R	80
620	R	9	980	R	45	1340	L	81
630	R	10	990	C	46	1350	R	82
640	C	11	1000	C	47	1360	R	83
650	C	12	1010	C	48	1370	R	84
660	C	13	1020	C	49	1380	R	85
670	C	14	1030	C	50	1390	R	86
680	C	15	1040	C	51	1400	L	87
690	C	16	1050	C	52	1410	R	88
700	C	17	1060	C	53	1420	R	89
710	C	18	1070	C	54	1430	R	90
720	C	19	1080	C	55	1440	R	91
730	C	20	1090	C	56	1450	L	92
740	C	21	1100	C	57	1460	R	93
750	C	22	1110	C	58	1470	R	94
760	C	23	1120	C	59	1480	R	95
770	C	24	1130	C	60	1490	L	96
780	C	25	1140	C	61	1500	C	97
790	R	26	1150	R	62	1510	C	98
800	C	27	1160	C	63	1520	C	99
810	C	28	1170	C	64	1530	C	100
820	C	29	1180	C	65	1540	C	101
830	C	30	1190	C	66	1550	C	102
840	C	31	1200	C	67	1560	C	103
850	C	32	1210	C	68	1570	C	104
860	C	33	1220	C	69	1580	C	105
870	C	34	1230	L	70	1590	R	106
880	C	35	1240	L	71	1600	R	107
890	C	36	1250	R	72			

C = Clear
R = Regional
L = Local

1.6.2 Channel Interference

No metropolitan area can support 107 AM broadcast stations simply because of the financial burden it would place on the commercial investors. Therefore, it is common to find a gap of 30 to 70 kHz between assigned AM stations in any locality. Table 1.4 shows the AM station assignments for two major cities in the Unites States.

Because of this wide channel separation, many AM broadcast stations extend their audio frequencies to provide better fidelity, taking advantage of the better equipment in today's market. The FCC regulations state that at 15 kHz on either side of a station's carrier frequency, the transmitted power must be reduced to 25 dB below the carrier level. The transmitter's characteristics are specified to 12 kHz; therefore, station operators run "proof of performance" tests to 12 kHz on each side of the carrier frequency to verify proper transmitter output performance.

Stations are rated according to class, which in turn sets the power limit that can be transmitted. A station's power and operating frequency are then plotted on an area map to determine proper separation from other stations on the same frequency. This will guarantee protection for the stations already on the air and provide for minimum interference. Any station that interferes with another station's protected zone is required by law to alter its performance to remove the interference. The station may (1) reduce the bandwidth of its transmitted signal, (2) lower its transmission power, (3) change the radiation pattern of its antenna, or (4) go off the air. Many broadcast stations that face this problem select option 2, 3, or 4 (option 3 is the most popular). Chapter 12 will describe how the transmission of radio waves differs from daytime to nighttime, when most interference occurs.

TABLE 1.4
Assigned standard AM broadcast radio stations

Chicago (Proper)						Phoenix (Area)					
560	WIND	News	1110	WMBI	Religious	550	KOY	Contemp.	1280	KHEP	Religious
670	WMAQ	Country	1160	WJJD	Country	620	KTAR	News	1310	KZZP	Rock
720	WGN	Variety	1240	WSBC	Foreign	740	KMEO	Music	1360	KRUX	Top 40
780	WBBM	News	1390	WVOX	Contemp.	860	KIFN	Spanish	1400	KXIV	Big Band
820	WAIT	Music	1450	WXOL	Blues	910	KJJJ	Country	1440	KXAM	Dance
890	WLS	Top 40	1490	WOPA	Ethnic	960	KARZ	Contemp.	1480	KPHX	Spanish
950	WJPC	Contemp.	1570	WBEE	Jazz	1010	KXEG	Gospel	1510	KDJQ	Modern
1000	WCFL	Contemp.	1590	WOXN	Foreign	1060	KKKQ	Oldies	1540	KASA	Religious
						1190	KRDS	Contemp.	1580	KNIX	Country
						1230	KFLR	Religious			

1.6.3 Definitions

Several new terms, some of which were introduced in the preceding paragraphs and in the tables, need to be defined here. The definitions are legal descriptions of an AM broadcast station's limits.

The **primary service area** is the area in which the transmitted signal has a measurable ground wave field strength between 10 and 50 μV/m (microvolts per meter) in the business and factory areas and a measurable ground wave field strength between 2 and 10 μV/m in the city residential areas. Signals in all rural areas in the winter (and northern U.S. rural areas in the summer) have a measurable field strength of 0.1 to 0.5 μV/m. Signals in the southern U.S. rural areas during the summer have a measurable field strength of 0.1 to 0.25 μV/m. The reason for this is that the moisture content changes the ground conductivity seasonally, affecting the radiation by geographical location.

Secondary service areas include all service areas that have a sky wave field strength of 0.5 mV/m (millivolts per meter) for 50% or more of the time.

A **clear-channel station** covers an extremely wide service area and is protected against objectional interference within this area. Originally, there where 60 clear channels, as noted in Table 1.3. However, as demands for air space became more numerous, the list was legally pared down to the present 25 clear-channel stations listed in Table 1.5

A **regional channel** renders primary service only to large cities and the surrounding rural areas. A regional channel may be occupied by several stations, with the primary service area of each limited by the interference patterns of the stations.

TABLE 1.5
Clear-channel standard AM broadcast radio stations

Frequency Class 1A Assignments (kHz)					
			830	WCCO	Minneapolis, MN
			840	WHAS	Louisville, KY
			870	WWL	New Orleans, LA
640	KFI	Los Angeles, CA	880	WCBS	New York, NY
650	WMS	Nashville, TN	890	WLS	Chicago, IL
660	WNBC	New York, NY	1020	KDKA	Pittsburgh, PA
670	WMAQ	Chicago, IL	1030	WBZ	Boston, MA
700	WLW	Cincinnati, OH	1040	WHO	Des Moines, IA
720	WGN	Chicago, IL	1100	WWWE	Cleveland, OH
750	WBS	Atlanta, GA	1120	KMOX	St. Louis, MO
760	WJR	Detroit, MI	1160	KSL	Salt Lake City, UT
770	WABC	New York, NY	1180	WHAM	Rochester, NY
780	WBBM	Chicago, IL	1200	WOAI	San Antonio, TX
820	WBAP	Fort Worth, TX	1210	WCAU	Philadelphia, PA

A **local channel** renders primary service to a city or town and the nearby suburban or rural areas. Its primary service area may be restricted by coexisting radiation interference patterns.

1.6.4 Effective Radiated Power

The power that is emitted into the atmosphere by the antenna is termed the **effective radiated power** (ERP). This rating takes into account (1) the transmitter output power, (2) cable-related and other losses incurred between the transmitter and the antenna, and (3) the antenna gain, if any. In most cases, the ERP of the system is stated on the station license as the limiting factor for that station.

1.6.5 Emission Code

The **emission code** is broken down into three basic categories, defined by the letter designations A, F, and P. The form of modulation studied in this chapter is amplitude modulation and carries the designation A. The other forms will be studied in later chapters, but the code is broken down in its entirety in Table 1.6.

This chapter has mainly covered transmissions that include a full carrier and both sidebands (double-sideband or DSB). The message signal studied has been voice or music. Therefore, letter designation A for such a signal would be followed by number 3 and no subscript (an A3 emission). Occasionally a bandwidth notation may accompany the emission code, for example, 5A3. The number 5 designates a 5 kHz bandwidth modulation of a double-sideband, full-carrier transmission.

The number designators subdivide the letter designators. The number 0 defines a transmission such as a radio beacon for aircraft. The carrier is on all of the time but with *no* modulation. The number 1 describes a carrier that is turned on and off, such as Morse code for dots and dashes, with no carrier modulation. The number 2 means that the carrier is on all of the time and is modulated with an on/off tone signal, again in the form of Morse code. The 3 designation is covered in Chapters 1, 2, and 3 of this text. Number 4 covers facsimile transmissions, which include charts, photographs, fingerprints, and other nonmoving pictures. A designation of 5 is the standard broadcast television signal, in which the full upper sideband, but only a part of the lower sideband, is transmitted. Designation 5 is termed "vestigial-sideband transmission."

Category 9 is set aside for any form of modulation that cannot be included in any of the other categories. Frequency modulation (FM), included under the letter F, is covered in depth as a separate subject in Chapter 5. The letter P includes all forms of pulse modulation, such as pulse amplitude, pulse width, and pulse position modulation. The letter P also carries a lettered subscript such as d, e, f or g.

TABLE 1.6
Emission code

Letter	A	Amplitude modulation
	F	Frequency modulation
	P	Pulse modulation
Number	0	Carrier on only, no message (radio beacon)
	1	Carrier on/off, no message (Morse code, radar)
	2	Carrier on, keyed tone on/off (code)
	3	Telephony, message as voice or music
	4	Facsimile, nonmoving graphics (slow-scan TV)
	5	Vestigial sideband (commercial TV)
	6	Four-frequency diplex telegraphy
	7	Multiple sideband, each with different message
	8	
	9	General (all others)
Subscripts	none	Double sideband, full carrier
	a	Single sideband, reduced carrier
	b	Double sideband, no carrier
	c	Vestigial sideband
	d	Carrier pulses only, pulse amplitude modulation (PAM)
	e	Carrier pulses only, pulse width modulation (PWM)
	f	Carrier pulses only, pulse position modulation (PPM)
	g	Quantized pulses, digital video
	h	Single sideband, full carrier
	j	Single sideband, no carrier
Variations	$35A3_h$	Amplitude-modulated, single sideband with full carrier and message frequency to 35 kHz
Other abbreviations	DSB	Double sideband
	DSSC	Double sideband suppressed carrier
	SSSC	Single sideband suppressed carrier (also SSB)
	FSK	Frequency shift keying

QUESTIONS

1. What does the abbreviation AM mean? (a) alternate mode, (b) amplitude modulation, (c) army manual, (d) altitude measurements, (e) arithmetic means.
2. Amplitude modulation results when two signals of different frequencies are applied to a (linear) (nonlinear) device.
3. Name the two signals that are applied to the modulating device in Question 2.
4. What is a good practical ratio for the two frequencies that contribute to the modulation process?
5. What is the name of the signal that has the higher frequency in the modulation process?
6. Name at least two ways to depict the AM wave.

7. List two advantages of the modulation process compared to no modulation.
8. How many characteristics of a sine wave are variable?
9. Which of the characteristics of the sine wave, when varied, would result in an amplitude-modulated wave?
10. The percentage of modulation of the carrier signal is (directly) (indirectly) proportional to the voltage amplitude of the message signal.
11. What are the theoretical upper and lower limits of the audible frequency range?
12. What is the absolute upper frequency limit for the message signal?
13. Is transmission noise considered to be a message signal?
14. What are the advantages of modulating an audible signal as opposed to not modulating the signal?
15. How many signals are required to effect modulation?
16. System noise is considered to be a message signal. (T) (F)
17. Describe the relationship between the carrier frequency and the upper and lower side frequencies.
18. What is the relationship between the modulation factor and the percentage of modulation?
19. When looking at an AM wave on a service oscilloscope, in what domain is the signal displayed?
20. Draw a time domain representation of an AM signal showing E_{max} = 83.5 V and E_{min} = 35.8 V. Make the carrier frequency ten times the message frequency.
21. What is the modulation factor of the wave in Question 20?
22. What is the percentage of modulation for the wave in Question 20?
23. What is the peak-to-peak voltage of the carrier in Question 20?
24. What is the peak-to-peak voltage of the message signal in Question 20?
25. Take the values from Question 20 and increase the percentage of modulation to 80%. Find the new values of E_{max} and E_{min}.
26. At 100% modulation, the sum of the voltages in both sidebands is equal to _____ % of the unmodulated carrier voltage. (a) 100 (b) 75 (c) 50 (d) 25 (e) 0
27. At 100% modulation, the sum of the power in both sidebands is equal to _____ % of the unmodulated carrier power. (a) 100 (b) 75 (c) 50 (d) 25 (e) 0
28. Signals are modulated in order to (a) separate the different transmissions at the receiver end, (b) provide a vehicle by which intelligence may be transmitted over great distances, (c) allow for practical antenna sizes, (d) all of the above.

A carrier voltage of 102.44 V_{rms} is applied to a 72 Ω antenna at a carrier frequency of 1250 kHz.

29. What is the total radiated voltage at 0% modulation?
30. What is the total radiated power at 0% modulation?
31. What is the total radiated voltage at 100% modulation?
32. What is the total radiated power at 100% modulation?
33. What is the total radiated voltage and power at 82% modulation?

34. What is the total radiated voltage and power at 25% modulation?
35. What is the ratio of the total voltage at 100% modulation to the voltage in the unmodulated carrier?
36. What is the ratio of the total power at 100% modulation, to the power in the unmodulated carrier?
37. How many times larger is the peak voltage in a 100% modulated wave compared to the peak voltage in the unmodulated carrier?
38. How many times larger is the rms power in a 100% modulated wave compared to the rms power in the unmodulated carrier?
39. At 0% modulation, the total power radiated from an antenna is transmitted as (a) the upper and lower sideband power, (b) the lower sideband power only, (c) the upper sideband power only, (d) the carrier power plus the upper and lower sideband power, (e) the carrier power only.
40. All of the message power is transmitted in the (a) carrier, (b) upper sideband only, (c) lower sideband only, (d) upper and lower sidebands, (e) the carrier plus both sidebands.
41. What is the total rms voltage in the transmitted wave when a 500 V_{rms} carrier is 100% modulated?
42. What is the power in the radiated wave for a 500 V_{rms} carrier applied to a 50 Ω antenna?
43. What is the total power radiated from an antenna when a 5000 W carrier is 100% modulated?
44. Name the three frequencies that are transmitted when a 1250 kHz carrier is modulated with a 5000 Hz tone.
45. What is the highest frequency that will be transmitted when a 780 kHz carrier is modulated by a message signal with a range of 300 Hz to 3 kHz?
46. At 100% modulation, what is the ratio of the voltage in one sideband compared to the voltage in the carrier?
47. At 100% modulation, what is the ratio of the power in one sideband compared to the power in the carrier?
48. The term RF, as used in the transmission of signals, refers to (a) random frequencies, (b) radiation factor, (c) ratio of frequencies, (d) rate of failure, (e) radio frequencies.
49. How many characteristics of the carrier wave may be modified to cause modulation?
50. Which characteristic of the AM carrier is held constant? (a) amplitude, (b) frequency, (c) phase
51. Overmodulation causes (a) distortion, (b) excessive bandwidth, (c) interference with other transmissions, (d) all of the above.
52. Modulating and mixing mean the same thing. (T) (F)
53. What is the range of frequencies set aside for use by the commercial AM broadcast industry?
54. How many channels are contained in the AM broadcast band?
55. What is the bandwidth of each of the channels in the AM broadcast band?

56. What is the carrier frequency of the lowest assignable station in the commercial AM broadcast band?

57. What is the carrier frequency of the highest assignable station in the commercial AM broadcast band?

58. What emission code is used to identify an AM wave with full carrier and with upper and lower sidebands?

Using Figure 1.1 for values of E_{max} and E_{min}, determine the following:

59. Find the percentage of modulation when $E_{max} = 132$ V p-p, $E_{min} = 28$ V p-p.

60. Find the percentage of modulation when $E_{max} = 35$ V p-p, $E_{min} = 5$ V p-p.

61. What is the peak-to-peak value of the carrier in Question 59?

62. What is the peak-to-peak value of the carrier in Question 60?

63. When the carrier voltage is 42 V p-p, what values of E_{max} and E_{min} would be displayed on an oscilloscope for 30% amplitude modulation?

64. What values of E_{max} and E_{min} would be displayed on the oscilloscope when a 134 V p-p carrier is amplitude-modulated to 64 percent?

65. What would be the peak-to-peak values of E_{max} and E_{min} when a 96 V p-p carrier is amplitude-modulated to 100%?

66. A 1120 kHz carrier is modulated to 75% by a 3 kHz tone signal. What are the three frequencies that would be transmitted in this signal?

67. What should the bandwidth of the transmitter be in order to transmit a carrier of 680 kHz modulated by a message signal of 7500 Hz?

68. What would be the highest frequency in the transmitted signal when a 1.6 MHz carrier is modulated by a 5 kHz sine wave message signal?

CHAPTER TWO

AM TRANSMITTERS

2.1   INTRODUCTION

A **transmitter** is an electronic system that will convert a message signal into a form that may be sent via an antenna system into the surrounding atmosphere.

A transmitter must be able to:

1. produce a carrier signal that will distinguish it from all other local transmitters.
2. code the carrier signal with the message signal before it is sent to the receiver.
3. supply enough power (energy) to the coded carrier so that it may travel the distance between the transmitter and the receiver.

These requirements apply to all transmitters, regardless of their power ratings. This chapter focuses on medium- and moderate-power AM transmitters but will touch on some extreme cases as required.

2.2   THE CIRCUITS

There are as many different kinds of transmitter circuits as there are people who design them and patents that govern their use. However, the functions of the circuits follow a standard pattern. The three major subdivisions of a basic transmitter are shown in Figure 2.1 within the dashed lines. The carrier signal is generated in the **exciter section.** It is this group of circuits that maintains the fidelity of the station's carrier frequency. The **power section** sets the level of energy to be launched by the system, without affecting signal frequency.

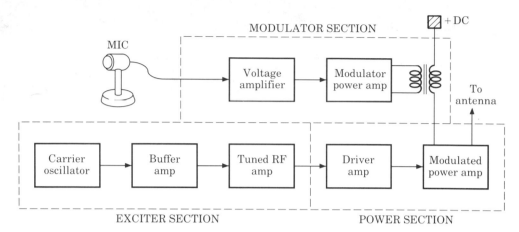

FIGURE 2.1
Basic high-level AM transmitter block diagram.

Modulation can take place in this section. The **modulator** processes the message signal and supplies the *modulating power,* designated earlier as *sideband power.*

2.2.1 Forming the Modulated Signal

Transmitters are categorized according to where the modulation takes place. The two basic categories are high-level and low-level modulation. It is rare to find both types of modulation within one transmitter.

High-level modulation is a process whereby the message signal controls the amplitude of the carrier signal by delivering to the final power amplifier an AC voltage that is *in series* with the DC power supply voltage. **Low-level modulation** describes a process in which the message signal controls the amplitude of the carrier signal at some location within the chain of carrier amplifiers. This control may take place at the base of the final power amplifier or at any previous point in the system, beyond the buffer amplifier. Figure 2.2a is the block diagram of a high-level AM transmitter. A low-level AM transmitter is shown in Figure 2.2b, with the modulator placed immediately after the buffer amplifier.

The advantage of high-level modulation is that all of the carrier signal amplifiers after the buffer can be operated as class C amplifiers for maximum efficiency, and 100% modulation is easily achieved. The disadvantage is that high levels of message power are required to fully modulate the carrier. Consider a transmitter with 5000 W of unmodulated carrier power. At 100% modulation, the sideband power is 50% of the carrier power, or 2500 W, which the modulator would need to supply.

The advantage to low-level modulation is that a small amount of message power will fully modulate the carrier. Modulation takes place in a part of the

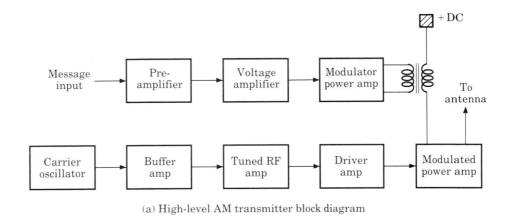

(a) High-level AM transmitter block diagram

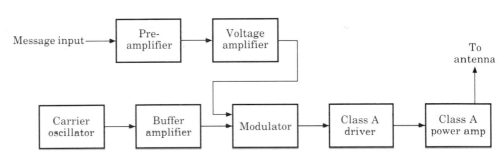

(b) Low-level AM transmitter block diagram

FIGURE 2.2
Block diagrams for high- and low-level modulated AM transmitters.

system where voltage is the prime consideration. The power requirements are taken into account at the final power amplifier, further on in the system. The disadvantage to low-level modulation is that once the carrier has been modulated, all of the following amplifiers in the circuit must be linear class A amplifiers, which suffer from low efficiencies. Occasionally, class B push-pull amplifiers are used to improve efficiency, at the price of a higher component count.

It may be wise at this point to briefly review the difference between a voltage amplifier and a power amplifier. A voltage amplifier provides good voltage gain into a high-impedance load, which demands a small current drain from the power supply. Since $E \times I$ = power, low currents result in low power. A power amplifier provides voltage gain into a low-impedance load, which draws heavy currents from the power supply, and large currents result in high power. Thus, the difference is low load impedance and high currents for power amplifiers, and large load impedance and low currents for voltage amplifiers.

2.2.2 The Exciter

The oscillator in the exciter panel may be any sine wave oscillator that is adaptable to crystal-controlled circuitry. The accuracy of the oscillator frequency is the major concern in any transmitter. Frequency tolerances of 20 parts per million (0.002%) are the rule. That is, a 1 MHz carrier signal would be considered illegal if it drifted more than 20 Hz off its assigned frequency. Penalties of several thousands of dollars or two years imprisonment, or both, are associated with violation of this regulation. Occasionally, penalties are exacted on a "per day" basis, that is, for each day that the violation is allowed to continue.

A crystal-controlled Pierce oscillator and crystal oven are shown in Figure 2.3. Recall that the crystal equivalent circuit is a parallel-resonant or series-resonant tank circuit; this will be the frequency-determining element for the oscillator. The Pierce oscillator uses the crystal in its series-resonant mode.

The oscillator frequency is set by the inductor, L_1, and capacitor C_1 (denoted by asterisks). Once oscillation starts, the crystal, Y_1, assumes full control of the frequency stability. The frequency stability of any oscillator may be improved by the adoption of the following steps, in whole or in part, listed in order of effectiveness.

1. Use a crystal to control frequency.
2. Regulate the DC supply voltage to the oscillator.

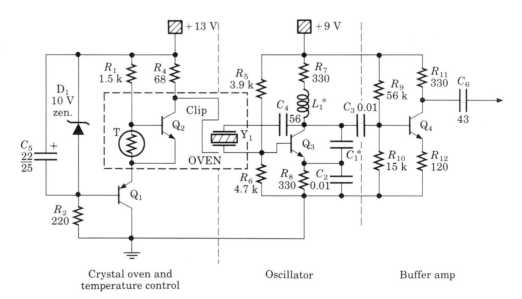

FIGURE 2.3
Pierce crystal oscillator, temperature oven, and buffer amplifier.

3. Hold a constant load at the output of the oscillator. The input impedance of the circuit that follows the oscillator should have a steady, high value.

4. Control the temperature of the crystal by placing it in an oven at an elevated temperature. Maintain the oven temperature at 175° Fahrenheit (80° Celsius).

The Buffer   Following the generation of the carrier signal, some amplification is needed. The amplifier that follows the oscillator is called the **buffer.** The name comes from its function of *buffering* (or isolating) the oscillator from the circuits that follow. The buffer amplifier has a high input impedance and low output impedance. It acts as a low-gain, impedance-matching amplifier. The buffer amplifier of Figure 2.3 has an input impedance of 7.2 kΩ (β = 150) and a voltage gain of about 2.70. Notice that to achieve these goals, the emitter resistor of Q_4 is not bypassed.

The Tuned RF Amplifier   The amount of gain provided by the RF amplifiers will depend on the signal strength leaving the buffer and the signal level needed to drive the final power amplifier at the selected conduction angle. These circuits may be tuned RF amplifiers with extremely narrow bandwidth or frequency multipliers (doublers or triplers as needed for the transmitting frequency). In commercial applications, only one transmitter is used in conjunction with possibly millions of receivers. Therefore, transmitters must reflect *quality* and *reliability*. It is common to find two or three amplifiers in a transmitter doing the corresponding work of one amplifier in a receiver. The gain per stage is far less important than the quality of performance at the transmitter.

2.2.3 The Power Amplifier

The **power amplifier** takes the energy drawn from the DC power supply and converts it to the AC power that is to be radiated. The need for high efficiency becomes apparent when the power to be transmitted exceeds a few watts. A 48 V DC supply to an amplifier that draws 10 mA of current will result in an input power of $E \times I$ = 48 V $\times$ 0.01 A = 0.48 W. If the amplifier is 60% efficient, then 0.48 W $\times$ 0.6 = 0.288 W will be converted to signal energy; the balance of the power, 0.122 W, is lost through heat dissipation. Another amplifier may draw 2.5 A from the same 48 V DC supply for an input power of 48 V $\times$ 2.5 A = 120 W. If the same 60% efficiency is taken into account, then 120 $\times$ 0.6 = 72 W will be radiated and 48 W wasted as heat in the transistor and surrounding circuit. This 48 W could be more power than is required by the rest of the transmitter. For this reason, the final power amplifier is usually a class C amplifier (for highest efficiency) for a high-level modulated transmitter and, very often, a class B push-pull power amplifier for use in a low-level modulated transmitter. The choice of amplifier type depends greatly on the output power intended for the transmitter. Transmitters having a carrier output power of 250 W or less are considered *low-power* transmitters. *Medium-power* transmitters are those with a rated carrier power between 250 and 5000 W. Transmitters that radiate carrier power greater than 5000 W are considered to be *high-power* units.

2.2.4 The Driver Amplifier

In the circuit of a high-level modulated transmitter, the final power amplifier is operated in class C. This means that it will conduct for less than one-half of the signal cycle. It also means that the final transistor is cut off for the majority of the cycle; during this period, no base current flows. Thus, the circuit that supplies the signal to the base of the final power amplifier will see a high input impedance for most of the cycle. During the part of the cycle when the final amplifier is conducting, heavy base current flows, which makes the final power amplifier input appear as a very low impedance. During this period of low resistance and heavy base current flow, the **driver amplifier** must supply a small amount of base power to the final power amplifier. For this reason, the driver amplifier must be a low-power amplifier (0.5–5 W) with low output impedance to match the low input impedance of the final amplifier during the conduction period.

2.2.5 The Modulator

The circuit for a **modulator** in a high-level modulated transmitter is depicted in Figure 2.4. Here we see the message power amplifier with its output transformer secondary connected *in series* with the DC supply voltage for the final power amplifier. The positive peaks of the message signal at the secondary of the transformer will *add* to the DC voltage for the final power amplifier and cause its gain to increase, thus increasing the amplitude of the output signal. The negative peaks of the message signal at the secondary of the modulation transformer will *oppose* the DC supply voltage and decrease the output signal of the final power amplifier. In this fashion, the peak-to-peak message signal alters the amplitude of the carrier signal to form the modulated wave.

2.2.6 The Audio Processor

The circuits that precede the modulator power amplifier fall into one of several categories. The **audio processor** of Figure 2.5 will be used as an example; other transmitters may use video processors, data processors, or multiplexers instead. The audio processor is part of the studio equipment, not part of the transmitter.

The audio console that houses the audio processer is made up of microphone preamplifiers, audio mixers, tone controls, compressors, and line amplifiers. Here many source signals are blended to form the message signal that will ultimately modulate the carrier. Each source signal is monitored for level and tonal values that it will contribute to the total message signal. An audio engineer is positioned at the console during transmission to ensure flawless operation and to act in a responsible manner in case of program change or interruption.

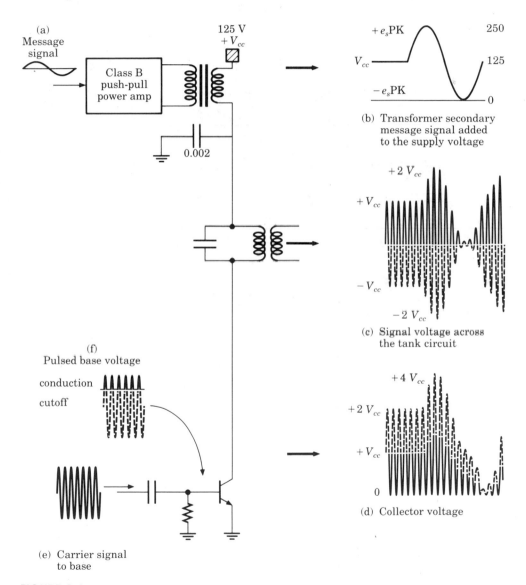

(a) Message signal

125 V $+V_{cc}$

Class B push-pull power amp

0.002

(b) Transformer secondary message signal added to the supply voltage

(c) Signal voltage across the tank circuit

(f) Pulsed base voltage

conduction
cutoff

(e) Carrier signal to base

(d) Collector voltage

FIGURE 2.4

Class B modulator and class C modulated power amplifier with signals that correspond to 100% modulation. Note: ———— ΔE_C due to base input drive signal, -------- ΔE_C due to energy stored in the tank circuit.

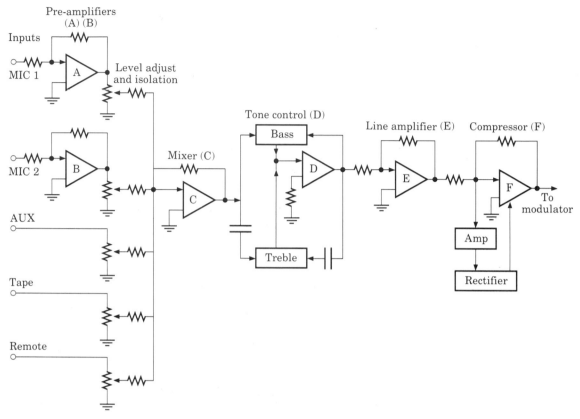

FIGURE 2.5
Typical audio processor block diagram.

Preamplifiers Preamplifiers are needed to raise the amplitude of the signal leaving the pickup device to a level that will facilitate signal processing while preventing system noise from affecting the quality of the information. Transducer output signals (from a microphone or a phono or tape head) are often in the range from a few microvolts to a few millivolts. When the microphone cables are shorter than about 15 feet, then a high-impedance (10 kΩ or higher) microphone will perform well. When longer cables are needed, more electrical noise gets into the system and the quality suffers. Therefore, low-impedance microphones (150 Ω) are almost universally used with long cables. An impedance-matching transformer is used at the input to the preamplifier to convert the low microphone impedance to the high amplifier input impedance.

Compressors A **compressor** is an amplifier whose gain is controlled by a level sensor at the input. If the average input signal changes, the sensor directs

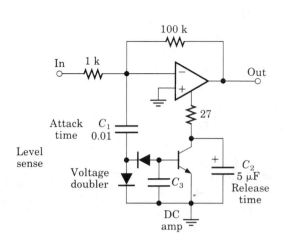

FIGURE 2.6
Compressor amplifier.

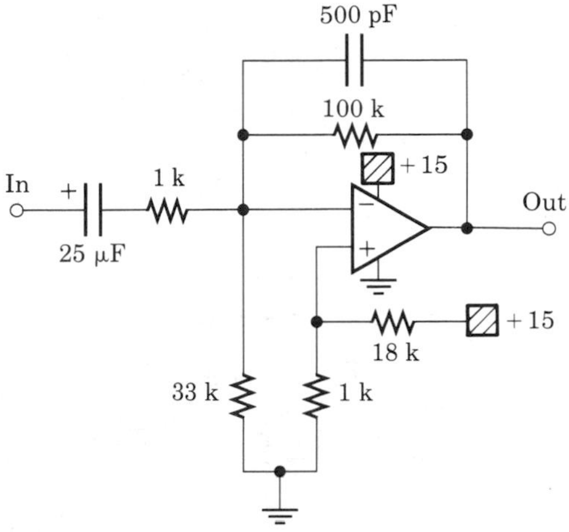

FIGURE 2.7
Line amplifier.

a DC control voltage to the amplifier that alters the gain in such a manner as to hold the average output level constant over a period of time. This keeps the percentage of modulation stable and helps to prevent overmodulation. The sound quality, known as *presence,* is adversely affected when a compressor is used. Loud sounds are softened, and soft sounds are intensified; the listener does not hear the same sounds as he or she would in the "presence" of the performing artist.

The compressor *attack time* is set by capacitor C_1 in Figure 2.6, and the *release time* is controlled by capacitor C_2. The attack and release times are relatively long so that the sound level does not change with each cycle of the audio message, but only with a change in the average signal strength. Typical attack time is 10 ms, with 50 ms release time.

Line Amplifiers The tone controls, mixers, and compressors will produce losses in the system. Figure 2.7 is a line amplifier used to recover these losses and ensure a level suitable for the modulator. Line amplifiers are *quality* voltage amplifiers (wide frequency response and low distortion).

2.3   HIGH-LEVEL MODULATION

The transistor amplifier in Figure 2.8 is a class C amplifier, as indicated by its bias arrangement. There is no DC supply voltage to the base for biasing. However, the base must be reverse-biased for the transistor to be held in cutoff for more than one-half of the input signal cycle.

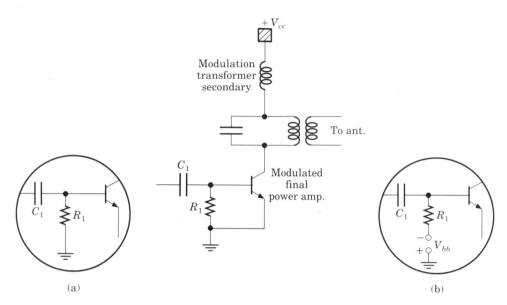

FIGURE 2.8
Modulated final class C amplifier. (a) Self-bias. (b) Fixed bias.

The two most common bias arrangements are shown in the inserts. The method shown in insert (a) is used for all low-power transmitters and is called *self-bias, signal bias,* or *base leak bias.* The negative DC voltage that holds the base in cutoff is developed by the signal charge across the coupling capacitor C_1 and is defined by the approximation:

$$-V_b = -V_{pk} + 0.6 \text{ V} \tag{2.1}$$

A 5 V_{pk} signal will place the base voltage at -4.4 V DC. On the positive peak of the input signal cycle, the coupling capacitor is charged through the base-emitter diode so that the voltage across the capacitor is negative toward the base. The base is negative with respect to the emitter and with respect to ground. On the negative half cycle, the base-emitter diode is reverse-biased and the capacitor will discharge through the base resistor R_b, which has a much longer RC time. The base will be held at a negative DC voltage with respect to the emitter.

The disadvantage of self-bias is that if the exciter should fail, the bias will be removed from the final amplifier, and the amplifier may go into very heavy conduction and self-destruct. For this reason, a negative bias supply ($-V_{bb}$) is often connected as shown in insert (b) of Figure 2.8. With this arrangement, if the exciter fails, the battery bias will hold the amplifier in cutoff and prevent burnout. Fixed negative DC bias is more common in higher-power transmitters.

2.3.1 Class C Power Amplifiers

Recognizing the important role played by the class C power amplifier in RF carrier transmitters is essential to understanding all types of RF power transmissions.

The following transmitter conditions will be used to examine the behavior of a high-level amplitude-modulated class C amplifier. The output power rating is 100 W (unmodulated carrier power) into a 50 Ω antenna at 27 MHz ($\pm 0.002\%$). The amplifier is assumed to be 80% efficient when working from a 125 V DC supply voltage, as seen in Figure 2.4. Efficiency is the amplifier's ability to take power from the DC supply, called input power, or $E \times I$, and convert it to signal power, called output power. 100 W output power divided by the efficiency (0.80) equals 125 W input power. Power divided by voltage (125 W/125 V) results in 1 ampere of DC current flow through the final amplifier.

The peak signal voltage at the final power amplifier cannot exceed the DC supply voltage, so that when these voltages match and the maximum power of 125 W is known, we can determine the peak signal current as

$$I_{\text{pk}} = \frac{P_{\text{in}}}{0.5V_{cc}} \tag{2.2}$$

$$= \frac{125 \text{ W}}{(0.5)(125)\text{V}} = 2 \text{ A}$$

Figure 2.9 shows a family of collector characteristic curves for a 2N5241 npn silicon transistor with the following ratings:

Voltage (C-E) (max) = 400 V DC
Collector current (max) = 5 A
Collector dissipation (max) = 125 W
Base current (max) = 2 A
DC β = 50

The load line drawn in Figure 2.9 between the supply voltage of 125 V DC and the peak current of 2 A identifies the load impedance as

$$Z_L = \frac{V_{\text{ceq}}}{I_{\text{max}}} \tag{2.3}$$

$$= \frac{125 \text{ V}}{2 \text{ A}} = 62.5 \text{ } \Omega$$

The output power is found as

$$P_o = \frac{(V_{cc} - V_{\text{sat}})^2}{2Z_L} \tag{2.4}$$

$$= \frac{(125 - 13)^2}{2(62.5)} = 100.352 \text{ W}$$

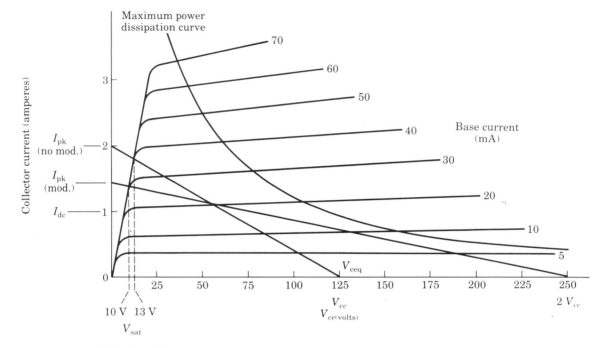

FIGURE 2.9
Collector characteristic curve for a 2N5241 power transistor.

Figure 2.9 also shows that the collector voltage should not drop below the saturation voltage (V_{sat}) of 13 V, where the transistor stops working like an amplifier, and it should never be operated in the shaded area, where it will dissipate over the maximum collector power rating. This area is found by solving for all values of collector current, expressed as maximum power divided by all values of DC voltage from 0 V to $2V_{cc}$:

$$I = \frac{P}{E} = \frac{125 \text{ W}}{\text{DC V}} \quad \text{(in amperes)} \qquad (2.5)$$

Figure 2.10 compares the change in collector voltage (E_c) and the change in collector current (I_c) over one half cycle of the carrier signal, for the 140-degree period over which the final power amplifier is conducting. When the current is maximum, the voltage is minimum. This comparison serves to show that the maximum collector power dissipation (P_c) does not occur when the current is maximum, but rather at points one-third and two-thirds of the way through the conduction cycle.

The amplifier is turned off (and cooling) for more than one-half of each cycle. Therefore, the power dissipated by the collector is a function of the con-

FIGURE 2.10
Voltage, current, and power relationships for the collector.

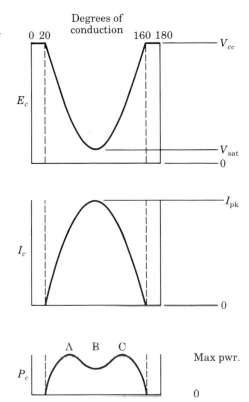

duction angle and the rms values of collector voltage and current. A close approximation of collector dissipation may be found as

$$\text{Power dissipation (collector)} = \sin\frac{\theta}{2}\left(\frac{E_{dc} \times I_{dc}}{4}\right) \tag{2.6}$$

$$= \sin\frac{140}{2}\left(\frac{125 \text{ V} \times 1 \text{ A}}{4}\right) = 29.4 \text{ W}$$

Adding the power dissipated to the desired output power of 100 W gives the total power drawn from the DC supply as 129.4 W. The efficiency is $P_{out}/P_{in} \times 100 = 77.3\%$, rather than the 80% predicted at the beginning of this discussion. The DC current equals total power divided by total voltage (129.4 W/125 V) = 1.035 A, instead of the 1 ampere predicted.

The Collector Load The impedance of the tuned circuit of the collector in Figure 2.4 or Figure 2.8 represents a parallel-resonant circuit where the inductive element is the primary of the output transformer used to couple energy to the antenna (see Figure 2.11). By rearranging the equation for resonant frequency and using a frequency of 27 MHz, the values of L and C may be determined:

$$LC = \left(\frac{0.159155}{f}\right)^2 = 3.4747 \times 10^{-17}$$

When $L = C$, then either value is $\sqrt{(3.4747 \times 10^{-17})} = 5.8946 \times 10^{-9}$. It is good to have a large L-to-C ratio (pick 2500:1). This ratio results in $50 \times 5.8946 \times 10^{-9} = 0.29473$ μH of inductance and $5.8946 \times 10^{-9}/50 = 118$ pF of capacitance. The closest standard capacitance is 120 pF, so L is recalculated to be 0.29 μH of inductance resonating with 120 pF.

The Base Circuit   Using signal bias to hold the final power amplifier's period of conduction to the desired percentage of the carrier cycle requires examination of the base characteristics for the selected device, in this case the 2N5241 npn transistor. The prescribed action is to charge the coupling capacitor, through the base-spreading resistance (r'_b) and the base-emitter diode, to the peak value of the 27 MHz carrier voltage.

The base-spreading resistance has a value between 25 Ω and 150 Ω for small-signal transistors but can be quite small for power transisters, often from 3 Ω to 25 Ω. The transistor equivalent circuit is shown in Figure 2.12b, and the base resistance can be found from the base current curve in Figure 2.12c. Select the most linear part of the curve and compare the *change* of base-emitter voltage to the *change* in base current. Then, from Ohm's law,

$$R = \frac{E}{I} = \frac{1.0 \text{ V} - 0.6 \text{ V}}{42 \text{ mA} - 15 \text{ mA}} = 14.8 \ \Omega$$

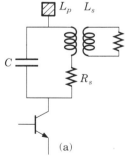

(a)

Measurable values
of collector load
impedance. R_s is
the coil resistance.

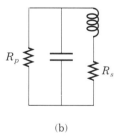

(b)

R_p is the load
that the primary
will see when the
secondary is
loaded with R_L.

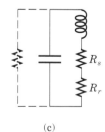

(c)

R_p is converted
to its series
equivalent (R_r),
which adds to R_s.

FIGURE 2.11
Electrical equivalent of the collector load impedance for a tuned resonant transformer with loaded secondary.

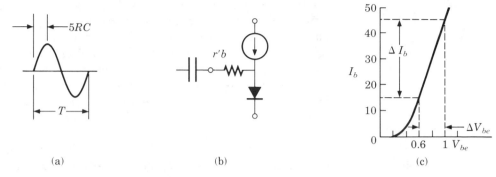

FIGURE 2.12
Base characteristics.

The time for one cycle at the carrier frequency of 27 MHz is shown in Figure 2.12a to be 0.037 μs. The coupling capacitor must charge to the peak value of the input wave within one-quarter of one cycle, which is equal to $5\,RC$ time constants:

$$\frac{T}{4} = 5RC$$

Since time is the reciprocal of frequency and $R = r'_b$, we can solve for the value of C as follows:

$$C = \frac{1}{20fR} \tag{2.7}$$

$$= \frac{1}{(20)\,(27 \times 10^6)\,(14.8)} = 125.1 \text{ pF}$$

We will use a standard value of 120 pF.

Applying the carrier signal to the input terminals will charge the coupling capacitor through the base-emitter diode and the base-spreading resistance r'_b, producing a value of "almost" V_{pk}. This voltage will be negative on the base side of the capacitor with respect to the emitter (ground). It is important that the base remain at a negative voltage in order to hold the transistor in cutoff for more than one-half of the cycle. For this to happen, the RC discharge time for the coupling capacitor must be longer than the charge time by several orders of magnitude. The base-emitter diode is reverse-biased during the negative peak of the input cycle, and the coupling capacitor discharges through the base resistance R_b. Typically, the base resistance is ten times the base-spreading resistance, $R_b = 10r'_b$. For this example,

$$R_b = 10(14.8) = 148 \ \Omega \qquad \text{(use 150 } \Omega\text{)} \tag{2.8}$$

Conduction Angle The selection of a conduction angle is a compromise between the greatest output power and the best efficiency. It is understood that when the conduction angle is 180°, the rms value of a half-wave peak signal will be 0.7071 times the peak voltage. When the angle is made smaller than 180°, the area under the curve will be reduced; the rms value will be less than 0.7071 by a factor of sin θ/2. Thus, as the conduction angle decreases, the rms output power also decreases. On the other hand, when the conduction angle is made smaller, the transistor is on for a smaller part of the input cycle, and the efficiency increases. The best compromise between efficiency and output power is an angle somewhere between 120° and 150°. The present example will use an angle of 140°.

In Figure 2.9, the load line crosses the position of V_{sat} at a base current of 40 mA. This information can be used to determine the required input signal.

$$V_{pk(in)} = \frac{V_{sat} - 0.05V_{cc} + 0.6}{1 - \cos 140/2} \tag{2.9}$$

$$= \frac{13 - 6.25 + 0.6}{1 - \cos 70} = \frac{7.35}{0.658} = 11.17 \ V_{pk}$$

The DC base voltage is, then,

$$-V_{DC} = -V_{pk} + 0.6 + V_{sat} - 0.05V_{cc} = -3.82 \ V \ DC \tag{2.10}$$

Care must be taken to ensure that the positive peak base voltage never approaches the collector saturation voltage, V_{sat}. In this case, V_{pk} is about 11.17 V positive from the DC base voltage of -3.82 V, so the most positive base voltage is 7.35 V, almost 6 V lower than the saturation voltage of $+13$ V. Always try to keep the peak base signal voltage lower than V_{sat} by at least 5% of V_{cc}. A close approximation of the conduction angle is

$$\theta = 2 \ \cos^{-1}\left[1 - \left(\frac{V_{sat} - 0.05V_{cc} + 0.6}{V_{b(pk)}}\right)\right] = 140° \tag{2.11}$$

2.3.2 Applying the Modulating Signal

When the message signal is fully applied through the modulator, one of two extreme cases can exist:

1.   When the peak message signal on the secondary of the modulation transformer is equal in value but opposite in polarity to the DC supply voltage, the two voltages will cancel. V_{cc} to the final amplifier will become zero, and the output signal will drop to zero.
2.   When the peak message signal on the secondary of the modulation transformer is equal in value and of the same polarity as the DC supply voltage, the two voltages will add together. V_{cc} will become $2V_{cc}$ and the output signal will be twice as great as the carrier signal alone.

Doubling the voltage across a circuit will normally double the current and quadruple the power. However, we know that this is not the case for 100% amplitude modulation. Examining the output load impedance is the easiest way to explain why this is so.

An 85 V_{rms} carrier voltage is set up across a 72.25 Ω antenna load. The load current is found to be 1.1765 A_{rms} and, by any of the power equations, the carrier power is estimated at 100 W.

When the voltage to the final power amplifier is $2V_{cc}$ (at the crest of the modulating wave), the voltage across the load doubles to 170 V_{rms}. This follows the principle that the sum of the voltages in the lower sideband, the carrier, and the upper sideband equals two times the carrier voltage alone (42.5 + 85 + 42.5 = 170). However, although the current doubles, it does not add directly as the voltage does. The total current is the square root of the sum of the squares of the individual currents in the lower sideband, the carrier, and the upper sideband:

$$I_{tot} = \sqrt{I_l^2 + I_c^2 + I_u^2} \qquad (2.12)$$

$$= \sqrt{0.58825^2 + 1.1765^2 + 0.58825^2} = 1.44 \text{ A}$$

Thus, $I^2 R = 1.44^2 \times 72.25 = 150$ W. This supports the principle that the output power at 100% modulation increases to 1.5 times the unmodulated carrier power.

2.3.3 The Modulator Amplifier

A class B push-pull amplifier is shown in Figure 2.13 as a typical power modulator. This circuit must supply the 50 W required for 100% modulation of the carrier. Class B amplifiers are about 60% efficient; therefore, this system will consume about 84 W from the 125 V DC power supply, drawing about 1 A of DC current.

Figure 2.13 shows the collector characteristic curves for the 2N3055 transistor. They are in push-pull configuration and are therefore shown one inverted relative to the other, so that as one transistor turns on, the other turns off. Also notice that the curves overlap slightly. Figure 2.13 shows the waveform for the two transistors in push-pull, each zero-biased and conducting for exactly 180° of the input cycle.

Notice the crossover distortion. To overcome this, each of the transistors must be forward-biased slightly to provide a trickle of collector current flow. The 7.2 kΩ and 43 Ω voltage divider biasing resistors set a base voltage of 0.74 V for each transistor, causing 40 mA of collector current to flow, thus eliminating the crossover distortion. Class B amplifiers are identified by the high ratio of the base biasing resistors. For R_7 and R_8 in Figure 2.13, this ratio is 7200/43 > 100/1. Class A amplifiers have a resistance ratio of 5 to 1.

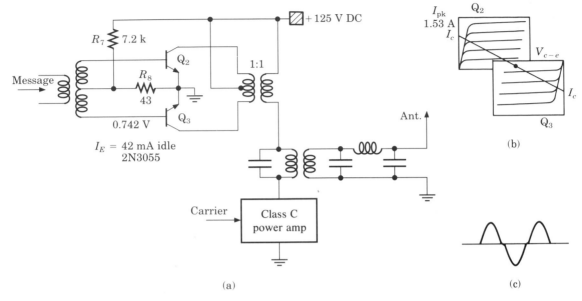

FIGURE 2.13
Class B modulator.

2.3.4 Heising Modulation

The modulation transformer for the preceding example would weigh 10 pounds for a 100 W output. For a 100 kW transmitter, a modulation transformer of 150 cubic feet (5′ × 5′ × 6′) and weighing several thousands of pounds would not be unrealistic. Eliminating this bulk and cost would be a distinct advantage. R.A. Heising, who is credited for inventing the collector modulation system in the 1930s (then called plate modulation), also developed a system to eliminate the modulation transformer. Named after its inventor, **Heising modulation** is outlined in Figure 2.14.

In this diagram, the modulated power amplifier (Q_6) and the power modulator (Q_5) are connected in parallel, and both draw their collector current through the series-connected RF choke L_3. As modulation is applied, the collector current of Q_5 increases and causes an increase in the voltage drop across the RF choke. This increased voltage drop lowers the collector voltage of both the modulated amplifier and the modulator. The decrease in the collector voltage of the modulated amplifier lowers its collector current. The sum of the two currents will remain constant, with or without modulation; hence, this configuration is also referred to as **constant-current modulation.**

Originally, a serious disadvantage to Heising modulation was that only about 80% modulation could be realized. This setback was overcome by adding a resistor R_3 in series with the modulated amplifier such that at zero modulation, the collector voltage of Q_6 is 80% of the supply voltage. In this manner,

FIGURE 2.14

Heising modulation or constant-current modulation.

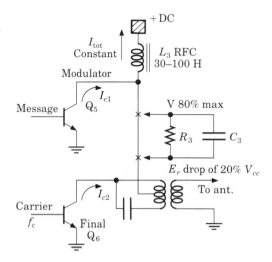

when modulation is applied, the 80% change in the collector voltage of Q_6 corresponds to 100% modulation. Capacitor C_3 bypasses the message signal around R_3 for 100% modulation. The loss of power in resistor R_3 was never overcome, limiting the use of Heising modulation to low-power transmitters.

2.3.5 Modulated Driver Amplifiers

Low-power transmitters, such as those used in citizens band radios, are outlined in Figure 2.15. The message signal applied to the base of modulator Q_7 causes its emitter to change according to the change in message voltage. The emitter of Q_7 is also the voltage supply point for the final power amplifier Q_9. Altering the supply to Q_9 in this manner is similar to modulating the collector of the final amplifier in a Heising-modulated system.

Modulator transistor Q_7 now acts like the RF choke in the Heising modulator and thus suffers some voltage loss. To counteract this, the emitter of Q_7 is also the supply for driver amplifier Q_8. Because of the resistive voltage divider network, Q_8's supply voltage does not change as much as Q_9's supply voltage.

The system works as follows: a message signal is fed to Q_7, causing its emitter to move away from ground. The collector current of Q_9 increases and increases the voltage drop across Q_9's collector load impedance, lowering the voltage at Q_9's collector (i.e., a larger output signal). The emitter voltage of Q_7 also increases the collector voltage of Q_8, but by a lesser amount. The rise in collector voltage of Q_7 is coupled to the base of Q_9 and causes the collector of Q_9 to move closer to ground. This increases the voltage drop across Q_9's collector impedance, producing an increase in signal output voltage. This circuit is similar in performance to the Heising modulator in some respects, and has proven effective in low-power transmitters.

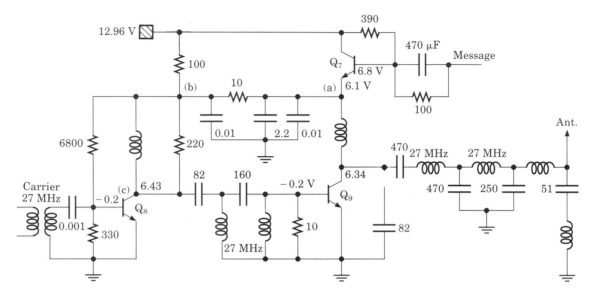

FIGURE 2.15
Modulated power amplifier and modulated driver amplifier.

2.3.6 Parallel Output

It is not uncommon to find power amplifiers connected in parallel to double the current-handling capacity of a system. Figure 2.16 shows such a circuit configuration. Ferrite beads are placed on the base wires to minimize parasitic oscillations. The beads may be anywhere on the wire and act as low-cost but effective series inductances. The resistors in series with the collectors serve the same purpose. The resistors are low-value but, because of the high currents, must be high-wattage. One to 2 Ω and 3 to 5 W are commonplace. Doubling the current is equivalent to quadrupling the power.

2.3.7 Heat Sinking

Transistors that are configured to dissipate more than 1–2 W without assistance are doomed to a short life. No one would dare touch a 25 W pencil-point soldering iron. Neither should a transistor (with a much smaller area) be expected to tolerate such high temperatures without additional provisions. **Heat sinks** range in size from small snap-on fan blades to large, finned metallic blocks. Each is designed to provide the greatest surface area for heat transfer to the surrounding air by convection. Often, the collector is the *case* of the transistor, so when a heat sink is used, the case must be insulated from ground. A thin mica washer set between the transistor and the heat sink is best for this use. Such a washer is thin enough to allow good heat transfer, yet is one of the best electrical insulators. Silicon grease is applied to both sides of the mica washer. The washer may be omitted if the transistor is bolted directly (electrically connected) to the heat sink, but then the heat sink must be insulated from ground. Again, silicon grease is used.

The heat sink size is determined from the transistor specifications. The 2N5241 transistor is rated with a maximum junction temperature of 150° C, and the collector thermal resistance is 0.7° C per watt. The transistor ratings are established at room temperature of 25° C (77° F). Say an installation dictates a maximum operating temperature of 52° C (125° F). In the 100 W transmitter we have used as an example, the collector dissipated 29.4 W. Allowing a small safety factor (up to 32 W dissipation), we can select a heat sink using the following equation:

$$P_{\text{diss}} = \frac{T_j - T_m}{\text{factor}} \tag{2.13}$$

where P_{diss} = a set level of dissipation
factor = combined thermal resistance of the transistor,
   the mica washer, and the heat sink
T_j = maximum junction temperature
T_m = maximum ambient temperature

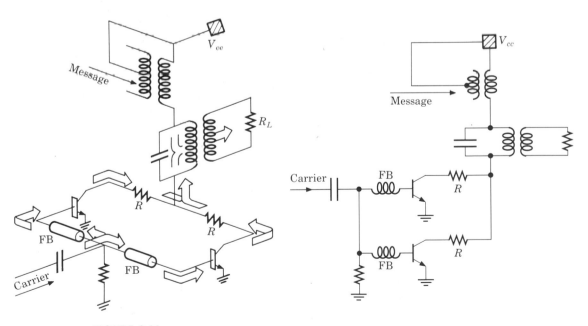

FIGURE 2.16
Parallel-connected class C power output amplifiers.

For the present example,

$$32 \text{ W} = \frac{150 - 52}{\text{factor}} \quad \text{or} \quad \text{factor} = \frac{150 - 52}{32 \text{ W}} = 3.06°\text{C/W}$$

The transistor has a thermal factor of 0.7 and the mica washer a factor of 0.5; therefore, the heat sink must have a thermal factor of $3.06 - 0.7 - 0.5 = 1.86°$ C/W. Suppose manufacturer A has a product that has a factor of 0.465° C per watt per inch. We divide 1.86 by 0.465 and find that 4 inches of the product are needed for a heat sink.

2.3.8 Progressive Series Modulation

Possibly the most attractive system to be developed in recent years is the **progressive series** high-level modulation arrangement. It boasts the advantages of eliminating the modulation transformer and overcoming the 20% power loss of the Heising system, plus it offers automatic power supply voltage selection depending on the percentage of modulation. Both steps increase the overall transmitter efficiency.

The circuit of Figure 2.17 can be analyzed in three steps. First, see how the circuit behaves for zero modulation. Transistor Q_1 conducts and holds the parallel pair of transistors Q_3 and Q_4 firmly, but just barely, in cutoff. (Henceforth Q_3 and Q_4 will be referred to as Q_{3-4}.) Transistor Q_2 is in a conducting state that holds Q_5 in a fully on (saturated) condition. With Q_{3-4} turned off, the 140 V DC supply is disconnected from the system. The 71 V DC supply is the power source for the carrier power amplifier. One volt is dropped across the diode and 2 V are dropped across transistor Q_5, leaving 68 V DC (at 18.383 A) to run the carrier amplifier. Thus, the input power to the final amp is 1250 W.

On the negative half cycle of the audio signal, Q_2 drives Q_5 toward cutoff, dropping the full 70 V across Q_5 and leaving 0 V as the carrier power amplifier voltage. The antenna signal is at the trough of the modulated wave. After the negative signal has bottomed out, it starts to climb again. It passes through the reference, where Q_5 is again saturated, the DC voltage to the final amp is again 68 V, and the modulated wave is again at the carrier's unmodulated power level.

Moving into the positive half cycle of audio, Q_5 remains saturated and Q_{3-4} is brought out of cutoff. As soon as Q_{3-4} begins to conduct, its emitter surpasses 70 V DC, diode D_1 is reverse-biased, the 70 V DC supply is disconnected, and the system runs on the 140 V DC supply for the full positive half cycle of the audio signal. At the crest of the positive half cycle, Q_{3-4} and Q_5 are saturated, about 2 V DC is dropped across each transistor, and the remainder, 136 V DC, is applied to the carrier final power amplifier. The power of the fully modulated carrier wave is 1875 W. The positive half cycle ends by turning off Q_{3-4}. The 140 V DC supply is disconnected, and the 70 V DC supply takes over for the start of the next cycle. (1875 W = 1.5 × 1250 W = 100% mod.)

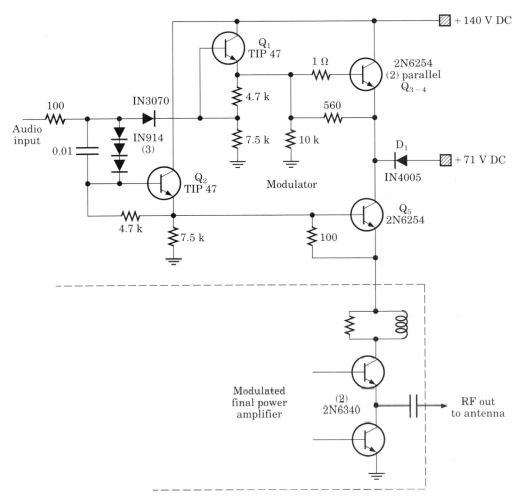

FIGURE 2.17
Progressive series modulation. (Courtesy of the Harris Corporation, Broadcast Division, Quincy, Illinois)

2.4   LOW-LEVEL MODULATION

Modulation may be implemented at any location in the carrier system after the buffer amplifier. When modulation takes place at or before the base circuit of the final power amplifier, then the transmitter is classed as a low-level modulation system. There are a great number of different circuits that can be used to perform this function. One common circuit is shown in Figure 2.18. Here, the RF carrier signal is transformer-coupled to the base of the modulator, while the modulating message signal is capacitively coupled to the emitter.

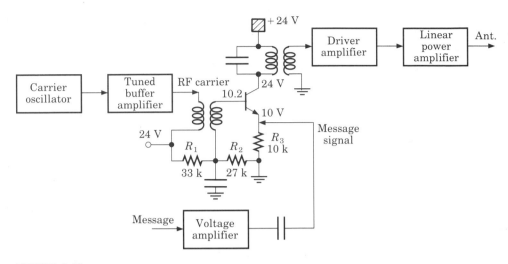

FIGURE 2.18
Low-level amplitude modulator.

The emitter resistor is larger than in the general amplifier case (10 kΩ) to hold the transistor slightly in conduction for nonlinear mixing (modulating). The voltage divider bias network R_1 and R_2 set the base voltage at 10.2 V DC above ground, causing about 1 mA of emitter current to develop 10 V of DC emitter voltage. Basic transistor theory for voltage gain states

$$A_v = \frac{r_c}{R_e + r'_e} \tag{2.14}$$

where r_c = the AC collector load impedance
R_e = the external emitter resistance
r'_e = the emitter diode internal resistance

Changing the emitter resistance changes the gain. The same effect is accomplished through changing the emitter voltage by feeding the message signal to the emitter. Good modulation results can be achieved when the message signal is ten times larger than the carrier signal, provided the two signals do not cause clipping. The tuned collector circuit and the tuned circuits in the RF amplifiers that follow the modulator will pass only the modulated carrier with its sidebands, filtering out all of the harmonics of the modulated wave.

2.4.1 Base Modulation

When the base of the final class C power amplifier is modulated as shown in Figure 2.19, the circuit acts almost the same as the low-level modulation scheme in the preceding discussion with one exception. Note that the base has a

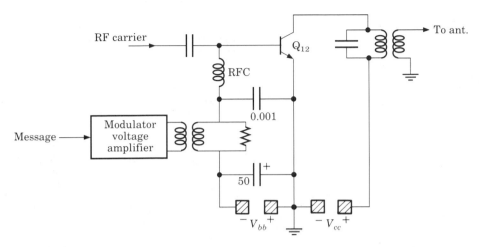

FIGURE 2.19
Base modulation (always use fixed bias).

fixed bias voltage from a negative DC supply $(-V_{bb})$. The base will not modulate when signal bias is used alone because of the small impedances used with signal bias. The base modulation circuit requires larger impedances.

2.5   VACUUM TUBES

Vacuum tubes are still extremely popular in high-power transmitters, where the operating frequencies range from 500 MHz to several tens of gigahertz. All of the circuit functions discussed so far were originally developed using vacuum tubes.

Collector modulation was developed as high-level *plate* modulation. The circuit for a plate-modulated *triode* appears in Figure 2.20a. The output tube could be the *pentode* or *beam power tube* of Figure 2.20b, for more efficient operation and greater power handling ability. The *screen grid* of the pentode could be modulated, as shown in Figure 2.20b, or the *suppressor grid* could be modulated, as in Figure 2.20c. All of the circuits in Figure 2.20 are variations of high-level series modulation for vacuum tubes. Each has a slight advantage in some area and a shortcoming in another area. Note the directly heated cathode, used to reduce internal capacitances, in each part of the figure.

One should be familiar with the elements of vacuum tubes and have at least an introductory exposure to vacuum tube parameters. Highly beneficial is a working knowledge of vacuum tubes coupled with the ability to interrelate vacuum tube values with those of their transistor counterparts. Figure 2.21 is a comparison diagram of the equivalent elements for FETs, bipolar transistors, and the varieties of tubes. Element (a) in one sketch, for example, is the

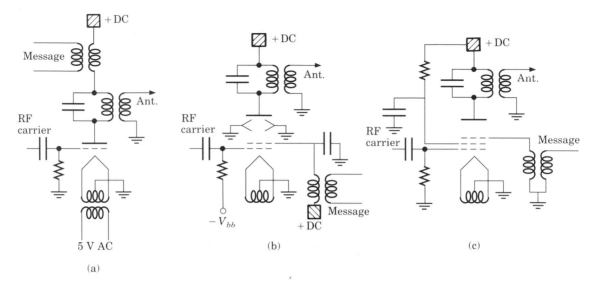

FIGURE 2.20
Series-modulated class C power amplifier tubes. (a) Plate-modulated triode.
(b) Screen-grid-modulated beam power pentode. (c) Suppressor-grid-modulated
pentode.

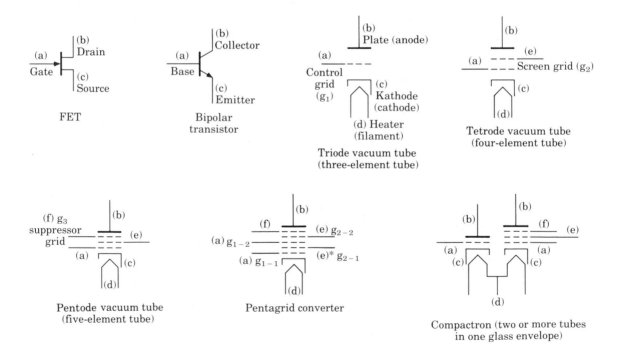

FIGURE 2.21
Elements of active devices. (Lowercase letters a, b, c, etc. denote similar functions in
each case.) Note: Pentagrid converter is a triode and pentode in series. The third element up acts as a plate for the triode at the bottom, and a cathode for the pentode
at the top.

TABLE 2.1
Comparison of parameters for bipolar transistor, FET, and vacuum tube technologies

	Bipolar	FET	Tubes	Units
Amplification factor (ratio of output to input)				
h_{fe}	$\beta = \dfrac{\Delta I_c}{\Delta I_b}$	$A = \dfrac{\Delta V_D}{\Delta V_{gs}}$	$u = \dfrac{\Delta E_p}{\Delta E_g}$	None
Transconductance (input voltage change to output current change)				
h_{oe}	$g_m = \dfrac{I_c}{(I_b r'_b) + 0.6}$	$g_m = \dfrac{I_D}{V_{gs}}$	$g_m = \dfrac{I_p}{E_g}$	Mhos (amps/volts)
Dynamic resistance (drain-collector-plate)				
h_{re}	$r_c = \dfrac{v_c}{i_c}$	$r_{Ds} = \dfrac{e_D}{i_D}$	$r_p = \dfrac{E_p}{I_p}$	Ohms

counterpart to element (a) in another. Table 2.1 draws a parameter comparison between the three different varieties of active devices. No attempt will be made to instruct in this area, so the table should serve only as a reminder of the similarity of the active devices.

A possible advantage of tubes over solid state devices is the physical size of the plate compared to the size of the collector. In a high-power vacuum tube, the plate may be 40 square inches, compared with 2 square inches of collector area in a power transistor. The large size makes it easier to conduct heat away from the amplifier. In vacuum tubes, the elements are separated by greater distances, and this reduces the interelectrode capacitances, which is beneficial for high-frequency work.

The heating element of the vacuum tube has always been the strongest disadvantage. One local TV station uses four 6166A tubes in the power stages of its transmitter. One 6166A is the power driver, and three 6166As are connected in parallel for the final power amplifier. In the final stage, a 6000 V DC supply draws 6 A (2 for each tube), for an input power of 36 kW. Since the transmitter operates at 78% efficiency, 28 kW are fed to the antenna with 10.5 dB gain, for an ERP of 315 kW. The final power amplifier dissipates 8000 W in heat. To this must be added the power loss for heating the filaments. Each filament requires 5 V at 181 A, or 905 W heater power *per tube*. Thus, four tubes use 3620 W for the filaments, added to 8000 W lost in the conversion of the AC signal in the final stage for a total power loss of 11620 W. Considering the present rate of technological advance, a permanent solid state replacement for the vacuum tube seems inevitable.

2.6   CLASS E AMPLIFIERS

The waveforms for the class C amplifier of Figure 2.10 should be reviewed at this time. Recall that power dissipated equals $E \times I$, that collector voltage is high when the current is low, and that current is high when the voltage is low;

this implies that almost no power is dissipated. At points A and C in Figure 2.10, the collector voltage is still fairly high and the current is well on its way toward peak. It is at these two points that the power dissipation is greatest. At point B the instantaneous current has reached its maximum, but the voltage is at a minimum; the product EI is less than at points A and C. In class C, D, and E amplifiers, the collector dissipation is averaged out over the entire cycle and is considerably less than the peak power at any instantaneous point during the cycle. When extreme care is taken in the design of a class C power amplifier, such as accelerating the rate of decrease in collector voltage while retarding the rise in collector current, the power dissipated by the collector will be less than normal. This controlled timing of the collector voltage and current has earned the name "high-efficiency tuned switching power amplifier." Efficiencies of 90% are claimed by patent 3,919,656.

2.6.1 Neutralization

One part of the circuit for power amplifiers has been deliberately omitted up to this point to avoid coverage of too many new principles at one time. Most of the circuits viewed in this Chapter so far normally include a neutralizing feedback capacitor. **Neutralization** of a circuit means canceling the effects of the inter-electrode capacitances. *All* active devices have interelectrode capacitances. The capacitance that is most offensive is that between the base and collector (grid and plate for tubes). Most amplifiers have 180° phase shift from base to collector, but an additional phase shift can occur through the reactance of the small value of base-collector capacitance, allowing a feedback signal to the base that is close enough in phase to cause parasitic oscillations. Commonly, these oscillations are at frequencies higher than the RF signal being amplified. Parasitic oscillations drain power from the amplifier and cause overheating and distortion.

Hazeltine Neutralization   One method of canceling these capacitances is shown in Figure 2.22 and was developed by Hazeltine. The bridge circuit is Hazeltine's cancellation network. The inductor of the tuned circuit is center-tapped and connected to the DC supply (AC ground). The signal voltage across the inductor will have a polarity at one end opposite to the polarity at the other end. When the collector (a) is positive compared to AC ground (b), then point c will be negative compared to ground (b). When C_n is adjusted to equal C_{BC}, the feedback voltages will be equal and opposite, canceling each other. Hazeltine is the first of four circuits commonly used to neutralize high-frequency amplifiers.

Rice Neutralization   This method of neutralization uses base inductance. When the input transformer center tap is grounded and the feedback signal is attached to the end away from the base, the circuit is neutralized. Figure 2.23 shows the *Rice* or *base (grid) neutralization* system.

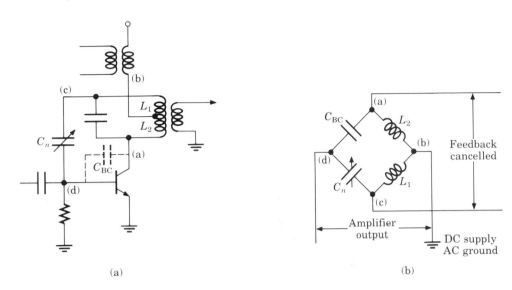

FIGURE 2.22
Hazeltine neutralization. (a) Collector neutralization. (b) Equivalent bridge circuit.

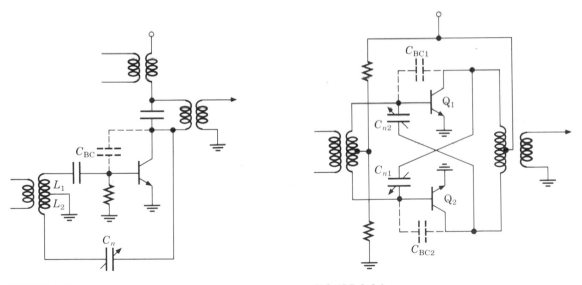

FIGURE 2.23
Rice or base neutralization.

FIGURE 2.24
Cross-neutralization of push-pull RF amplifier.

Cross-neutralization Another neutralization circuit cancels the interelectrode capacitances in a push-pull amplifier. Here there is no need to create a phase shift. The signal at one collector is already at the correct phase for the base of the other transistor. Simply cross-connect the base of one transistor to the collector of the other and neutralization is complete. *Cross-neutralization* is shown in Figure 2.24.

FIGURE 2.25
Direct or inductive neutralization. C_n
is a large-value capacitor and has
a short-circuit effect at RF frequencies.

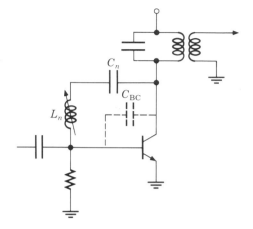

Direct Neutralization A very popular high-power, single-ended neutralization circuit is shown in Figure 2.25 and is called *direct* or *inductive neutralization*. It was determined that an inductor in parallel with C_{BC} would form a
parallel-resonant circuit and a high impedance at resonance for the offending
parasitic signal frequency. Figure 2.25 shows the neutralizing inductor L_n for
this system. Any large-value capacitor in series with the inductance will prevent the DC at the collector from affecting the base DC voltage but will have no
effect on the resonant circuit.

Interelectrode capacitance values for tubes range from about 1 to 5 pF.
For transistors the range is about 3 to 7 pF. The neutralizing capacitor need not
be large, only variable, so that it may be adjusted to exactly match the interelectrode capacitances. C_n is usually adjustable over the range of 1 to 15 pF.
Figure 2.26 presents oscilloscope views of varying degrees of parasitic oscillation.

Neutralization is most often used in amplifiers meeting any of the following conditions.

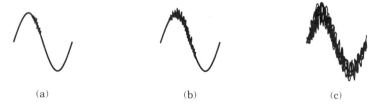

(a)   (b)   (c)

FIGURE 2.26
Various degrees of parasitic oscillations. (a) Causes overheating only. (b) Causes
overheating and distortion. (c) Severe overheating and distortion.

1. in tuned RF amplifiers, where the input and output are tuned to the same frequency
2. at frequencies generally over 25 MHz
3. in circuits when the active device has three elements (transistors and triodes)

These conditions are general rules of thumb only. Untuned audio power pentodes, for example, also have been known to exhibit parasitic oscillations.

2.7 TRANSMITTER ADJUSTMENTS

All transmitter adjustments will be pointless if the carrier oscillator is off frequency. Therefore, the carrier frequency must first be measured using a reliable frequency counter placed at any point in the system after the buffer amplifier. The carrier frequency may be adjusted only by someone holding an FCC license or by someone under the direction of the license holder. Remove the frequency counter before doing any other alignment steps.

2.7.1 Adjusting Neutralization

1. Disconnect the DC voltage from the amplifier to be neutralized. Do *not* remove the drive signal or bias.
2. Misadjust the neutralizing capacitor to either extreme of its range (C_{max} or C_{min}).
3. Connect an RF reading device (scope or RF meter) at the output of the amplifier. Loosely couple an RF coil near the output resonant tank circuit, and connect the RF indicator to the coil. A neon lamp will substitute in emergencies.
4. Tune the output resonant tank circuit for maximum output voltage. With the DC disconnected, the RF signal reaches the tank circuit through the neutralizing capacitor C_n.
5. Adjust the neutralizing capacitor for minimum output voltage using the RF reading device, signifying total cancellation of the interelectrode capacitances.
6. Disconnect the RF measuring devices and reestablish DC power.

2.7.2 Intermediate RF Stages

Adjusting the intermediate amplifiers involves simply tuning the RF circuits for maximum output signal at the carrier frequency. Any measuring device is likely to cause some loading effects when connected to these RF amplifiers during tuning, and since the final measurements will be made at the antenna terminals, time will be saved by taking this set of output readings from the antenna terminals also.

2.7.3 Loading and Output Power

Several of the adjustments to the final stage are interrelated and may require repeating some steps two or three times, each repetition bringing the circuit closer to perfect alignment. This outline defines only one pass to avoid repetition. The circuit of Figure 2.27 encompasses the entire power amplifier circuit for a high-level, high-power transmitter.

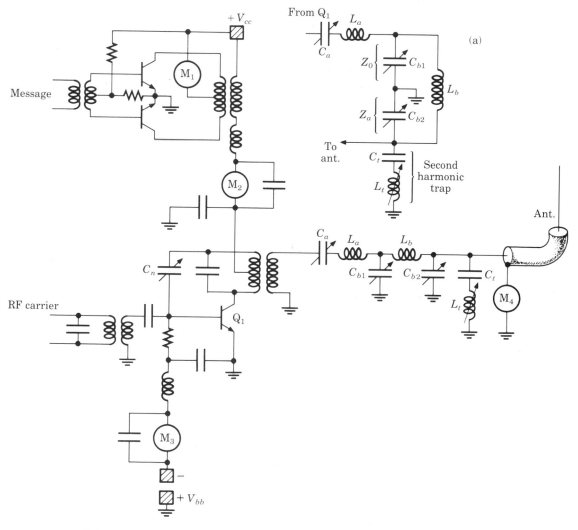

FIGURE 2.27
Output stage of transmitter showing antenna matching network. (a) Antenna network rearranged.

During major alignment and overhaul adjustments, the antenna is disconnected and a *dummy load* is connected in its place. A dummy load is a shielded, noninductive resistor of equal ohmic value and power rating as the intended antenna. The antenna current meter may still be used. High-power transmitters (100,000 W) may require a dummy load that is immersed in a container of pure mineral or fish oil; a water cooling coil is then piped through the oil chamber to conduct heat away from the load.

After adjusting the neutralization and frequency-tuning the final power amplifier, measure the DC supply voltage and set it to the specified value.

The base adjustments are as follows. Set the base bias supply $(-V_{bb})$ to turn off the final power amplifier. Disable (or turn down) the RF carrier signal to the final stage. While watching the collector current meter, M_2, advance the base bias voltage $(-V_{bb})$ until the meter moves *just off* zero current. Next, while watching the antenna current meter, increase the RF signal to the base of Q_1. As the RF signal gets stronger, the antenna current will increase. A point will be reached where an increase in RF carrier signal will no longer cause an increase in antenna current. This is the knee of the curve in Figure 2.28 and indicates that the final power amplifier is in saturation. Try varying this adjustment several times to ensure an exact setting. Too much drive signal causes positive carrier shift and distortion. Too little drive signal causes negative carrier shift, low output power, low efficiency, overmodulation, and distortion. The antenna current should remain at the knee of the curve.

The antenna coupling network may now be adjusted to correctly load the collector circuit for a specified power output. This is done by observing the DC supply voltage and the collector current meter, M_2, while adjusting the antenna coupling network for a specified value of DC collector current. The impedance matching (antenna coupling) network is shown in Figure 2.27 in its schematic form and rearranged in insert (a) to further specify the circuit action. In the normal schematic, C_a and L_a form a series-resonant circuit to present the lowest impedance to the desired frequency signal only, retarding all other frequencies. Coil L_b with capacitors C_{b1} and C_{b2} form a typical π-type low-pass filter. The desired signal falls within the passband, while higher frequencies are faced with greater attenuation as the frequency increases. C_t and L_t form a series-resonant circuit for the *second harmonic* ($2f_c$) of the carrier signal. The

FIGURE 2.28
Graph of antenna current vs. power amplifier signal drive voltage.

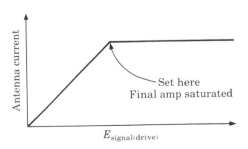

second harmonic contributes the most interference of the spurious radiation signals, so the series resonant *trap* sets an impedance to ground that is almost a short circuit across the antenna terminals at this frequency.

   Rearranging the drawing of the circuit components (see part (a) of Figure 2.27) shows how C_{b1} and C_{b2} form a parallel-resonant circuit with L_b. Adjusting C_{b1} affects the impedance seen by the secondary of the power output transformer and alters the DC current in the final power output amplifier. Adjusting C_{b2} sets the antenna impedance to the value specified by the matching network.

2.8   THE TRAPEZOID PATTERNS

An attractive alternative to the standard AM wave oscillograph is the **trapezoidal** wave pattern. The circuit and initial setup are shown in Figure 2.29.

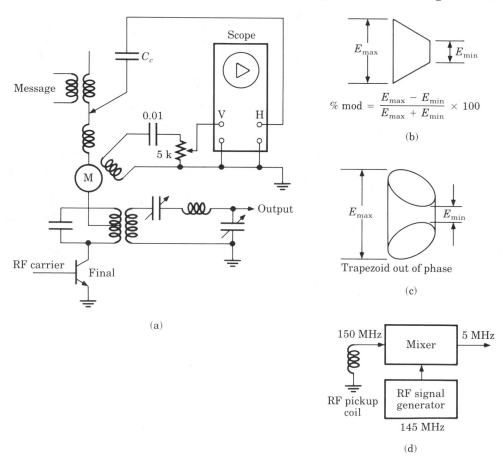

$$\% \ mod = \frac{E_{max} - E_{min}}{E_{max} + E_{min}} \times 100$$

(b)

Trapezoid out of phase

(c)

(a)

(d)

FIGURE 2.29
The trapezoidal wave as a measuring system.

The modulated carrier is applied to the vertical input of the oscilloscope by means of a loosely coupled RF coil and an RC phase shifting network. The phase shifting network will restore the true phase relationship between the modulated RF wave and the modulating wave. The trapezoidal wave pattern (Figure 2.29c) is essentially like a cardboard mailing tube cut at two angles. If rotated, the tube looks like two joined ovals. The correct rotation will display the side view, and the ovals will not be apparent. The message signal is taken from the output of the modulation transformer and applied through any large coupling capacitor to the Horizontal input of the oscilloscope.

There are several advantages to the trapezoid compared to the standard AM wave. While both permit measurement of the percentage of modulation, the standard wave is useful for sine wave modulation only. The trapezoidal wave allows observation of any complex modulating signal and provides a gauge for measuring the percentage of modulation. This is because the same signal that modulates the carrier wave also causes instantaneous horizontal deflection on the oscilloscope. Secondly, the taper of the trapezoid is a measure of the *linearity* of the modulated wave. Study Figure 2.30 for interpretations of variations of the trapezoidal wave, where distortion in some forms becomes more obvious.

Oscilloscopes that do not exceed the 20 MHz frequency response of the vertical amplifier are becoming rare. However, if your oscilloscope does not respond to the high frequency of the circuit under test, then some alternate arrangement must be used to view the output. A circuit that has had some success is the mixer circuit of Figure 2.29d. The purpose of this circuit is to lower the carrier frequency to match the response of the scope. The RF coil is positioned as before, except that it feeds the pickup signal to the base of a nonlinear mixer. The signal generator, set at 5 MHz above or below the carrier frequency, is supplied to the emitter of the mixer. Only the difference frequency of 5 MHz is within the passband of the scope, so the modulated wave is seen on the scope at a reduced carrier frequency. Only two precautions need be observed:

1. Be careful not to overdrive the base of the mixer. This condition is ascertained by moving the RF coil closer to or further from the output tank while checking the behavior of the signal on the scope. If the signal amplitude on the oscilloscope does not change when the pickup coil is moved closer to the tank circuit, then the coil is already too close.

2. The signal generator voltage should be at least five times larger than the RF pickup signal. This is usually not a problem, and, to examine the effects, the generator voltage may be increased or decreased while the scope pattern is being observed. Decrease the generator voltage until the modulated wave on the oscilloscope begins to flatten out. This is the point of too small a generator signal voltage. Increase the level until no flattening is observed.

Time domain	Remarks	Trapezoid

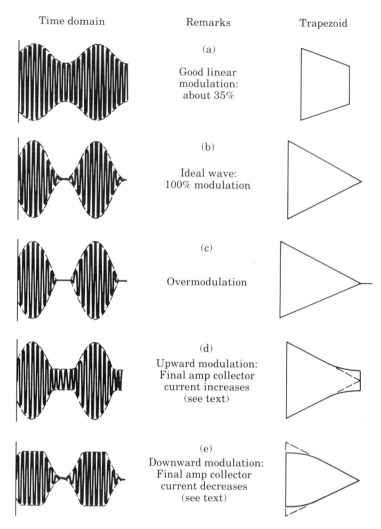

(a)

Good linear
modulation:
about 35%

(b)

Ideal wave:
100% modulation

(c)

Overmodulation

(d)
Upward modulation:
Final amp collector
current increases
(see text)

(e)
Downward modulation:
Final amp collector
current decreases
(see text)

FIGURE 2.30
Interpretation of the trapezoidal wave.

2.9   ANTENNA CURRENT

The antenna current is a convenient and accurate measure of the percentage of modulation. A meter that reads current through a known antenna impedance can be calibrated to indicate both power and percentage of modulation.

A 50 Ω antenna connected to a 1250 W transmitter (rated power is unmodulated carrier power) indicates an antenna current of 5 A. We know that at 100% modulation by a sine wave signal, the output power will increase by 50% to 1875 W. In this case the antenna current meter will read 6.1237 A through the 50 Ω load. This represents a 22.4745% increase in antenna current, which can be used to indicate 100% modulation. In technical circles, this value is rounded off to state that a 22.5% change in current will represent 100% modulation by a sine wave signal.

The antenna current meter is not as useful when the carrier is modulated by a voice or music signal because of the wide range of frequencies and amplitude variations in the message signal. Also, for such transmission, the meter is damped to respond to slow average changes in current and will not be a true measure of the rapid changes of the frequency range in the message signal.

QUESTIONS

1. A transmitter must be able to (a) transmit a distinctive carrier signal, (b) code the carrier signal, (c) supply transmit power, (d) all of the above.
2. The major subdivisions of an AM transmitter are (a) the exciter, (b) the power amplifier, (c) the modulator, (d) all of the above.
3. The fidelity of the carrier frequency is maintained by the (a) exciter section, (b) power amplifier section, (c) modulator section, (d) all of the above.
4. The energy level to be transmitted is determined by the (a) exciter section, (b) power amplifier section, (c) modulator section, (d) all of the above.
5. If misadjusted, the power amplifier section can affect the frequency of the transmitted signal. (T) (F)
6. The modulator section will alter the instantaneous power of the power amplifier section. (T) (F)
7. The sideband power is supplied by the (a) exciter section, (b) power amplifier section, (c) modulator section, (d) all of the above.
8. List the categories of AM transmitter styles.
9. What type of AM transmitter has the modulating voltage added in series with the DC supply to the final power amplifier? (a) very low level, (b) low-level, (c) intermediate-level, (d) high-level, (e) very high level.
10. When the message signal is applied to the input terminal of the final power amplifier along with the carrier signal, the transmitter is categorized as (a) very low level, (b) low-level, (c) intermediate-level, (d) high-level, (e) very high level.
11. What section supplies sideband power to a low-level AM transmitter? (a) the exciter section, (b) the modulator section, (c) the power amplifier section, (d) all of the above, (e) none of the above.
12. A transmitter rated at 50 kW will radiate how much power when modulated to 100%? (a) 25 kW, (b) 50 kW, (c) 75 kW, (d) 100 kW, (e) 150 kW.
13. List the steps that may be taken to ensure the accuracy of the carrier frequency.
14. How far off the assigned frequency of 1550 kHz would an AM transmitter be allowed to drift without violating the standard transmission regulation? (a) 14 Hz, (b) 20 Hz, (c) 26 Hz, (d) 31 Hz, (e) 37 Hz.
15. All AM transmitter carrier oscillators are crystal-controlled. (T) (F)

16. The buffer amplifier should present a constant _____ impedance as the load on the carrier oscillator. (a) low, (b) medium, (c) high, (d) amount of impedance is not a factor.

17. From the circuit of Figure 2.3, the buffer amplifier is operated as (a) class A, (b) class B, (c) class C, (d) class D, (e) class E.

18. The tuned RF amplifiers that follow the buffer amplifier are (a) linear wideband amplifiers, (b) linear narrowband amplifiers, (c) nonlinear wideband amplifiers, (d) nonlinear narrowband amplifiers.

19. What is the efficiency of a power amplifier that puts 400 W into the antenna when the amplifier circuit draws 5.6 A from an 80 V DC power supply?

20. High-power transmitters are those that (a) add modulation in series with the final power amplifier, (b) transmit powers greater than 5 kW, (c) have class B push-pull power amplifiers, (d) transmit powers greater than 250 W.

21. The final power amplifier of a high-level modulated transmitter has (a) a low input impedance during the conduction cycle, (b) a high input impedance during the cutoff period, (c) low input impedance all of the time, (d) high input impedance all of the time, (e) both (a) and (b).

22. The driver amplifier (a) never supplies power to the final amplifier, (b) supplies small amounts of power to the final amplifier, (c) supplies large amounts of power to the final amplifier.

23. The collector signal voltage of the final power amplifier can reach a peak-to-peak value of (a) $0.5V_{cc}$, (b) V_{cc}, (c) $1.5V_{cc}$, (d) $2V_{cc}$, (e) $4V_{cc}$.

24. In high-level modulation, the message signal (a) adds to the DC collector voltage of the final power amplifier, (b) opposes the DC collector voltage of the final power amplifier, (c) maintains a steady DC collector voltage in the final power amplifier all of the time, (d) both (a) and (b).

25. The audio processor circuits are found in the (a) exciter section, (b) modulator section, (c) power amplifier section, (d) they are not part of the transmitter.

26. The audio mixer, tone controls, and compressor of Figure 2.5 (a) are all lossy circuits, (b) all have variable gain, (c) all alter the frequency response.

27. Microphone cables longer than 15 feet should be (a) high-impedance, (b) low-impedance, (c) single-ended connections, (d) balanced line connections, (e) both (b) and (d).

28. Class C power amplifiers are recognizable by their (a) collector load circuit, (b) emitter biasing circuit, (c) base biasing circuitry.

29. The base voltage of an npn class C amplifier, compared to the emitter voltage, is (a) always positive, (b) always negative, (c) may be either positive or negative.

30. The value of DC self-bias base voltage of a class C amplifier will change as the signal strength changes. (T) (F)

31. In a high-level modulated class C power amplifier, the DC collector current (a) increases as the percentage of modulation increases, (b) decreases as the percentage of modulation increases, (c) remains constant for all levels of modulation.

32. What is the major disadvantage of self-bias?

33. The input power to the final power amplifier is found as (a) base supply I DC $\times$ E DC, (b) base signal I^2_{rms}/Z_{in}, (c) collector supply I DC $\times E$ DC.

34. As the desired conduction angle of a class C amplifier is made smaller, the peak drive signal (a) will increase, (b) will decrease, (c) will remain constant, (d) is not a factor.

35. Determine the peak drive signal to saturate a class C power amplifier for a conduction angle of 60° when a base current of 75 mA must be drawn through a base-spreading resistance of 8.8 Ω.

36. In parallel-output power amplifiers, the condition most guarded against is (a) incorrect loading, (b) parasitic oscillations, (c) overdriven base signal, (d) insufficient base signal, (e) improper collector tuning.

37. Many transistors require assistance in dissipating heat when the collector dissipation exceeds approximately (a) 1 W, (b) 5 W, (c) 25 W, (d) 50 W, (e) 100 W.

38. A transformerless series high-level modulation system that overcomes the Heising effect with no sacrifice in power is called (a) class E modulation, (b) progressive series modulation, (c) suppressor grid modulation, (d) emitter modulation, (e) neutralized modulation.

39. In the modulator circuit of Figure 2.17, with no modulation, transistor _____ is saturated and transistor _____ is in cutoff. (Consider only Q_{3-4} and Q_5.)

40. On the negative half-cycle peak of the modulating signal, of the transistors in Figure 2.17, Q_{3-4} is _____ and Q_5 is _____. (Consider only saturation and cutoff.)

41. On the positive half-cycle peak of the modulating signal, of the transistors in Figure 2.17, Q_{3-4} is _____ and Q_5 is _____. (Consider only saturation and cutoff.)

42. The power amplifier in the progressive series modulation circuit of Figure 2.17 has a DC collector current that is the same at zero modulation as it is for 100% modulation. (T) (F)

43. In the low-level modulated circuit of Figure 2.18, the modulator is biased for _____ operation. (a) linear, (b) nonlinear.

44. What form of bias is recommended for a base-modulated power amplifier? (a) self-bias, (b) voltage divider, (c) signal bias, (d) emitter bias, (e) fixed bias.

45. Neutralization is a circuit arrangement used to (a) cancel the effects of interelectrode capacitances, (b) reduce gain, (c) offset the Miller effect, (d) widen the amplifier frequency response, (e) limit collector current.

46. Neutralization is usually found in circuits that have (a) a three-element active device, (b) input and output tuned to the same frequency, (c) a frequency range over 25 MHz, (d) all of the above.

47. The trapezoid modulation test pattern will (a) show nonlinearity in the modulated wave, (b) show upward modulation, (c) show downward modulation, (d) work well on program signals, (e) all of the above.

CHAPTER THREE

SIDEBAND TRANSMISSION

3.1 INTRODUCTION

We already know that in a double-sideband transmission, the two sidebands contain identical information. Thus, either could be transmitted by itself without loss of meaning to the receiver. **Single-sideband transmission** has obvious advantages, some implied by the name alone. For instance, such a transmission can occupy one-half the frequency space required for two sidebands, permitting better management of the frequency spectrum. Moreover, suppose an A3 double-sideband transmitter and a single-sideband transmitter are each capable of 150 W maximum output. From Chapter 1 we know that there are 50 W of message power in the double-sideband transmission, while the 150 W of the single-sideband signal is *all* message power. This simple introduction indicates the advantages and disadvantages that go with single-sideband radio signals.

3.1.1 Advantages

The advantages of single-sideband transmission are as follows.

1. It allows better management of the frequency spectrum. More transmissions can fit into a given frequency range than would be possible with double-sideband signals.
2. *All* of the transmitted power is message power; none is dissipated as carrier power. Since total power relates to distance transmitted, the carrier power is not wasted energy, but it contains no intelligence and therefore has limited use at the receiving end of the system.

3. The noise content of a signal is an exponential function of the bandwidth: the noise will decrease by 3 dB when the bandwidth is reduced by half. Therefore, single-sideband signals have less noise contamination than do double-sideband signals.

A fourth advantage, which is not so obvious, is that single-sideband transmission is less prone to *selective fading*. Transmitted signals arrive at the receiver in part by reflection off of atmospheric layers of different densities. One of these layers (described in Chapter 12) is called the ionosphere. Depending on the angle of transmission, higher frequencies will not reflect as well as lower frequencies, and the transmitted signal may arrive at the receiver with a different phase between the upper and lower sidebands than it had when it left the transmitter. In some cases, this phase change is sufficient to cause complete cancellation of the received signal. Single-sideband transmission is less subject to this kind of difficulty (called selective fading) due to the absence of the carrier frequency and the narrower bandwidth of the transmitted signal.

3.1.2 Disadvantages

In spite of the overwhelming advantages that the single-sideband signal has over the double-sideband transmission, the few disadvantages are serious enough to discourage a mass movement toward this mode of transmission.

1. The cost of a single-sideband receiver is higher than its double-sideband counterpart by a ratio of about 3 to 1.
2. The average radio user wants only to flip a power switch and dial a station. Single-sideband receivers require several precise frequency control settings to minimize distortion and may require continual readjustment during the use of the system.

The second disadvantage is slowly being overcome by improvements in automatic frequency control circuitry. However, more automatic circuitry adds to an already complex system. Nevertheless, as new bands of transmitting frequencies are opened up and as old bands are revised, regulations by the FCC are geared more and more toward single-sideband transmissions.

3.1.3 Types of Single-Sideband Signals

A quick review of the emission code categories in Chapter 1 will help identify the family of single-sideband transmissions, broken down in Table 3.1. Double-sideband signals are included as a comparison. Figure 3.1 shows the frequency domain response characteristics for the five amplitude-modulated categories of sideband signals.

TABLE 3.1

Code	Abbreviation	Energy Transmitted
A3	DSB	Double sideband: two sidebands and full carrier
A3$_b$	DSSC	Double sideband suppressed carrier: two sidebands with the carrier removed, or two independent sidebands
A3$_a$	SSB	Single sideband: one sideband with reduced carrier (10% carrier energy, also called pilot carrier)
A3$_j$	SSSC	Single sideband suppressed carrier: one sideband only, no carrier
A3$_c$	Vestigial sideband (VSB)	Full upper sideband and carrier, with only part of the lower sideband; standard commercial television (visual signal)

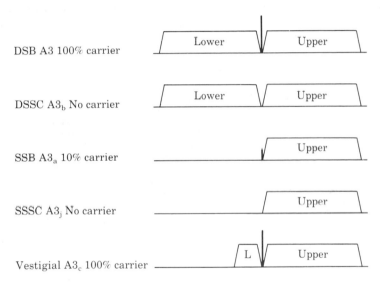

FIGURE 3.1
Transmission signals in the frequency domain.

3.2   SEPARATION OF SIDEBANDS

It is easy to see that the single-sideband signal takes only one-half the frequency bandwidth of the double-sideband signal. For a specified power level, more of the radiated signal can be used for message power. To see how this is done, we will first examine the functional block diagrams for the single-sideband transmitter and then look at the circuits that make up the functional blocks. Two methods for separating the sidebands are in common use: The **filter** method and the **phase shift** method.

3.2.1 Generating the Sidebands Only (the Filter Method)

The most popular circuit used to generate sideband signals is a **balanced modulator.** A balanced modulator requires two input signals: the message signal and the carrier signal. The function of the balanced modulator is to generate an amplitude-modulated signal having two sidebands, at the same time canceling out the carrier signal. The upper sideband range includes the carrier plus the message frequencies; the lower sideband range contains the carrier minus the message frequencies. This action is displayed in Figure 3.2.

Notice that the **separation** (the space where the carrier has been removed) between the lower and upper sidebands in the modulated signal is only 200 Hz in this example. The lowest frequency in the message signal is the factor that determines the separation. From the figure, it is clear that the separation is two times the lowest message signal frequency. The carrier and the sideband to be eliminated must be attenuated at least 40 dB below the level

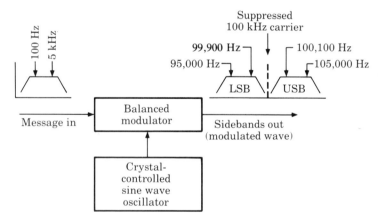

FIGURE 3.2
Balanced modulator with input and output signals.

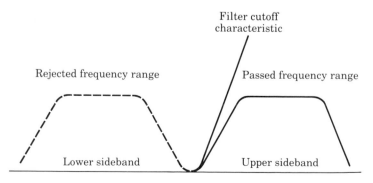

FIGURE 3.3
Performance curve of a sharp cutoff filter.

of the desired signal. This means a reduction to less than 1% of the original signal. Even the best *RLC* filters cannot block the unwanted signals only 200 cycles away from the desired signal this effectively. For this degree of control, special filters, called SAW filters, crystal filters, and mechanical filters, are most effective (in that order).

Figure 3.3 shows the action of a sharp cutoff filter. Notice that the balanced modulator has removed the carrier signal, and the sharp cutoff filter has removed the unwanted sideband (in this case, the lower sideband). Either sideband may be removed, and the effect on the transmission will be the same, since both sidebands contain the same information. Later we will see special cases where the choice of sideband may make a difference.

Raising the Sideband Frequency   In Figure 3.2, the double sidebands are modulated on a low carrier frequency of 100 kHz. The carrier frequency should be low to ensure that the ratio of the frequency spread between the sidebands to the carrier frequency does not place undue demand on the filter circuit. A ratio of 200 Hz to 100 kHz is a factor of 0.2%, whereas a ratio of 200 Hz to, say, 1 MHz would be only 0.02%, an impractical value even for sharp cutoff filters. The carrier frequency should be low enough not to involve critical circuitry, yet high enough to maintain a stable frequency and to allow the selection of practical component part size in the resonant circuit of the oscillator. This low frequency is seldom the frequency to be transmitted, however. In the example, the lower sideband is rejected while the upper sideband of 100.1 kHz to 105 kHz will be raised to the desired transmitting frequency by modulating the sideband to a higher carrier frequency level. This action is shown in Figure 3.4.

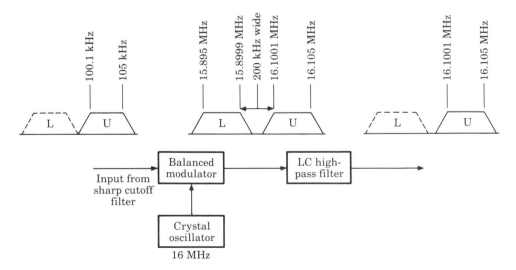

FIGURE 3.4
Balanced modulation of the lower sideband to raise the frequency up to the transmission frequency.

The balanced modulator circuit produces sideband frequencies that are the sum of and difference between the single-sideband input and the new carrier frequency of 16 MHz. The output has a new lower sideband of 15.895 MHz to 15.8999 MHz and a new upper sideband of 16.1001 MHz to 16.105 MHz. Notice that the space between the upper and lower sidebands is now 200 kHz rather than 200 Hz as before. These sidebands will be easy to separate with a simple *LC* filter because of this wider space. If the frequency is to be raised further, the same balanced modulation principles will be used and the space between the two sidebands will become wider, making it easier with each level to separate the sidebands with a simple *LC* filter. At each stage, either sideband may be selected for transmission. In this example, the final signal will be the upper sideband for a 16.1 MHz suppressed carrier. The lowest modulating frequency in the sideband will be 100 Hz above the suppressed carrier. Had the lower sideband been selected for transmission, the carrier frequency would be 15.9 MHz, since the highest frequency in the sideband (15.8999 MHz) is 100 Hz below the carrier frequency.

EXAMPLE 3.1

Let's take another example with different values to see how it works out. A message signal that ranges from 300 Hz to 3 kHz is to be the upper sideband of a suppressed 79.9 MHz carrier.

1. Select a low carrier frequency to derive the amplitude-modulated double-sideband carrier signal to be suppressed. A 150 kHz oscillator, balance-modulated by the message signal, will have sidebands at 147 kHz to 149.7 kHz and at 150.3 kHz to 153 kHz. Filtering out the lower sideband leaves the upper frequencies of 150.3 kHz to 153 kHz.
2. A second balanced modulator with an oscillator frequency of 2.85 MHz will produce a lower sideband range of 2.697 MHz to 2.6997 MHz and an upper sideband range of 3.0003 MHz to 3.003 MHz. Again, removing the lower sideband frequencies leaves the upper sideband of 3.0003 MHz to 3.003 MHz.
3. To have the transmitted signal appear as the upper sideband to a 79.9 MHz carrier, the third oscillator will be operated at a frequency that is *lower than the reference carrier by an amount equal to the sum of the two preceding oscillator frequencies.* For a desired reference carrier frequency of 79.9 MHz, the third oscillator will be set to 76.9 MHz.

Adding the sidebands out of the second filter to the frequency of the third oscillator will produce the desired sideband referenced to 79.9 MHz. The final sideband will include frequencies from 76.9 + 3.0003 MHz = 79.9003 MHz (300 Hz higher than 79.9 MHz) to 76.9 + 3.003 MHz = 79.903 MHz (3 kHz higher than 79.9 MHz). This example is pictorially represented in Figure 3.5.

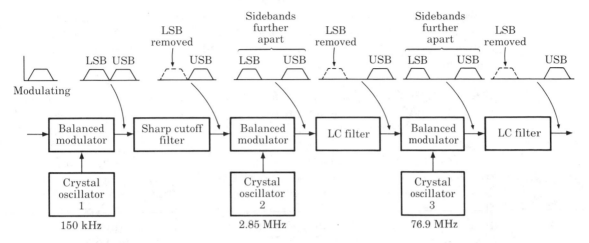

FIGURE 3.5
SSB transmitter block diagram with associated sideband positions (see Example 3.1 for sideband values).

Note that the reference carrier of 79.9 MHz does not appear in the transmitter diagram. Keep in mind, however, that the frequency of the reference carrier must be known since it is one of the prime considerations at the receiver end. This will be shown in the following chapter.

When the nature of the transmission is A3$_a$ (10% carrier SSB) and the carrier must be at 79.9 MHz, then the transmitter of Figure 3.5 must include the circuits of Figure 3.6. The carrier to be reinserted is the *sum* of all three oscillator frequencies. The frequencies of oscillators 1 and 2 are added in the first mixer for 3 MHz output frequency. This 3 MHz is added to the 76.9 MHz of oscillator 3 and to the sideband frequency range. The transmitter output will

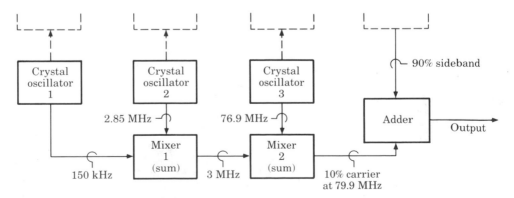

FIGURE 3.6
Generating and reinserting a 10% carrier at 79.9 MHz for A3$_a$ SSB transmission.

now consist of 10% carrier, at 79.9 MHz, and 90% upper sideband information, at 79.9003 MHz to 79.903 MHz. If either oscillator should drift in frequency for any reason, the net result would be a shift in sideband frequency and in carrier frequency in the same direction and by the same amount.

The Voltage Wave of the Single Sideband   Chapter 1 showed the voltage wave of the A3 double-sideband signal in several forms. We can adapt two of these forms to represent the energy in the single-sideband wave, as shown in Figure 3.7.

The sine wave constituting the top of the DSB wave in Figure 3.7 traces the *shape* of the carrier envelope, as does the dashed line at the bottom of the DSB wave. Neither line represents the sideband voltage; each merely defines the changing amplitude of the carrier wave. In Figure 3.7b, the carrier amplitude has been reduced, and the top and bottom waves are beginning to overlap. Complete overlap occurs in Figure 3.7 and represents the double-sideband suppressed-carrier wave. Note in particular the time for one cycle of the message signal in parts (a) and (c). In the DSB (A3) transmitter this action would cause overmodulation, as defined earlier. For the DSSC (A3$_b$) transmission, the waveform shown is not 100% accurate but is the best way to understand the

FIGURE 3.7
A3 full-carrier emission changed to A3$_b$ double-sideband suppressed carrier.

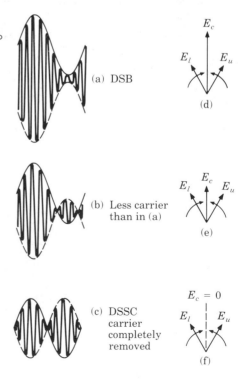

(a) DSB

(d)

(b) Less carrier than in (a)

(e)

(c) DSSC carrier completely removed

(f)

behavior of the wave. The vector diagrams in Figure 3.7d, e, and f serve to show the summation of the carrier vector voltage and the counter-rotating sideband vector voltages. In Figure 3.7d, the rotating sideband vectors add to the stationary carrier vector voltage, and in Figure 3.7f, the carrier is absent and only the rotating sideband vectors remain to be added together.

Power Comparisons   The most widely published FCC specifications for DSB and SSB transmitters of equal class are those for class D citizens band radio. The power limits are 4 W unmodulated carrier power for DSB (A3) emission and 12 W peak envelope power (PEP) for single-sideband (A3$_j$) emission (SSSC). For A3 emission at 100% modulation, the output power is $1.5 \times P_c$ (unmodulated), which is equal to 6 W. Of this 6 W, 4 W are carrier energy and 2 W are the message energy in the sidebands. For SSSC, the 12 W PEP may be reduced to true (or rms) power as follows:

$$P = \frac{E^2}{Z} \quad \text{or} \quad E = \sqrt{PZ}$$

Select any antenna impedance, say 50 Ω, and solve for voltage of a known power level:

$$E = \sqrt{PZ} = \sqrt{12 \times 50} = \sqrt{600} = 24.495 \text{ V}_{pk}$$

Convert the peak voltage to an rms voltage:

$$E_{rms} = 0.7071 \times E_{pk} = 0.7071 \times 24.495 = 17.3205 \text{ V}_{rms}$$

Now convert this rms voltage into true (rms) power:

$$P = \frac{E^2}{Z} = \frac{(17.3205)^2}{50} = 6 \text{ W}_{rms}$$

This shows that the A3 emission of 6 W at full modulation and the A3$_j$ emission at 12 W PEP are related to each other through the peak-to-rms voltage conversion. However, for the A3 emission only 2 W are message power, while the entire 6 W are message power in the SSB transmission. Thus, peak envelope power is two times the rms *total* power for equal-rated emissions. That is, PEP is two times the total power of a 100% modulated DSB transmission or three times the DSB unmodulated carrier power ($2 \times 6 = 12$ and $3 \times 4 = 12$ W).

3.2.2   The Phase Shift Method

In the **phase shift modulator,** two identical balanced modulators are used in parallel. The message signal to each modulator is identical in all respects except phase. The signal is fed directly to one modulator and delayed by 90° to the second modulator. Both modulators use the same carrier signal frequency, but the carrier is delayed by 90° when applied to the second modulator.

In Figure 3.8, modulator A receives the in-phase message and carrier signals, and modulator B receives the 90°-delayed message and carrier signals. The output voltages from the modulators are then added in a summing amplifier. The summing of these two signals results in the *cancellation of one sideband*.

At some point in time, the phase of the voltage from modulator A can be represented by the vectors in Figure 3.9a. The lower sideband, designated by the subscript *la*, is a clockwise-rotating vector and is shown lagging the suppressed carrier vector by 15°. The upper sideband (subscript *ua*) is a counter-rotating vector shown leading the suppressed carrier by 15°.

The suppressed carrier of modulator B lags the suppressed carrier of modulator A by 90°, as shown in Figure 3.9b. If modulator B were supplied with an in-phase message signal, its sidebands would be as indicated by the dashed

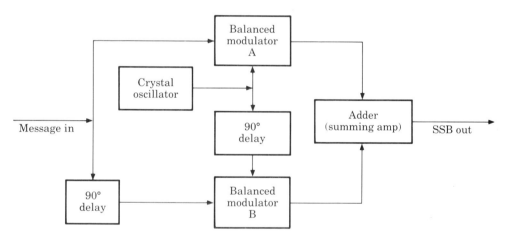

FIGURE 3.8
Phase shift-balanced modulation.

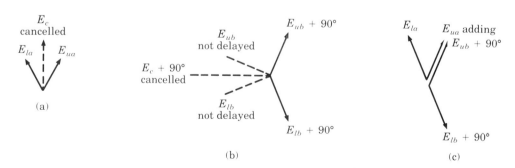

FIGURE 3.9
Voltage vectors for a phase shift-balanced modulator. (a) Output-balanced modulator A. (b) Output-balanced modulator B. (c) The vector sum of the output voltages.

vectors for E_{lb} and E_{ub}. However, they are altered by an additional 90°, placing them behind the dashed vectors as shown.

Figure 3.9c shows the sum of the sideband vectors. Note that the two lower sidebands are equal in amplitude but 180° out of phase and thus cancel. The two upper sidebands are in phase and of equal amplitude, giving an output voltage of $2E_u$.

Figure 3.10 depicts circuits used as phase-shifting networks for the carrier (3.10a) and for the modulating signal (3.10b). The modulating signal delay network shown is one used for telephone systems and is for the frequency range of 300 Hz to 3 kHz.

The advantages of the phase shift method can be summarized as follows.

1. Not only is the carrier removed from the modulated signal, but one of the sidebands is also canceled, producing true direct single-sideband modulation.
2. Switching the carrier oscillator signal to a leading or lagging phase shift allows easy selection of the lower or the upper sideband.
3. Since one sideband is already canceled, a sharp cutoff sideband filter is not required in this system; thus, it is no longer necessary to start at a low modulating frequency and heterodyne up to the desired carrier frequency. The intended carrier frequency can be presented to the first balanced modulator set, eliminating the need for more sets of balanced modulators to raise the frequency.

The only disadvantage to the phase shift method is its limitation of the message frequency range. Shifting the carrier oscillator signal by 90° is no problem because it is constant at one frequency, but the message signal covers a range of frequencies. It is extremely difficult to hold the phase shift constant

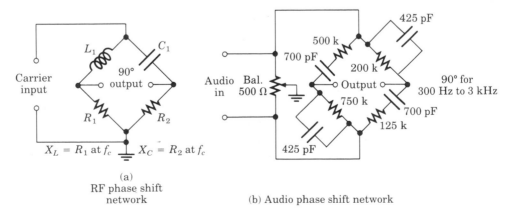

(a)
RF phase shift
network

(b) Audio phase shift network

FIGURE 3.10
Phase-shifting networks.

over the audio frequency range of 20 Hz to 20 kHz using an *RC* phase-shifting network. A circuit developed by this author (patent pending), shown in Figure 3.11, will maintain a 90° phase shift over the full audio range (or any selected range) of the message signal.

The message signal is first amplitude-modulated with full double side-bands onto a low carrier frequency, say, 1 MHz. The carrier with its sidebands (0.98 MHz to 1.02 MHz) is then shifted in phase by 90°, after which the message signal is recovered by a simple diode detector (see Chapter 4). The full-range audio signal of 20 Hz to 20 kHz (at +90°) is remodulated via the balanced modulator onto the 90°-shifted carrier, and the carrier is suppressed. The wide-range 90° SSSC signal is then added to the *zero* phase-shifted signal to produce a wideband true SSB signal, ready to be raised in power for transmission.

3.2.3 After the Modulators

Once the sideband has been generated and raised to the final transmission frequency, any further amplification must be through a *linear* system. The modulation in sideband transmitters takes place at low levels and therefore needs to be elevated to the required power for transmission. Distortion of the modulated wave cannot be tolerated, even if measures used to prevent it hamper efficiency. Unfortunately, even the distortion levels of class B push-pull amplifiers are too high. Push-pull amplifiers are rich in *odd-order* intermodulation harmonics (third- and fifth-order) that are so close to the desired frequencies that they cannot be easily separated. Therefore, class A amplifiers are used in such cases.

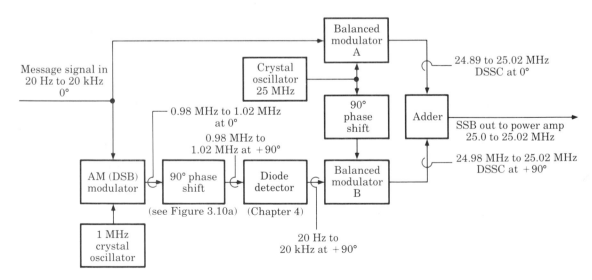

FIGURE 3.11
Wideband single-sideband (A3ⱼ) modulator (patent pending).

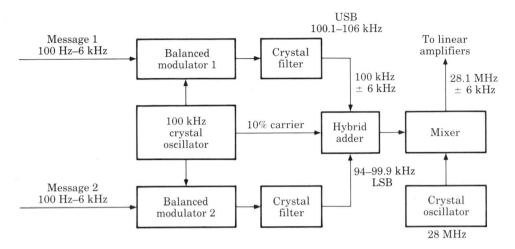

FIGURE 3.12
Independent sideband modulator (ISB).

3.2.4 Independent Sidebands (ISB)

We have discussed the generation of a lower sideband with suppressed carrier and an upper sideband with suppressed carrier as individual and distinct forms of transmission. It is also possible to take the lower sideband from one sideband generator and add it to the upper sideband from a different generator, combining them as one transmission. This arrangement, called **independent sideband transmission,** is shown in Figure 3.12. It is also possible to expand on this idea, to fit more than one sideband on each side of the removed carrier. This modulation class is identified by the emission code $A3_b$. *Multiple* independent sidebands are used for data transmission and in telephone and telegraph systems. This form of signal processing, called *multiplexing,* will be covered in a later chapter in depth.

3.3   BALANCED MODULATORS

The required output from a balanced modulator is an amplitude-modulated signal with two sidebands but no voltage at the carrier frequency. A **differential amplifier** is ideally suited to this function.

  In the differential amplifer in Figure 3.13a, transistors Q_1 and Q_2 constitute a matched set of amplifiers. The base of each transistor is pushed slightly above ground by the voltage drop across one-half of the secondary of the input transformer. With the emitters at ground, the transistors are biased slightly forward in a class B mode, and each conducts for 180° of the input cycle. A signal applied to the primary of the input transformer will at some point show a positive polarity at the top of the transformer compared to that at the bottom.

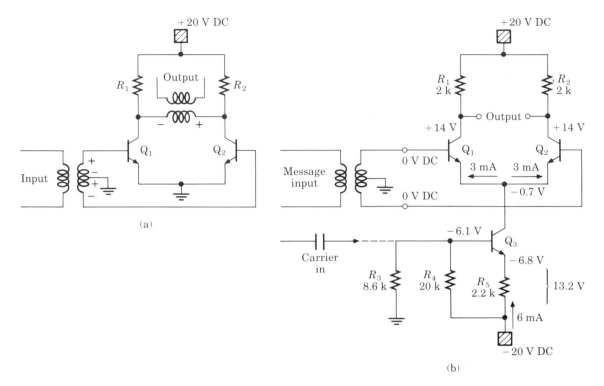

FIGURE 3.13
(a) Differential amplifier. (b) Differential amplifier with constant-current generator as a modulator for balanced modulation.

With these two ends equidistant from the grounded center tap, the base of Q_1 will be positive and the base of Q_2 negative. A positive potential on Q_1's base causes the collector current of Q_1 to rise and increase the voltage drop across the collector load resistor R_1, causing the collector voltage to decrease. A negative voltage on the base of Q_2 drives the transistor into cutoff, and its collector voltage increases. The potential *difference* between the collectors of Q_1 and Q_2 presents a voltage across the output transformer that is negative at the collector of Q_1 relative to the potential at the collector of Q_2.

The opposite condition will hold true when the input signal begins its negative half cycle. The base of Q_1 then goes in the negative direction, and its collector moves in the positive direction. The base of Q_2 now becomes more positive and raises Q_2's collector current, thus increasing the voltage drop across R_2, and lowers the collector voltage of Q_2. The voltage across the output transformer now has a positive polarity at the side toward the collector of Q_1 and a negative polarity toward the collector of Q_2.

In Figure 3.13b, a constant-current generator has been added in series with the emitter circuit of Q_1 and Q_2. The bias on the base of generator Q_3 is set by the resistive voltage divider R_3 and R_4 at -6.1 V DC. Assuming a 0.7 V base-emitter potential, the emitter will be set at -6.8 V DC. With this emitter voltage and the other end of the emitter resistor tied to -20 V DC, the voltage drop across the emitter resistor will be 13.2 V DC, causing an emitter current in Q_3 of 6 mA. The bases of Q_1 and Q_2 are at DC ground, which places their emitters (and the collector of Q_3) at -0.7 V DC. Transistors Q_1 and Q_2 and their collector resistors act as the load for Q_3. Dividing the current of Q_3 equally between the two transistors sets up a collector current through Q_1 and Q_2 of 3 mA each. Collector current of 3 mA causes 6 V DC drop across each collector load resistor, placing the collectors at $+14$ V DC each. With the collectors of Q_1 and Q_2 *both* at $+14$ V, the difference of their potentials is zero. A signal across the input transformer secondary that makes the base of Q_1 go positive and the base of Q_2 go negative will cause the collector current of Q_1 to increase to 4 mA. The resulting voltage drop across R_1 of 8 V will reduce the collector of Q_1 to 12 V DC. The same input signal causes the base of Q_2 to drop in voltage, decreasing the collector current of Q_2 to 2 mA and producing a voltage drop across R_2 of 4 V DC, which in turn sets the potential of the Q_2 collector at $+16$ V DC. The difference in potential between the collectors is 4 V p-p, with Q_2 positive compared to Q_1.

Suppose that a signal at the carrier frequency is now applied through a coupling capacitor to the base of Q_3. Every time the base of Q_3 is driven positive, the collector current of Q_3 increases, to 8 mA, for example, and the current through Q_1 and Q_2 increases to 4 mA each. This increased current, coupled with the signal already present at the base of Q_1 or Q_2, will result in the sum and difference of the two input signal frequencies, generating the sidebands. However, when *no* signal is applied to the base of Q_1 or Q_2 and only the carrier signal is applied to the base of Q_3, the collector current of Q_3 increases, the currents through Q_1 and Q_2 increase equally, and the voltage drops across the load resistors of Q_1 and Q_2 change equally. The difference in voltage at the collectors of Q_1 and Q_2 remains at zero because both collectors have changed equally in the same direction.

No difference in potential between the collectors of Q_1 and Q_2 for a carrier input at the base of Q_3 is a sure indication of carrier cancellation. The circuit of Figure 3.13 is slightly oversimplified but nevertheless factual. A circuit for an MC1495L balanced modulator is shown in Figure 3.14. It would be beneficial for manufacturers to supply detailed diagrams similar to this figure and wise for the technician to seek out such drawings. Understanding at this level may separate the chip changers from the craftsmen.

In Figure 3.14, the Darlington pair Q_1 and Q_2 form the input differential amplifier, with Q_{3a} and Q_{3b} the associated constant-current generator. Their balanced outputs are fed to a cross-coupled differential amplifier consisting of Q_4 with Q_8 and Q_5 with Q_7. The constant-current generator associated with this differential amplifier is transistor Q_{10}. Q_6 and Q_9 form the series differential modulator with Q_6 acting on Q_4 and Q_8, while Q_9 acts on transistors Q_5 and Q_7. Carrier symmetry is achieved by a balancing resistor between pins 5 and 6; the message signal is balanced by a resistor between pins 10 and 11. The actual

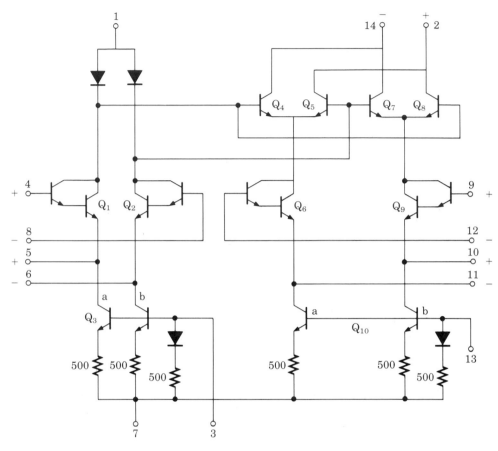

FIGURE 3.14
MC1495L four-quadrant multiplier to be used for balanced modulation.

circuit is shown in Figure 3.15, where the carrier is supplied to pins 4 and 8 and the message signal is input to pins 9 and 12. In practice, either signal can be fed to either set of input terminals with no ill effects because of the wide frequency response of the amplifiers.

The suppressed carrier output is taken from terminal 14. The same circuit can be used as a *demodulator* (to be explained later), in which case the output is taken from pin 2. Carrier suppression of 60 dB can be realized by applying a signal to one input and adjusting the opposite input null control, and then reversing the process for the second input. The advantages of this circuit are that the carrier and message frequencies can be close together or widely separated, and either signal may be applied to either input. Also, a high degree of carrier suppression is readily obtainable, and the circuit can be used as a DSSC modulator or demodulator.

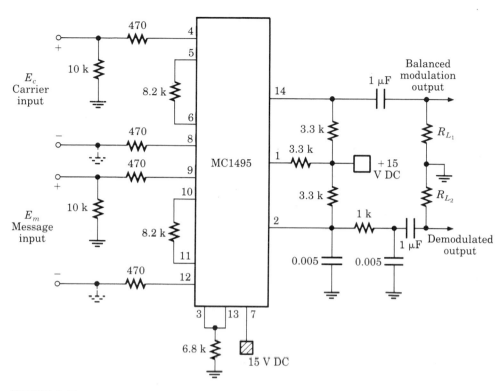

FIGURE 3.15
Working circuit for an integrated balanced modulator (or demodulator).

There are other popular balanced modulators in use that were developed for vacuum tube circuitry and later converted to solid state circuits. They were the proving ground for today's microcircuits. One such system is that of Figure 3.16. It is the circuit of a class B push-pull amplifier and performs in every way like a regular class B amplifier. A low-frequency message signal is applied through the input transformer and is amplified by Q_1 and Q_2 on alternate half cycles to produce a linear output signal. A third transformer is added between the center tap of the input transformer secondary and ground. A carrier signal applied to T_3 will at any instant have a polarity that is positive at the center-tap end of the T_1 secondary. At this time, both bases are raised in level, and the collector voltages of both transistors decrease, becoming equal in amplitude and phase. Therefore, the output signals cancel as they appear across each half of the output transformer primary. Modulation takes place when both signals are applied. When the message signal is positive, Q_1 conducts more than Q_2; when the message signal goes negative, Q_2 conducts more than Q_1. The sum and difference frequencies of the message and carrier signals are of unequal amplitude and do *not* completely cancel, causing a modulated output signal voltage. The carrier signal still cancels.

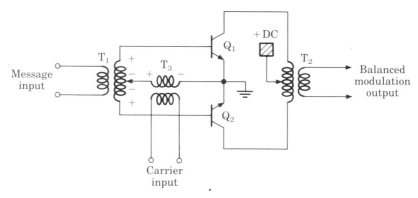

FIGURE 3.16
Discrete class B push-pull balanced modulator.

3.4   **RING MODULATORS**

The ring (or lattice) modulator of Figure 3.17 performs the same functions as
the balanced modulator, with the exception that it does not have gain. The
carrier voltage at the secondary of T_3 should be between 3 and 6 V_{rms} and the
carrier should be about ten times larger than the message signal voltage.

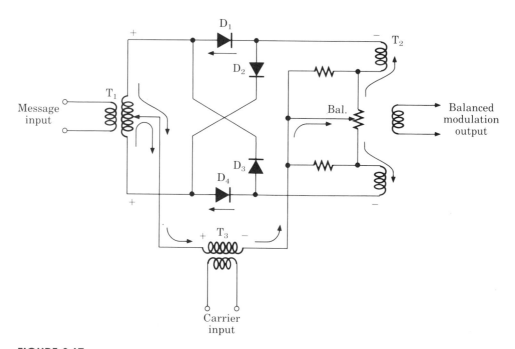

FIGURE 3.17
Ring modulator (full lattice).

When a carrier signal of the polarity indicated in the diagram is applied, diodes D_1 and D_4 will conduct and draw current through the output transformer primary, whose halves are of opposite polarity but equal amplitude. This develops voltages across the primary that are equal but opposite in nature, so they cancel each other. The balancing potentiometer in the center of the windings is needed to ensure complete cancellation of the carrier signal. When the carrier changes polarity, the second set of diodes, D_2 and D_3, conduct.

As a message signal is applied to the input transformer primary, depending on its polarity, the signal will add to or subtract from the carrier signal already present and cause the corresponding set of diodes to conduct more heavily, giving a larger output signal. The nonlinear operation of the diodes will generate the sum and difference frequencies to make up the sideband voltages, at the same time canceling out the voltage at the carrier frequency.

3.5   FILTERS

There are five families of filters in popular use for separating the sidebands. They are, in order of their Q values,

1. SAW (surface acoustical wave) filters,
2. crystal filters,
3. mechanical resonating filters,
4. ceramic filters, and
5. RLC filters.

The required Q of the circuit will establish which family of filter is best suited for the application (see Table 3.2).

The required Q is a function of (a) the amount of desired attenuation, (b) the number of cycles over which the attenuation is to take place, and (c) the center frequency around which (b) is stated. The equation is

$$Q = \frac{f_c}{\Delta f}\left(0.25\sqrt{\frac{1}{X}}\right) \qquad (3.1)$$

where $X = \log^{-1}(-\text{dB}/20)$.

TABLE 3.2

Device	Relative Q
SAW filter	Over 35,000
Crystal filter	20,000
Mechanical filter	10,000
Ceramic filter	2000
RLC filter	500

In the example used for Figure 3.2, the first sharp cutoff filter was to separate two sidebands 200 Hz apart and centered around 100 kHz. The lower sideband was to be removed by a filter providing 80 dB of attenuation within the frequency limits of 200 Hz. Thus,

$$X = \log^{-1}\left(\frac{-80}{20}\right) = 1 \times 10^{-4}$$

The required Q is

$$Q = \frac{100 \text{ kHz}}{200 \text{ Hz}}\left(0.25\sqrt{\frac{1}{1 \times 10^{-4}}}\right) = 12,500$$

In this instance, even the mechanical filter will not do the job properly. A crystal or SAW filter is required.

3.5.1 Octaves and Decades

Before getting too deeply involved in the filter circuits themselves, let's take the time to ensure a full understanding of the terms *octave* and *decade*. An **octave** is a change in frequency by a factor of either 2 or ½. One octave up from 10 kHz is 20 kHz. One octave down from 10 kHz is 5 kHz. A **decade** is a change in frequency by a factor of 10 or ¹⁄₁₀. One decade up from 10 kHz is 100 kHz. One decade down from 10 kHz is 1 kHz. A **slope** is defined as a decibel change over a unit of frequency, such as 6 dB per octave or 20 dB per decade. The units of frequency may or may not be consistent between descriptions of the same slope. For example, a response curve may be expressed as 12 dB/octave, and another reference may describe the same slope as 40 dB/decade. Therefore, the following relation should be memorized: 6 dB/octave and 20 dB/decade represent the *same* slope.

The slope may be approximated as

$$\text{Slope} = \frac{\text{dB } f_c}{\Delta f} \quad \text{(in decibels per octave)} \tag{3.2}$$

For the Figure 3.2 example,

$$\frac{(80)100 \text{ kHz}}{200} = 40,000 \text{ dB/octave}$$

3.5.2 Surface Acoustical Wave (SAW) Filters

Several members of the silicon-germanium family of oxides exhibit a piezoelectric effect that results in a rippling on the surface of the substrate. The ripples vary with the rate of the applied signal frequency but travel along the surface of the substrate at the speed of sound. It is from this velocity that the **surface**

acoustical wave filter takes its name. Figure 3.18a shows the ripple wave on the surface of the substrate, moving from the input transducer toward the output transducer.

Figure 3.18b shows that the length of the transducer "fingers" can be varied to control the frequency response of the filter. The input is termed a *broadband* transducer because the fingers are all of the same length. A transducer in which the fingers are of different lengths, as with the output transducer in this figure, is termed a *weighted finger array* and is highly frequency-selective. The frequency range is controlled by the space between the fingers. For high-frequency operation, the fingers are placed close together; here operation into the low gigahertz has been successfully demonstrated. The spacing is widened for lower frequencies. The surface of the transducers must be highly polished. A conductive coating is added to the bottom of the substrate and serves as the ground connection for the filter.

SAW filters are basically band-pass filters with extremely sharp cutoff skirts, so that the desired frequency falls within the passband while the frequencies to be attenuated fall outside the passband. A typical response curve for a SAW filter can be seen in Figure 3.18c. The input and output impedance is typically 75 Ω, but SAW filters are available with higher values.

Materials commonly used in the construction of SAW filters are silicon and bismuth germanium oxides, lithium tantalate and lithium niobate, thin

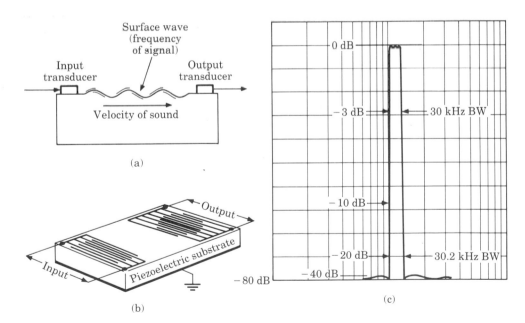

FIGURE 3.18
SAW filter. (a) Surface wave. (b) Input-output transducers. (c) Response curve.

film zinc oxides, and quartz. The finished substrate is placed inside a ceramic box with the five leads brought to the outside. A ceramic lid is placed over the substrate, and the entire package is sealed in epoxy. The completed package may measure only 0.32 cm (0.125 in) × 1 cm (0.375 in).

3.5.3 Crystal Filters

The quartz crystal has the unique property of distorting its shape when a voltage is applied across its terminals. An AC voltage will cause the crystal to warp first in one direction, then in the other direction—in effect, to mechanically oscillate. This is the basis for **crystal filters.**

The dimensions of the quartz material set the frequency of oscillation. The frequency is high when the crystal is thin and flexible. An approximate upper limit is 10 MHz; thinner crystals are too easily broken. However, crystals may be operated at harmonics of the fundamental frequency, in which case they are called *overtone crystals.* An overtone crystal operating at ten times the fundamental frequency is common. Thick crystals are stiffer and flex at a lower frequency. An approximate lower frequency limit is 100 kHz based on size and cost. This author has seen a 60-cycle crystal made for a very special purpose, but its size and cost place it outside typical system usage.

The size and shape of the crystal wafer can be equated to a series electrical *RLC* circuit. Figure 3.19b shows the series-resonant equivalent circuit. The mounting plates used for mechanical support of the wafer form a capacitance (C_h) that is in parallel with the crystal. The crystal thus has two resonant

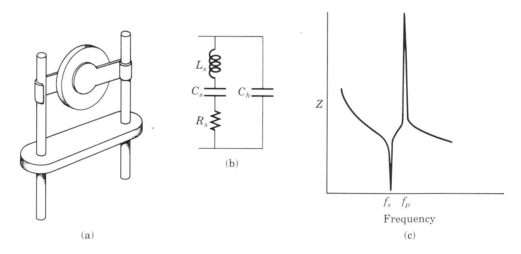

(a)   (b)   (c)

FIGURE 3.19
Quartz crystal. (a) Isometric view of crystal mounting without case. (b) Equivalent electrical circuit. (c) Crystal impedance curve.

frequencies. The series R_s-C_s-L_s of the crystal material itself sets the series-resonant frequency; holder capacitance C_h in parallel with the inductance L of the crystal forms the parallel-resonant circuit. The impedance characteristics of the crystal appear in Figure 3.19c, which shows the series-resonant frequency f_s with a very low impedance and the parallel-resonant frequency f_p with a very high impedance. The separation between the series-resonant frequency and the parallel-resonant frequency is called the **pole zero separation.** It is a function of the ratio between the series capacitance C_s and the holder capacitance C_h, where C_s is in series with C_h and the total capacitance is in parallel with the inductance L_s for parallel resonance:

$$f_p = f_s\left(1 + \frac{C_s}{2C_h}\right) \tag{3.3}$$

EXAMPLE 3.2
Given a crystal with $L_s = 1.2665$ μH, $C_s = 20$ pF, and $C_h = 0.01$ μF, the series-resonant frequency, the parallel-resonant frequency, and the pole zero separation will be:

$$f_s = \frac{1}{2\pi\sqrt{L_sC_s}} = 1 \text{ MHz}$$

$$f_p = f_s\left(1 + \frac{C_s}{2C_h}\right) = 1.001 \text{ MHz}$$

Pole zero separation = 1.001 MHz − 1 MHz = 1 kHz

Table 3.2 specified the Q of the crystal filter to be around 20,000, which implies that a high Q is usually associated with a narrow bandwidth. It is the extremely narrow bandwidth that makes the crystal so frequency-stable. Frequency stability of 2 parts per million per degree Celsius (0.0002%) is typical.

Figure 3.20 shows some possible crystal filter circuits. Because of the frequency-selective properties of the crystal, most crystal filters are *band-pass* filters. The lattice filter gives the sharpest cutoff frequency response curve, followed in order of sharpness by the π filter, the T filter, and finally, the simple band-pass filter. The response curve in Figure 3.20 shows that each crystal in the lattice network contributes to the overall response much the same as a stagger-tuned resonant circuit.

Although we can calculate the required Q and the dB/octave slope to select a filter, the value used to check the circuit after it is constructed is the **shape factor.** The shape factor is the ratio of the bandwidth at −60 dB to the bandwidth at −6 dB:

$$\text{Shape factor} = \frac{\text{BW at } -60 \text{ dB}}{\text{BW at } -6 \text{ dB}} \tag{3.4}$$

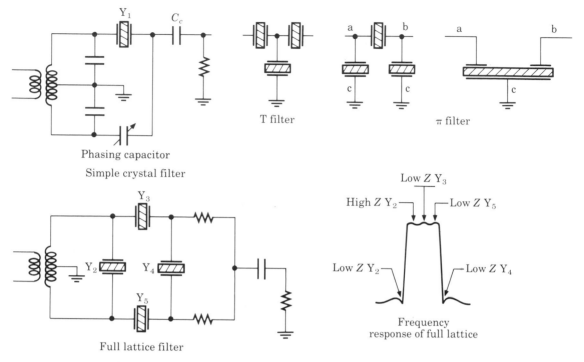

T filter

π filter

Phasing capacitor

Simple crystal filter

Full lattice filter

Frequency
response of full lattice

FIGURE 3.20
Circuit configurations for crystal filters.

For example,

$$\frac{\text{BW at } -60 \text{ dB}}{\text{BW at } -6 \text{ dB}} = \frac{28 \text{ kHz}}{25 \text{ kHz}} = 1.12$$

Thus, a very steep response curve will have a small shape factor, while a larger ratio defines a lower-Q, broader-bandwidth response curve. Low-Q resonant circuits may have a shape factor of 25.

The -6 dB and -60 dB bandwidths are measured values. From them, we can work backward to find the dB change per octave and the effective Q of the circuit. At the center frequency of 250 kHz, $\Delta f = (28 \text{ kHz} - 25 \text{ kHz})/2 = 1500$ Hz.

$$\text{dB/oct} = \frac{(60 - 6)(250{,}000)}{1500} = 9000$$

$$\text{Effective } Q = \frac{250 \text{ kHz}}{1500}\left(0.25\sqrt{\frac{1}{X}}\right) = 932.8$$

where $X = \log^{-1}(-54/20) = 1.9953 \times 10^{-3}$.

3.5.4 Mechanical Filters

Although the **mechanical filter** does not have as high a Q as the crystal filter, it has the advantage of ruggedness. In military or aircraft equipment, the mechanical filter has no equal. It is compact and has excellent band-pass frequency response characteristics. The mechanical filter is shown in Figure 3.21.

The input and output transducers are identical; either terminal may serve either function. The transducers are designed for 600 Ω at specific frequencies and act very much like high-frequency microphones and speakers, changing electrical energy into mechanical motion and mechanical motion back to electrical energy. A bias magnet, placed in the vicinity of the input transducer, will aid or oppose the flow of current through the input coil. The coil spring will induce mechanical vibrations in the disks. The disks are coupled to each other and to magnetostrictive rods, which have a piezoelectric property of lengthening and shortening as an AC voltage is applied to them. The opposite effect, changing mechanical motion to electrical energy, takes place at the output transducer. As energy is applied to each disk, the desired frequency vibrations are passed on to the next disk, and the unwanted frequencies are damped out. The size of the disk determines the resonant frequencies, and the number of disks sets the steepness of the skirts. The bandwidth is set by the spacing between the disks and by the thickness of the magnetostrictive rods.

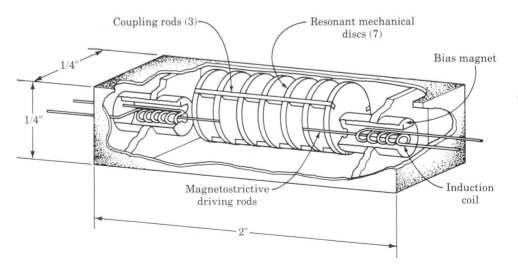

FIGURE 3.21
Mechanical resonant filter.

3.5.5 *RLC* Filters

There are five families of passive filters in common use, listed here in order of
the steepness of the attenuation skirt. Each family may be branched into low-

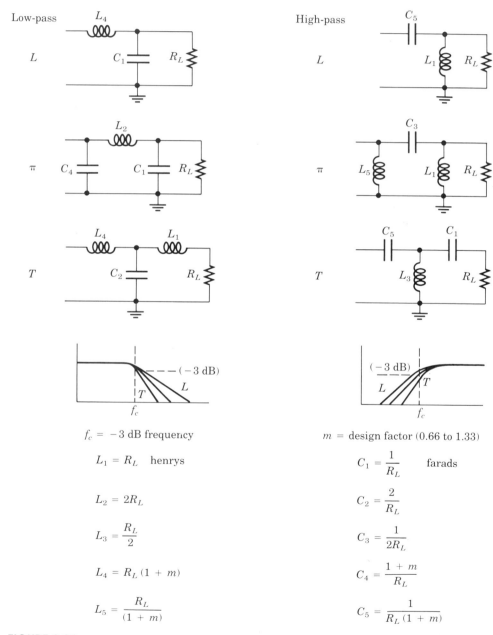

$f_c = -3$ dB frequency

$L_1 = R_L$ henrys

$L_2 = 2R_L$

$L_3 = \dfrac{R_L}{2}$

$L_4 = R_L\,(1 + m)$

$L_5 = \dfrac{R_L}{(1 + m)}$

$m = $ design factor (0.66 to 1.33)

$C_1 = \dfrac{1}{R_L}$ farads

$C_2 = \dfrac{2}{R_L}$

$C_3 = \dfrac{1}{2R_L}$

$C_4 = \dfrac{1 + m}{R_L}$

$C_5 = \dfrac{1}{R_L\,(1 + m)}$

FIGURE 3.22
Constant-*K* filters.

pass, high-pass, and band-pass filters, and each branch may have a number of circuit configurations such as L, π, and T filters. The main families are

1. constant-K
2. m-derived
3. Butterworth-Thompson
4. Butterworth
5. Chebychev

Each type may be single-pole (one section) or may use up to seven poles, cascaded to make up one filter. The greater the number of poles, the faster the rolloff response will be.

Constant-K Filter Using a simple L filter as an example, values of inductance and capacitance can be selected to make the product $XL \times XC$ a constant at all frequencies. Such a filter is called a *constant*-K *filter*. The intention is that the filter present a constant impedance at the input and output over the desired frequency range. See Figures 3.22 and 3.23.

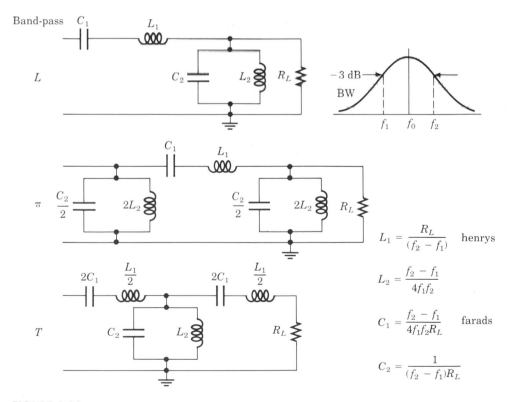

$$L_1 = \frac{R_L}{(f_2 - f_1)} \quad \text{henrys}$$

$$L_2 = \frac{f_2 - f_1}{4f_1 f_2}$$

$$C_1 = \frac{f_2 - f_1}{4f_1 f_2 R_L} \quad \text{farads}$$

$$C_2 = \frac{1}{(f_2 - f_1)R_L}$$

FIGURE 3.23
Constant-K band-pass filters.

***m*-derived Filter**   Another filter type is based on the ratio of the cutoff frequency (where the output drops to -3 dB) to the frequency where the attenuation is considered to be infinite. This ratio is termed the *m* factor and is generally between 2/3 and 4/3. The m-*derived filter,* shown in Figures 3.24 and 3.25, has a sharper rolloff than the constant-*K* filter. The improvement usually comes from adding to the branch of the filter a reactive component that provides infinite attenuation.

    The three remaining filter families are beyond the scope of this text as far as circuit descriptions and equations are concerned. These modern filters bear the names of the people who either developed the set of polynomial equations from some preselected curve shapes or broke down such equations into the component values for the circuits. Some of the major features of each family of filters are summarized next.

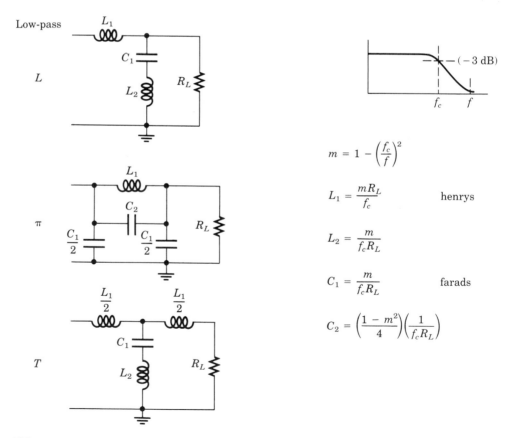

$$m = 1 - \left(\frac{f_c}{f}\right)^2$$

$$L_1 = \frac{mR_L}{f_c} \qquad \text{henrys}$$

$$L_2 = \frac{m}{f_c R_L}$$

$$C_1 = \frac{m}{f_c R_L} \qquad \text{farads}$$

$$C_2 = \left(\frac{1 - m^2}{4}\right)\left(\frac{1}{f_c R_L}\right)$$

FIGURE 3.24
m-derived low-pass filters.

Butterworth-Thompson Butterworth-Thompson filters are medium- to low-Q filter circuits. The lowest-Q filters were developed by Bessel and later modified by Butterworth and Thompson for higher Q values. The advantages to this filter type are absolute freedom from ripple across the passband and the best phase response (minimum phase change) of the three "elite" families. However, these are not sharp cutoff filters.

Butterworth Only one filter was really developed by Butterworth. It is a medium-Q band-pass filter with steep skirts, yet it has very little phase change and is free from ripple across the band. The input and output impedances are the same; however, the impedance changes with frequency.

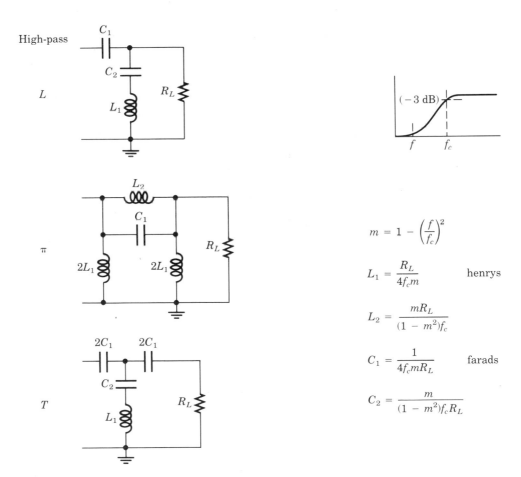

$$m = 1 - \left(\frac{f}{f_c}\right)^2$$

$$L_1 = \frac{R_L}{4f_c m} \qquad \text{henrys}$$

$$L_2 = \frac{mR_L}{(1 - m^2)f_c}$$

$$C_1 = \frac{1}{4f_c m R_L} \qquad \text{farads}$$

$$C_2 = \frac{m}{(1 - m^2)f_c R_L}$$

FIGURE 3.25
m-derived high-pass filters.

Chebychev   Chebychev developed an entire family of extremely good filters in the medium- to high-Q range. They offer the sharpest slope of the three elite filters but have relatively poor phase characteristics. The ripple across the passband is directly related to the steepness of the response curve, and one is

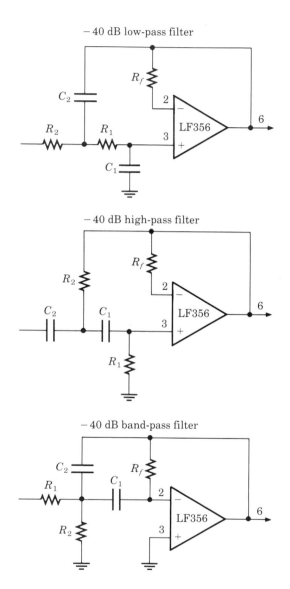

−40 dB low-pass filter

Select $R = R_1 = R_2$

$R_f = 2R_1$

$C_1 = \dfrac{0.159155}{f_c R_1}$ farads

$C_2 = 2C_1$

−40 dB high-pass filter

Select $C = C_1 = C_2$

$R_f = R_1$

$R_2 = \dfrac{R_1}{2}$

$R_1 = \dfrac{0.225}{f_c C_1}$

−40 dB band-pass filter

Select f_0 and BW

$Q = \dfrac{f_0}{\text{BW}}$

Select $C = C_1 = C_2$

$R_f = \dfrac{1}{2(\text{BW})C_1}$

$R_1 = \dfrac{R_f}{2 \times \text{gain}}$

$R_2 = \dfrac{R_f}{4Q^2 - 2 \times \text{gain}}$

FIGURE 3.26
Active filters.

generally traded off for the other. Ripple factors of 0.1 dB to 5 dB are possible, with the steepest slopes rendering the highest ripple. When an odd number of sections are cascaded together, the input and output impedances are equal. When an even number of sections are cascaded together, the input and output impedances are not equal, and the filter is used as an impedance-matching device.

3.5.6 Active Filters

The word *active* usually implies a circuit that has gain. The passive RLC filters we have looked at suffer a loss across the passband of -1 dB to -6 dB from input to output. In some active filters, the amplifier gain is only sufficient to overcome the circuit loss, called **insertion loss.** Figure 3.26 identifies some of the active filters, which provide the advantages of (1) fewer components and generally smaller part values, and (2) capacitive reactances, which means there are no inductances. The filtering action is accomplished by first building an amplifier that covers the desired frequency range. Then a frequency-selective signal voltage is fed from the output back to the input. This signal is opposite in phase to reduce the input signal voltage at the unwanted frequencies. For most practical purposes, operational amplifier active filters should have low Q values and operate at frequencies below 5 MHz. The circuits of Figure 3.26 work well within this range. Above this range, or at Qs above 100, the circuit values mushroom in both directions. Within the same circuit some values become very small while other values become extremely large.

QUESTIONS

1. What emission code is used to describe a transmission of one sideband only?
2. What emission code is used to describe a pilot carrier transmission?
3. What emission code is used to describe a transmission having two equal sidebands but no carrier?
4. Selective fading (is) (is not) a problem with SSB.
5. Name the two methods used to separate a sideband from the remainder of the signal.
6. What will be the frequency of a reinserted carrier for an $A3_j$ transmission that modulates the message signal up to the carrier level in three steps—using an oscillator at 150 kHz, one at 3.75 MHz, and a third at 87.1 MHz—when the message signal is limited between 300 Hz and 3 kHz?
7. What are the frequencies contained in the upper sideband of the transmission in the preceding question?
8. An $A3_j$ transmitter, rated at 450 W PEP, has how much carrier power?
9. What is the equivalent PEP output power for an $A3_a$ SSB transmitter that is comparable to a 300 W A3 transmitter?

10. How many times more message power does an A3$_a$ 450 W PEP transmitter have than a 300 W A3 transmitter?

11. An A3$_a$ phase shift SSB transmitter uses one in-phase balanced modulator and a second balanced modulator that is at $-90°$. Which sideband is transmitted?

12. How many signals are required at the input to a balanced modulator?

13. Name the signals that are contained in the output voltage of a balanced modulator.

14. At the output of a balanced modulator, what factor is affected by choosing the lowest message frequency?

15. Why is the first balanced modulator in an SSB transmitter operated at a low frequency (say, 100 kHz) rather than a high frequency (such as 75 MHz)?

16. A balanced modulator is basically a class B push-pull amplifier. (T) (F)

Identify the following forms of emission.

17. DSSC   (a) A3$_h$
18. VSB   (b) A3$_a$
19. DSB   (c) A3$_j$
20. SSB   (d) A3$_b$
21. SSSC   (e) A3$_c$

Independent sidebands in a single transmission must be:

22. on different sides of the suppressed carrier. (T) (F)

23. of the same bandwidth. (T) (F)

24. of the same signal form (data, voice, etc.). (T) (F)

25. limited to only one sideband on either side of the suppressed carrier. (T) (F)

26. What emission code is ISB?

27. What electrical characteristic is shared by a balanced modulator using transistors and one using diodes?

28. Which is more desirable in a filter, a large or a small slope factor?

29. In Figure 3.15, the carrier must be applied to (a) terminals 4 and 8, (b) terminals 9 and 12, (c) either set of terminals.

30. A 60 dB carrier suppression is equivalent to (a) 10% E_o, (b) 5% E_o, (c) 1% E_o, (d) 0.5% E_o, (e) 0.1% E_o.

31. A filter-type SSB transmitter has a low carrier oscillator of 200 kHz and a high carrier oscillator of 80 MHz. The message signal covers the frequency range of 50 Hz to 10 kHz.
    a. What is the frequency range in the transmitted USB?
    b. What is the required Q of the first band-pass filter?
    c. What is the required Q of the second band-pass filter?

32. What are the minimum filter types for (a) the first filter in the preceding question and (b) the second filter in the preceding question?

33. What are the approximate upper and lower frequency limits of a crystal operating in its fundamental mode of oscillation?

34. How many fundamental resonant frequencies does a crystal have?

35. Name the fundamental mode(s) from the preceding question.

36. If Question 34 has more than one mode, which has the highest frequency? (Answer "none," if applicable.)
37. What are the approximate frequency limits of the mechanical filter?
38. Describe the typical input and output signals (by name and frequency relationship) of a balanced modulator.
39. What are the advantages and disadvantages of a phase shift ($A3_j$) SSB generator?
40. In an active filter, 60 dB/decade attenuation is comparable to how many dB/octave?
41. Find the required Q of a filter that will convert a DSSC signal to an SSB signal when the separation between sidebands is 500 kHz, the suppressed carrier is at 48 MHz, and the attenuation is to be 80 dB.

CHAPTER FOUR

AM RADIO RECEIVERS

4.1 INTRODUCTION

In the commercial radio business, there are several million receivers for every transmitter, whereas in the industrial market, the ratio may be as small as 1:1. Regardless of the ratio, anyone in the field of electronics is likely to encounter a receiver considerably more often than he or she will a transmitter. A great deal of emphasis is placed on the study of receivers for this reason.

The simplest receiver contains the four component circuits shown in Figure 4.1. They are

1. the antenna,
2. a tuned circuit (fixed or variable),
3. a diode rectifier, and
4. a reproducing device to convert the electrical impulses back to a work function.

The basic circuit of Figure 4.1 will work, but it may leave much to be desired in terms of quality, noise, and sensitivity.

FIGURE 4.1
The minimum required circuits for a radio receiver.

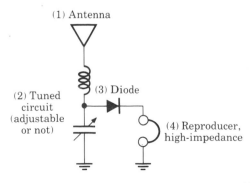

(1) Antenna

(2) Tuned circuit (adjustable or not)

(3) Diode

(4) Reproducer, high-impedance

The radio receiver has come a long way from its inception to the design seen today, namely, the superheterodyne concept of receiver circuitry. The reader can refer to other books to study the history, but there are two basic receiver forms that should not be omitted from this initial discussion. The *tuned radio frequency* receiver (TRF) of Figure 4.2 was technically the most advanced system for many years and was still popular into the mid-1930s. A disadvantage was that *all* the RF (high-frequency) amplifiers needed adjustment as the radio was tuned across the band. This caused squealing sounds when the set was not accurately aligned. Some receivers required several tuning adjustments to change stations. A further complication was that when the tuning direction was changed, often a station could not be found at the same place on the dial.

The *regenerative* (or *super-regenerative*) receiver of Figure 4.3 presented an improvement in that it used the same amplifier for several major functions.

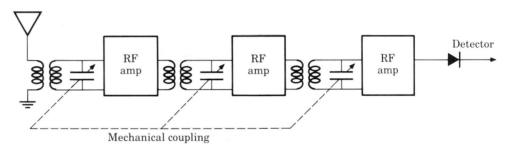

FIGURE 4.2
Functional block diagram of a tuned radio frequency (TRF) receiver.

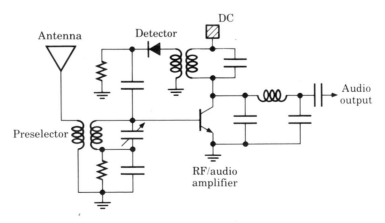

FIGURE 4.3
One version of a super-regenerative AM receiver.

In the days of tubes, this was a distinct advantage. The one tube was a wide-band RF amplifier that was biased in a nonlinear region to rectify the signal; the recovered audio was then fed through the same amplifier at the audio frequencies. The super-regenerative (miniature) receiver did a good job where size was a limiting factor.

4.2   THE SUPERHETERODYNE RECEIVER

During World War I, Major Edwin H. Armstrong was trying to find a way to detect the ignition spark from German aircraft by means of a radio receiver. In the process, he developed a circuit he called a **superheterodyne receiver.** This basic system did not achieve popularity until the middle or late 1930s, but since then, it has become the model for *all* receivers, AM, FM, amateur radio, CB, television, radar, satellite systems, etc.

We will study the superheterodyne receiver of Figure 4.4 in block diagram form first and then examine the circuits that can be used for each function.

The antenna intercepts the electromagnetic waves as they travel from the transmitter. When these waves cut through a cross section of the antenna, they induce a voltage in the antenna that ranges from a few tenths of a microvolt to several thousand microvolts. This signal strength depends on the power of the transmitter and how far the transmitter is from the receiver.

4.2.1 The RF Amplifier

The antenna signal is fed to the RF amplifier. The signal at the antenna has the lowest signal noise level that will be found anywhere in the receiver. All

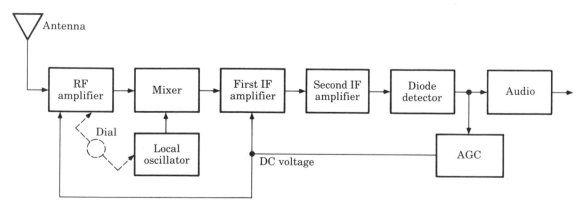

FIGURE 4.4
Typical block diagram of an AM radio receiver.

following amplifiers add some noise to the signal, so a major function of the RF amplifier is to provide gain at the point of lowest noise in the system. In many receivers the RF amplifier also participates in selecting the signal to be processed through the receiver. Some receivers have more than one RF amplifier stage, and other receivers have none. In strong-signal areas, the difference between a receiver with no RF amplifier or one with two RF amplifiers may not be noticeable. The difference in performance will show up in rural or weak-signal areas, where the receiver with one or more RF amplifiers will outperform the receiver with no RF amplifiers. The RF amplifier is often a tuned-input, tuned-output amplifier with a typical voltage gain of just over 3 (+10 dB).

4.2.2 The Mixer/Oscillator

The output signal from the RF amplifier is coupled to a **mixer** amplifier. The word *mixer* automatically indicates that there is more than one signal fed into this section. Indeed, there are two input signals, with the second coming from a **local oscillator.** Although this circuit is called a mixer, in reality it is a true modulator. It has a high- and a low-frequency input signal, it is biased nonlinearly, and it will produce the sum and difference frequencies just as the modulator at the transmitter does. The term *mixer* was probably first used to avoid association with the transmitter. In the receiver, the effort of the mixer is to *down*shift the carrier signal to an **intermediate frequency** rather than to shift it up. The intermediate frequency (called IF) for low-carrier-frequency AM receivers is an industry standard of 455 kHz. Other services, such as FM radio or television, may use different IF standards, to be specified later.

The local oscillator in low-frequency systems is always operated at a frequency higher than the incoming RF signal by an amount equal to the IF. In higher-frequency service bands, the local oscillator may be operated above or below the frequency of the incoming RF signal at the option of the designer.

For example, say the RF antenna signal is 600 kHz. Therefore, the local oscillator must operate at 600 kHz + 455 kHz = 1055 kHz for this carrier frequency. Note that in Figure 4.4 the RF amplifier tuned circuit and the local oscillator tuned circuit are linked by the tuning dial assembly. When the station selection is changed, say to 1 MHz, the RF amplifier tuning is changed to 1 MHz and the local oscillator frequency is changed to 1455 kHz. The frequency difference between the RF signal frequency and the oscillator frequency will *always* be 455 kHz, regardless of where the station selector is set. The two circuits *track,* so they are always 455 kHz apart in frequency.

In some systems, one transistor serves as *both* the local oscillator and the mixer. When this type of circuit is used, it is called a **converter.** Sometimes the oscillator/mixer circuit is referred to as the *first detector* because it changes the carrier to a lower frequency.

4.2.3 The Intermediate Frequency (IF) Amplifiers

The mixer, being a true modulator, has an output signal that contains the two original input signal frequencies plus the sum and difference frequencies of the two input signals. From the preceding example these would be an RF of 600 kHz, an oscillator frequency of 1055 kHz, a sum frequency of 1655 kHz, and a difference frequency of 455 kHz, plus some low-level harmonics of these signal frequencies. The IF amplifiers are tuned circuits that select the difference frequency of 455 kHz *only*. All other frequencies are attenuated severely. The IF circuit may have 1, 2, or 3 separate amplifiers, all tuned to the same frequency. We know that single-frequency tuned amplifiers, when cascaded, can furnish a gain as high as 80,000 (98 dB). The IF amplifiers contribute 2/3 or more of the *total* gain of the receiver. This is one big advantage of the superheterodyne receiver. The high gain is due to the gain-bandwidth product of IF amplifiers, since the bandwidth of the IF system is only 10 kHz (± 5 kHz).

4.2.4 The Detector

There are several circuits for the detection of the AM wave. All use a diode rectifier and a low-pass filter. The **diode detector** recovers the negative half cycle of the modulated wave in some cases and the positive half cycle in other cases. The low-pass filter allows the message signal to continue on to the next circuit, while it bypasses the high RF carrier signal to ground. The output voltage of the detector contains a DC component of the wave plus the message signal. Because the detector does a job at the receiver that is opposite to that of the modulator at the transmitter, it might be suggested that the detector be renamed the demodulator. This is wrong. In the strictest technical sense, the word *demodulator* can only be applied to a detector that requires the input of a carrier signal to recover the modulation. This will be clear later, when single-sideband demodulators are discussed.

4.2.5 Automatic Gain Control (AGC)

The DC component of the wave recovered by the detector is further filtered and fed back to the *first* IF amplifier as a bias voltage to modify the gain of the amplifier relative to the signal strength at the antenna. **Automatic gain control** works as follows. When a strong signal is received, the detector DC output level is high, and a large DC voltage is sent back to the first IF to *reduce* the gain. When the incoming signal is weak, a smaller DC voltage from the detector is sent back to the IF amplifier to *increase* the gain. The net result is that the signal level at the output of the detector is relatively constant (within limits) for either a weak or a strong signal at the antenna. The circuit of the IF amplifiers will decide if a negative or a positive voltage is required for AGC bias to alter the gain. The polarity of voltage then dictates which way the detector diode is connected to establish the positive or negative DC component. It should be remembered that the modulated wave is symmetrical; the top half and the bottom half of the modulated wave each contain the same message envelope. Thus, the diode can be configured in either direction without affecting the message signal. During servicing, testing, or alignment, it is common practice to disconnect the AGC line and hook up a battery in its place so that the AGC voltage will be fixed.

4.2.6 Output Reproducer

For radio receivers, the output reproducer is an audio circuit connected to a speaker to change the message signal into sound. The signal level from the detector into the audio amplifier will range from about 0.1 to 2.0 V_{rms} on the average, depending on the receiver design. The audio amplifiers increase the strength of the signal to be presented across the speaker. For instance, 6 V_{rms} of audio across an 8 Ω speaker will result in 4.5 W of audio power. The load for the detector is often a variable resistor that functions as a volume control.

Reexamine the block diagram for the superheterodyne receiver, Figure 4.4. Compare the circuit descriptions to the block diagram. Repeat this procedure until you are confident that you can draw a block diagram from memory, name all of the circuits, and describe their functions.

4.3   DOUBLE-CONVERSION RECEIVERS

Receivers in the high-frequency RF range are more likely than low-frequency receivers to incorporate **double-conversion** front-end circuitry. This is due in part to the wide frequency spread between the RF signal and the standard IF value (455 kHz) and in part to the highly sensitive receivers (1 μV) used in high-frequency work. Figure 4.5 displays a double-conversion system. The frequencies and input levels of citizens band radio lend themselves to double conversion and so are used to illustrate this process.

The antenna input signal at 27.005 MHz ($\pm$5 kHz) is amplified by the RF amplifier and fed to the first mixer. The first local oscillator may be higher or lower than the RF antenna signal in frequency. Figure 4.5 shows the oscillator set lower than the incoming signal frequency. The output frequencies of the first mixer are the two original input frequencies, 27.005 MHz and 16.31 MHz, plus their sum and difference frequencies, 43.315 MHz and 10.695 MHz. The crystal band-pass filter between the two mixers will pass only the difference frequency, 10.695 MHz ($\pm$5 kHz). The first local oscillator and the RF amplifier are linked through the channel selector, so that as one is adjusted, the other is also. In this way, the signal passed through the filter between the two mixers will be at 10.695 MHz for *all station selections*. Mixer 2 always sees the same signal frequencies at its inputs, 10.695 MHz from mixer 1 and 10.24 MHz from

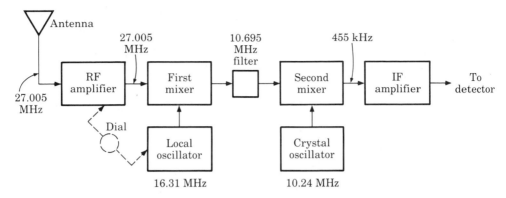

FIGURE 4.5
Double-conversion RF section. Frequencies of channel 4 in the CB band.

fixed crystal oscillator 2. The difference frequency between these two signals, then, is 455 kHz. The system is called *double conversion* because the RF frequency is reduced to the intermediate frequency in two steps.

There are several advantages to double conversion. For example, the RF signal is reduced to a very low intermediate frequency, which is less problematic and which provides gain-bandwidth product features better suited to stable circuits. Also, double-conversion systems have a better **image rejection** ratio than single-conversion circuits. The only disadvantages are a slightly higher cost and a few more steps in the alignment procedure.

4.3.1 Image Frequency Rejection

It is possible for an unwanted RF signal to get through the RF amplifier, mix with the oscillator signal, and appear along with the desired signal at the intermediate frequency to be amplified. An **image** is a signal that is the same distance from the local oscillator frequency as the desired signal, but in the opposite direction. When the oscillator operates at a frequency higher than the desired RF signal, the image frequency is

$$\text{Image frequency} = \text{RF} + 2\text{IF} \qquad\qquad \textbf{(4.1a)}$$

When the local oscillator operates at a frequency below the RF signal, then the image frequency is

$$\text{Image frequency} = \text{RF} - 2\text{IF} \qquad\qquad \textbf{(4.1b)}$$

A single-conversion receiver in the standard broadcast AM radio band that is tuned to 540 kHz and that has a local oscillator working above the RF signal has an image frequency at

$$540 \text{ kHz} + 2(455 \text{ kHz}) = 1450 \text{ kHz}$$

Notice that the image signal is still within the AM band.

The rejection of an image signal is dependent on the ratio of the wanted to unwanted signal frequencies and on the Q of the resonant circuits that precede the mixer amplifier. The attenuation in decibels of an image frequency signal may be found by

$$\text{Image rejection (dB)} = 20 \log \sqrt{Q_t^2 F^2} \tag{4.2}$$

where Q_t is the total Q of the system. In a preselector receiver, in which no RF amplifier is used, Q_t represents the Q of the preselector tuned circuit. In circuits that have a tuned-input, tuned-output RF amplifier, the total Q is found by

$$Q_t = \sqrt{Q_i^2 + Q_o^2} \tag{4.3}$$

F is a factor that compares the ratio of the image frequency to the desired RF frequency, and is found by

$$F = \frac{\text{image}}{\text{RF}} - \frac{\text{RF}}{\text{image}} \tag{4.4}$$

for cases where the local oscillator operates higher than the RF, or by

$$F = \frac{\text{RF}}{\text{image}} - \frac{\text{image}}{\text{RF}} \tag{4.5}$$

for cases where the local oscillator works at a frequency lower than the RF.

EXAMPLE 4.1

Given a Q_t of 50 and the frequencies used in the image example above,

$$F = \frac{1450 \text{ kHz}}{540 \text{ kHz}} - \frac{540 \text{ kHz}}{1450 \text{ kHz}} = 2.313$$

$$\text{Image rejection} = 20 \log \sqrt{50^2 (2.313)^2} = 41.26 \text{ dB}$$

The image is attenuated by more than 40 dB below the level of the desired signal strength. This represents good image rejection.

4.3.2 Signal-to-Noise Ratio (S/N)

A relative measure of the desired signal power P_s compared to the noise power P_n at a given point in the system is known as the **signal-to-noise ratio** and is derived as follows:

$$S/N(\text{dB}) = 10 \log \left(\frac{P_s}{P_n} \right) = 20 \log \left(\frac{E_o}{E_n} \right) \tag{4.6}$$

where $E_o = E_s + E_n$

4.3.3 Sensitivity

The ability of the receiver to amplify weak signals is called **sensitivity.** Sensitivity identifies the *minimum* input signal at the antenna that will result in a signal 20 dB greater than the noise at the detector output terminals. Sensitivity is usually defined in microvolts at the low end, middle, and high end of the receiver range of frequencies. The sensitivity specification includes the system's signal-to-noise ratio, for example,

$$\text{Sensitivity} = 3 \ \mu\text{V for } S/N = 20 \text{ dB (20 dB of quieting)}$$

Different input impedances will alter this measurement, so the trend is toward a reference to a power level that will facilitate comparison between receivers. Thus, the sensitivity voltage is related to the input impedance and compared in dB to 1 fw (1 femtowatt or 10^{-15} W), a relation abbreviated as "dBf." For example, 10 μV input across 1 kΩ is an input power of 10^{-13} W. The dBf value is found by

$$\text{Sensitivity} = 10 \ \log \left(\frac{10^{-13}}{10^{-15}} \right) = 20 \text{ dBf} \quad (+20 \text{ dB compared to 1 fW})$$

4.3.4 Selectivity

Selectivity describes the ability of the receiver to accept the desired signal frequency while rejecting all closely adjacent disturbances. Selectivity is usually poorer at the high-frequency end of the band and improves as the frequency goes lower. As noted in Figure 4.6, the bandwidth of the RF amplifier is a compromise between good selectivity and good fidelity; therefore, the bandwidth is made wider than would be optimal for good selectivity. The steep skirts of the IF amplifier's response curve contribute heavily to improved selectivity.

FIGURE 4.6
Selectivity curve. The response is continued until the input voltage is 80 dB greater than at f_0 or until the bandwidth reaches ±100 kHz, whichever occurs first.

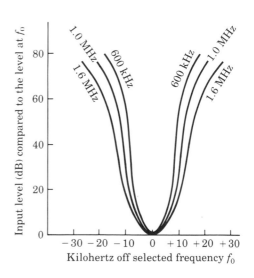

Selectivity and image rejection are related actions; both reduce the strength of signals away from the selected resonant-frequency signal. Image rejection attenuates alien signals at a fair distance from the selected carrier, while selectivity reduces interference closer to the carrier signal frequency. The degree of selectivity is stated as a decrease in response at noted deviations from the center frequency, f_0. In the testing phase, the receiver output is recorded when the signal generator and receiver are set to the same frequency. The generator frequency is changed, and the difference in receiver output level is monitored. Typically, receiver specifications are stated as follows:

-6 dB at ± 2.5 kHz off center frequency
-60 dB at ± 7.5 kHz off center frequency

Note again the -6 and -60 dB reference levels, used in Chapter 3 in the discussion of the slope of a response curve.

4.3.5 Dynamic Range

A receiver that falters under strong-signal conditions is as problematic as one that has excessive noise in a weak-signal area. The **dynamic range** of a receiver can be defined as the difference in input level, in decibels, between the minimum discernible signal and the overload condition that causes compression. Although this may be the single most important performance characteristic, it seems to appear only in the specifications for military equipment or very high quality commercial products.

The minimum discernible signal (MDS) detectable by some measuring instrument is considered to be an output signal voltage plus noise voltage that is 3 dB greater than the noise signal alone. This level is referred to as the *noise floor*.

The minimum input noise signal to a receiver stems from the thermal agitation currents of the first resistor encountered by the signal. A single isolated resistor may have a thermal noise of approximately -126 dBm (decibels compared to 1 mW). Fortunately, the receiver's input is not resistive, but rather reactive; therefore, the noise floor at the input terminals to the receiver are closer to -135 dBm (0.18 μV across 1 kΩ) depending on the Q of the input circuit. The two-generator test setup indicates that an input level of -55 dBm (1800 μV across 1 kΩ) causes compression of the signal in the RF stages. It is therefore concluded that an input signal level between -135 dBm and -55 dBm will pass through the receiver without adverse effects; in other words, the dynamic range is 80 dB. This corresponds to a voltage ratio of 10,000 to 1. A dynamic range of 100 dB is said to be the highest possible. A poor dynamic range causes blocking (desensitizing) in the RF amplifier and results in cross modulation and severe intermodulation distortion at weak signal levels.

4.4   RECEIVER CIRCUITS: THE FRONT END

4.4.1 The RF Amplifier

It is important to be able to look at a diagram of an unfamiliar circuit and pick out the salient features. For example, from Figure 4.7a we can establish the following conclusions. This is a high-frequency amplifier, as evidenced by the

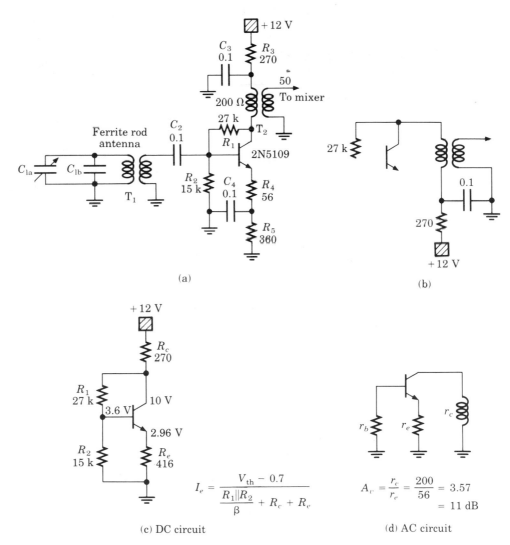

(a)

(b)

(c) DC circuit

$$I_e = \frac{V_{th} - 0.7}{\dfrac{R_1 \| R_2}{\beta} + R_c + R_e}$$

(d) AC circuit

$$A_v = \frac{r_c}{r_e} = \frac{200}{56} = 3.57$$
$$= 11 \text{ dB}$$

FIGURE 4.7
RF amplifier configurations.

tuned-input and tuned-output circuits (tuned circuits are *not* used in low-frequency amplifiers). It is a small-signal RF amplifier, as indicated by the ferrite rod antenna at the input side. The schematic does not give values for the tuned circuits, but if the diagram is marked as an AM radio receiver, then we know that this RF amplifier must operate in the 540 kHz to 1600 kHz range of frequencies.

Note the very different appearance of parts a and b of Figure 4.7. Figure 4.7a is the formal textbook schematic form showing the positive voltage (or supply) at the top of the diagram and ground (or common) at the bottom. Figure 4.7b is the standard industry method of showing the same circuit. You should be able to mentally interchange these two methods of presentation. Also, you should learn to visually identify the DC bias circuit for the amplifier, shown in Figure 4.7c.

This circuit has a bias arrangement that is a cross between collector feedback bias and true voltage divider bias. Because of the small value of R_c (270 Ω) compared to the sum of R_1 and R_2 (42 kΩ), the collector feedback bias equation can be used to find the emitter current (I_e) to a close approximation. If β is not known, use a value of 50 for wide-frequency, small-signal RF amplifiers and a value of 150 for low-frequency audio or video amplifiers. From this, you can determine the collector, base, and emitter voltages. Before going into these calculations, identify the amplifier class by the values of the resistors in the DC circuit. Compare the value of R_1 to R_2. If R_2 is one-tenth to one-half the value of R_1, the amplifier is class A. If R_2 is much smaller than one-tenth of R_1, then check for an emitter resistor. If there is no emitter resistor, R_2 can be $(1/100) R_1$ before class B operation takes over.

When there is a need to know the voltages on the transistor, for testing, analyzing, or repairing, then solve for the value of emitter current (I_e) using the equation of Figure 4.7c. At this point, assume that the emitter and collector currents have the same value, and solve for the emitter and collector voltages first, then the base voltage. Knowing the voltages, you can make a definite statement about the class of amplification: this is a class A amplifier. When the voltages appear on the schematic, they are the DC operating voltages under conditions of *no* signal. Finally, to know the approximate gain of the amplifier, you should be able to convert the diagram to an AC equivalent circuit. First note that C_3 places the DC side of the output transformer primary at AC ground. The AC signal will encounter only the primary of the output transformer in the collector circuit. Note next that the junction between the two emitter resistors is at AC ground through C_4, which means that the emitter will see only 56 Ω to ground. The top of R_1 is almost at AC ground through C_3, and the bottom of R_2 is at ground, so that R_1 and R_2 are in parallel for the AC circuit of the base. The impedance of the output transformer primary is given as 200 Ω, so the gain, from the equation of Figure 4.7d, is 3.57 or 11 dB. The concluding observations regarding Figure 4.7 are:

1. The circuit is a tuned-input, tuned-output, common-emitter, class A, small-signal RF amplifier.
2. The input is tunable by the variable capacitor across the antenna tank circuit over the frequency range of 540 kHz to 1600 kHz. Fixed tuning of the output is through the inductance of the output transformer, having a gain of about 11 dB.
3. The emitter current is about 7 mA, which sets up bias voltages of $V_e =$ 2.91 V, $V_b = 3.57$ V, and $V_c = 10.0$ V.

4.4.2 The Mixer

Carefully examine the circuit of Figure 4.8 and try to analyze it before reading this section.

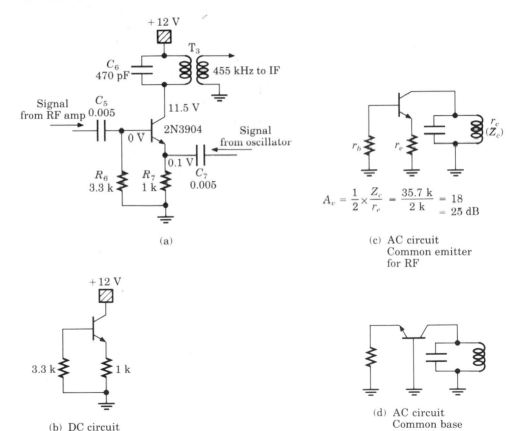

$$A_v = \frac{1}{2} \times \frac{Z_c}{r_e} = \frac{35.7 \text{ k}}{2 \text{ k}} = 18 = 25 \text{ dB}$$

(a)

(c) AC circuit
Common emitter
for RF

(b) DC circuit

(d) AC circuit
Common base
for oscillator

FIGURE 4.8
The mixer circuit configuration.

The first observation is that this circuit has two input signals. One signal comes from the RF amplifier and will mix with the signal coming from the local oscillator. The output is tuned to the intermediate frequency. The circuit performs as a common-emitter amplifier for the RF signal and as a common-base amplifier for the oscillator signal. Most distinctive in this circuit is the DC biasing. Note that there is *no* DC voltage applied to the base of this amplifier. The base is connected through resistor R_6 to ground, placing the base at 0 V and just at cutoff. The DC voltages on the transistor with both signals removed are $V_e = 0$, $V_b = 0$, and $V_c = V_{cc}$. When an RF signal is present, the base goes more negative due to the base leak bias capacitor C_5. These are all conditions for class C amplification. To further enhance this class of amplification the signal from the RF amplifier will charge C_5 with a voltage that is negative on the base side and that will hold the mixer in class C cutoff. The RF signal at the base will cause current pulses in the collector tuned circuit each time the input pulses are sufficiently positive to turn the transistor on. The resonant tank in the collector will restore the missing half cycle, just as the final power amplifier did for the transmitter example in Chapter 2.

The oscillator signal at the emitter will drive the mixer into conduction on the negative half cycle of the wave. Driving the emitter negative is the same as driving the base positive. Both will cause conduction and a flow of collector current. Collector and emitter current flow will raise the emitter slightly above ground by the voltage drop across the emitter resistor.

The AC circuit for the mixer (relative to the RF signal) is shown in Figure 4.8c, where the gain is the collector impedance divided by the emitter impedance. However, this amplifier only conducts for one-half of the cycle (or less); therefore, the gain is one-half of the value derived by the standard equation.

The collector inductance of 260 μH (from the values $C = 470$ pF and $f = 455$ kHz) is used to find the collector impedance at resonance:

$$Z_c = \frac{L}{RC} = 35.7 \text{ k}\Omega$$

where $R = 15.5\ \Omega$ (measured). The gain is found as

$$A_v = \frac{1}{2}\left(\frac{Z_c}{r_e}\right) = \frac{1}{2}\left(\frac{35.7 \text{ k}\Omega}{1 \text{ k}\Omega}\right) = 17.85$$

dB gain = 20 log 17.85 = 25 dB

When the AC equivalent circuit for the oscillator signal (Figure 4.8d) is examined, it is found to be the common-base configuration. The gain equation is the same as for the common-emitter configuration; therefore, the two signals are treated exactly alike.

Another method of injecting a signal into the mixer is shown in Figure 4.9. Here the RF signal is transformer-coupled to the base of the mixer, with C_8 acting as a DC blocking capacitor so that the base is not at DC ground through

T_4's secondary. The local oscillator signal is also applied to the base of the mixer through capacitor C_9, which is kept small to reduce *oscillator pulling* (changing the tuning of T_4 when the oscillator frequency is changed). Because of the small value of C_9, oscillator signal voltages as great as 8 V p-p may be needed. This configuration has a poorer dynamic range than the emitter-injected mixer because of the rigid bias voltage created by the voltage divider bias network (R_8 and R_9).

Thevenizing the base voltage and base resistance and estimating a β of 150, we use the emitter current equation of Figure 4.9b to find

$$I_e = 0.38 \text{ mA}$$
$$V_e = 0.38 \text{ V}$$
$$V_b = 1.09 \text{ V} \qquad (V_{b-e} = 0.7 \text{ V})$$
$$V_c = 11.5 \text{ V}$$

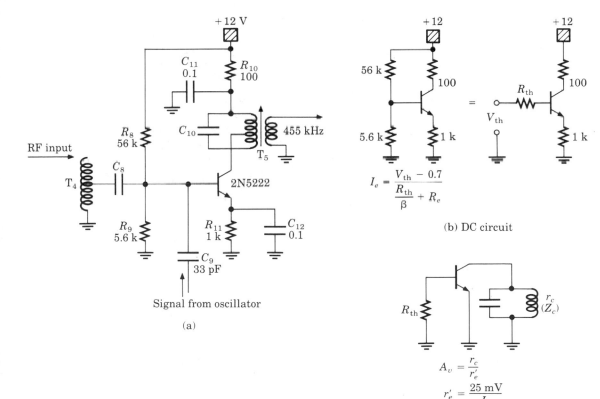

(a)

(b) DC circuit

$$I_e = \frac{V_{th} - 0.7}{\dfrac{R_{th}}{\beta} + R_e}$$

(c) AC circuit

$$A_v = \frac{r_c}{r_e'}$$

$$r_e' = \frac{25 \text{ mV}}{I_e}$$

FIGURE 4.9
Alternate mixer circuit.

Examining the ratio of R_9 to R_8 plus the calculated voltages, we conclude that this amplifier would greatly distort signal levels greater than 0.7 V_{pk} because its operation is right on the border between class A and class B. The same values effect a low input impedance of about 1 Ω. This mixer circuit would typically be employed when *no* RF amplifier were used, and the input transformer would be a ferrite rod antenna.

4.4.3 The Local Oscillator

The local oscillator found in the typical AM radio receiver is a sine wave oscillator. It makes little difference which sine wave oscillator circuit is used (Colpitts, Hartley, Armstrong, etc.); however, the Hartley and Armstrong circuits are the most common because it is easier to vary the capacitance for tuning than to vary the inductance. (Inductive tuning, by the way, was popular in many auto radio receivers with pushbutton selectors and when the IF amplifier frequency was 262.5 kHz.)

The circuit of Figure 4.10 is a Hartley sine wave oscillator, which is a common-base amplifier (base at AC ground through C_{13}). The input signal is capacitively coupled to the emitter. The collector signal is transformed with zero phase change to the secondary of T_6, where a portion of the secondary signal is capacitively fed back to the emitter through C_{14}. The in-phase feedback plus gain results in sustained oscillations that will have a maximum peak-to-peak signal at the collector equal to V_{cc}. The tapped-down secondary signal is also capacitively coupled through C_{16} to the mixer. The DC biasing is the standard base bias configuration. With a 0.7 V DC base-emitter potential and a β of 150, the loop equation as given in Figure 4.10b is set up to solve for emitter current (I_e) and produces the following values:

$$I_e = 1.30 \text{ mA}$$
$$V_e = 7.25 \text{ V}$$
$$V_b = 7.95 \text{ V} \qquad (V_{b-e}) = 0.7 \text{ V}$$
$$V_c = 12.0 \text{ V}$$

Examination of these values identifies a class A amplifier. However, when the circuit is constructed and the voltages verified, the values are found to be:

$$I_e = 1.16 \text{ mA}$$
$$V_e = 6.50 \text{ V}$$
$$V_b = 2.80 \text{ V} \qquad (V_{b-e}) = -3.7 \text{ V}$$
$$V_c = 12.0 \text{ V}$$

These measured voltages, with the base voltage much lower than the emitter voltage, identifies class C operation.

Most sine wave oscillators are designed with DC biasing for class A amplification. However, with in-phase feedback sufficient to cause sustained oscillation, they will self-bias (signal bias) into class C operation. Measuring the DC

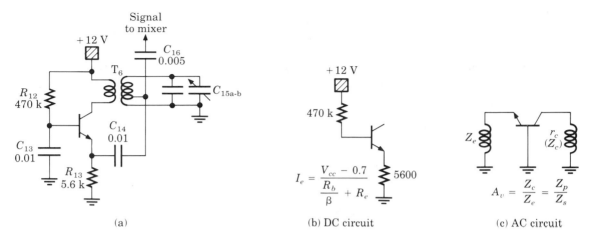

FIGURE 4.10
Hartley oscillator circuit configuration.

voltages with a high-impedance voltmeter will determine if the circuit is oscillating. When the frequency of oscillation is low, most oscilloscopes will allow
you to observe the output waveform directly. However, when the frequency of
the oscillator is above the limits of the scope, the DC voltages may be the only
means available to validate circuit operation.

The AC equivalent circuit of the oscillator is shown in Figure 4.10c. The
tapped secondary of T_6 is in parallel with the emitter resistor R_e, so that the
emitter impedance is the impedance of the secondary at the tap, designated Z_s.
The collector impedance is the primary impedance of the oscillator transformer
T_6, designated Z_p. The gain is thus Z_p divided by Z_s, as seen from Figure 4.10c.
Oscillations will be sustained as long as Z_p is larger than Z_s and the gain is
greater than 1.

4.4.4 Tuning Capacitors

A capacitor is formed by any two conducting materials separated by an insulator. The electrical value of capacitance is a function of the *size* of the conductors, the *distance* between the conductors, and the *insulating material* that
separates the conductors. Changing any of these three variables will change
the value of capacitance. The equation is

$$C = \frac{\epsilon_o \epsilon_r A}{d} \quad \text{(in farads)}$$

where ϵ_o = permittivity of a vacuum = 8.85×10^{-12} F/m
$\quad\epsilon_r$ = permittivity of insulator compared to a vacuum (air = 1.006)
$\quad A$ = area of one plate
$\quad d$ = distance between the plates (in same units as A)

Figure 4.11 shows an example of capacitance calculation. The most common way to change the capacitance value is to change the size of the plates. This is demonstrated in Figure 4.12 by changing the overlap area of the plates. A second method is to change the spacing between the plates. In the trimmer capacitor in Figure 4.12 a slice of mica insulation is squeezed between the plates. Greater pressure moves the plates closer together and raises the capacitance.

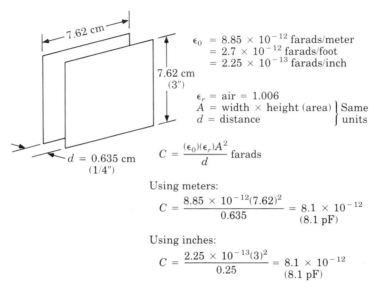

$$\epsilon_0 = 8.85 \times 10^{-12} \text{ farads/meter}$$
$$= 2.7 \times 10^{-12} \text{ farads/foot}$$
$$= 2.25 \times 10^{-13} \text{ farads/inch}$$

$\epsilon_r = \text{air} = 1.006$
$A = \text{width} \times \text{height (area)} \big\}$ Same
$d = \text{distance}$ units

$$C = \frac{(\epsilon_0)(\epsilon_r)A^2}{d} \text{ farads}$$

Using meters:
$$C = \frac{8.85 \times 10^{-12}(7.62)^2}{0.635} = 8.1 \times 10^{-12}$$
$$(8.1 \text{ pF})$$

Using inches:
$$C = \frac{2.25 \times 10^{-13}(3)^2}{0.25} = 8.1 \times 10^{-12}$$
$$(8.1 \text{ pF})$$

FIGURE 4.11
Determination of capacitance values.

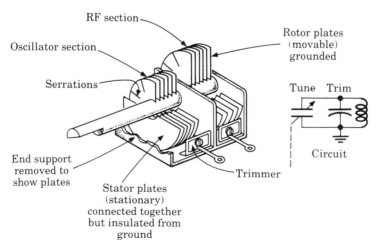

FIGURE 4.12
Variable air capacitor.

The trimmer capacitor is in parallel with the main tuning capacitor. The shape of the tuning plates defines how the dial scale is distributed across the face of the dial. The scheme of Figure 4.13a is most common in receivers in which space is a prime consideration. This plate arrangement occupies the smallest space. From the capacitance equation, it is clear that the capacitance is directly proportional to the overlapping plate area. From the resonant frequency equation,

$$f = \frac{0.159155}{\sqrt{AL}}$$

it is seen that the frequency is modified by the inverse square root of the area. As the area doubles, the frequency goes down by $1/\sqrt{2} = 0.7071$.

Thirty degrees of dial rotation at one end of the scale will *not* produce the same change in frequency as 30° at the other end of the scale. The scale is nonlinear, and stations at the high end of the scale are harder to tune than stations at the low end. To make the scale spread even across the dial, the plate

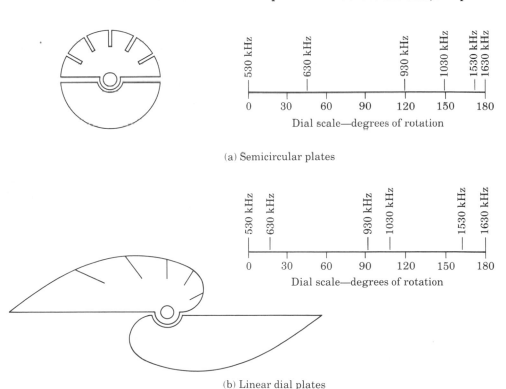

(a) Semicircular plates

(b) Linear dial plates

FIGURE 4.13
The two extreme cases of tuning capacitor plate shapes. (a) Semicircular plates take less room but have a badly distorted dial scale. (b) Linear dial plates are strangely shaped, require more extensive calculations, and take much more room, but have an accurate linear dial scale.

shape must be distorted (made nonround) so that 30° of rotation at the low end of the dial will cause the same frequency change as 30° at the high end. This is shown in Figure 4.13b, which depicts a *linear scale* capacitor plate.

4.4.5 High- or Low-Frequency Oscillator?

The electronic industry involves many tradeoffs. None of the systems we have is perfect, and the differences in circuitry are based on the opinions of the decision makers. Rarely do we encounter such a clear-cut case as the decision to make the local oscillator in a low-frequency receiver run at a frequency higher than or lower than the RF signal. Technically, it could be either.

When the oscillator frequency is higher than the frequency of the signal being received, it must be so by an amount equal to the intermediate frequency. In AM receivers, this is 455 kHz. An AM oscillator will generate a sine wave from 995 kHz (540 kHz + 455 kHz) to 2055 kHz (1600 kHz + 455 kHz). The ratio of the highest to lowest frequency is 2055/995 = 2.065 to 1. Using the resonant frequency equation and a fixed value of inductance (L), the necessary change in capacitance is the frequency ratio (2.065) squared, or 4.265 to 1. The tuning capacitor value must be approximately 4¼ times as large at the lowest frequency as at the highest frequency. This is certainly an achievable goal.

When the oscillator frequency is lower than the carrier frequency being received, the circuit must oscillate at 85 kHz (540 kHz − 455 kHz) at the low end and 1145 kHz (1600 kHz − 455 kHz) at the high end of the band. The frequency ratio is now 1145/85 = 13.47 to 1. Again, the capacitance ratio is the frequency ratio (13.47) squared, or 181.45 to 1. It would be impractical to construct a capacitor with a ratio of 181:1 when one with the smaller ratio of 4.25:1 works as well. For this reason, all low-carrier-frequency receivers are constructed so that the local oscillator operates at a frequency *higher* than the receiver's RF signal frequency.

When the center frequency of the receiver's band divided by the spread of frequencies covered by the band exceeds about 5, then the oscillator frequency can be selected either above *or* below the incoming RF signal with no apparent problems. The AM band's center frequency of 1070 kHz divided by its spread of 1060 kHz equals about 1, so the oscillator must operate at a higher frequency than the RF signal.

4.4.6 Front-End Circuits

The receiver is divided into three major parts: the high-frequency section, the intermediate-frequency section, and the low-frequency section. The high-frequency section is referred to as the **front end.** Figure 4.14 is a composite of Figures 4.7, 4.8, and 4.10 and shows how these circuits interconnect.

The waveforms associated with the front-end circuitry of Figure 4.14, at locations R, S, and T, are shown in Figure 4.15a. The wave at R is a 1 MHz

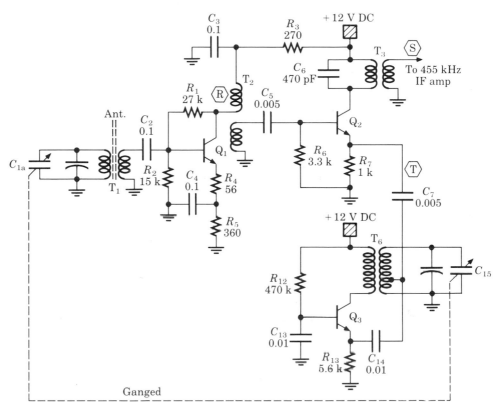

FIGURE 4.14
Complete front-end circuitry (Figures 4.7, 4.8, and 4.10 combined).

amplitude-modulated carrier with a 5 kHz modulating sine wave. The wave at T is the local oscillator sine wave of constant amplitude at 1.455 MHz, and the wave at S is the carrier *after* it has been reduced to the 455 kHz signal. Observe wave S in particular. Note that the shape of the modulated envelope has *not* changed; only the carrier frequency that supports the envelope has been reduced. This is further illustrated in Figure 4.15b, where the frequency domain display locates the RF frequencies of the carrier and its two sidebands on a scale with the local oscillator signal. Observe the quantitative difference in frequency between the oscillator signal and the three components of the RF wave. It is this quantitative difference in frequency that constitutes the IF signal, as shown in Figure 4.15c.

Not all receiver front ends are as refined as the one in Figure 4.14. The complexity of the system depends on the intended service area and the cost vs. sensitivity tradeoffs of the manufacturer. A modified front end is shown in Figure 4.16, where the RF amplifier has been omitted. The frequency-selecting

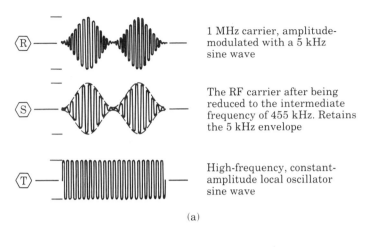

(R) — 1 MHz carrier, amplitude-modulated with a 5 kHz sine wave

(S) — The RF carrier after being reduced to the intermediate frequency of 455 kHz. Retains the 5 kHz envelope

(T) — High-frequency, constant-amplitude local oscillator sine wave

(a)

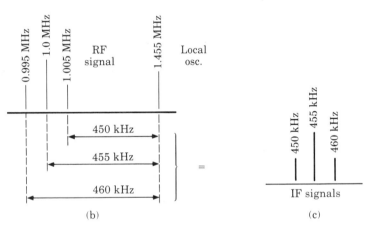

(b)

(c)

FIGURE 4.15
Conversion frequency spectrum.

circuits of the RF amplifier have been replaced with a preselector. The primary disadvantages of removing the RF amplifier are:

1. loss of sensitivity,
2. reduced selectivity,
3. poorer signal-to-noise ratio, and
4. increased signal radiation from the oscillator to the antenna.

The first three are self-explanatory, but the fourth needs some discussion. The signal from the oscillator is connected to the emitter of the mixer and can be fed

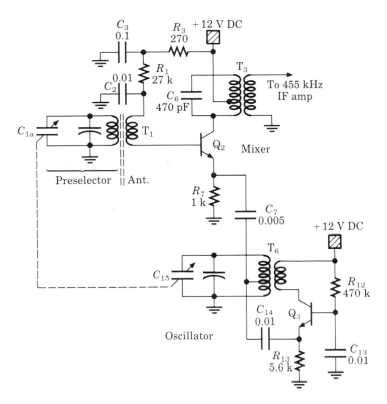

FIGURE 4.16
Mixer-oscillator. Preselector takes the place of a "no gain" RF section.

through the base to the antenna coil and be transmitted into the atmosphere. The FCC has rigid standards on the amount of energy that may be emitted from a receiver. The regulation is intended to reduce interference in other receivers.

Figure 4.17 shows a further reduction in the front-end circuitry. Here the RF amplifier is omitted, and the mixer is combined with the local oscillator. When the mixer and oscillator are the same transistor, it is called a **converter.** Any time one circuit is called upon to perform several functions, it will do so at the expense of quality. Ideal alignment may be more difficult or impossible to achieve, and the circuit may drift out of adjustment more easily. Tuning stations may require more precision, and component tolerances will have a greater effect on performance depending on age and temperature. Any investment in the receiver front end is a justifiable cost.

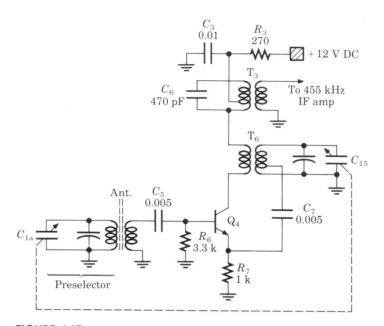

FIGURE 4.17

A converter. One transistor takes the place of the oscillator and mixer, with a prese-lector and no RF amplifier.

4.5   INTERMEDIATE-FREQUENCY AMPLIFIERS

It is impossible to tell the difference between an IF amplifier and an RF ampli-fier from a schematic if no further information is given. The IF amplifier is, in every sense of the word, an RF amplifier. It has a tuned input circuit and a tuned output circuit, and it amplifies a modulated carrier signal. The IF ampli-fier takes its name from its function within the receiver system, not from the frequency values it amplifies.

The IF amplifier diagram in Figure 4.18 can be examined in the same way as all of the other circuit diagrams. By this time, some of the main features of an amplifier should stand out and require less effort to identify. This transistor has voltage divider bias, with R_{14} and R_{15} between the 12 V supply and the bias source of -1 V. (The bias source will be dealt with later.) Through voltage divider theory, a DC base bias voltage of 2.7465 V is found. Considering that this voltage is referenced to -1 V, we must subtract 1 V from the DC base bias calculated voltage, for a value of 1.7465 V. Thevenizing the base resistance by placing R_{14} in parallel with R_{15}, we find the equivalent base resistance is 11.83 kΩ. Using these values and an estimated β of 150 in the equation in Figure 4.18b will result in an emitter current of 2.58 mA. From this

$$I_e = 2.58 \text{ mA}$$
$$V_e = I_e R_e = 0.843 \text{ V}$$
$$V_b = V_e + 0.7 = 1.543 \text{ V} \qquad (V_{b-e} = 0.7 \text{ V})$$
$$V_c = V_{cc} - I_e R_c = 11.75 \text{ V}$$

In all of the cases so far, the gain has been found by means of an AC circuit like the one in Figure 4.18c. Gain calculation for this circuit should be no different, except that we know little about the collector load impedance. We do

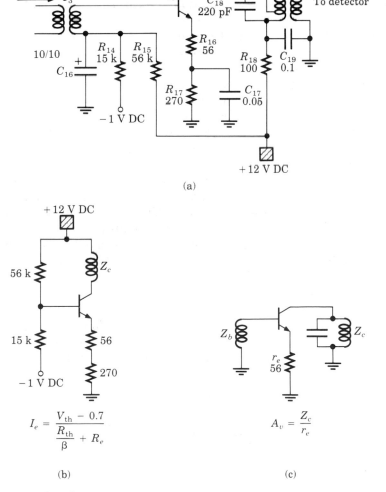

(a)

(b)   (c)

FIGURE 4.18
Intermediate frequency amplifier.

know, however, that the collector is tied to a tap on the output transformer primary. This is done to match the output impedance of the transistor to the impedance of the transformer at the tap. To determine gain, all we need to know is the output impedance of the transistor. Z_o is thus E divided by I, or $V_{c\text{-}e}/I_e = 10.9 \text{ V}/2.58 \text{ mA} = 4225 \text{ } \Omega$. From this,

$$A_v = \frac{Z_o}{R_e} = \frac{4225}{56} = 75.45$$

$$\text{dB gain} = 20 \log 75.45 = 37.54 \text{ dB}$$

We have seen how the resistors of the circuit establish the static DC operating Q point on the family of $V_{c\text{-}e}$ vs. I_c characteristic curves and how the bypass capacitors place AC ground at selected points to aid the collector impedance calculation in the determination of gain. However, the key to high-performance IF amplifiers is the prudent use of tuned transformers. It is the transformer that dictates the frequency response pattern of the amplifier and regulates the selectivity of the receiver.

There is such a wide variety of tuned transformer shapes and sizes that it would be impractical (if not impossible) to show all of the configurations. Three general styles are depicted in Figure 4.19. Part a of the figure shows two π windings, core-tuned and separated by distance d. The distance is adjusted during the design stage to set the degree of coupling between the primary and secondary. Figure 4.19b shows a bifilar wound transformer, which consists of two parallel wires wrapped alternately (primary-secondary-primary). Bifilar wound coils have a high coefficient of coupling. The style shown in Figure 4.19c is a primary coil wound on the outside of the secondary coil and uses a large shell core to surround the transformer. The large shell core provides a high-permeability flux path while maintaining small size.

In preliminary testing of transformers, you should ensure the continuity of each winding (P to P' and S to S' in Figure 4.19), measure the DC resistance of each winding, and verify the absence of shorts between the windings. You could measure the inductance and Q of each winding and the mutual inductance between the primary and secondary, then test to see if the transformer will tune over the desired frequency range. The mutual inductance may be measured on the same test device used to measure the primary and secondary inductance, using the format described in Figure 4.19. The mutual inductance (L_m) is

$$L_m = \frac{L_{\text{tot(aiding)}} - L_{\text{tot(opposing)}}}{4} \tag{4.7}$$

$$= \frac{1021 \text{ } \mu\text{H} - 991 \text{ } \mu\text{H}}{4} = 7.5 \text{ } \mu\text{H}$$

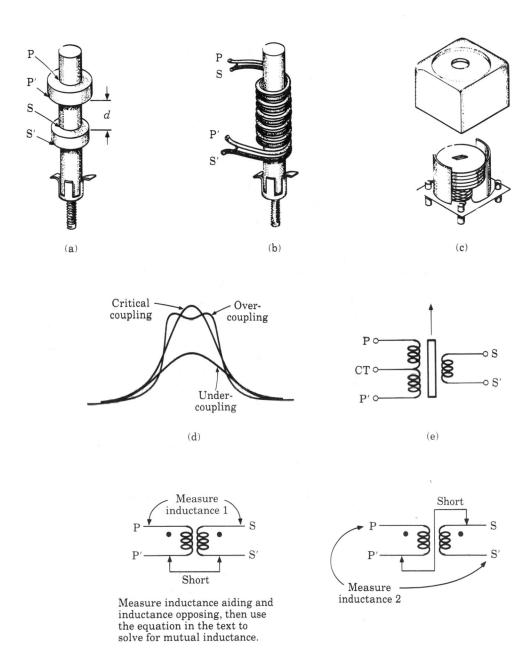

(a)

(b)

(c)

Critical coupling

Over-coupling

Under-coupling

(d)

P
CT
P'
S
S'

(e)

Measure inductance 1

P   S
P'   S'

Short

Measure inductance aiding and inductance opposing, then use the equation in the text to solve for mutual inductance.

Short

P   S
P'   S'

Measure inductance 2

FIGURE 4.19
General forms of RF transformers with the degree-of-coupling curves and transformer symbol.

This is a small value of mutual inductance and indicates loose coupling. The coefficient of coupling (k) is

$$k = \frac{L_m}{\sqrt{L_p L_s}} \qquad (4.8)$$

Typical values of k range from 0.01 to 0.5, with smaller values being more desirable. Our sample has $k = 0.015$ ($L_p = 408$ μH, $L_s = 613$ μH).

At resonance, the reactance of the secondary is made small compared to the impedance of the load on the secondary. This will make the primary see the secondary as a pure resistive load.

Inductive coupling begins with the primary and secondary windings far apart, where $k = 0$ and $L_m = 0$. As the coils are moved closer together, k and L_m increase proportionally. As the current (I_p) flows in the primary winding, the magnetic field around the primary induces a small voltage (E_s) into the secondary winding. This small secondary voltage causes a small secondary current (I_s), which in turn produces a magnetic field around the secondary winding. This field is coupled back into the primary and induces a feedback voltage in the primary of a polarity opposite that of the original primary voltage. This reflected voltage is called *counter emf*. The counter emf reduces the primary current and has the effect of adding series resistance to the primary coil. More primary resistance lowers the Q and widens the bandwidth. Eventually, a point is reached where a further increase in k causes the primary current to decrease faster than the mutual inductance can increase, and the secondary induced voltage begins to fall off. This point is called the *critical coefficient of coupling*.

We can summarize by saying that when k increases, L_m increases, the secondary output voltage increases, the impedance reflected back into the primary increases, Q gets smaller, and the bandwidth gets wider. When selectivity is desired, k is kept small, Q is high, and the bandwidth is narrow, with steep walls on the response curve. This is called *loose* coupling.

When high secondary voltage is the prime consideration, k is made larger (*tight* coupling), but now the reflected impedance is large, Q is low, and the bandwidth is wide. Usually a tradeoff is made to have a fair amount of secondary voltage, a medium Q, and a tailored bandwidth to enhance selectivity.

We can use the circuit of Figure 4.18 as an example to show the calculations for such a circuit. The value of C across the primary winding is 220 pF, so L_p is found to be

$$L_p = \frac{\left(\dfrac{0.159155}{f_0}\right)^2}{C} = 556.15 \text{ μH} \qquad (4.9)$$

When the transformer is tuned to 455 kHz, the measured inductance is 556 μH with a DC resistance of 14.5 Ω. The primary is a universal wound coil with 175 turns of #7 × 44 litz wire on a ¼ in (6 mm) ceramic form. The reactance at 455 kHz is 1590 Ω. The untuned secondary is 157 turns of #7 × 44 litz wire with 11.5 Ω of DC resistance wound on the same form. At resonance, the secondary inductance is 1285 Ω.

The resistance reflected back into the primary by the mutual inductive coupling is found to be

$$R_{\text{ref}} = \frac{X_{Lm}^2}{R_s} = \frac{(2\pi f L_m)^2}{11.5} = 39.89 \ \Omega \qquad (4.10)$$

(Note that $L_m = 7.5 \ \mu\text{H}$.) The reflected resistance is added to the DC resistance of the primary, so that $R_{p(\text{tot})} = R_{\text{coil}} + R_{\text{ref}} = 54.28 \ \Omega$. The loaded Q of the primary is then

$$Q_{\text{loaded}} = \frac{X_{Lp}}{R_{\text{tot}}} = \frac{1590}{54.28} = 29.29 \qquad (4.11)$$

When the Q is known, the -3 dB bandwidth can be found as

$$\text{BW}_{-3\text{dB}} = \frac{f_0}{Q} = 15.54 \ \text{kHz} \qquad (4.12)$$

The desired bandwidth of the IF amplifiers in an AM radio receiver is typically 10 kHz. That is, the transmitter is modulated with a 5 kHz message to produce the two sidebands, for a bandwidth of 10 kHz.

Reexamine Figure 4.18 and note that transformers T_3 and T_7 are both tuned to 455 kHz. When this happens, the bandwidth shrinks. As the signal goes through T_3, the sidebands at 5 kHz on either side of the carrier are amplified at a level 3 dB lower than the carrier signal. The sidebands arrive at T_7 already suppressed by 3 dB and encounter the bell-shaped gain characteristics of T_7, which reduces the sidebands another 3 dB. The total sideband suppression is, then, $0.7071^2 = 0.5 = -6$ dB. The bandwidth shrinkage factor may be determined by

$$\text{BW}_n = \text{BW}_1 \ \sqrt{2^{1/n} - 1} = 15.54 \ \text{kHz} \ (0.6436) = 10 \ \text{kHz} \qquad (4.13)$$

The factor can also be taken from Table 4.1 if you know the bandwidth of one tuned circuit (BW_1) and the number (n) of tuned circuits to process the signal at the same carrier frequency.

TABLE 4.1
Bandwidth shrinkage factor

Number of Tuned Circuits	Factor
1	1.0
2	0.6436
3	0.5098
4	0.4350
5	0.3856
6	0.3500

The shrinkage factor plays another important role in the shape of the tuned amplifier response curve, namely, it steepens the skirts. The side frequencies will fade at a faster rate when more tuned circuits are used, and this will improve the selectivity of the receiver.

Center taps are regularly used in RF transformers. This is done to match the output impedance of the transistor to the primary impedance of the tuned transformer. The transformer is designed to resonate at a selected frequency and to have a desirable bandwidth and sharp frequency rolloff above and below resonance. The primary impedance is of less importance than Q or BW and is found as

$$Z_p = \frac{L_p}{R_{tot}C} = \frac{556\ \mu H}{(54.28)(220\ pF)} = 46.56\ k\Omega \qquad (4.14)$$

This value does not match the collector impedance calculated earlier in the chapter. Therefore, a tap on the primary is used to match the output impedance of the transistor to the tank circuit. The tap value may be found by

$$\frac{Z_p}{Z_c} = \left(\frac{N_p}{N_{tap}}\right)^2 \quad \text{or} \quad N_{tap} = N_p\sqrt{\frac{Z_c}{Z_p}} \qquad (4.15)$$

In our example, the tap is at 55 turns on the primary to match the collector impedance of 4231 Ω.

4.6   AM DETECTORS

After a signal is received and strengthened, the modulated wave must be broken down into its original parts for use at the receiver. The detector is the receiver's circuit that does this job.

Other common names for the AM detector are diode detector, peak envelope detector, square law detector, and second detector. The term *second detector* indicates that there is a detector ahead of this one; the reference is to the oscillator-mixer at the front end of the receiver. The oscillator-mixer circuits and the AM detector act similarly in terms of changing frequency. The oscillator-mixer down-converts the signal from an RF to an IF, and the second detector down-converts the signal from the IF to the message frequencies.

The AM detector has sometimes been called a *demodulator,* based on the idea that it does the opposite of the modulator at the transmitter. However, a demodulator is a special form of detector that has two input signals, one of which is a carrier signal (see the discussion of product detectors later in this chapter). Therefore, a diode detector in itself is *not* a true demodulator.

There is very little difference between the circuitry of an AM detector and that of a low-voltage DC power supply circuit. The circuit format and principles of operation are identical. The only difference is the frequency of operation; as a result, the component values are smaller for the AM detector circuit. The DC power supply works at 60 Hz or 120 Hz, whereas the AM detector center frequency is 455 kHz (for broadcast radio).

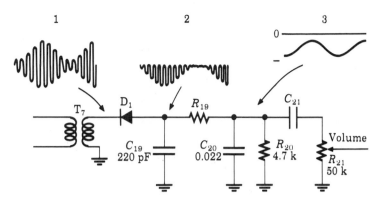

FIGURE 4.20
AM diode detector circuit. Waveform 1 is the secondary signal (open-circuited).
Waveform 2 is the rectified wave with no filter. Waveform 3 is the output wave after
filtering.

The basic circuit of Figure 4.20 shows the diode as a half-wave rectifier followed by a low-pass filter. The signal from the IF amplifier is a modulated envelope that is symmetrical about its axis, so that as far as the audio is concerned, either polarity of the IF signal may be recovered. The polarity of the recovered signal will be determined by the polarity of the AGC voltage needed as feedback DC bias voltage to the IF amplifiers (see the section on AGC later in this chapter). After the wave is rectified by the diode, the high-frequency carrier is filtered out by a low-pass RC filter, and the remaining audio signal is processed forward to the audio amplifiers.

This simplified description does not tell the whole story. A second approach to the AM detector is needed to explain the nonlinear effects of the diode and the location of the frequencies at the output of the detector. This is shown in Figure 4.21, which also shows that the diode resistance equals the change in voltage across the diode divided by the change in current through the diode. Remember that the IF signal contains the carrier with upper and lower sideband frequencies. Assume for the moment a 455 kHz IF carrier modulated by a 5 kHz tone, so that the IF signal contains three frequencies: 450 kHz, 455 kHz, and 460 kHz. As these frequencies encounter the nonlinearity of the diode curve, a heterodyning effect generates the original signals plus the sum and difference frequencies. From these suggested frequencies, the detector output by-products, as shown in Figure 4.21, will be

$$\text{Sum frequencies: } 450 \text{ kHz} + 455 \text{ kHz} = 905 \text{ kHz}$$
$$450 \text{ kHz} + 460 \text{ kHz} = 910 \text{ kHz}$$
$$455 \text{ kHz} + 460 \text{ kHz} = 915 \text{ kHz}$$
$$\text{Difference frequencies: } 460 \text{ kHz} - 455 \text{ kHz} = 5 \text{ kHz}$$
$$455 \text{ kHz} - 450 \text{ kHz} = 5 \text{ kHz}$$
$$460 \text{ kHz} - 450 \text{ kHz} = 10 \text{ kHz}$$

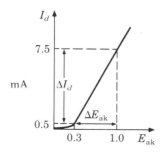
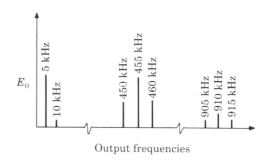

Output frequencies

$$R_d = \frac{\Delta E_{ak}}{\Delta I_d} = \frac{0.7}{0.007} = 100\ \Omega$$

FIGURE 4.21
A diode slope showing diode resistance of 100 Ω and the output frequencies that result from a 455 kHz carrier with ±5 kHz side frequencies.

Although there are two low frequencies in the output signal, the 10 kHz signal is at least 15 dB lower in amplitude than the 5 kHz coming out of the diode. The low-pass filter further attenuates the 10 kHz signal, so it will be omitted from this discussion.

The components of Figure 4.20 have been replaced by their equivalent resistances in Figure 4.22 at the two critical frequencies of 5 kHz and 455 kHz, where we will examine the attenuation. Coupling capacitor C_{21} is a short circuit at these frequencies, and that places the 50 kΩ volume control in parallel with the 4700 Ω DC return resistor, to make R_L 4300 Ω. Figure 4.22a shows the circuit configuration with equivalent values at the audio frequencies and shows the signal output from the diode (at A) to be 0.94 of the input signal. If we add on the second section, the output to the volume control (at B) will be 0.655 of the input signal, or 3.67 dB lower at 5 kHz. Figure 4.22b has the same components at 455 kHz and shows the diode output to be 0.788 of the input signal (at A). Adding on the rest of the circuit reduces the output to the volume control to 0.02585 of the input signal, for a carrier loss of 31.75 dB; that is, the carrier will be 28 dB lower than the audio signal. The carrier represents less than 4% of the total output signal of the detector.

A second look at the component value selection will show how to avoid distortion of the message signal through the detector. The controlling factors are the resistance of the diode and the input capacitor on the charging cycle. The capacitor must be able to charge up fully (5 RC time constants) in one-quarter of one cycle at the carrier frequency, as shown in Figure 4.23b.

$$C = \left(\frac{1}{4}\right)\left(\frac{1}{f}\right)\left(\frac{1}{5R}\right) = \frac{1}{20fR} \quad \text{(in farads)} \qquad \text{(4.16)}$$

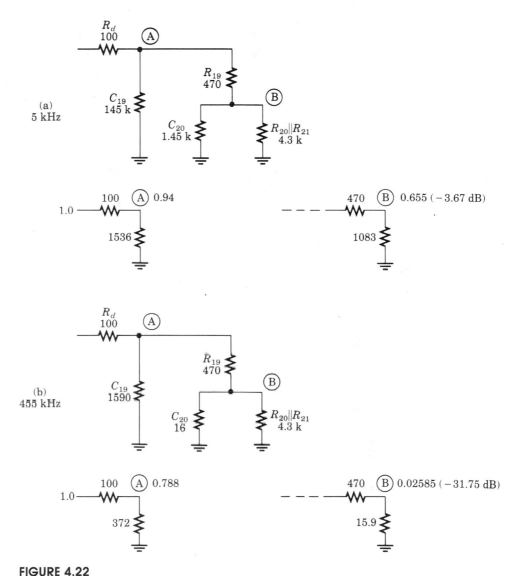

FIGURE 4.22
The impedances of the detector circuit reduced to simple resistances at 5 kHz and 455 kHz.

From Figure 4.20 the diode resistance was found as 100 Ω. The time for one quarter cycle at 455 kHz is 0.55 μs. Therefore, $C = 0.55 \times 10^{-6}$ divided by $5R =$ 1100 pF $= C_{\max}$.

The capacitor discharges through the load resistance in one half cycle of the highest audio frequency. In Chapter 1, the proof-of-performance test frequency was 12.5 kHz. If we use this frequency here, we will ensure good

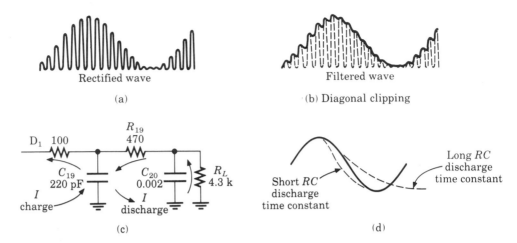

FIGURE 4.23
Effects of long and short discharge time constants.

performance under worst-case conditions. The time for one half cycle at 12.5 kHz is 40 μs = $5RC$. With a C value of 1100 pF, R_{max} is calculated at 7.27 kΩ. R_{max} is the sum of R_{19} and R_L in Figure 4.23. In order to make R_L large, C_{max} (C_{19}) in Figure 4.23 is reduced to 220 pF to make the reactance high at the audio frequency. As a result, R_L could be larger than 7.27 kΩ (when C decreases, R_L may increase).

When the input capacitor is too large compared to the resistance of the diode, the capacitor cannot charge fast enough at the carrier frequency, and the output from the detector will be less than the highest achievable value. When the load resistance is too large compared to the input capacitor of the filter, the capacitor cannot discharge fast enough at the highest audio frequency, and diagonal clipping will result.

4.7   AUTOMATIC GAIN CONTROL (AGC)

Figure 4.24 is the composite drawing of Figures 4.18 and 4.20. Here we see that the average DC voltage out of the detector, which is dependent on signal strength, is a DC feedback bias used to automatically alter the gain of the IF amplifiers. R_{14} and C_{16} act as additional filtering to remove the audio component and present only a pure DC voltage as bias. C_{16} must be large to remove even the lowest audio frequency. The negative DC voltage to R_{14} is the −1 V source in Figure 4.18.

When a strong signal arrives at the antenna, the audio signal level at the detector output becomes large. This results in a large negative DC feedback voltage to the base of the IF amplifier, which lowers its gain. The opposite conditions hold when the signal at the antenna is weak. The audio signal from the detector is small, the negative DC detector voltage is less, and the IF ampli-fier gain would increase to compensate for the weaker antenna signal.

The transconductance curve of a transistor is shown in Figure 4.24 with an operating Q point on the linear part of the curve. Because the curve is more round above and below Q, the gain of the amplifier will be less when Q is moved in either direction. The AGC voltage changes the Q point and therefore changes the gain. It is typical for the Q point to be away from center so that the AGC bias voltage change either increases or decreases the gain depending on whether the AGC voltage becomes more positive or more negative. The direc-tion in which the detector diode is connected will control only the polarity of the AGC voltage and the IF amplifier gain and will have no effect on the recovered audio.

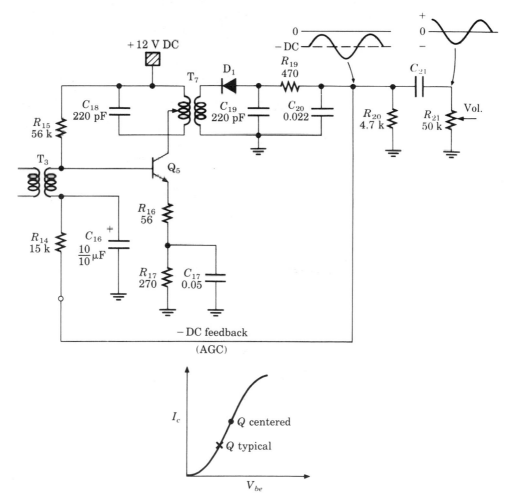

FIGURE 4.24
IF amplifier and detector circuit with AGC feedback.

4.8   AUDIO AMPLIFIERS

The audio frequencies (defined earlier as those frequencies detectable by the human ear) from the detector output are amplified and applied to a speaker. The detector output level is typically greater than 100 mV$_{\text{rms}}$ but generally no larger than 2.0 V$_{\text{rms}}$.

A discrete-component, low-power audio amplifier is shown in Figure 4.25 and contains a class A preamplifier (Q_3) and a class A power output amplifier (Q_4). The preamplifier has a gain of 10 with no feedback and an adjusted gain of 5 with feedback through R_{23}. The feedback resistor supplies an out-of-phase signal to the base of Q_3 to decrease its gain and expands the frequency response of the amplifier. The feedback resistor R_{23} serves the second function as one of the biasing resistors for Q_3. The two amplifiers are direct-DC-coupled to reduce the number of reactive components and improve the low-frequency response. Both amplifiers are operated in class A and consume some power from the supply even when no signal is present. The DC biasing voltages are somewhat self-adjusting. The base voltage of Q_3 is derived from a voltage divider in the emitter of Q_4. This sets the collector voltage of Q_3, which is also the base voltage for Q_4. The latter sets the emitter voltage of Q_4 and the bias voltage returned to the base of Q_3.

The output power may be found even with the few data this circuit diagram provides. The emitter current of Q_4 is the emitter voltage divided by the emitter resistance, or 8.8 mA. The input power from the supply is $V_{\text{dc}} \times I_{\text{dc}} = 90 \times 0.008 = 800$ mW. Assume Q_4 (class A) to have efficiency of about 50% $\times$ 800 mW = 400 mW power output. The formal equation may also be used:

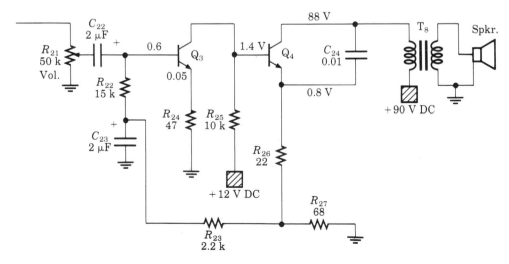

FIGURE 4.25
Class A audio preamplifier and power output amplifier.

$$P_{\text{load}} = (0.5)\, \frac{V^2_{\text{ceq}}}{r_c + r_e} \quad \text{(in watts)} \qquad \textbf{(4.17)}$$

where V_{ceq} is the difference between the emitter and collector voltages, 87 V. r_c is V_{ceq} divided by the emitter current, or 9880 Ω, which is added to r_e (90 Ω). Thus, 0.5 times 87 V squared divided by 9970 Ω equals 400 mW from Equation 4.17.

Figure 4.26 has a preamplifier (Q_5-Q_6 Darlington pair) feeding a push-pull power amplifier (Q_7-Q_8 Darlington pair with Q_9-Q_{10} Darlington pair). Again, this entire circuit is direct-DC-coupled to reduce the use of reactive components. It is noteworthy to observe at this time the preparation for IC circuits. It is not uncommon for manufacturers to test a discrete circuit like this one to develop an integrated circuit with a minimum of external parts.

There are two feedback loops in this amplifier. The first has R_{30} supplying the DC voltage for the base of Q_7 and Q_9 through the speaker voice coil from +DC. R_{30} also supplies AC feedback, to reduce gain and widen the bandwidth, and provides DC voltage to the collectors of Q_5 and Q_6. The second feedback is from the collector of Q_{10} to the bases of the drivers, Q_5 and Q_6. This first feedback is called a *bootstrap* circuit because it multiplies the input impedance of the push-pull amplifier. The voltage across R_{30} is due to the Q_9 base signal voltage at one end and the speaker signal voltage at the other. The voltage across R_{30} is about 1/100 of the signal from the base of the output push-pull pair to ground. R_{30} draws about 1/100 of the Q_9 base current and therefore multiplies the input resistance (for AC) about 100 times its value of 1.5 kΩ, for a final value of 150 kΩ for Z_{in} of the push-pull amplifier.

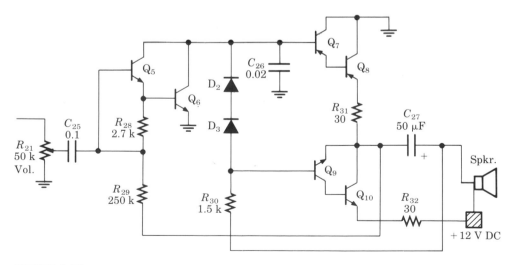

FIGURE 4.26
Complementary-pair preamplifier and push-pull complementary-pair power output amplifier.

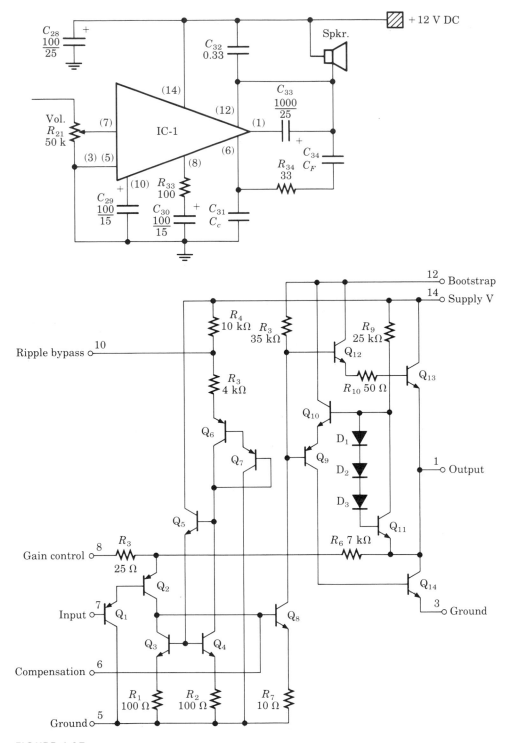

FIGURE 4.27
Solid state audio amplifier. Five watts output with large heat sink.

The bases of the push-pull amplifier are separated by a set of diodes to establish 2×0.7 V = 1.4 V DC difference between the base voltages. When the collector of Q_5-Q_6 moves in a positive direction, Q_7 and Q_8 are placed in cutoff while Q_9 and Q_{10} are biased on. Q_9-Q_{10} conducts for this half cycle. As the collector of Q_5-Q_6 swings in the negative direction, Q_7 and Q_8 are moved into conduction while Q_9 and Q_{10} are biased into cutoff. The two output transistor pairs are both in very light conduction when no signal is present, called *trickle bias*, and draw a minimum of current from the DC supply (about 10 mA). As the signal level increases, the DC current drain from the power supply increases to about 350 mA. The product $I^2 \times R$ places 4.9 W into a 40 Ω speaker.

The function of the audio amplifiers is to raise the signal level at the output of the detector to a power level strong enough to operate the speaker while keeping the noise level down to an acceptable value below the signal level. At the same time, the fidelity of the signal must be maintained. A measure of fidelity is the percentage of **total harmonic distortion** (THD). The weakest link in the system is the loudspeaker.

A high-output-power, low-distortion IC audio amplifier is shown in Figure 4.27. The amplifier is contained in a 14-pin minidip package and uses a minimum of external parts. C_{28} is a decoupling capacitor to keep the audio out of the power supply. C_{30}, C_{31}, C_{32}, and C_{34} along with R_{33} and R_{34} are frequency response tone control components, while C_{33} is a DC blocking capacitor and provides AC coupling from the chip output to the speaker.

The frequency response covers the range from 50 Hz to 40 kHz, at 2 W output with less than 1% THD across the frequency range. The power may be increased to 3 W with light heat sinking and to 5.5 W with heavy heat sinking. The distortion rises to 3.5% at 3 W and to about 10% at 1 kHz and 5.5 W. Temperature derating should be considered when the circuit is operating at these elevated power levels.

Note the striking similarities between the discrete amplifier of Figure 4.26 and the internal circuit of the IC chip.

4.9   RECEIVER SCHEMATICS

Figure 4.28 is a very simple schematic for a typical pocket-sized AM radio receiver. The ability to read a schematic is essential in all electronic work, and to make it easy, most schematics follow the same standard layout as the English printed page, that is, in order from left to right and then down.

At the start of a project, your main concerns may be directed toward a systems concept. The second step would be to narrow your scope to a rack, panel, or small product. The third step would involve the functional operation within the device. Finally, you direct your attention to the individual circuits and components. The same approach is taken whether the subsystem is to be built and tested, analyzed and evaluated, or tested and repaired. Whatever your goal, you need to know

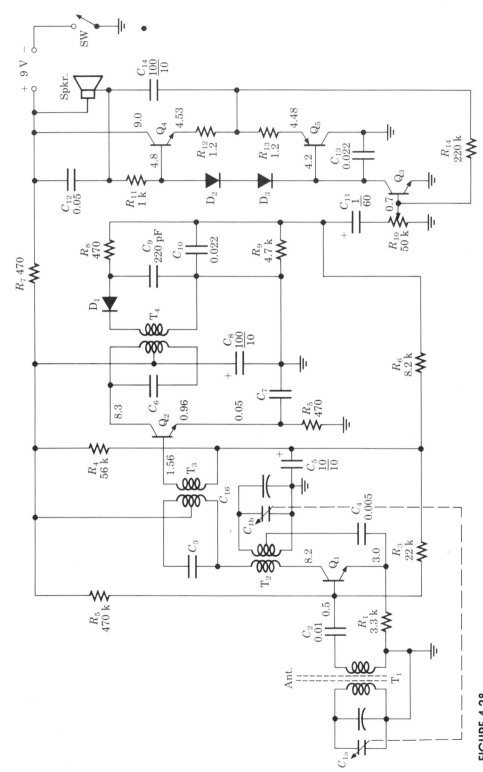

FIGURE 4.28
Simple AM radio receiver schematic.

1. what this device is intended to do,
2. what functional blocks are used to obtain the end results,
3. how each circuit attains the end results, and
4. how the individual components support the circuit operation.

We begin our analysis of Figure 4.28 knowing that the circuit is an AM radio receiver; thus, it will accept any selected amplitude-modulated carrier signal in the frequency range of 540 kHz to 1.6 MHz. It will amplify the selected signal, remove the message from the carrier, and convert it to a power level that will operate a speaker.

The schematic is subdivided by identifying the detector (D_1) and recognizing that everything before the detector is RF while everything after the detector is used for low frequencies. Starting at the left, capacitor C_{1a} and the primary of T_1 form a resonant circuit called a **preselector** that determines which carrier frequency will pass through the tuned circuit to be amplified. The signal that appears on the secondary of T_1 is capacitively coupled to the base of Q_1 by capacitor C_2. Q_1 has a collector load and has its emitter connected to ground through R_1, making it a common-emitter amplifier. The signal goes in the base and out the collector, where it encounters two load impedances. The primary of T_2 is in series with the signal going to T_3. The energy is coupled to the secondary of T_2, where C_{1b} and the secondary inductance establish the frequency of the energy that is returned through capacitor C_4 to the emitter of Q_1 to support oscillations. T_2's secondary inductance and capacitance C_{1b} are adjusted to oscillate at a frequency 455 kHz higher than the frequency of the energy in the primary of T_2.

The signal continues on past T_2 to the primary of T_3, which is tuned to 455 kHz. Here the energy at 455 kHz passes from primary to secondary of T_3 and is connected to the base of Q_2. The bottom side of T_3's secondary is at AC ground through capacitor C_5. The Q_2 emitter is grounded through R_5 and AC-bypassed by capacitor C_7. The collector load is transformer T_4. This is a common-emitter amplifier with the signal entering the base and leaving by the collector. The signal is inductively coupled to the secondary of T_4, rectified by diode D_1, and filtered so that only the shape of the modulated envelope (audio) is passed through coupling capacitor C_{11} to the volume control R_{10}. The level of the signal to the base of Q_3 is set by the position of R_{10}'s center arm. The signal then enters the base of Q_3, leaves via the collector, and is applied equally to the bases of Q_4 and Q_5. Q_4 and Q_5 operate on alternate half cycles to drive the speaker.

When only the audio circuit is to be examined, the speaker should be replaced by a 10 Ω resistor of appropriate wattage, a signal generator should be connected to the positive side of C_{11}, and an oscilloscope should be used to view the signal at each of the three active devices. Connect the scope to the base of Q_3 and set the signal generator to any audio frequency (1 kHz is a common standard). Then set the signal generator output signal (and volume control) for about 0.5 V p-p at the base of Q_3. Follow the signal through the complete

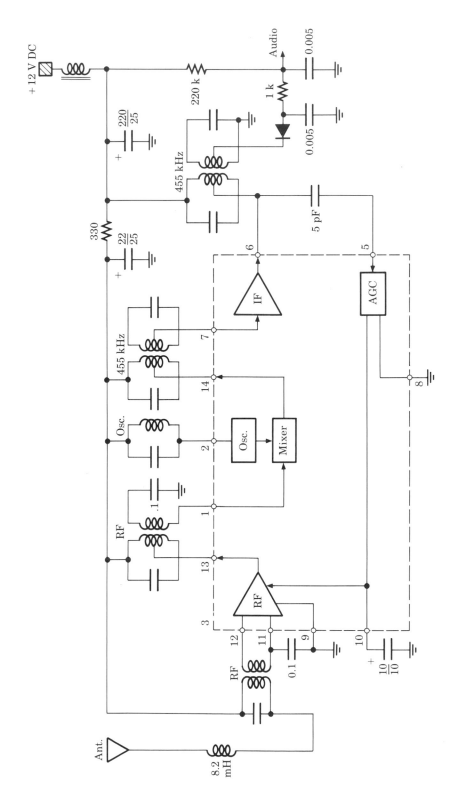

FIGURE 4.29
Single-chip RF radio receiver for AM.

circuit. You should see a signal at the collector of Q_3 with gain, but not at the emitter of Q_3, which is at ground. The same signal level seen at the collector of Q_3 should be seen at the bases of Q_4 and Q_5. A signal slightly smaller than the one seen at the bases should be seen at the common emitter between R_{12} and R_{13}, and again at the positive side of C_{14}. There should not be a signal at the collector of Q_4 because the collector connects to the DC supply line, which is AC ground. Nor should a signal appear at the collector of Q_5 because it is grounded.

Conditions normally tested for in audio amplifiers are: gain at 1 kHz, the frequency response of the amplifier, maximum power output and THD at a specified frequency, and distortion across the audio response passband. The speaker is then replaced and the system is subjected to listening tests. Testing of the RF and IF sections are covered in the section of this chapter on alignment.

Figure 4.29 shows one version of a single-chip RF section for an AM broadcast band radio receiver. Note that all of the sections found in the discrete circuits are still in the chip circuitry, except now they are confined to composite blocks. All of the tuning is external to the chip, as is the detector circuit. When the single-chip audio amplifier of Figure 4.27 is added on to this circuit, the result is a two-chip complete radio receiver.

4.10   LOUDSPEAKERS

As shown in Figure 4.30, a good speaker should have:

1. a large magnet (1 lb or more),
2. a large voice coil (1 in or larger),
3. a large stiff spider (a 3 in spider is average for a 1 in voice coil),
4. very flexible edge webbing (annular surround), and
5. a quality cone stock.

A large magnet and voice coil allow for high power to the speaker. As power is applied to the speaker, the voice coil moves in and out of the permanent magnet and makes the cone move large volumes of air. The voice coil should not bottom out when moving inside the magnet, nor should it rub against the side of the magnet. The spider holds the center for the voice coil to prevent rubbing. The large, flexible annular surround gives freedom of movement to the cone. Large speakers move great amounts of air and are best for low frequencies. Smaller, more rigid speaker cones are better for high frequencies.

Figure 4.31 shows the frequency response and the percentage of distortion curve for an amplifier such as the one shown in Figure 4.26. The output level is set to 4.5 W at 1 kHz, and the input level is noted. This input will be kept constant for all other frequencies in the determination of a frequency response curve. The output voltage or power are measured, along with the percentage of

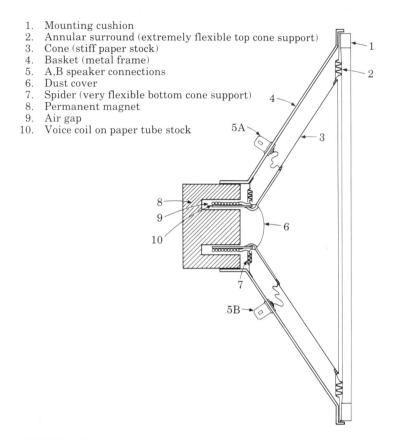

1. Mounting cushion
2. Annular surround (extremely flexible top cone support)
3. Cone (stiff paper stock)
4. Basket (metal frame)
5. A,B speaker connections
6. Dust cover
7. Spider (very flexible bottom cone support)
8. Permanent magnet
9. Air gap
10. Voice coil on paper tube stock

FIGURE 4.30
Permanent magnet speaker.

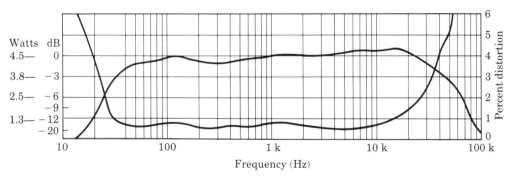

FIGURE 4.31
Frequency response curve and distortion for the 5 W amplifier in Figure 4.27.

distortion at each frequency during this test. The frequency response for this curve is described as *flat* from 50 Hz to 25 kHz ± 1 dB and it drops off to −3 dB at 30 Hz and 50 kHz. The distortion is less than 1% from 40 Hz to 15 kHz.

4.11   PILOT CARRIER RECEIVERS

A typical pilot carrier receiver is shown in Figure 4.32. Such a receiver may sense the presence of a reduced carrier signal level and use it for frequency synchronization with the transmitter. This double-conversion receiver takes the 143.07 MHz RF signal down to 10.6 MHz in the first conversion by mixing it with a 132.47 MHz local oscillator. The second converter mixes a 10.8 MHz signal with the 10.6 MHz from the first conversion, to produce the 200 kHz second intermediate frequency. Note that the second local oscillator frequency is generated by a 200 kHz crystal-controlled reference oscillator and then multiplied by 54 to derive the 10.8 MHz signal.

Keep in mind that during the reception, mixing, and amplifying of the signals, the circuits are working with a carrier signal frequency that is only 10% of the total voltage being received. After the second mixer, the IF signal is split into two parallel paths. One path is to the 200 kHz IF amplifier, which increases the signal amplitude and then connects the signal to the product detector. The second circuit path is a sharply tuned carrier-only band-pass amplifier that increases the carrier amplitude to a voltage greater than that of the sideband signal before it is applied to the product detector.

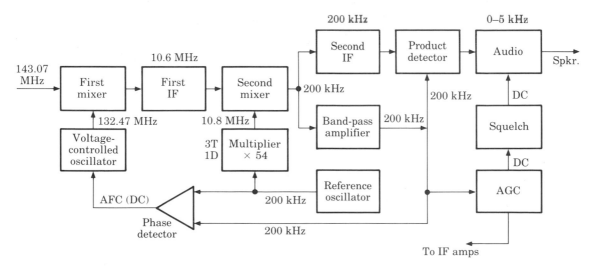

FIGURE 4.32
Pilot carrier DSSC receiver block diagram.

Our example of a pilot carrier system uses an RF carrier frequency of 143.07 MHz with message frequencies 5 kHz above and below, resulting in a band of frequencies from 143.065 to 143.075 MHz. Again, the carrier frequency of 143.07 MHz is only 10% of the voltage in the transmission. The signal is reduced to 200 kHz ± 5 kHz = 195–205 kHz by the IF amplifier in two steps. At the product detector, the 195–205 kHz signal is heterodyned with a 200 kHz carrier to recover only the sidebands of 0 to 5 kHz.

The 200 kHz crystal reference oscillator also sends a signal to the phase detector, where it is compared to the signal frequency coming out of the band-pass amplifier. If either the transmitter or the receiver drifts off frequency for any reason, the VCO acting as the first local oscillator will change in the same direction and correct the frequency drift.

4.11.1 Squelch

The squelch circuit acts as a controllable antenna level input switch and turns the audio amplifier on or off. The audio cannot be partially on or partially off; it is subject to either a go or a no-go condition. For this reason, the squelch amplifier is a switching transistor that knows no middle ground; it must be fully on or fully off.

The squelch circuit of Figure 4.33 shows the AGC voltage, which is a DC voltage with a value dependent on signal strength, connected through a level-set control to the base of the squelch amplifier. When the base voltage is low, the transistor is cut off and its collector voltage is high. The collector voltage is connected to the emitter of one of the audio amplifiers. As the audio amplifier emitter voltage goes high, the audio transistor is shut off. (Low antenna sig-

FIGURE 4.33
Squelch amplifier.

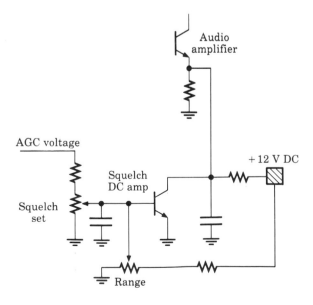

nal = low DC from detector = low base, DC squelch amplifier = high collector, DC squelch amplifier = high emitter, audio amplifier = audio cutoff.)

As the antenna signal strength increases, the detector's DC output level goes up, as does the DC base voltage on the squelch amplifier. At some level of DC base voltage, the collector voltage will snap toward zero, lowering the audio amplifier emitter voltage and allowing the audio amplifier to pass the message signal.

The purpose of a squelch switch is evident if you have operated a communications receiver for several hours. Unlike broadcast receivers, there is not just one steady transmission being received at a time. There can be any number of transmitters and almost all possible signal strengths (due to varying distance) on the air at one time. The constant hissing and crackling of meaningless noise signals are irritating to the radio operator. A squelch control setting such that the audio will be turned on (opened up) when a strong signal arrives at the antenna will allow reception of clean signals and block out all weak, noisy transmissions.

4.12   INDEPENDENT SIDEBAND RECEIVERS

Figure 4.34 is the block diagram of an independent sideband receiver (ISB) that uses frequency synthesis and individual sideband selection. Remember that in $A3_b$ transmission, the upper side frequency may contain a different message than the lower side frequency.

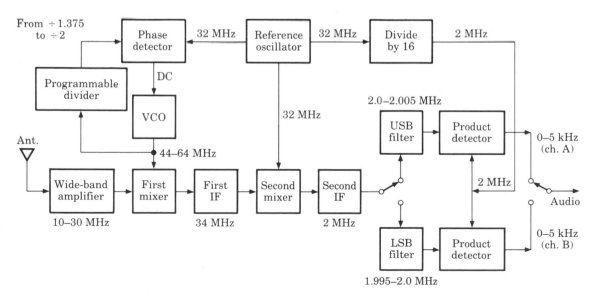

FIGURE 4.34
ISB receiver with frequency synthesizer.

The antenna intercepts the transmitted wave and connects it to a wide-band RF amplifier that is shaped to amplify signals between 10 MHz and 30 MHz. The RF amplifier is fixed for this band of frequencies.

The first mixer is unique in that it is an *up*-converted. That is, when a 10 MHz signal is to be selected from the incoming signal range, the first local oscillator (VCO) is set to 44 MHz, so that the output of the first mixer is the sum and difference of the two input signals, 54 MHz and 34 MHz. The difference frequency is amplified by the first IF section, and the sum, 54 MHz, is rejected. To tune the receiver to 18 MHz, the VCO is set to 52 MHz so that the difference frequency will still be 34 MHz. The input signals to the second mixer are the 34 MHz from the first IF section and the second local oscillator frequency signal of 32 MHz. The second mixer output signals are the two original input signals plus the sum and difference frequencies: 2 MHz, 32 MHz, 34 MHz, and 66 MHz. The second IF amplifiers are tuned to pass only the 2 MHz signal (plus its sidebands) and attenuate all higher frequencies.

It should be understood that the transmitted ISB signal has a totally suppressed carrier; only the sidebands are sent. When the message signal is a 5 kHz tone, only signals at 10.005 MHz and 9.995 MHz are transmitted (for a 10 MHz transmission).

The sidebands of the 2 MHz second IF signal may be switched to a lower sideband filter or an upper sideband filter, and then to one of two separate product detectors. The sideband filters can be any of the high-*Q* filters discussed earlier in the context of SSB transmitters, such as mechanical, crystal, or SAW filters. The upper sideband filter for this receiver sharply attenuates all frequencies below 2 MHz and passes only the 2.005 MHz signal. The lower sideband filter sharply attenuates all frequencies above 2 MHz and passes only the 9.995 MHz signal.

A pure sine wave carrier signal at 2 MHz must now be mixed with the output of the filters to recover the difference frequencies of 5 kHz for channel A and 5 kHz for channel B. The product detector performs the true function of *demodulation*.

4.12.1 Frequency Synthesis

The frequency synthesizer in Figure 4.35 shows three separate signal frequencies, all dependent on the frequency of one oscillator, the reference oscillator. This is a 32 MHz crystal-controlled, temperature-stable oscillator. The 32 MHz sine wave output voltage is used directly as the local oscillator signal for the second mixer. The same 32 MHz sine wave signal from the reference oscillator is divided by 16 (four divide-by-2 networks) to produce the 2 MHz sine wave signal supplied to the product detector.

The 32 MHz signal is also applied to the first local oscillator, which is basically a phase-locked loop (PLL). In Figure 4.35, the 32 MHz is fed into a preset divider containing ten divide-by-2 networks (a total division of 2^{10} or 1024). The 32 MHz (into pin 7) divided by 1024 equals 31.25 kHz (out of pin 14),

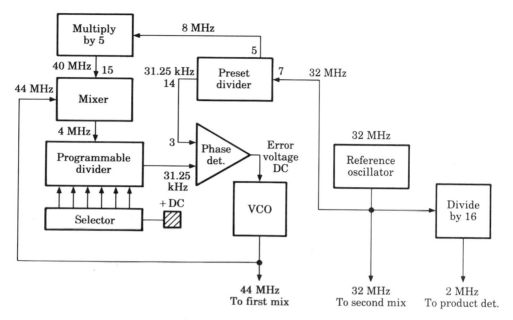

FIGURE 4.35
Frequency synthesizer (synthetic frequency generator). VCO output shown for 44 MHz only (adjustable from 44 to 64 MHz).

which is applied to the phase detector (at pin 3). A second output signal from the preset divider, at pin 5, is the voltage at the output of the second divide-by-2 network in the chain of ten. Two divide-by-2s is a division by 4, making the signal frequency out of pin 5, 8 MHz. This 8 MHz is then multiplied by 5, for a total of 40 MHz into pin 15 of the mixer. The output of the voltage-controlled oscillator at 44 MHz is also applied to the mixer, so that the output signal frequency from the mixer is the difference of 4 MHz.

The 4 MHz signal from the mixer is fed through a programmable divider, which delivers an output signal frequency of 31.25 kHz to the second input of the phase detector. The two inputs to the phase detector are of the same frequency, so that the error voltage out of the phase detector locks the VCO output at 44 MHz. If you wish a different channel carrier frequency, increase the division factor of the programmable divider by switching its variable DC voltages to make its output frequency slightly lower. The lower frequency into the phase detector will cause the error voltage to go up by a fraction of a volt, causing the VCO output frequency to increase slightly. The higher frequency from the VCO is fed back to the mixer, whose output frequency rises slightly, so that the larger division factor will again produce a frequency of 31.25 kHz, bringing the phase detector back into lock. The overall effect is that the VCO output frequency is higher than before the carrier change.

To maintain the 34 MHz IF center frequency, an incoming signal of higher frequency would heterodyne with the VCO first local oscillator to produce the difference frequency of 34 MHz. The DC error voltage from the phase detector need only change by 3 mV for a VCO frequency change of 10 kHz. With the programmable divider set to change the error voltage by 3 mV per step, this receiver could have 2000 channels in the 20 MHz bandspread 10 kHz apart. Considering the ability to choose an upper or lower sideband for each channel, there is room for 4000 communications channels in this band.

4.12.2 Product Detectors

A detector circuit in which an externally generated carrier signal must be added to the modulated wave before the sidebands can be recovered is called a **demodulator. Product detector** is another name for this type of circuit. All of the transmissions that contain sidebands only, namely, $A3_a$, $A3_b$, and $A3_j$, fall into this category, along with the subcarrier systems for stereo and the color signals for television. Figure 4.36 shows a simple circuit capable of acting as a product detector. Figure 3.15 shows the MC1495 chip circuit, which may be either a balanced modulator or a balanced demodulator, depending on where the output signal is taken from. Carefully compare the circuits of Figure 3.15 and Figure 4.36. Can you recognize the similarities and subtle differences between the two circuits without reading further? If you can see the similarities, your progress is better than average. If you can discern the differences, then you are well above average.

Note first that the circuit of the balanced modulator and the circuit of the balanced demodulator (product detector) are the same. Both are class B push-pull amplifiers. The major difference lies not in the circuit, but rather in the signals that are applied to the circuit, which dictate if it is to act as a modulator or a demodulator. In the modulator system, the two input signals are the carrier and the message signals. The output signal is the sum of the two inputs, the modulated sidebands. In the demodulator system, the two input signals are the carrier and the modulated sidebands. The output is the difference between the two inputs, the message signal. Thus, if you know the basic balanced modulator circuit operation, you can determine what you need to put into the system based on your desired output results.

4.13   AM STEREO

Although prospects for AM stereo were dealt a damaging blow by the controlling federal agency in 1982, there is still potential for this mode of transmission. There were five proposed systems under consideration by the commission, all of which appeared to have equal merit. The commission decided to let the industry set the standards. Proposed systems were submitted by The Harris Corporation (Quincy, Ill.); Belar Electronics Laboratory, Inc.; Kahn Communications, Inc.; Magnavox Consumer Electronics Co.; and Motorola, Inc. To conserve space here, only the Harris system will be discussed. This is the only

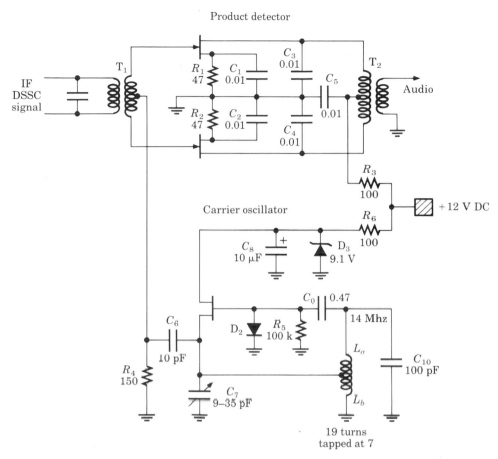

FIGURE 4.36
Product detector with carrier oscillator.

proposed system within the limits for radiated sideband power in the transmission spectra as set forth by the FCC. Also, it exhibits the least amount of monophonic envelope detector distortion.

The Harris Compatible Phase Multiplex (CPM) system is a linear-additive quadrature modulation system. The CPM system amplitude-modulates two carrier signals separated in phase by 30°. The left channel signal amplitude-modulates a carrier that lags the transmitted resultant by 15°, and the right channel signal amplitude-modulates a carrier that leads by 15°. The two signals are linearly combined (added) to form the CPM signal, making Harris the only linear system. One method of generating the CPM signal is shown in Figure 4.37.

A matrix circuit adds the left and right audio signals to produce the L + R signal and a reduced-level L − R signal. A low-frequency, low-level stereo

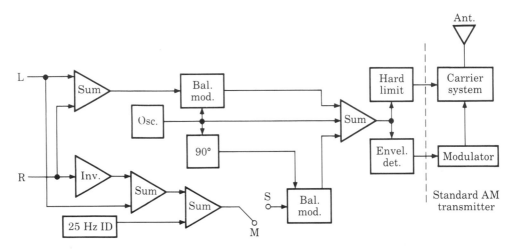

FIGURE 4.37
Harris matrixing system for stereo AM transmission.

identification (ID) tone of 25 Hz for AM stereo signaling is added to the L − R signal. The tone is not heard on stereo broadcasts because it appears out of phase on the stereo channel, and it will not be heard on monaural receivers because it is on the L − R channel only.

The L − R signal is applied to a balanced modulator along with a 90° leading phase-shifted carrier. The L + R signal is applied to a balanced modulator along with a carrier that is *not* phase-shifted. The output signals from the two balanced modulators are summed to make up the CPM signal. To interface the CPM signal to an AM transmitter, it is separated into an envelope and a phase-modulated component. The CPM signal is passed through a limiter, to remove any amplitude changes, and is then used in place of the crystal-controlled oscillator signal for the AM transmitter. The L + R signal is passed through an envelope detector and used as the audio input for the AM transmitter.

The receiver for this system is shown in Figure 4.38. The reduced gain in the L − R channel is the key to providing compatibility in the monaural receiver using envelope detection. Distortion of 0.5% results due to the presence of the quadrature sidebands. If the L − R signal were not reduced in transmission, the distortion would be about 11%. The stereo receivers in the Harris CPM system use synchronous, rather than envelope, detectors to eliminate any distortion in the stereo and monaural modes. Figure 4.38 illustrates how the CPM signal is recovered. Equal phase and amplitude characteristics are not required for the front end of the receiver to keep distortion low, as they are in other systems.

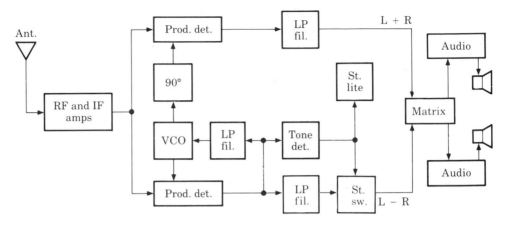

FIGURE 4.38
Harris stereo AM receiver block diagram.

The output of the IF amplifier is applied to a synchronous detector, which serves two purposes. First, it works as a phase detector in the phase-locked loop, where the VCO is locked to the IF and oscillates 90° out of phase from the incoming IF signal. Second, the balanced modulator directly demodulates the quadrature L − R part of the signal. The VCO's output signal is shifted 90° and used to demodulate the in-phase L + R component of the signal. Low-pass filters remove all content of the carrier signal coming out of the detectors. The L − R and L + R signals are combined in a simple audio matrix to recover independent right and left channel signals. The stereo ID tone activates the stereo/monaural switch and turns on a stereo indicator.

Stereo systems were not mentioned before the receiver section of this text because all of the stereo processing occurs in the audio sections of the transmitter and receiver. Furthermore, to understand stereo operation it is necessary first to understand simple amplitude modulation as well as sideband modulation using balanced modulators and product detectors.

4.14 AM RECEIVER TESTING

When testing receivers for repair purposes, many of the following procedures may be helpful. When testing receivers in a production or laboratory situation, all may be required. It is always good practice to make a record of the test equipment used to perform the measurements in a laboratory notebook. List the make, model number, and serial number of each piece of test equipment used in the measurements.

4.14.1 Test Setup

Figure 4.39 shows the test setup used for all of the tests to be discussed. There are sufficient differences in receiver format to warrant a whole "menu" of setup arrangements. To meet your particular requirements, select one circuit from group 1, one from group 2, and one from group 3 in Table 4.2. Use the table as a guide to the tests indicated in the figure.

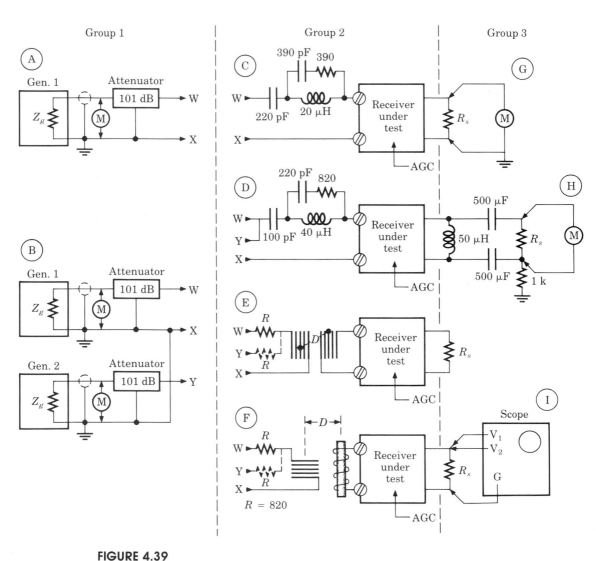

FIGURE 4.39
Generator, attenuator, impedance matching, receiver, load, and output measurements for AM receivers.

TABLE 4.2

Group	Circuit	Description
1	A	One generator, meter, and attenuator
	B	Two generators, meters, and attenuators
2	C	Dummy antenna for one-generator setup, coupled to receiver with antenna input terminals
	D	Dummy antenna for two-generator setup, coupled to receiver with antenna input terminals
	E	Induction coil (Figure 4.40) coupled to receiver with flat loop antenna, for one- or two-generator application
	F	Induction loop (Figure 4.40) coupled to receiver with ferrite rod antenna, for one- or two-generator application
3	G	Output load and metering for receivers with no DC current through the speaker
	H	Output load and metering for receivers with DC current through the speaker
	I	Single- or dual-trace scope measurements of output signal conditions

When distance D in Figure 4.39e and f equals 24 inches, the voltage induced into the receiver antenna is:

$$E_a = 0.05E_g$$

as measured on the input meter, minus any losses inserted by the input attenuator.

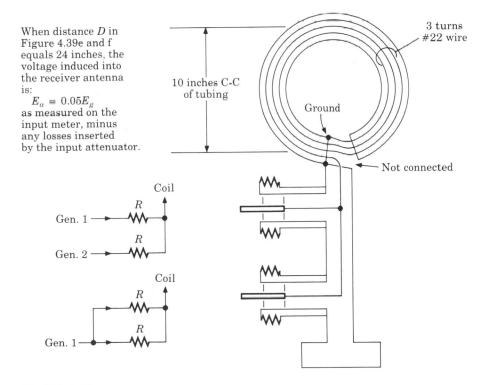

FIGURE 4.40
Radiation coupling coil to be used between the generator and the receiving antenna.

4.14.2 Alignment

Simply stated, aligning the IF amplifiers means tuning the interstage coupling transformers to obtain the maximum output signal at the IF center frequency. Although there are several generally accepted methods, the one described here yields highly accurate results and is basically simple to perform.

A. Disable the local oscillator. The easiest way is to remove the oscillator feedback path (C_4 in Figure 4.28). If removal of a part is impractical, then add a large value of capacitance (1 µF) in parallel with the oscillator tuning circuit.

B. Establish a fixed AGC value for the IF amplifiers. IF transformer tuning is not affected by the AGC bias, but reading the output indicator is easier and more accurate when the gain is not changing at the same time you are trying to induce more voltage through the transformer. Solder a wire to each side of a 1.5 V penlight battery. Check the polarity of the AGC voltage at the base of the IF amplifier, and connect the battery from the AC ground point of the base (junction of R_3, R_4, R_6, and C_5 in Figure 4.28) to the DC ground point. In Figure 4.28, the negative side of the battery is grounded.

C. A DC voltmeter is connected across the detector load resistor (R_9 in Figure 4.28) and set to read less than 2 V DC. Some procedures call for a scope or AC voltmeter to be connected across the speaker for this step, but then the RF signal must be modulated, and an audible tone will be present during the entire alignment procedure.

D. Feed an RF signal through a coupling capacitor (any medium value, such as 0.01, 0.05, 0.1, 0.5, or 1 µF) to the base of the mixer amplifier. Keep the generator output voltage level low and use a frequency counter to set the output signal frequency for an unmodulated 455 kHz carrier. Turn the receiver power on (tune the dial to the high end of the band, volume to zero) and set the signal generator output amplitude just high enough for the DC voltmeter at the detector to read about 1.0 V DC.

E. Using a *non-metallic* adjustment tool, start at the detector and tune each transformer for maximum signal output on the DC meter. If the DC meter voltage advances higher than 1 V, turn down the level of the input signal. *Hints:* (1) Some transformers have top *and* bottom tuning adjustments. In such cases, tune each one (bottom one first) for maximum output at the meter. (2) Do *not* force the tuning slug. If it cannot be turned easily, put a drop of penetrating oil into the opening at the top of the transformer, allow a few minutes, and try again. If it still cannot be turned, it must be replaced. (3) When the transformer tuning slug can be turned to the top or bottom of its movement without passing through a maximum voltage point, this usually indicates a cracked slug. Replace the slug, if possible, or replace the entire transformer. Repeat the tuning procedure several times to ensure that the transformers are set to maximum.

F.   Enable the oscillator by removing the short from step A, and remove the fixed bias from step B. Remove the coupling capacitor installed in step D.

RF Alignment

G.   Using a frequency counter, set the signal generator for 1610 kHz unmodulated. Feed the generator signal through a 0.01 μF capacitor to the base of the RF amplifier. Set the main tuning capacitor fully open (dial set to the high end), and adjust the oscillator coil tuning slug for maximum output reading at the detector DC voltmeter. Lower the RF signal input voltage if the meter moves above 1.0 V. The oscillator should now be 455 kHz higher than the RF signal, or 2065 kHz.

H.   Turn the main tuning capacitor to 1400 kHz on the dial. Set the RF signal generator to indicate maximum output on the DC voltmeter (keep the level below 1.0 V) and then adjust the oscillator trimmer capacitor to the highest output voltage on the DC meter.

I.   Set the main tuning capacitor to maximum capacitance (dial at the low end of the band). Construct a home-made coil of three turns of #22 hookup wire 10 in in diameter, as shown in Figure 4.40. Connect the signal generator to this coil (see A and F in Figure 4.39) and place it 24 in from the antenna rod. Adjust the signal generator to a frequency (about 540 kHz) that produces the maximum reading on the meter. Always keep the meter reading below 1.0 V. Tune the RF trimmer capacitor for the highest reading on the meter.

J.   Remove the DC meter from the detector, turn the signal generator off, and move it away from the receiver. Turn up the volume control and tune the receiver to a station near the center of the dial. Adjust for quiet listening level. Remove the wedge between the antenna coil and the antenna rod. Using a pencil or plastic tuning wand, position the antenna coil along the antenna rod for the loudest output signal. Replace the wedge to lock the antenna coil in place.

4.14.3 Noise Measurements

All noise, regardless of its origin, manifests itself as random amplitude changes in voltage and current through the system. Some noises are man-made, and others are natural due to the physics of electronics. Some noise arrives via the antenna terminals of the receiver under test, and some noise is caused by the current moving through the components within the receiver. At this time, our concern is the noise within the receiver only.

Although there are special instruments to measure noise, they are expensive and offer few other general uses. Both the analog and the digital devices are difficult to read because of the continuous changes in the noise voltage. It is uncertain that a valid set of readings of sufficient accuracy could be duplicated. The following procedure demonstrates the use of a dual-trace oscilloscope, found at every work station, to allow meaningful measurements that can be repeated with the degree of accuracy common in industry.

1. Cancel the receiver's ability to pick up radio waves by placing a 75 Ω resistor in parallel with the preselector.
2. Replace the speaker with an equivalent-value resistor, and turn the volume to maximum.
3. Turn the receiver on, but do not tune to a station.
4. Using a dual-trace oscilloscope with two input ports, connect both vertical input cables to the same place across the speaker load resistor.
5. Momentarily set the bottom trace mode switch to ground, and position the bottom trace in line with one of the scope's lower graticule lines.
6. Set both vertical input mode switches to AC input, and adjust both vertical sensitivity controls equally to view the noise pattern on each trace separately (see Figure 4.41a).
7. *Slowly* move the top trace toward the bottom trace until you see no separation between the traces (see Figure 4.41b).
8. Set both channel input mode switches to ground, and observe the two traces (see Figure 4.41c). The distance between the two traces is twice the rms noise voltage as determined by the vertical input sensitivity control settings.

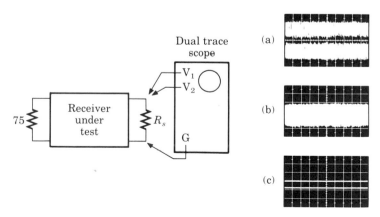

FIGURE 4.41
Setup and display for dual-trace scope measurements of noise.

EXAMPLE 4.2

The two traces are 4 cm apart, and the vertical sensitivity controls are set at 2 mV per cm. The distance between the lines is 4×2 mV = 8 mV; therefore, the rms noise voltage is one-half of 8 mV, or 4 mV$_{rms}$.

4.14.4 Receiver Sensitivity Measurements

All receivers must be tested under exactly the same conditions if the input sensitivity is to have a valid basis for comparison. These measurements should provide a numerical value so that we can judge the performance of one product against another.

The common denominator is the receiver's noise limiting. The term used to describe this noise control is "dB of quieting." The measuring procedure is as follows:

1. Terminate the audio output to its proper impedance with a resistive load.
2. Connect the test setup of Figure 4.41 with the signal generator turned *off*.
3. Turn the receiver *on*, and set the main tuning capacitor to maximum capacitance (minimum frequency, but *not* tuned to a station) and the volume control to maximum output.
4. Measure and record the noise at the load using the dual-trace technique.
5. Turn *on* the signal generator, set for 30% modulation at the carrier frequency to match the receiver tuning dial setting. Once the tuning is verified to be correct, turn *off* the modulation, but leave the RF signal *on*.
6. Adjust the input signal from the generator until the noise at the output decreases to a value 10 dB below the noise measurement from step 4. (A realistic and desirable value for dB of quieting is 20 dB, to allow for better signal-to-noise ratio. However, since some published figures are at 10 dB S/N, their sensitivity values look better. Measure and record *both* values.
7. Find the input signal level to the receiver. Measure the signal level at the output of the signal generator and reduce it by the amount of dB loss in the test setup attenuator. Remember that the voltage divider in the impedance-matching network adds 6 dB to the losses.

EXAMPLE 4.3

The RF generator output is 500 µV. The setup attenuator shows 28 dB attenuation, added to which is the 6 dB of the matching network, for a total of -34 dB = 0.02×500 µV = 10 µV. The sensitivity would be read as "10 µV sensitivity for 10 dB of quieting." This level is typical for AM radio receivers that have no RF amplifiers.

4.14.5 Selectivity Measurements

1. Connect a DC voltmeter from the detector load resistor to ground. Set the meter at 1.0 V.
2. Turn the receiver on and to the center of the dial (off of a station). Set the volume control to minimum.
3. Connect an RF signal generator through the correct matching network to the antenna.
4. Tune the *unmodulated* RF signal generator to the same frequency as the receiver, as indicated by the output voltage on the DC meter. Set the input level so that the DC output reads 1 V on the meter.
5. Measure the AGC voltage. Disable the AGC circuit by replacing the AGC voltage to the IF amplifier with a voltage from an external source. This voltage will remain unchanged for the balance of this test.
6. Record the RF input signal required to develop 1 V DC on the output meter. Use a frequency counter to provide accurate frequency settings of the RF signal generator from this point to the end of the test.
7. The balance of this test will depend on the bandwidth of the typical transmitted signal within the range of the receiver under test. In the AM broadcast band, each channel is theoretically allotted 10 kHz of bandwidth. This test should cover three times the channel bandwidth on *each side* of the selected center frequency, or 60 kHz total. Recommended steps are: ±1 kHz, ±2 kHz, ±4 kHz, ±6 kHz, ±10 kHz, ±15 kHz, ±20 kHz, ±25 kHz, and ±30 kHz away from the center.
8. With the receiver set as in step 4, change the generator frequency as specified in step 7, using a frequency counter at each setting. Adjust the signal generator output level to obtain a receiver output voltage of 1 V DC at each new frequency setting. Record the RF signal generator output voltage for each step.
9. Plot the results of this test on semilog graph paper to produce a set of curves similar to the graph in Figure 4.6.

4.14.6 Dynamic Range Measurements

1. Using Figure 4.39 as a guide, set up for a two-generator input test (B) and the dual-trace output noise measurement of Figure 4.41. Make sure you use the correct input signal connection for your receiver (C–F in Figure 4.39). Both generators should be turned off.
2. Turn the receiver on, with the volume control to maximum, normal AGC, and a dial setting near the center of the band (*not* on a local station). (The geometric center frequency for AM is 930 kHz.)
3. Turn on generator 1, with 30% amplitude modulation and a carrier frequency to match the dial setting. Watching the scope for maximum signal output will establish the correct frequency settings. You may

adjust the receiver and the generator so that they are at identical frequencies and at a convenient number, but away from any local station. Using a frequency counter, measure and record the generator carrier frequency (example: 930 kHz). Turn *off* generator 1.

4. Measure and record the receiver's output noise using the dual-trace scope method. Do *not* change the receiver's settings from those established in step 3. The output noise should be checked with *no* signal input, receiver off of a station, and at maximum gain (example: $e_{noise} = 2$ mV$_{rms}$).

5. Set attenuator 1 for maximum attenuation, and turn generator 1 *on*. Set the voltage at the generator to 500 μV, 30% modulation at 930 kHz (from step 3). Decrease the attenuation, watching the output scope, until the output voltage rises by 3 dB above the noise level measured in step 4. Using our sample values,

    A. Convert 2 mV$_{rms}$ to 5.657 mV p-p.
    B. Increase to 3 dB greater than 5.657 mV, or 8 mV p-p.
    C. Decrease the generator 1 setup attenuator so that the receiver output on the scope reads 8 mV p-p.
    D. Find and record the signal level at the output of attenuator 1 (example: $E_{gen} = 500$ μV, atten. = 59 dB).

    $$-59 \text{ dB} = 20 \log \frac{E_{s1}}{500 \text{ μV}} \qquad E_{s1} = 0.561 \text{ μV}$$

6. Turn on generator 2 (full 101 dB attenuation). Using a frequency counter, set for an unmodulated carrier 20 kHz higher *or* lower than the carrier frequency of generator 1 at step 3 (example: 910 kHz). Set the generator output voltage to 100,000 μV (0.1 V).

7. While observing the receiver output signal on the scope (from generator 1), decrease the attenuation of generator 2 until the receiver output signal on the scope decreases by 1 dB (example: 0.89125 × 8 mV = 7.13 mV p-p).

8. Find and record the signal level at the output of attenuator 2 (example: $E_{gen} = 100,000$ μV, atten. = 18 dB).

    $$18 \text{ dB} = 20 \log \frac{E_{s2}}{0.1 \text{ V}} \qquad E_{s2} = 12.59 \text{ mV}$$

9. The dynamic range is now found as the difference in decibels between the two input signals. For our example,

    $$\text{Dynamic range} = 20 \log \frac{E_{s2}}{E_{s1}}$$

    $$= 20 \log \frac{12.59 \text{ mV}}{0.561 \text{ μV}} = 87 \text{ dB}$$

The minimum discernible signal of this receiver is an antenna signal of 0.561 μV. The largest antenna input signal before compression (blocking) occurs at 12.59 mV, a range of 87 dB. Sixty to 80 dB is considered acceptable. A value over 80 dB is thought to be very good, with the ceiling considered to be 100 dB dynamic range.

QUESTIONS

1. Name the four circuits that a minimal receiver must have.
2. What do the initials TRF mean?
3. What operation is performed exclusively by a super-regenerative receiver?
4. What form of receiver circuitry is most popular in communications systems? (a) TRF receivers, (b) super-regenerative receivers, (c) superheterodyne receivers, (d) diversity receivers, (e) doppler receivers.
5. What function does the antenna serve for the receiver?
6. The noise signal will be lower at the antenna than anywhere else in the receiver. (T) (F)
7. A radio receiver will function without an RF amplifier. (T) (F)
8. The function of the RF amplifier is to: (a) amplify the signal more than the noise, (b) afford signal gain at the lowest noise level, (c) match the antenna impedance to the receiver, (d) amplify the RF and AGC voltages, (e) amplify the oscillator signal.
9. A mixer in a radio receiver is a (linear) (nonlinear) device.
10. A mixer in a radio receiver is a frequency converter. (T) (F)
11. The mixer circuit requires at least _____ input signals. (a) 1, (b) 2, (c) 3, (d) 4, (e) 5.
12. In the standard AM commercial radio broadcast band, the local oscillator frequency will always be (higher than) (lower than) (the same as) the frequency of the RF signal at the antenna.
13. A single active device that does the work of the mixer and the oscillator is called (a) an oscillating mixer, (b) a converter, (c) an IF amplifier, (d) an RF amplifier, (e) a frequency changer.
14. The frequency difference between the local oscillator and the incoming RF signal will always be (a) two times the RF frequency, (b) one-half the RF frequency, (c) the IF frequency, (d) the same as the RF frequency, (e) two times the IF frequency.
15. Double conversion in a radio receiver means (a) you can receive stereo, (b) there is a first detector and a second detector, (c) the RF signal is reduced to the IF signal in two steps.
16. The standard IF frequencies for a double-converted AM receiver are _____ and _____.
17. The advantage(s) of double conversion are (a) better image rejection, (b) less noise, (c) more stable low-frequency IF amplifiers, (d) more gain, (e) both a and c.

18. What is the image frequency for a receiver tuned to 820 kHz when the oscillator operates 455 kHz above the incoming RF signal?

19. What is the attenuation of the image signal in the preceding question when the tuned circuits have a combined Q of 75?

20. What would happen to the attenuation of the image if the Q of the circuit were reduced?

21. What would be the attenuation of the image signal if the Q dropped to 35?

22. Define "signal-to-noise ratio."

23. Define "sensitivity."

24. What unit of measurement is used so that all sensitivity measurements have the same reference level?

25. Define a "femtowatt."

26. Define "selectivity."

27. Name at least three electrical characteristics that describe an RF amplifier.

28. Name at least three electrical characteristics that describe a mixer amplifier.

29. Another name for a mixer is a (a) linear adder, (b) nonlinear adder.

30. The signals from the oscillator and the RF signal into a mixer must always be applied to two different mixer terminals. (T) (F)

31. The local oscillator in a radio receiver must always be a sine wave oscillator. (T) (F)

32. Oscillators developed from an LC tuned amplifier circuit are DC-biased for class (a) A, (b) B, (c) C, (d) D, (e) E.

33. After oscillations are sustained, the circuit bias is class (a) A, (b) B, (c) C, (d) D, (e) E.

34. The capacitance of an air capacitor is _____ the closed area of the capacitor plates. (a) directly proportional to, (b) inversely proportional to, (c) proportional to the square of, (d) inversely proportional to the square of, (e) proportional to the square root of, (f) inversely proportional to the square root of.

35. An air capacitor with half-round tuning plates requires a (a) linearly calibrated dial scale, (b) nonlinearly calibrated dial scale compressed at the high-frequency end, (c) nonlinearly calibrated dial scale compressed at the low-frequency end.

36. How many different frequency ranges will the signal experience while passing through the receiver? (a) 1, (b) 2, (c) 3, (d) 4, (e) 5.

37. The receiver front end *must* contain:
    a. an RF amplifier. (T) (F)
    b. a mixer. (T) (F)
    c. an oscillator. (T) (F)

38. What safeguard measurement does the FCC require when no RF amplifier is used in the receiver front end? (a) sensitivity, (b) selectivity, (c) RF radiation, (d) dynamic range, (e) signal-to-noise ratio.

39. More front-end noise will result (with) (without) an RF amplifier.

40. An IF amplifier is an RF amplifier. (T) (F)

41. The intermediate frequency of every radio receiver will be the same. (T) (F)

42. The standard IF for the commercial AM broadcast radio band is (a) 540 kHz, (b) 455 kHz, (c) 450 kHz, (d) 995 kHz, (e) 535 kHz.

43. The major function of the IF amplifier circuit is to provide (a) improved signal-to-noise ratio, (b) high gain with selectivity, (c) constant gain, (d) variable gain with signal strength, (e) low gain and wide bandwidth.

44. Increasing the number of tuned transformers at one set frequency will (a) increase the bandwidth, (b) decrease the bandwidth, (c) not change the bandwidth.

45. When the degree of coupling (coefficient of coupling, k) increases, the mutual inductance (a) increases, (b) decreases, (c) does not change.

46. The detector in an AM receiver carries out the following two circuit functions: (a) clamper and high-pass filter, (b) low-pass filter and clipper, (c) limiter and low-pass filter, (d) rectifier and low-pass filter, (e) rectifier and high-pass filter.

47. The AM detector in a standard broadcast radio receiver is the same as a demodulator. (T) (F)

48. The polarity of the diode determines if a high- or low-pass filter is to be used. (T) (F)

49. The component value most critical in the AM detector circuit is the (a) diode, (b) input capacitor, (c) R of the π filter, (d) output capacitor, (e) load resistance.

50. The charging time of the detector filter is based on the time for (a) one cycle at the carrier frequency, (b) one half cycle at the carrier frequency, (c) one quarter cycle at the carrier frequency, (d) one half cycle at the lowest audio frequency, (e) one quarter cycle at the highest audio frequency.

51. The discharge time for the detector filter is based on the time for (a) one cycle at the carrier frequency, (b) one half cycle at the carrier frequency, (c) one quarter cycle at the carrier frequency, (d) one half cycle at the highest audio frequency, (e) one quarter cycle at the lowest audio frequency.

52. The AGC voltage is a(an) (a) AC voltage, (b) DC voltage, (c) RF voltage, (d) IF voltage, (e) audio voltage.

53. The audio amplifiers cover the frequency range of (a) 1 Hz to 1 MHz, (b) frequencies above 25 kHz, (c) frequencies below 25 kHz, (d) frequencies above 1 MHz.

54. The function of the audio amplifiers is to _____ .

55. The audio circuits contain both a voltage amplifier and a power amplifier. (T) (F)

56. To increase the output power of the audio circuit, only the power amplifier need be changed. (T) (F)

57. Why must the collector dissipation be guarded when higher output powers are operative?

58. How does a heat sink help collector dissipation?

59. Name the four circuits every radio receiver must have.

60. Draw a block diagram of a single-conversion AM superheterodyne radio receiver and describe the function of each block. There should be seven functional circuits (minimum).

61. Can there be more than one RF amplifier? (a) yes, (b) no.

62. What is the minimum number of RF amplifiers found in a radio receiver? (a) zero, (b) one, (c) two, (d) three.

63. The function of the RF amplifier in a radio receiver is to (a) provide gain at the lowest noise point in the receiver, (b) help select a station, (c) isolate the antenna from the local oscillator, (d) all of the above.

64. The front-end mixer circuit is a _____ mixing amplifier. (a) linear, (b) nonlinear.

65. The two input signals to the mixer are the _____ and the _____ signals.

66. The combined signal at the output of the mixer contains the _____ and _____ frequencies of the two input signals.

67. The selected output signal frequency from the mixer is: (a) 540 kHz, (b) 1600 kHz, (c) 455 kHz, (d) 995 kHz, (e) 2055 kHz.

68. One amplifier circuit made to act like both an oscillator and a mixer is called (a) oscillodyne, (b) mixalator, (c) combiner, (d) collator, (e) converter.

69. Define the following four terms:
   a. sensitivity
   b. selectivity
   c. signal-to-noise ratio
   d. dynamic range
   State the units of measurement in the definition and include the qualifying conditions. Give an example of each term as it would appear on a product specification sheet.

70. The intermediate frequency for standard AM commercial broadcast radio is (a) 455 kHz, (b) 540 kHz, (c) 995 kHz, (d) 1600 kHz, (e) 2055 kHz.

71. The minimum number of signal frequencies that result from mixing two signals is (a) one, (b) two, (c) three, (d) four, (e) five.

72. The IF amplifier is a true RF amplifier. (T) (F)

73. The IF amplifier contributes about _____ of the total gain of the receiver. (a) 20%, (b) 40%, (c) 60%, (d) 80%, (e) 100%.

74. The diode detector is also sometimes called the (a) envelope detector, (b) AM detector, (c) second detector, (d) peak detector, (e) all of the above.

75. The diode detector in a broadcast band AM radio receiver is a demodulator. (T) (F)

76. The AGC allows for remote adjustment of the volume level. (T) (F)

77. The AGC action is a function of (a) the number of IF stages, (b) the presence or absence of an RF amplifier, (c) the gain of the IF amplifier, (d) the strength of the antenna signal, (e) the volume control setting.

78. The AGC is a(an) (a) AC voltage, (b) DC voltage.

79. Name the two circuit blocks that contribute to selectivity.

80. Which of the two circuits in the preceding question has the greater effect on selectivity?

81. The selectivity will be the same at the high end and at the low end of the receiver band of frequencies. (T) (F)

82. A meaningful selectivity measurement should contain what two statements of facts?

83. Define "dynamic range," as applied to receiver measurements.

84. An oscillator must first be an amplifier. (T) (F)
85. The gain of an oscillator must be greater than the losses of the feedback circuit. (T) (F)
86. The feedback signal of the oscillator is an (in-phase) (out-of-phase) voltage.
87. Should the local oscillator operate above or below the RF carrier frequency for a receiver in the 2.7 MHz to 3.0 MHz band? (a) above, (b) below, (c) either.
88. The coefficient of coupling will increase when the secondary of the transformer is moved (closer to) (further from) the primary of the transformer.
89. An untuned secondary of an RF transformer will look like _____ to the tuned primary of the transformer. (a) a capacitance, (b) an inductance, (c) a resistance, (d) an open circuit, (e) a short circuit.
90. When good selectivity is a desired feature, the coefficient of coupling is made (a) small, (b) medium, (c) large.
91. The quality of the sound will not be affected by the direction in which the diode is connected. (T) (F)
92. The magnitude of the AGC voltage is a function of the signal strength. (T) (F)
93. The AGC voltage is used as (a) audio voltage, (b) bias for the IF amplifiers, (c) DC supply for the detector, (d) oscillator frequency control voltage, (e) antenna bias voltage.
94. The efficiency of the audio output amplifier is found by _____.
95. "THD" means (a) the highest distortion, (b) twice highest distortion, (c) total hypothetical delay, (d) theoretical harmonious delight, (e) total harmonic distortion.
96. What electrical measurement of the transistor changes when the output power increases? (a) V_{b-e}, (b) V_{c-e}, (c) V_{b-c}, (d) I_c, (e) I_b.

Which of the following speaker construction details would improve low-frequency sounds?

97. Magnet: (a) 2 lb, (b) 8 oz.
98. Voice coil: (a) ½ in, (b) 3 in.
99. Spider: (a) 1½ in, (b) 5 in.
100. Cone: (a) stiff, (b) soft.
101. Annular surround: (a) flexible, (b) rigid.
102. Overall diameter: (a) 15 in, (b) 4 in.
103. A pilot carrier receiver is best suited for reception of (a) A3, (b) $A3_a$, (c) $A3_b$, (d) $A3_j$, (e) $A3_h$.
104. Single-sideband receivers (are) (are not) superheterodyne receivers.
105. The detectors for SSB signals use a (a) diode detector, (b) peak detector, (c) product detector, (d) envelope detector, (e) second detector.
106. A DSSC transmission reduced to an IF frequency range of 490–510 kHz would (a) require what reinserted carrier frequency? (b) produce what demodulated signal frequencies?
107. The squelch circuit acts as (a) an amplifier, (b) a high-pass filter, (c) a low-pass filter, (d) a level-activated switch, (e) a frequency changer.
108. The squelch system is (a) AC-controlled, (b) DC-controlled.

CHAPTER FIVE

FREQUENCY MODULATION PRINCIPLES

5.1   INTRODUCTION

Frequency modulation (FM) presents several attractive improvements over amplitude modulation. It offers the advantages of almost total immunity from noise interference, and it all but eliminates the problem of fading, which is so pronounced in amplitude modulation. Both of these advantages hold even when the signal at the receiver is extremely weak. You can conduct a simple demonstration of this effect by removing the antenna connections to your television receiver. The quality of the amplitude-modulated picture will be greatly affected, while almost no change can be detected in the frequency-modulated sound signal. These results are even more dramatic when you consider that the picture carrier power is six times greater than the sound carrier power. Another advantage is that the power in the transmitted FM signal is constant for *both* the modulated and unmodulated conditions of the carrier. The major disadvantage is that FM transmissions usually require a wider bandwidth for a message signal compared to AM.

In the expression for a sine wave, $E_c \sin (2\pi f_c)t$, it was stated earlier that for amplitude modulation, the voltage (E_c) changes while the frequency (f_c) remains constant. In angle modulation, the carrier frequency (f_c) changes while the voltage remains constant. The term **angle modulation** includes the two forms of frequency modulation:

1. *Direct* frequency modulation
2. *Indirect* frequency modulation (phase modulation)

There is only a subtle difference between the resulting modulated waveforms. The major difference lies in how and where the carrier is modulated and in the advantages of each process. By definition, **direct frequency modulation** is a system in which the modulating signal circuits act on the carrier oscillator to *directly* change the oscillator frequency. **Indirect frequency modulation** is a system in which the modulating signal circuits act on the carrier *signal*, at a point shortly after the carrier oscillator, to shift the signal phase and *indirectly* change the carrier frequency.

5.2   THE MODULATED WAVE

A time domain display of both the frequency-modulated (FM) carrier and the phase-modulated (PM) carrier is shown in Figure 5.1. It would take a trained eye to detect the difference between the frequency-modulated wave and the phase-modulated wave if either were displayed on an oscilloscope alone. FM receivers cannot tell the difference and treat both waves the same.

With no modulation applied, the carrier oscillator for the FM wave (and eventually the transmitted carrier signal) is at its rest frequency. The **rest frequency** is the center frequency assigned by the FCC as a station's transmitting frequency. The rest frequency will use the symbol f_c and is shown at point A in the wave of Figure 5.1. Point B for the FM wave in the figure is a frequency that is higher than the rest frequency. Point C denotes a frequency that is lower than the rest frequency. The carrier changes frequency between these two extremes.

In Figure 5.2, the FM wave is shown in the frequency domain. The carrier is seen at the center (rest) frequency at point A, changing upward by an amount $+\Delta f_c$ above the rest frequency at point B. It then changes downward to a level at point C (or $-\Delta f_c$) below the carrier rest frequency.

5.2.1 Frequency Deviation

The change in carrier frequency is called the **frequency deviation** and uses the symbol Δf_c. (Δ is the Greek letter delta and is used to symbolize change.) For a sample transmitter with an assigned rest frequency of 100 MHz deviated by ± 25 kHz, the carrier changes frequency with modulation between the limits of 99.975 MHz and 100.025 MHz. The total frequency change of 2×25 kHz = 50 kHz is called the **carrier swing.**

Deviation Limits   A logical question at this time is, "How far can the carrier change frequency?" There is no technical limit to the frequency change. A carrier oscillator of 5 MHz could change down to *zero* cycles and up to *two times* 5 MHz (10 MHz) without distorting the modulated signal. It is easy to see that a deviation of ± 5 MHz (a carrier swing of 10 MHz) for one station would be an undesirable waste of the frequency spectrum. Moreover, if all stations deviated down to zero cycles, they would have overlapping frequency bands, and it would be impossible to separate them.

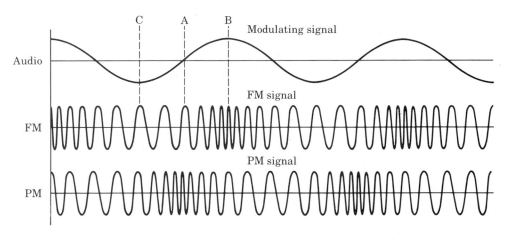

FIGURE 5.1
The relationship between the audio signal, the frequency-modulated carrler signal, and the phase-modulated carrier signal.

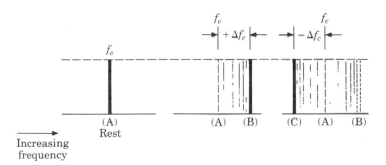

FIGURE 5.2
Frequency-modulated carrier wave shown in the frequency domain. Note the *constant* amplitude and the carrier frequency *change* at points A, B, and C.

The FCC has set legal limits of deviation for each of the different services that use FM as the form of modulation. Table 5.1 indentifies the transmission bands that use FM as the primary form of modulation and states the deviation limits for each category. The deviation limits are based on the quality of the intended transmissions, where wider deviation usually results in higher fidelity. The deviation limit is the term used to express 100% modulation of the FM carrier signal.

TABLE 5.1
Specifications for transmission of FM signals

Service Type	Frequency Assignment	Channel Bandwidth	Maximum Deviation	Highest Audio
Commercial FM radio broadcast	88.0 MHz to 108.0 MHz	200kHz	±75 Khz	15 kHz
Television sound	4.5 MHz above the picture carrier frequency	100 kHz	±25 kHz monaural ±50 kHz stereo	15 kHz
Public safety: police, fire, ambulance, taxi, forestry, utilities, transportation, and government, etc.	50 MHz and 122 MHz to 174 MHz	20 kHz	±5 kHz	3 kHz
Amateur and CB class A and business band radio	216 MHz to 470 MHz	15 kHz	±3 kHz	3 kHz
Wireless mics., wireless telephones	The same as commercial FM broadcast, but limited in power to less than 1 W			
Video tape recorders	All functions are within a closed system and are not restricted, except to radiation into the air. System specs may vary with each manufacturer. Typically: the carrier is at 3.4 MHz, sync tips cause a frequency change to 3.0 MHz, the white level to 4.0 MHz, with a typical bandwidth of 4.0 MHz.			
Satellites	See FSK and data communications (special)			
Military	Intermingled with public safety, and extending to microwave frequencies			

5.2.2 Percentage of Modulation

For the commercial FM broadcast radio band, 100% modulation is established as a carrier frequency change of ±75 kHz around the carrier rest frequency (a carrier swing of 150 kHz). The modulating voltage that causes the carrier frequency to change has both a linear and a proportional relationship to the percentage of modulation. When the modulating voltage drops to one-half of the original value, the percentage of modulation also drops to one-half of the original value; ±37.5 kHz represents 50% modulation. The percentage of modulation, then, is simply the ratio of the carrier frequency change to the maximum allowable carrier frequency change:

$$\% \text{ modulation} = \frac{\text{measured carrier frequency change} \times 100}{\text{max. allowable carrier frequency change}} \quad (5.1)$$

EXAMPLE 5.1 ═══

An FM broadcast carrier is deviated by ±30 kHz. Find the percentage of modulation.

$$\frac{30 \text{ kHz} \times 100}{75 \text{ kHz}} = 40\% \text{ modulation}$$

(It was known from Table 5.1 that the maximum carrier frequency change for FM broadcast radio is ±75 kHz.)

You may have noticed in the preceding statements that the FM carrier modulation process is a voltage-to-frequency conversion. That is, the voltage of the modulation signal causes a frequency change in the carrier generator. Thus, it may be stated that when the amplitude of the modulating signal changes the carrier generator frequency, it also changes the percentage of modulation. This characteristic of frequency modulation is the same as for amplitude modulation.

5.3   THE FM RADIO FREQUENCY BAND

The range of frequencies from 88.0 MHz to 108.0 MHz was set aside by the FCC for FM radio broadcasting in 1945. Each station in this band is allowed ±75 kHz deviation (150 kHz carrier swing); a 25 kHz guardband is then added above and below the carrier frequency swing. Thus, a total bandwidth of 200 kHz is assigned to each station in this frequency band. There is, then, room for 100 stations within the 20 MHz span between 88 to 108 MHz, as shown in Figure 5.3.

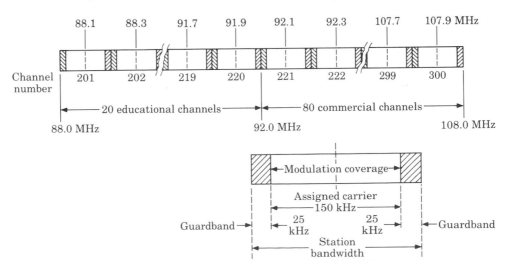

FIGURE 5.3
Standard FM radio broadcast band.

The AM radio channel numbers end at 107. The FM channels could have continued from that number, but the FCC decided that a whole new service deserved a whole new set of numbers. The FM channel numbering system starts at channel 201 and ends at channel 300. Channel 201 has a carrier rest frequency at 88.1 MHz, so that the deviation ($-\Delta f_c$) of 75 kHz and the 25 kHz guardband at the low end places the bottom end of channel 201 at 88.0 MHz. The upper sideband has the same range for modulation and guardband. With each channel assigned a 200 kHz bandwidth, the next higher channel rest frequency will be 200 kHz higher than 88.1 MHz, or 88.3 MHz for channel 202. As the process continues up the band, carrier rest frequencies will be found at all *odd* 0.1 MHz intervals, that is, at 88.1 MHz, 88.3 MHz, 88.5 MHz, and so on, all the way up to channel 300, at 107.9 MHz.

The first 20 channels in this band are reserved for educational broadcasts (nonprofit), and the remaining 80 channels are for commercial broadcast services. Channels 201 (at 88.1 MHz) through 220 (at 91.1 MHz) are educational and have a maximum ERP of 10 W (see Table 5.2). Channels 221 (at 92.1 MHz) through 300 (at 107.9 MHz) are commercial channels.

5.3.1 Radiation Standards

We will examine the FM transmitted signal shortly and discover that it is possible to radiate a bandwidth of energy beyond the ±75 kHz deviation limits. Therefore, to ensure an interference-free service area, a set of guidelines have been established that take into account: (1) the station ERP, (2) the assigned rest frequency, and (3) the distance between stations.

A service area is generally considered to be a circle having a radius of 75 miles. This is typical for line-of-sight transmissions, such as FM signals. Table 5.2 lists the specifications for each station class by power, antenna height, and

TABLE 5.2

Transmitter power and antenna height by class

Class	Power	Antenna Height	Area
A	3 kW 100 W (min.)	300 feet above the average terrain	Small communities, cities, towns, rural surroundings
There are 20 class A channels: 221, 224, 228, 232, 237, 240, 244, 249, 252, 257, 261, 265, 269, 272, 276, 280, 285, 288, 292, and 296			
B	50 kW 5 kW (min.)	500 feet above the average terrain	Large communities in zone I or zone IA
C	100 kW 25 kW (min.)	2000 feet above the average terrain	Large communities in zone II
D	10 W (max.)		Noncommercial channels

TABLE 5.3
Minimum mileage separation between FM stations

	Class A				Class B				Class C			
	Carrier Separation (in kilohertz)				Carrier Separation (in kilohertz)				Carrier Separation (in kilohertz)			
	Co-channels	200	400	600	Co-channels	200	400	600	Co-channels	200	400	600
Class A	65	40	15	15	—	65	40	40	—	105	65	65
Class B	—	—	—	—	150	105	40	40	170	135	65	65
Class C	—	—	—	—	—	—	—	—	180	150	65	65

area coverage. Table 5.3 sets the minimum distance (by class) between stations having the same rest frequency assignment (co-channels) and minimum distance from the first, second, and third closest channel rest frequencies. Each FM transmission service group has its own set of guidelines, which may be obtained from the National Bureau Office.

The terms used in Table 5.2 are defined as follows:

1. *Effective radiated power (ERP):* the product of the antenna input power and the antenna gain.
2. *Antenna input power:* transmitter output power minus transmission line losses.
3. *Average terrain:* The average elevation between points 2 miles and 10 miles from the antenna measured along eight radials, each separated by 45°.
4. *Zone I:* The northeastern part of the United States.
5. *Zone IA:* Puerto Rico, Virgin Islands, and California south of the 40° latitude.
6. *Zone II:* Parts of the United States not in zone I or zone IA, including Alaska and Hawaii.
7. *Zone III:* Florida and the gulf coast through Texas.

5.4 DIRECT FREQUENCY MODULATION

The FM transmitter has three basic sections (see Figure 5.4):

1. the **exciter,** which contains the carrier oscillator, the modulator, and a buffer amplifier;
2. the **multiplier section,** which features several frequency multipliers; and
3. the **power output section,** which includes a low-level power amplifier, the final power amplifiers, and the impedance-matching network to properly load the power section with the antenna impedance.

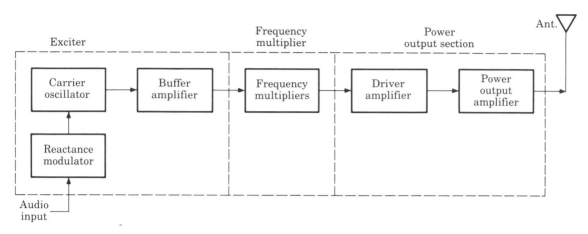

FIGURE 5.4
Basic direct frequency-modulated transmitter block diagram.

Existence of the power supply is assumed, and it receives little or no mention in transmitter discussions. The audio (or message) section is considered to be part of the studio network rather than part of the transmitter, since the same audio signals could be used for AM or FM transmissions (see Chapter 2).

The essential function of each circuit in the FM transmitter may be described as follows:

1. The exciter:
    a. The function of the **carrier oscillator** is to generate a stable sine wave signal at the rest frequency, when no modulation is applied. It must be able to linearly change frequency when fully modulated, with no measurable change in amplitude.
    b. The **buffer amplifier** acts as a constant high-impedance load on the oscillator to help stabilize the oscillator frequency. The buffer amplifier may have a small gain.
    c. The **modulator** acts to change the carrier oscillator frequency by application of the message signal. The positive peak of the message signal generally lowers the oscillator's frequency to a point below the rest frequency, and the negative message peak raises the oscillator frequency to a value above the rest frequency. The greater the peak-to-peak message signal, the larger the oscillator frequency deviation.
2. Frequency multipliers are tuned-input, tuned-output RF amplifiers in which the output resonant circuit is tuned to a multiple of the input frequency. Common frequency multipliers are 2X, 3X and 4X multiplication. A 5X frequency multiplier is sometimes seen, but its extremely low efficiency forbids widespread usage. Note that the mul-

tiplication is by whole numbers *only*. There cannot be a 1.5X multiplier, for instance.

3. The final power section develops the carrier power to be transmitted and often has a low-power amplifier to drive the final power amplifier. The impedance-matching network is the same as for the AM transmitter and matches the antenna impedance to the correct load on the final power amplifier.

The performance of the FM transmitter circuits can best be understood through the use of an example.

EXAMPLE 5.2

The FM transmitter of Figure 5.5 has a carrier oscillator with a rest frequency of 3.5 MHz. The oscillator shifts frequency by ±1.6 kHz when a 3.6 V p-p message signal is applied. The frequency multiplier section has three frequency triplers. Find the carrier rest frequency (f_c), the deviation (Δf_c), and the percentage of modulation *at the antenna*.

SOLUTION Look at the multiplier section first. Three triplers represent a factor of $3 \times 3 \times 3 = 27$. Thus, the carrier oscillator frequency is multiplied by 27 when passing through the multipliers: 3.5 MHz $\times$ 27 = 94.5 MHz.

The final power amplifier does not change the frequency; therefore, the transmitted carrier frequency (f_c) is 94.5 MHz (channel 233). The carrier oscillator is deviated by ±1.6 kHz by the message signal. When the deviation is multiplied in the multiplier section, the signal to the final amp is 27 $\times$ ±1.6 kHz = ±43.2 kHz. Again, the final section does not change the frequency, so the deviation at the antenna is ±43.2 kHz. The percentage of modulation is 43.2 kHz times 100 divided by 75 kHz, or 57.6% for the commercial FM broadcast radio band.

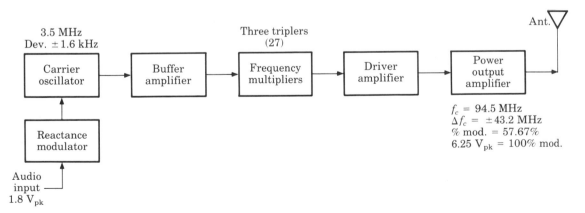

FIGURE 5.5
Sample working FM transmitter.

The next challenge may be to determine what peak-to-peak voltage of the message signal will cause 100% modulation. To do this, you need to divide the known voltage at the modulator by the percentage of modulation it caused, as

$$100\% \text{ modulation} = \frac{3.6 \text{ V p-p}}{0.576} = 6.25 \text{ V p-p}$$

A message signal of 6.25 V p-p would fully modulate the 94.5 MHz carrier signal to ±75 kHz deviation at the antenna.

EXAMPLE 5.3 ══════════════════════════════════════

A carrier oscillator at 5.05 MHz is deviated ±3.125 kHz by a 2.85 V p-p message signal. Determine the radiated carrier frequency, the deviation, and the percentage of modulation at the antenna, and find the message voltage needed to cause 100% modulation when the multiplier section has one doubler and two triplers.

SOLUTION Following Example 5.2, we find that the multipliers produce a signal 18 times f_c, or 90.9 MHz (channel 215), the deviation is ±56.25 kHz for 75% modulation, and 3.8 V p-p is needed to fully modulate the transmitter.

5.4.1 Deviation Sensitivity

More often than not, the voltage required to change the oscillator frequency is stated in terms of "frequency change per volt," called the **deviation sensitivity.** The deviation sensitivity is a characteristic of the individual FM modulator and defines the *peak* voltage required for a *peak* frequency change. The sensitivity of the modulator in Example 5.2 is 888.8 Hz/V. Therefore, at 3.6 V p-p (1.8 V_{pk}), the carrier oscillator changes frequency by ±1.8 V $\times$ 888.8 Hz $= \pm1.6$ kHz. For 100% modulation, 6.25 V p-p (±3.125 V_{pk}) $\times$ 888.8 Hz $= \pm2.778$ kHz, which is multiplied by 27 through the frequency multiplier section, for a total of ±75 kHz at the antenna.

5.4.2 Carrier Frequency Tolerances

The 88–108 MHz FM broadcast transmitters are required to hold the radiated carrier rest frequency to within 2000 Hz of the assigned station carrier rest frequency. Other services have similar tolerances, which could be greater or smaller than 2 kHz depending on the band of operating frequencies. Although the actual limits vary slightly, an FM transmitter is generally expected to hold its carrier rest frequency at the antenna to within 0.002% of the assigned carrier.

5.4.3 The Effects and Limits of the Audio Frequency

The *frequency* of the message signal has been omitted from the discussions to this point. The audio frequency limits, like the deviation limits, are different for each category of service and were established as legal boundaries by the FCC. The audio for FM broadcast radio and for television sound is contained between 50 Hz and 15 kHz. Public safety services are limited to between 50 Hz and 5 kHz, while some amateur classes are restricted to the range between 100 Hz and 3 kHz.

The message signal frequency determines the rate of change of the carrier oscillator, that is, how many times per second the carrier oscillator will go back and forth between its lowest and its highest frequencies. The message signal frequency has no control over how far the oscillator deviates from rest, an effect caused by the message signal amplitude. Example 5.2 described a 3.5 MHz carrier oscillator changing frequency by ±1.6 kHz. The frequency changed from 3.5 MHz up to 3.5016 MHz, then downward past the rest frequency to 3.4984 MHz (1.6 kHz below MHz), and then back to rest at 3.5 MHz. The oscillator does this for *each cycle* of the message signal. For the FM broadcast radio band, the oscillator could repeat this pattern from 50 times per second to 15,000 times per second.

The carrier oscillator signal changing frequency over the range indicated in Example 5.2 is depicted in the three-dimensional diagram of Figure 5.6. From time 0 to A_1, the carrier is at rest at 3.5 MHz. From time A_1 to time B, the carrier has increased frequency to 3.5016 MHz. It returns to rest at time A_2. Between time A_2 and time C, the carrier swings lower in frequency, to 3.4984 MHz, and from time C to time A_3, the carrier again swings upward, returning to rest at 3.5 MHz. The amplitude of the carrier signal remains constant, but the frequency changes through time in the shape of the sine wave modulating signal. The dashed line shows how the carrier would respond to an increase in audio amplitude, which would increase the *amount* of carrier frequency change but would *not* change the shape or the time of one cycle of the modulation signal.

5.4.4 Modulation Index

The frequency of the modulating message signal has a more important role in FM than simply determining the rate of change of the carrier signal frequency. It is used in evaluating the voltage and power distribution in the modulated FM wave. The relationship between the carrier deviation and the audio frequency, measured after the multipliers (usually at the antenna), is a factor called the **modulation index** (m_f), and may be found by:

$$m_f = \frac{\text{measured deviation at the antenna}}{\text{modulating signal frequency}} \quad \text{(in radians)} \qquad (5.2)$$

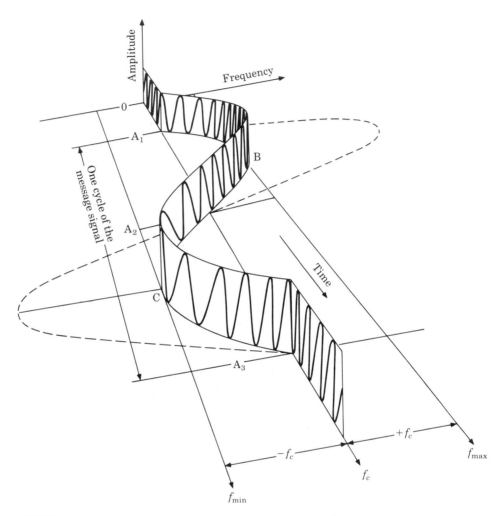

FIGURE 5.6
A three-dimensional view of the FM carrier wave, as it changes frequency away from
the rest frequency for a distance proportional to audio amplitude and a rate equal
to the audio frequency.

The modulation index was originally known as the *modulation factor;* hence
the symbol m_f. The modulation index is a measure of *radian* phase shift of the
modulated FM signal compared to the phase of the unmodulated carrier alone.
In Example 5.2, the deviation was ±43.2 kHz. Had a message signal frequency
of 7.2 kHz been specified, the modulation index would be

$$m_f = \frac{43.2 \text{ kHz}}{7.2 \text{ kHz}} = 6 \text{ rad}$$

The modulated signal, at the crest of its departure from rest, would be shifted in phase 6 radians (343.8°) from the carrier's phase when unmodulated. Over the years, the trend has been to disregard the phase relationship and simply state the modulation index as a dimensionless number, in this case 6.

5.4.5 The FM Wave Equation

We will assume that the carrier signal and the message signal are both sine waves for discussion and testing of the frequency modulated wave. The expression for the instantaneous voltage in the carrier wave, where one sine wave frequency modulates another sine wave, is stated as

$$e_c = E_c \sin \{(2\pi f_c t) + m_f \sin (2\pi f_s t)\} \tag{5.3}$$

where E_c = peak amplitude of the radiated carrier wave (constant)
f_c = radiated carrier rest frequency
m_f = modulation index (radians)
f_s = message signal frequency

There are now three terms $(f_c, m_f, \text{and } f_s)$ affecting the sine of a sine in Equation 5.3. It is best to bypass the analysis of this equation for three reasons.

1. Solving this equation would take more room than this text allows.
2. It is not our intent to develop equations, only use them.
3. This work has already been done, so there is no reason to do it again.

Before the age of computers, Mr. Fred Bessel expanded Equation 5.3 to the form shown as Equation 5.4.

The voltage in the:

$$
\begin{aligned}
e_c = E_c \{ \ & J_0(m_f) \sin (\omega_c t) && \leftarrow \text{carrier} \\
+ \ & J_1(m_f) \sin (\omega_c + \omega_s)t - \sin (\omega_c - \omega_s)t && \leftarrow \text{first SB pair} \\
+ \ & J_2(m_f) \sin (\omega_c + 2\omega_s)t + \sin (\omega_c - 2\omega_s)t && \leftarrow \text{second SB pair} \\
+ \ & J_3(m_f) \sin (\omega_c + 3\omega_s)t - \sin (\omega_c - 3\omega_s)t && \leftarrow \text{third SB pair} \\
+ \ & J_4(m_f) \sin (\omega_c + 4\omega_s)t + \sin (\omega_c - 4\omega_s)t && \leftarrow \text{fourth SB pair} \\
+ \ & J_n(m_f) \dots\dots\dots\dots\dots\dots\dots \ \} && \leftarrow n\text{th SB pair}
\end{aligned}
\tag{5.4}
$$

The expansion was accompanied by the resulting value chart, called the Bessel function chart, which has been reprinted in every text since the original work was completed. An updated chart that expands the values to four decimal places and increases the number of modulation indices is presented in Table 5.4, called the modulation function chart.

Equation 5.4 has a very large number of terms and can look overwhelming at first glance. However, a closer examination reveals several very important facts about the FM wave.

TABLE 5.4
Modulation function chart

m_f	J_0	J_1	J_2	J_3	J_4	J_5	J_6	J_7	J_8	J_9	J_{10}	J_{11}	J_{12}	J_{13}	J_{14}	J_{15}	J_{16}	J_{17}	J_{18}
0.00	1.000																		
0.10	0.9975	0.0499	0.0013																
0.20	0.9900	0.0995	0.0050	0.0001															
0.25	0.9845	0.1241	0.0078	0.0002	0.0001														
0.30	0.9776	0.1484	0.0111	0.0006	0.0001														
0.40	0.9604	0.1961	0.0197	0.0013	0.0001														
0.50	0.9385	0.2423	0.0306	0.0026	0.0002	0.0001													
0.60	0.9120	0.2867	0.0437	0.0044	0.0004	0.0001													
0.70	0.8812	0.3290	0.0588	0.0069	0.0006	0.0001													
0.80	0.8463	0.3689	0.0758	0.0103	0.0011	0.0001													
0.90	0.8075	0.4060	0.0946	0.0144	0.0017	0.0002													
1.00	0.7652	0.4400	0.1150	0.0195	0.0025	0.0003	0.0001												
1.25	0.6459	0.5107	0.1711	0.0369	0.0059	0.0008	0.0001												
1.50	0.5119	0.5579	0.2321	0.0610	0.0118	0.0018	0.0003												
1.75	0.3690	0.5802	0.2940	0.0919	0.0209	0.0038	0.0006	0.0001											
2.00	0.2239	0.5767	0.3529	0.1289	0.0340	0.0070	0.0012	0.0002											
2.50	-.0484	0.4971	0.4461	0.2166	0.0738	0.0195	0.0043	0.0008	0.0002										
3.00	-.2601	0.3391	0.4861	0.3091	0.1320	0.0430	0.0114	0.0026	0.0005	0.0001									
3.50	-.3801	0.1374	0.4586	0.3868	0.2044	0.0805	0.0255	0.0068	0.0016	0.0004	0.0001								
4.00	-.3972	-.0661	0.3642	0.4302	0.2812	0.1320	0.0491	0.0152	0.0040	0.0010	0.0002								
4.50	-.3206	-.2311	0.2179	0.4247	0.3484	0.1947	0.0843	0.0301	0.0092	0.0025	0.0006	0.0002							
5.00	-.1776	-.3276	0.0466	0.3649	0.3913	0.2612	0.1311	0.0534	0.0184	0.0055	0.0015	0.0004	0.0001						
5.50	-.0069	-.3415	-.1174	0.2562	0.3967	0.3209	0.1868	0.0866	0.0337	0.0113	0.0034	0.0009	0.0002						
6.00	0.1507	-.2767	-.2429	0.1148	0.3577	0.3621	0.2458	0.1296	0.0565	0.0212	0.0070	0.0021	0.0006	0.0002					
6.50	0.2601	-.1539	-.3074	-.0354	0.2748	0.3736	0.2999	0.1802	0.0881	0.0366	0.0133	0.0043	0.0013	0.0004	0.0001				
7.00	0.3001	-.0047	-.3014	-.1676	0.1578	0.3479	0.3392	0.2336	0.1280	0.0589	0.0236	0.0084	0.0027	0.0008	0.0002				
7.50	0.2664	0.1353	-.2303	-.2580	0.0239	0.2835	0.3542	0.2832	0.1744	0.0889	0.0390	0.0151	0.0053	0.0017	0.0005	0.0002			
8.00	0.1714	0.2345	-.1131	-.2912	-.1053	0.1858	0.3376	0.3206	0.2235	0.1263	0.0608	0.0256	0.0097	0.0033	0.0011	0.0003	0.0001		
8.50	0.0417	0.2729	0.0222	-.2627	-.2078	0.0672	0.2867	0.3376	0.2694	0.1694	0.0895	0.0410	0.0167	0.0062	0.0021	0.0007	0.0002		
9.00	-.0906	0.2451	0.1447	-.1810	-.2655	0.0552	0.2043	0.3275	0.3051	0.2149	0.1247	0.0622	0.0274	0.0108	0.0039	0.0013	0.0004	0.0001	
9.50	-.1944	0.1609	0.2275	-.0656	-.2692	-.1614	0.0992	0.2868	0.3234	0.2578	0.1651	0.0897	0.0427	0.0182	0.0070	0.0025	0.0008	0.0003	
10.0	-.2454	0.0438	0.2549	0.0584	-.2196	-.2339	-.0145	0.2167	0.3179	0.2919	0.2075	0.1231	0.0634	0.0290	0.0120	0.0045	0.0016	0.0005	0.0002
10.5	-.2369	-.0791	0.2215	0.1633	-.1286	-.2612	-.1202	0.1237	0.2850	0.3108	0.2477	0.1611	0.0898	0.0442	0.0195	0.0079	0.0029	0.0010	0.0003
11.0	-.1709	-.1770	0.1384	0.2267	-.0156	-.2390	-.2019	0.0184	0.2249	0.3087	0.2804	0.2010	0.1216	0.0643	0.0304	0.0130	0.0051	0.0019	0.0006
11.5	-.0687	-.2301	0.0258	0.2361	0.0953	-.1720	-.2460	-.0851	0.1421	0.2823	0.2996	0.2390	0.1576	0.0898	0.0454	0.0207	0.0087	0.0033	0.0012
12.0	0.0490	-.2244	-.0871	0.1928	0.1802	-.0746	-.2445	-.1711	0.0448	0.2305	0.3005	0.2705	0.1953	0.1202	0.0651	0.0316	0.0140	0.0057	0.0022
12.5	0.1537	-.1608	-.1711	0.1106	0.2256	0.0334	-.1988	-.2257	0.0547	0.1560	0.2790	0.2899	0.2314	0.1543	0.0896	0.0464	0.0218	0.0094	0.0038

1. The $J_0(m_f)$ factor in the first term represents the percentage of voltage in the modulated carrier for a known modulation index compared to the voltage in the unmodulated carrier.
2. The $J_1(m_f), J_2(m_f), J_3(m_f)$, etc. factors in the following terms represent the percentage of voltage in each of the sideband pairs compared to the voltage in the unmodulated carrier. That is, the $J_1(m_f)$ term is part of the expression for the first sideband pair, the $J_2(m_f)$ term is part of the expression for the second sideband pair, and so on for as many sideband pairs as exist for this value of m_f.
3. The first sideband pair is separated in frequency from the carrier by $\pm f_s$ (the audio frequency). The second sideband pair is separated from the carrier by $\pm 2f_s$, the third sideband pair by $\pm 3f_s$, and so on for each of the numbered sideband pairs.
4. Notice the polarity of the upper and lower sidebands in each of the sideband expressions. The upper sideband is expressed as a positive value, while the lower sidebands are alternately positive and negative. The algebraic process states that all even-numbered sidebands contribute directly to the total voltage in the modulated wave. The odd-numbered sidebands are in *phase quadrature* relative to the carrier component of the modulated wave. Thus, the odd-numbered sidebands are dropped from consideration of contributing factors in the determination of the total voltage in the modulated wave.

Determination of the J factor values in Equation 5.4 has been omitted from this treatment for the same reasons as was the analysis of Equation 5.3.

A simple program in Basic is included here to aid in the solution of the J terms for *any* entered modulation index value. The program in Table 5.5 was written for the TRS-80 but may be translated into other machine languages.

If every value of modulation index from 0 to 25 were to be solved for using Equation 5.4, the result would be the graph shown in Figure 5.7. J_0 would represent the changing voltage of the carrier, and the changing sideband voltages would be represented by the terms J_1 through J_{25}. This graph could be used as a substitute for Table 5.4, and it may be easier to visualize which terms are negative and which are positive for any given value of modulation index.

Figure 5.8 shows the carrier and sideband voltages of a frequency-modulated wave with a modulation index of 2.5. The carrier has a negative phase at the instant depicted; the first upper and lower sideband voltages are in phase quadrature to the phase of the carrier (as are the third and fifth sideband pairs), while the even-numbered sideband voltages are direct contributors to the total radiated voltage in the modulated wave.

The spectrum analyzer, used to display a frequency domain pattern, cannot distinguish vector phases, but simply renders *all* sideband voltages as positive amplitude vectors of the correct magnitude and position, as seen in Figure 5.9. Figure 5.9 shows the relative carrier and sideband voltages for modulation indices of 0.5 , 1.0 , 2.5, and 4.0. Here we see that the number of sidebands increase as the modulation index increases.

Note in the figure that the carrier voltage decreases to a point and then starts to increase again and that the sideband voltages do *not* always decrease as they get farther from the carrier frequency. Not shown in the figure is the fact that the separation between the sidebands is an amount equal to the message signal frequency.

TABLE 5.5
TRS 80 modulation index program

20	CLS DEFDBL N,X: DIM J(100)
30	PRINT "THIS PROGRAM WILL CALCULATE THE MAXIMUM POWER FACTORS"
40	PRINT "OF THE FM SIDEBANDS. THE LIMITATIONS OF THE TRS 80 MAKE"
50	PRINT "IT NECESSARY TO LIMIT THE MODULATION INDEX 0<MF<15."
60	PRINT LP = 0
	INPUT "ARE YOU USING THE LINE PRINTER FOR OUTPUT" :Q\$: Q\$ = LEFT \$ (Q\$,1): CLS: IF Q\$="Y" OR Q\$="Y" THEN LP =1:PRINT: PRINT "MAKE SURE LINE PRINTER IS ON."
100	PRINT: INPUT"WHAT IS VALUE OF MF" ;MF
220	N=0: PRINT "MODULATION INDEX =" MF: PRINT: PRINT "SIDE FREQUENCIES MAXIMUM AMPLITUDE": PRINT: IF LP THEN L PRINT "MODULATION INDEX = "MF: L PRINT"": L PRINT "SIDE FREQUENCIES MAXIMUM AMPLITUDE": L PRINT""
230	GO SUB 300: PRINT USING "## #.#####"; N,J(N): IF LP THEN L PRINT USING "## #.#####"; N,J(N):
240	IF ABS (J(N))<1E−5: PRINT: PRINT: IF L PRINT THEN L PRINT"": L PRINT""GO TO 100 ELSE GO TO 100
250	N = N+1: GO TO 230
300	CALCULATE MAXIMUM AMPLITUDE WITH APPROPRIATE BESSEL FUNCTION
310	***CALCULATE HEAD (X∧N)/(2∧N*X!)***
320	J(N) = (MF∧N)/(2∧N): X=N: K=0: GO SUB 500
330	J(N) = J(N)/X1 : GO SUB 350
340	J(N) = J(N)*B: RETURN
350	**** CALCULATE BODY ****
360	DE = 0: B = 1
370	DE = DE+1
400	CALCULATE NUMERATOR (INCLUDE SIGNS + −)
410	NM = MF∧ (DE)*((2*INT(DE/2)−DE)*2+1)
420	DIVIDE NUMERATOR BY DENOMINATOR
430	NM = NM/2∧ (DE*2): X=DE: K=0: GO SUB 500
440	NM = NM/X1: NM = NM*MF∧ (DE): X=DE: K=N GO SUB 500
460	IF ABS (NM) < IE-4 THEN RETURN
470	GO SUB 370
500	****X! ROUTINE****
510	X1 = 1: FORL = 1 TOX: X1 = *(K+L): NEXT RETURN

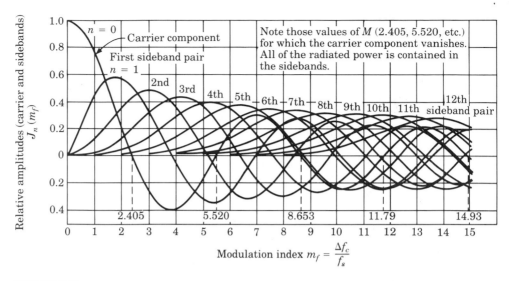

FIGURE 5.7
The variations of the carrier and sideband amplitudes as a function of the modulation index, m_f. Only the coefficients $J_n (m_f)$ greater than 0.01 are considered.

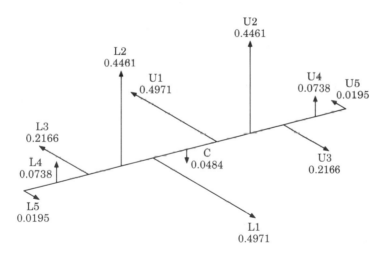

FIGURE 5.8
Carrier and sideband voltages for $m_f = 2.5$. Odd-numbered sidebands are shown in phase quadrature.

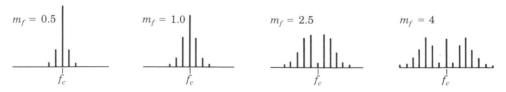

FIGURE 5.9
Spectrograms of energy levels for several different modulation indices.

183

5.4.6 The Bandwidth of the FM Wave

If the sidebands are separated by an amount equal to the message signal frequency and, from the modulation function chart, you know the number of sidebands, you can almost guess the bandwidth of the transmitted signal. Every calculated sideband must be considered. Notice that the sideband values are smallest farthest from the carrier frequency. A point is reached farthest from the carrier where the power in the sidebands contributes negligible power. The modulation function chart draws a zig-zag line to denote where the voltage in the sideband constitutes less than 1% of the total voltage in the wave.

Carson's rule for bandwidth takes into account only those sidebands that support 96% of the total radiated power. The bandwidth is then found as

$$BW = 2(\Delta f_c + f_s) \tag{5.5}$$

A change of carrier frequency (deviation) equal to ±60 kHz modulated by a 5 kHz tone would have a bandwidth of 2(60 kHz + 5 kHz), or 130 kHz.

5.4.7 Deviation Ratio

The **deviation ratio** is an expression of the worst-case modulation index. The deviation ratio tells what the transmitter is capable of doing, in terms of the maximum carrier deviation and the highest audio signal frequency:

$$\text{Deviation ratio} = \frac{\text{maximum allowable deviation}}{\text{highest allowable message frequency}} \tag{5.6}$$

For the broadcast FM radio band, the deviation ratio is 75 kHz divided by 15 kHz, or 5. For television sound, the deviation ratio is 25 kHz/15 kHz = 1.67. For public safety services, the deviation ratio is 5 kHz/3 kHz = 1.67. Amateur band radio has 3 kHz deviation and 3 kHz message frequency, for a deviation ratio of 1. A higher deviation ratio usually defines a higher-quality FM transmitter (see Table 5.1 for data).

The deviation ratio is used to determine the worst-case bandwidth for a given transmitter. For example, FM broadcast radio has a deviation ratio of 5. From the modulation function chart, it is found that that an index of 5 has eight upper and eight lower sidebands separated by 15 kHz each. Therefore, the radiated sidebands cover 2 × 8 × 15 kHz = 240 kHz. This signal would exceed the 200 kHz allowed per channel and cause interference in the adjacent channels. However, two redeeming factors must be considered:

1.   The percentage of the time that the highest message frequency is present is small; when present, this frequency would need to have an amplitude large enough to fully modulate the carrier, again small.
2.   Even when the preceding condition exists, the power in the two extreme upper and lower sidebands will add up to less than 5% of the total power in the transmitted wave.

It was concluded from these findings that the interference caused by such wide bandwidths would be negligible and within industry standards. The educated professional in FM systems should know that these overlap conditions do exist and should be willing to accept the minimal duration and repetition rates as noninterfering. Excluding the two outer sidebands (for $m_f = 5$) would leave a bandwidth of $2 \times 6 \times 15$ kHz $= 180$ kHz, which agrees with Carson's rule ($2 \times (75$ kHz $+ 15$ kHz$) = 180$ kHz).

5.4.8   The Voltage Distribution in the FM Wave

The sidebands generated when a carrier signal is frequency-modulated share the voltage distribution in such a manner that the sum of all of the voltages in the modulated carrier and sideband frequencies will be equal to the voltage of the unmodulated carrier alone (review Figures 5.8 and 5.9). This supports the statement that the total output voltage of an FM transmission will remain constant under condition of *no* modulation to *full* modulation.

An example will best prove this. A 150 V broadcast radio carrier is frequency-modulated to 40% by a 12 kHz tone. The deviation is 0.4×75 kHz $= 30$ kHz. The modulation index is 30 kHz divided by 12 kHz, or 2.5, which indicates that there are five upper and five lower side frequencies. The voltage at the carrier frequency is $J_0 E_c = -0.0484 \times 150$ V $= -7.26$ V. The voltage in the first side frequency *pair* is $2J_1 E_c = 2 \times 0.4971 \times 150$ V $= 149.13$ V (74.565 V in each side frequency). The second side frequency pair has $2J_2 E_c = 2 \times 0.4461 \times 150$ V $= 133.83$ V (66.915 V in each side frequency). The third side frequency pair has $2J_3 E_c = 2 \times 0.2166 \times 150$ V $= 64.98$ (32.49 V in each side frequency). The fourth side frequency pair has $2J_4 E_c = 2 \times 0.0738 \times 150$ V $= 22.14$ V (11.07 V in each side frequency), and the fifth side frequency pair has $2J_5 E_c = 2 \times 0.0195 \times 150$ V $= 5.85$ V (2.925 V in each side frequency).

The voltages may be summed as the *square root of the sum of the squares,* for a total of 150 V, keeping in mind that there are 11 terms to be squared, summed, and rooted. Alternatively, only the carrier and the contributing vectors may be added together:

$$E_{\text{tot}} = -7.26 + 2(66.915) + 2(11.07) = 149 \text{ V}$$

When the voltage in any one side frequency is sought, simply multiply the J term by the unmodulated carrier voltage.

5.4.9   The Power in the Modulated FM Wave

The power in the carrier and each of the side frequencies of a frequency-modulated signal is related to the power in the unmodulated carrier signal by $J_n{}^2 P_c$.

$$P_{\text{tot}} = J_0{}^2 P_c + 2J_1{}^2 P_c + 2J_2{}^2 P_c + 2J_3{}^2 P_c + 2J_4{}^2 P_c + 2J_5{}^2 P_c \qquad (5.7)$$

For example, if the 150 V carrier from the preceding section were applied across a 75 Ω antenna, the power would be $E^2/R = 300$ W. At 40% modulation, the power at each frequency would be $P_{\text{tot}} = 0.703$ W + 148.256 W + 119.403 W + 28.149 W + 3.268 W + 0.228 W = 300.016 W. This shows that the power remains constant for all levels of modulation in the frequency-modulated wave.

This calculation supports the bandwidth theory by showing that the two outermost side frequencies contain slightly more than 1% of the total power in the modulated wave and thus may be ignored.

5.4.10   Narrowband FM

By definition, narrowband FM has a modulation index not exceeding $\pi/2 = 1.57$ rad = 90°. It is intended to have a frequency-modulated carrier with only *one* set of side frequencies, similar to AM. The purpose is to save space in the frequency spectrum while providing a better-quality signal than AM for noise-free voice transmissions.

Wideband FM is reserved for the broadcast industry and narrowband FM for television sound, public safety services, and amateur radio. From Table 5.1, television sound has a deviation ratio of 25 kHz/15 kHz = 1.67. The resulting bandwidth is 2(25 kHz + 15 kHz) = 80 kHz. Although this is a wider bandwidth than would normally be considered for narrowband FM, its deviation ratio fits the definition for narrowband.

The carrier deviation for the public safety band is ±5 kHz, with a message frequency of 3 kHz and a deviation ratio of 1.67. The bandwidth for public safety services is 2(5 kHz + 3 kHz) = 16 kHz. For amateur radio, a carrier change of 3 kHz and a message frequency of 3 kHz gives a deviation ratio of 1. The resulting bandwidth is 2(3 kHz + 3 kHz) = 12 kHz.

Because of the small carrier deviations, most narrowband FM transmitters use indirect frequency modulation (see Section 5.5). Emphasis is placed on the ERP for narrowband FM transmitters, typically limited to 50 W. The structure of Equation 5.8 for narrowband FM is strikingly similar to that of Equation 1.6 for the AM wave. A notable exception is the reversal of the polarity of the lower sideband.

$$
\begin{aligned}
e = E_c \sin\,(2\,\pi f_c t) &- \frac{mE_c}{2} \cos\,[2\pi(f_c - f_s)t] \\
&+ \frac{mE_c}{2} \cos\,[2\pi(f_c + f_s)t]
\end{aligned}
\qquad (5.8)
$$

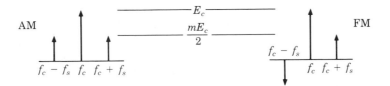

FIGURE 5.10
Frequency and amplitude spectra of the AM wave and FM narrowband.

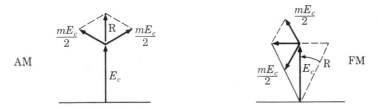

FIGURE 5.11
The vector relationship of the AM and narrowband FM waves, showing the angular positions of the carrier and sideband voltages at some point in the modulation cycle. R: resultant, E_c: carrier vector, $mE_c/2$: sideband vectors

Figure 5.10 compares the frequency domain voltages for the AM wave to those of the wave for narrowband FM and shows the lower sideband in reverse polarity.

Examine the vector diagram of the AM wave in Figure 5.11, and note that if you simply reverse the polarity of either sideband vector, you get the vector diagram of the narrowband FM wave. Here it is seen that the algebraic sum of the sideband vectors in the narrowband FM wave is in quadrature phase with the carrier (one quarter cycle = 90°).

5.4.11 Preemphasis (Deemphasis)

This section describes a system for improving the signal-to-noise ratio of the recovered audio at the receiver. No amount of filtering will remove the noise from the RF circuits, but noise control in the low-frequency (audio) amplifiers is achieved through a high-pass filter at the transmitter (preemphasis) and a low-pass filter at the receiver (deemphasis). The measurable noise in low-frequency electronic amplifiers is most pronounced over the frequency range of 1 kHz to 12 kHz. At the transmitter, the audio circuits are tailored to provide a higher level of audio signal at the upper audio frequency range. For a fixed noise level, the greater signal voltage yields a better signal-to-noise ratio. At the receiver, when the upper audio frequency signals are attenuated to form a flat frequency response, the associated noise level is also attenuated.

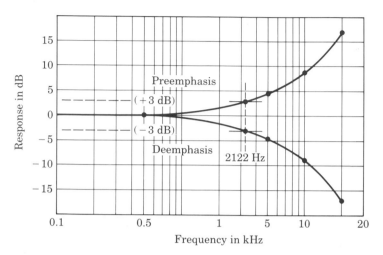

FIGURE 5.12
The standard EIAA 75 μs emphasis curves.

     The response curves are shown in Figure 5.12 for the EIAA standard time constant of 75 μs. A time constant was selected rather than a set of component values for the curve for the preemphasis and deemphasis circuit to allow product designers a free choice of component values to match either the RC or the L/R time. L/R circuits are less common because of the large inductance values and small resistance values required to achieve the 75 μs time. For example,

$$L/R = 75 \ \mu s \qquad L = 75 \times 10^{-6} \times 1500 \ \Omega = 112.5 \ \text{mH}$$
$$RC = 75 \ \mu s \qquad R = 75 \times 10^{-6}/0.01 \ \mu F = 7500 \ \Omega$$

The 3 dB frequency is determined by

$$f_1 = \frac{1}{2\pi RC} \quad \text{(in hertz)} \tag{5.9}$$

Since RC is equal to time and time is set at 75 μs, then the 3 dB point is

$$f_1 = \frac{1}{2\pi 75 \ \mu s} = 2122 \ \text{Hz}$$

The dB increase for *any* frequency over the range can be found as

$$\text{dB gain} = 20 \log \sqrt{1 + \left(\frac{f}{f_1}\right)^2} \tag{5.10}$$

where f_1 is the frequency at the 3 dB point and f is the frequency under investigation.

EXAMPLE 5.4

Find the dB gain at 100 Hz, 2122 Hz, 6 kHz, 10 kHz, and 15 kHz. *Note: f_1* remains at 2122 Hz for all calculations involving a 75 µs time constant. Only the 100 Hz and 10 kHz calculations are shown here.

$$\text{At 100 Hz,} \quad \text{dB gain} = 20 \log \sqrt{1 + \left(\frac{100}{2122}\right)^2} = 0.02 \text{ dB}$$

$$\text{At 10 kHz,} \quad \text{dB gain} = 20 \log \sqrt{1 + \left(\frac{10 \text{ k}}{2122}\right)^2} = 13.66 \text{ dB}$$

The deemphasis characteristic curve for 75 µs is determined in a similar fashion from Equation 5.11.

$$\text{dB loss} = 20 \log \frac{1}{\sqrt{1 + \left(\frac{f}{f_1}\right)^2}} \tag{5.11}$$

Equation 5.11 will yield dB values identical to those of Equation 5.10, but of opposite sign.

The dB improvement in the signal-to-noise ratio at any frequency is found as a power ratio:

$$\text{dB} = 10 \log \left(\frac{1}{3}\right)\left(\frac{f}{f_1}\right)^2 \tag{5.12}$$

Several circuit configurations for preemphasis and deemphasis are shown in Figure 5.13. The European and Australian TV sound uses 50 microseconds.

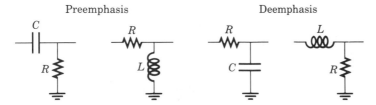

FIGURE 5.13
Sample preemphasis and deemphasis networks. $T = 75$ µs (North America FCC); $T = 50$ µs (Europe CCIR); $T = 25$ µs (Dolby). Select a value of C and solve for R in $R = T/C$. (CCIR: Comité Consultatif International Radio.)

5.5   INDIRECT FREQUENCY MODULATION (PHASE MODULATION)

The system diagram of Figure 5.14 describes the principle of operation of an Armstrong phase modulator. It should be noted first that the output signal from the carrier oscillator is supplied to circuits that perform the task of modulating the carrier signal. The oscillator does *not* change frequency, as is the case for direct FM. This points out the major advantage of phase modulation (PM), or indirect FM, over direct FM. That is, the carrier oscillator is *crystal-controlled* for frequency stability.

The crystal-controlled carrier oscillator signal is directed to two circuits in parallel. This signal (usually a sine wave) is established as the reference phase carrier signal and is assigned a value of 0°.

The balanced modulator is an amplitude modulator used to form an envelope of double sidebands and to suppress the carrier signal (DSSC). This requires two input signals, the 0° carrier signal and the modulating message signal. The output of the balanced modulator is connected to the adder circuit; here the 90° phase-delayed carrier signal will be added back to replace the suppressed carrier. The act of delaying the carrier phase by 90° does not change the carrier frequency or its waveshape. This signal is identified as the 90° carrier signal.

In the discussions on amplitude modulation, a phasor diagram of the carrier and sideband vectors was used to create a visual image of the modulation process. This format is carried one step further here to describe the output of the adder circuit (see Figure 5.15). The adder has two input signals, the zero-referenced double-sideband AM envelope and the 90° carrier signal. A vector diagram of the adder output shows the effects of adding the two input signals. The 90° carrier is labeled E_c and the vector sum of the two sidebands (E_u and E_l), denoted E_{sb}, is shown 90° from E_c. As the two sideband vectors counterrotate, their resultant (E_{sb}) will always be 90° from E_c but will change amplitude and polarity from $+E_{sb}$ to $-E_{sb}$.

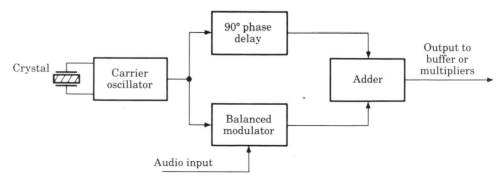

FIGURE 5.14
Armstrong phase modulator.

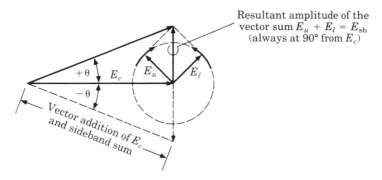

Resultant amplitude of the
vector sum $E_u + E_l = E_{sb}$
(always at 90° from E_c)

FIGURE 5.15
Vector analysis of the Armstrong phase modulator.

The vector addition of E_{sb} and E_c in Figure 5.15 will form the hypotenuse of the triangle that changes phase through $\pm\theta$ as the sideband amplitude of E_{sb} changes from $+E_{sb}$ to $-E_{sb}$. The hypotenuse represents the output voltage (E_o) of the adder. As the angle changes from $+\theta$, through 0, to $-\theta$, the length of the hypotenuse (E_o) changes, and since this is the output of the adder, an undesirable amount of amplitude modulation appears at the adder output.

The amount of AM that is acceptable in the PM signal is a matter of how much can be controlled (or eliminated) in later circuits. There is no industry standard, but 10% seems to be about the maximum amount found in such systems. Assuming 10% to be the AM limit, Equation 1.3, for the percentage of modulation of the AM carrier, may be rearranged by setting $E_c = E_{min} = 1$ and $E_o = E_{max}$. We can then solve for E_o at a known percentage of modulation:

$$E_o \text{ (as a factor of } E_c) = \frac{1 + \% \text{ modulation}}{1 - \% \text{ modulation}} \qquad (5.13)$$

For 10% AM, $E_o = 1.222 \times E_c$. Knowing E_o and E_c, we can find the angle as cos $\theta = 1/1.222$, or $\theta = \pm35.1°$ phase shift. The angle is then converted to radians (35.1 divided by 57.3 equals 0.6125 rad).

The carrier frequency change at the adder output is a function of the output phase shift and is found by:

$$f_c = \Delta\theta f_s \quad \text{(in hertz)} \qquad (5.14)$$

where θ is the phase change in radians and f_s is the lowest audio modulating frequency. In most FM radio bands, the lowest audio frequency is 50 Hz. Therefore, the carrier frequency change at the adder output is 0.6125 $\times$ 50 Hz = ±30 Hz. Since 10% AM represents the upper limit of carrier voltage change, then ±30 Hz is the maximum deviation from the modulator for PM.

The 30 Hz deviation is multiplied in the frequency multiplier section to determine the deviation of the radiated signal. Herein lies the major disadvantage of phase modulation, that is, the small value for carrier frequency change at the modulator compared to that for direct FM.

The preceding paragraphs discussed two different phase shift networks that may appear to behave inconsistently. The 90° phase shift network does not change the signal frequency because the components and resulting phase change are constant with time. However, the phase of the adder output voltage is in a continual state of change brought about by the cyclical variations of the message signal, and during the time of a phase change, there will also be a *frequency change.*

In Figure 5.16, during time (a), the signal has a frequency f_1 and is at the zero reference phase. During time (c), the signal has a frequency f_1 but has changed phase to θ. During time (b), when the phase is in the process of changing from 0 to θ, the frequency is less than f_1.

A carrier signal frequency change of 30 Hz out of the modulator is acceptable for narrowband FM such as amateur radio, where 100% modulation (after the multipliers) is only ±3 kHz. Here a multiplier of 100 increases the deviation from 30 Hz to 3000 Hz. However, for commercial FM, where 100% modulation is ±75 kHz, a multiplication of 2500 is required. Although this is not impossible (seven frequency triplers will do it), it does point out another undesirable condition. That is, the carrier oscillator would need to be at a frequency of about 40 kHz, which is much too low for a crystal-controlled oscillator. Crystals operate best in the frequency range of 100 kHz to 10 MHz as determined by crystal thickness. For a frequency below 100 kHz, the crystal needed would be thick and bulky. For frequencies above 10 MHz, the crystal material would be thin and fragile. (Crystals that operate at frequencies above 10 MHz are operating at harmonics of the crystal fundamental frequency and are called "overtone crystals.")

The Armstrong modulator is joined to the remainder of the transmitter in Figure 5.17. Here a carrier oscillator frequency of 400 kHz has been chosen, and the deviation out of the modulator is ±14.468 Hz. The 400 kHz carrier is frequency-multiplied to 32.4 MHz, then down-converted to 1.4078 MHz, and again multiplied to 90.1 MHz. If the carrier were not down-converted, the rest frequency at the antenna would be 400 kHz × 5184 = 2.0736 GHz. However, a large multiplying factor is needed to increase the small deviation of ±14.468 Hz to ±75 kHz. The basic principle that should not be overlooked here is that when a modulated carrier is frequency-converted in a mixer-

FIGURE 5.16

In the process of changing phase, with constant input frequency, the output experiences a frequency change.

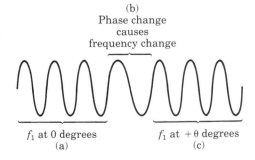

(b)
Phase change
causes
frequency change

f_1 at 0 degrees
(a)

f_1 at $+\theta$ degrees
(c)

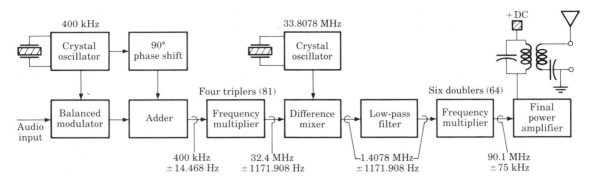

FIGURE 5.17
FM transmitter.

oscillator circuit, the carrier frequency is changed (up or down) but the modulation is not changed in any respect.

Following the signal conditioning through the transmitter of Figure 5.17, we see that the output of the modulator (adder) is 400 kHz ±14.468 Hz. This signal is multiplied by four frequency triplers ($3 \times 3 \times 3 \times 3 = 81$) so that at the mixer input, the carrier is at 32.4 MHz and the deviation is at ±1171.9 Hz. The second crystal-controlled oscillator may operate higher or lower than the carrier frequency at this point. In this example, it is at a higher frequency than 32.4 MHz (33.8075 MHz). The signal out of the mixer is the difference between the two RF signals, that is, 33.8075 MHz − 32.4 MHz = 1.4078 MHz. The deviation is *unchanged,* at 1171.9 Hz. The second set of frequency multipliers has six frequency doublers, or a factor of $2 \times 2 \times 2 \times 2 \times 2 \times 2 = 64$.

The carrier signal of 1.4078 MHz × 64 = 90.1 MHz is passed through the power amplifier to the antenna. The deviation through the second set of frequency multipliers is 1171.9 Hz × 64 = ±75 kHz at the antenna. The power amplifier does not change the frequency content of the signal, so whatever signal is at the output of the second set of frequency multipliers will be transmitted. The frequency of the second crystal-controlled oscillator may be changed to change the radiated rest frequency to any of the rest frequencies of the 100 FM stations. Changing the second oscillator in 3125 Hz steps will correspond to the standard FM station assigned frequencies.

5.6   THE CARRIER PHASE IN THE FREQUENCY-MODULATED WAVE

One example of the carrier phase shift as a result of modulation is given here to introduce the concept. An in-depth explanation is included in Appendix A for the user who needs a detailed description of the workings of the carrier phase shift.

The carrier phase is determined by the summation of any number of vector voltages, similar in philosophy to advanced trigonometry. The advantage to the FM sideband voltage vectors is that each progressive vector is at right angles to its predecessor and advanced by 90° for positive values of the J term. The sideband vectors are in phase quadrature with the carrier vector. For a simple modulation index of 1 radian ($m_f = 1$) $J_0 = 0.7652$, $J_1 = 0.4401$, $J_2 = 0.1149$, and $J_3 = 0.0196$.

Figure 5.18 identifies the unmodulated carrier at 100% amplitude and at zero phase as a heavy dashed line. When the modulation index is 1 radian, the voltage in the carrier is reduced to 76.52%, identified in Figure 5.18 as the J_0 vector. The first sidebands *each* have a magnitude of 44.01% (88.02% total) and form a right-angle vector to the carrier advanced 90°, shown as (1) J_1 and (2) J_1. The second sidebands each have a value of 11.49% (22.98% total) and form a vector advanced another 90°, or 180° from the carrier; these are labeled (1) J_2 and (2) J_2. The third sidebands each have a value of 1.96% (3.92% total) and form still another right angle at the end of the J_2 vector length. The J_3 vectors advance still another 90° or 270° from the carrier. The length of the two J_3 vectors should just meet the circumference of the 100% circle. When a radius line is drawn to the point where the last J term touches the circumference, an angle is formed with the carrier vector that equals the value of the modulation

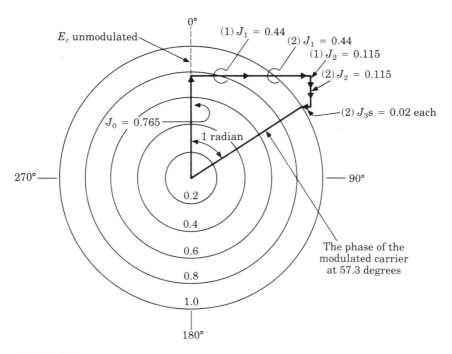

FIGURE 5.18
The phase shift of a modulated FM carrier with a modulation index of 1 radian.

index in radians. In Figure 5.18, this radius is drawn as a heavy solid line and forms an angle of $57.3° = 1$ rad $= m_f$.

QUESTIONS

1. In frequency modulation, the carrier changes frequency when it is modulated. (T) (F)

2. The range of frequencies assigned for use as the FM radio broadcast band is _____.

3. The bandwidth of a single FM broadcast channel (including the guardbands) is _____.

4. The carrier rest frequency of the lowest channel in the FM broadcast band is _____.

5. The carrier rest frequency of the lowest commercial channel in the FM broadcast band is _____.

6. The amplitude of the frequency-modulated wave _____ with modulation. (a) remains constant, (b) increases, (c) decreases, (d) both increases and decreases.

7. One hundred percent modulation of the carrier for the broadcast FM radio band is achieved when the (a) carrier envelope changes by ±100%, (b) carrier frequency changes by ±75 kHz, (c) carrier frequency changes by ±100%, (d) audio frequency is 15 kHz.

8. The highest audio frequency allowed for use in commercial FM radio broadcasting is _____.

9. Eighty percent modulation of the carrier in FM broadcast radio is achieved when the deviation reaches (a) ±15 kHz, (b) ±30 kHz, (c) ±60 kHz, (d) ±75 kHz, (e) ±90 kHz.

10. In frequency modulation, what characteristic of the audio signal determines the percentage of modulation? (a) power, (b) frequency, (c) amplitude, (d) phase.

11. In frequency modulation, what characteristic of the audio signal determines the rate of change of the carrier signal? (a) power, (b) frequency, (c) amplitude, (d) phase.

12. The primary oscillator of an FM transmitter has a rest frequency of 5.25 MHz, and 3.0 V_{rms} of audio signal deviates the oscillator by ±4 kHz. When the transmitter has one frequency doubler and two frequency triplers,
    a. What is the radiated carrier rest frequency?
    b. What is the deviation at the antenna?
    c. What is the percentage of modulation of the radiated signal?

13. What is the minimum distance in miles between two class C FM stations operating on the same frequency?

14. What is the minimum distance in miles between two class C FM stations operating on adjacent channels?

15. What is the minimum distance in miles between a class A station and a class B station operating on the same frequency?

16. What is the highest carrier rest frequency of a class D FM station?

17. The carrier oscillator of a directly modulated FM transmitter is usually crystal-controlled. (T) (F)

18. The carrier oscillator of a directly modulated FM transmitter generates a (a) square wave, (b) sine wave, (c) triangular wave, (d) ramp wave, (e) pulse wave.

19. The input of a buffer amplifier is (a) high-impedance, (b) low-impedance.

20. The output frequency of a frequency multiplier can be 2.6 times the input frequency. (T) (F)

21. The final power amplifier increases the carrier frequency. (T) (F)

22. The modulation index is a ratio of the _____ compared to the _____.

23. The modulation index is measured at (or calculated for) the (a) carrier oscillator, (b) buffer amplifier, (c) frequency multipliers, (d) driver amplifier, (e) antenna terminals.

24. The modulation index is a measure of (a) the number of sideband pairs in the modulated wave, (b) the voltage and power in the modulated wave, (c) the bandwidth of the modulated wave, (d) the radian phase shift of the modulated carrier, (e) all of the above.

25. The deviation ratio is a measure of the worst-case modulation index. (T) (F)

26. What two factors are compared in the deviation ratio?

27. What is the bandwidth of a transmitted signal that is 72% modulated by a 9 kHz tone?

28. The advantage of phase modulation over direct FM is that (a) more multipliers can be used, (b) the deviation is smaller, (c) the oscillator is crystal-controlled, (d) greater deviation to center frequencies is obtainable.

29. In a phase-modulated transmitter, modulation takes place (a) before the master oscillator, (b) at the master oscillator, (c) just after the master oscillator, (d) at the final power amplifier.

30. A deemphasis circuit is a (a) high-pass filter at the transmitter, (b) low-pass filter at the transmitter, (c) high-pass filter at the receiver, (d) low-pass filter at the receiver.

31. Preemphasis and deemphasis circuits are used to (a) make the voice sound clearer, (b) increase the percentage of modulation, (c) increase the modulation index, (d) improve the signal-to-noise ratio at the receiver, (e) increase the trebles.

32. A frequency multiplier is a tuned RF amplifier with (a) the output voltage larger than the input voltage, (b) a divide-by-N flip-flop, (c) the output tuned to an integer value of the input frequency, (d) a heterodyne oscillator, (e) none of the above.

33. A frequency multiplier multiplies the carrier frequency but not the deviation. (T) (F)

34. A frequency multiplier multiplies the deviation but not the carrier frequency. (T) (F)

35. What is the purpose of neutralizing an amplifier? (a) to cancel the gain, (b) to reduce frequency multiplication, (c) to maintain a constant output power, (d) to cancel interelectrode capacitances, (e) to cancel negative feedback.

36. When is neutralization likely to be required? (a) at frequencies over 25 MHz, (b) when a three-element active device is used, (c) with single-frequency RF tuned amplifiers, (d) all of the above.

37. Neutralization is (a) negative feedback, (b) positive feedback.

38. The effect of parasitic oscillations on an RF amplifier circuit is to (a) cause over-heating, (b) reduce output power, (c) cause distortion in the signal, (d) all of the above.

39. The maximum deviation at the output of a phase modulator is approximately (a) ±25 Hz, (b) ±250 Hz, (c) ±1 kHz, (d) ±100 kHz, (e) ±2.5 MHz.

40. Only direct frequency-modulated signals are transmitted in the commercial FM broadcast band. (T) (F)

CHAPTER SIX

FREQUENCY-MODULATED TRANSMITTERS

6.1   INTRODUCTION

In electronic theory, the functional behavior of a circuit indicates how the circuit is to treat the signals in question. A particular treatment calls for a specific functional circuit. However, the component selection and arrangement can encompass a wide range of possibilities. This chapter will examine some recent-vintage FM transmitter circuits and some second-generation circuits, explaining their operation and comparing their tradeoffs.

If engineering time is the chief concern, then the search for a stable frequency carrier oscillator circuit offering identical characteristics over a wide frequency range would take top priority. The efficiency and linearity of the final power amplifiers would receive secondary attention. The design of the modulator would have third priority, with all other circuits assigned smaller time allotments.

6.2   THE CARRIER OSCILLATOR

The heart of the FM transmitter, the carrier oscillator, was once thought to require a clean sine wave output signal, as does the AM transmitter carrier oscillator. Recent investigations have shown that a square wave oscillator signal (for direct frequency modulation) has two advantages:

1.   The *symmetry* of the carrier wave is more critical than the linearity of the wave. Symmetry here pertains to *duty cycle* only, where a 50% duty cycle square wave is the ideal. It is easier to control the symmetry of a square wave than it is for a sine wave.

2. A square wave oscillator will remain symmetrical over a much wider frequency range than will a sine wave oscillator. In the discussion of phase modulation, it was noted that the frequency-modulated oscillator should limit amplitude modulation of the FM wave. A square wave oscillator that switches output voltage between 0 V (ground) and the +DC supply voltage will also act to limit amplitude changes.

After the carrier is modulated, it will pass through tuned circuits and be restored to its fundamental sine wave frequency.

The carrier oscillator circuit in Figure 6.1 differs from the description in Chapter 5 in that here the carrier frequency is generated directly by the oscillator; frequency multipliers are not needed. The VCO of Figure 6.1 is part of a phase-locked loop (PLL) and has a rest frequency that can be varied from 87.5 MHz to 107.9 MHz. The VCO frequency must be capable of deviation, so it cannot be crystal-controlled directly. Rather, it is controlled by way of a phase-locked feedback loop that *is* crystal-controlled. The 10 MHz crystal oscillator is divided by 3200, to 3125 Hz, at the input to the phase detector. For the PLL to be in a locked condition, the phase detector input from the programmable divider must also be at 3125 Hz. Using the highest division factor of the programmable divider and working backward to the output of the VCO gives 3125 Hz $\times$ 2158 $\times$ 16 = 107.9 MHz at the VCO output (channel 300). Jumper cutting and/or strapping the terminals on the programmable divider will set the division to the desired count so that the VCO can operate on any channel carrier fre-

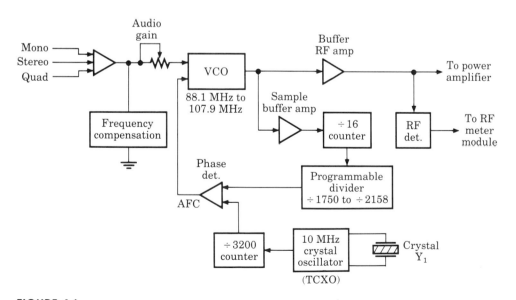

FIGURE 6.1
VCO master oscillator for FM transmitter in the carrier frequency range from 88.0 MHz to 108.0 MHz. (Courtesy of The Harris Corporation, Broadcast Division, Quincy, Illinois)

quency. Programming a division by 1750 results in a VCO output at the lowest channel, 87.5 MHz. Each time the programmed division factor is adjusted by a count of 4, the VCO will be on the carrier rest frequency of the next channel. Divide-by-1750 produces 87.5 MHz, which is out of the band. The "on" channel count starts with divide-by-1762 and sets the VCO at 88.1 MHz (channel 201). Four counts higher is divide-by-1766, which sets the carrier at 88.3 MHz (channel 202), and so on up the FM band. When the VCO generates a signal directly at the transmission carrier rest frequency, its tolerance is limited by law to ±2000 Hz.

The temperature-compensated crystal-controlled oscillator (TCXO) is guaranteed to within 2 ppm (parts per million) per year. At 10 MHz, this amounts to ±20 Hz per year. At the phase detector input, after a division by 3200, the guaranteed change is 0.00625 Hz or less. The VCO output would be controlled to within $0.00625 \times 2158 \times 16 = 215.8$ Hz, about 10% of the allowable carrier drift.

The directly controlled parameter in a phase-locked loop is not frequency, but phase. Since frequency represents the rate change of phase, by controlling the phase with a finite constant error, the frequency can be controlled with no error with respect to the reference signal. The phase-locked loop indirectly controls frequency by accurately controlling phase.

The phase detector compares the phase of the two input signals. If they are exactly in phase, the error voltage out of the phase detector will be one-half its DC supply voltage. When the phase advances, the error voltage increases. When the phase is delayed, the error voltage decreases. When the carrier signal is divided to 3125 Hz, its phase may be different than that of the 3125 Hz from the reference oscillator. If so, an error voltage will result and attempt to match the phase of the two signals. The error voltage difference, when applied to the VCO, will establish the oscillator frequency. An error voltage change of 0.1 V DC will shift the VCO frequency by 200 kHz (one FM channel separation).

At 98.0 MHz, the geometric center frequency of the FM band, the phase comparator's error output voltage is 7.5 V DC. As the VCO frequency increases, the phase error increases, and the DC voltage out of the phase detector increases to maintain the increased frequency of the VCO. The output voltage of the phase detector is put through a low-pass filter and emerges as a relatively pure DC voltage. This DC error voltage is coupled back to the AFC (automatic frequency control) input to the VCO to maintain phase lock and hold the VCO to center frequency.

The circuit for the voltage-controlled oscillator of Figure 6.1 is detailed in Figure 6.2. Q_1 is the JFET Colpitts oscillator, Q_2 and Q_3 are a cascode RF amplifier, Q_4 regulates the load at the output of the oscillator, and Q_5 sets the DC bias voltage.

CR_1, CR_2, CR_3, and CR_4 are varactor diodes, technically defined as *voltage variable capacitors*. The DC voltage across the reverse-biased diodes institutes a value of capacitance that can be changed by simply changing the voltage across the diode. The capacitance is formed by the area of the anode and the

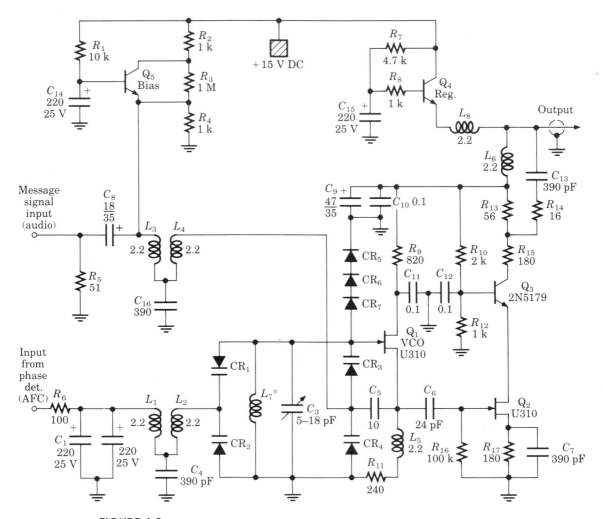

FIGURE 6.2
The circuit of the VCO in Figure 6.1. (Courtesy of The Harris Corporation, Broadcast Division, Quincy, Illinois)

area of the cathode, which are separated by a depletion layer, exactly like the base-emitter junction of a transistor. As the reverse bias is made stronger, the depletion layer widens and makes the effective capacitance decrease. The effective capacitance of the diode junction becomes a function of the voltage across the diode. The diode capacitance is part of the tuned circuit that determines the oscillator frequency. Thus, the oscillator is voltage-controlled. Two conditions must be observed at all times:

1. The diode *must* be reverse-DC-biased
2. The peak AC signal to be superimposed on the DC bias should not be capable of forward-biasing the diode.

The characteristic curve for the varactor diodes used in Figure 6.2 is shown in Figure 6.3 as a capacitance vs. reverse bias voltage curve.

The center contact between diodes CR_3 and CR_4 is connected through L_3 and L_4 to the emitter of the Q_5 bias transistor. The emitter current of Q_5 is found using the standard equation for I_e of a base-biased transistor.

$$I_e = \frac{V_{cc} - 0.7 \text{ V}}{R_c + R_e + \left(\dfrac{R_b}{\beta}\right)} = 6.97 \text{ mA}$$

The emitter voltage of Q_5 (7 V) is applied to the junction of CR_3 and CR_4. At 7 V DC, the capacitance of these diodes is 16 pF each from Figure 6.3. Therefore, the sum of CR_3 and CR_4 in series is 8 pF. C_3 at optimum tuning has a capacitance of 8.5 pF. The total capacitance of CR_3-CR_4 and C_3 in parallel is 16.5 pF.

The 50 Ω coaxial line (L_7) (denoted by an asterisk in Figure 6.2) is equivalent to an inductance of 0.1 μH (see Chapter 8). At 88.1 MHz, with L equal to 0.1 μH, the total required tuning capacitance is 32.64 pF. Subtract the values of C_3 and the combination of CR_3 and CR_4 to find the effective capacitance of CR_1 and CR_2. That is, $32.64 - 8 - 8.5 = 16.14$ pF. Because 16.14 pF applies to CR_1 and CR_2 in series, each must be $2 \times 16.14 = 32.28$ pF. From the curve of Figure 6.3, a DC voltage of 2.5 V must be supplied as AFC from the phase detector to form a capacitance of 32.28 pF and tune the VCO to 88.1 MHz.

At 107.9 MHz, the capacitance totals 21.757 pF. Subtracting the effects of C_3 plus CR_3 and CR4 leaves 5.075 pF as an effective capacitance to tune the VCO to 107.9 MHz. Since CR_3 and CR_4 are in series, each must be $2 \times 5.075 = 10.15$ pF. This value corresponds in Figure 6.3 to 12.5 V DC, which represents the AFC voltage required for tuning to 107.9 MHz.

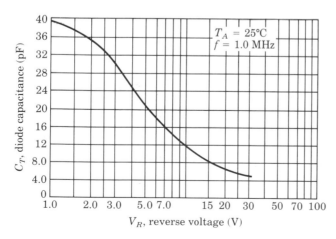

FIGURE 6.3
Capacitance vs. reverse bias voltage curve for the MV109 varactor diode.

The AFC voltage changes from 2.5 to 12.5 V DC, representing a 10 V change to tune across the FM band of 100 stations. This constitutes a 0.1 V DC AFC voltage change per station (200 kHz).

A truth table is supplied with the transmitter. During installation, the appropriate jumpers on the programmable divider are cut or installed to set the transmitter VCO to the proper frequency, where it will remain throughout its service life. The frequency is measured for accuracy and C_3 is adjusted, if needed, to ensure a correct carrier rest frequency.

6.2.1 The Reactance Modulator

The varactor diodes in the voltage-controlled oscillator of Figure 6.2 are formally known as **reactance modulators.** They are the functional devices that respond to a voltage change and convert it to a capacitance change to cause a frequency change. With the exception of the integrated chip circuit, the varactor diode (also called *varicap*) is the discrete device most commonly used as a reactance modulator.

The capacitance change over the active range of the varicap is nonlinear with frequency; however, the voltage that causes the capacitance change does change linearly with frequency. In Figure 6.2, the audio signal is applied through coupling capacitor C_1 and superimposes the AC signal onto the DC bias voltage. This causes a fluctuation in the capacitances of diodes CR_3 and CR_4, which causes a frequency change. Because the oscillator experiences a 200 kHz frequency change for each 0.1 V DC voltage change, a ratio can be set up to show how much signal is needed to fully modulate the FM carrier, that is, what voltage change will cause a 75 kHz frequency change if 0.1 V causes a 200 kHz frequency change.

$$\frac{x}{75\text{ kHz}} = \frac{0.1}{200\text{ kHz}} \qquad x = \frac{0.1(75\text{ kHz})}{200\text{ kHz}} = 37.5\text{ mV}_{\text{pk}} \qquad \textbf{(6.1)}$$

A signal voltage of 37.5 mV$_{\text{pk}}$ or 75 mV p-p will fully modulate the FM carrier in this transmitter.

Inductive or Capacitive Modulators   The reactance modulator takes its name from the fact that the impedance of the circuit acts as a reactance (capacitive or inductive) that is connected in parallel with the resonant circuit of the oscillator. The varicap can only appear as a capacitance that becomes part of the frequency-determining branch of the oscillator circuit. However, other discrete devices can appear as a capacitor or as an inductor to the oscillator, depending on how the circuit is arranged. A Colpitts oscillator uses a capacitive voltage divider as the phase-reversing feedback path and would most likely employ a modulator that looks capacitive. A Hartley oscillator uses a center-tapped coil as the phase-reversing element in the feedback loop and most commonly uses a modulator that appears inductive.

The theory behind the reactance modulator shown in Figure 6.4 is to add a phase-shifting network consisting of a series R and C. The modulator's active device and the series RC network are both in parallel with the resonant circuit of the carrier oscillator. The oscillator signal voltage that is present across the resonant tank circuit also appears across the transistor and the series circuit of R and C. The coupling capacitor C_4 and the RF bypass capacitor C_2 can be any large value to allow near-zero reactance to the oscillator frequency; as such, they are removed from consideration in this analysis.

The capacitive reactance of C at the oscillator frequency is made five to ten times larger than the value of series resistance. In this case, the 100 pF capacitor has 500 Ω of reactance at 3.5 MHz. The value of R, then, is chosen as 100 Ω. The RC circuit now appears to be purely capacitive. In a purely capacitive circuit, the current leads the voltage by 90°. Thus, the current through capacitor C leads the oscillator signal voltage across the capacitor by 90°. The same leading current flows through resistor R, where E and I are in phase, so the voltage across the resistor leads the oscillator signal voltage by 90°. The transistor collector current (I_c) is in phase with the base voltage (the voltage across R) so if E_R leads E_{osc} by 90°, then I_c also leads E_{osc} by 90°. The current through the transistor leads the voltage across the transistor by 90°, making the circuit look like a capacitor. An audio signal applied to the base of the modulator causes a change in the collector current. The vector result of changing I_c relative to E_{osc} causes a change in the value of equivalent capacitance, which in turn causes a change in the oscillator frequency.

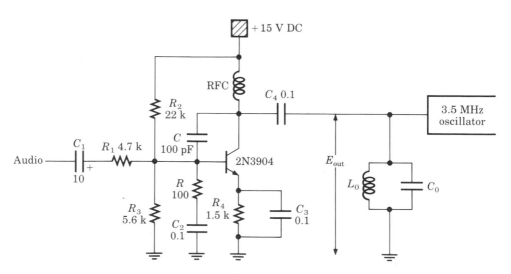

FIGURE 6.4
A reactance modulator as an equivalent value of capacitance.

This reactance circuit looks capacitive to the oscillator by an amount equal to

$$C_{\text{equiv}} = \frac{RC}{Z_{\text{in}}} \quad \text{(in farads)} \tag{6.2}$$

When a field effect transistor is used as the modulator, the same circuit configuration will result in an equivalent capacitance of

$$C_{\text{equiv}} = g_m RC \quad \text{(in farads)} \tag{6.3}$$

When it is desirable to make the modulator look *inductive,* simply reverse the positions of the R and C values; connect the resistor from collector to base and the capacitor from base to ground. This results in a current through R that is in phase with the oscillator voltage. This in-phase current also flows through the capacitor, where the capacitor voltage drop *lags* the current through it by 90°. A lagging base voltage controls a lagging collector current. When I_c lags E_{osc}, the transistor looks inductive (see Figure 6.5).

All other circuit conditions remain the same; that is, X_c is still five times larger than R at the oscillator frequency, and the coupling capacitor is any large value needed to appear as a short circuit for the oscillator signal. The resulting values of L are as follows. For the bipolar transistor,

$$L_{\text{equiv}} = \frac{Z_{\text{in}}RC}{\beta} \quad \text{(in henrys)} \tag{6.4}$$

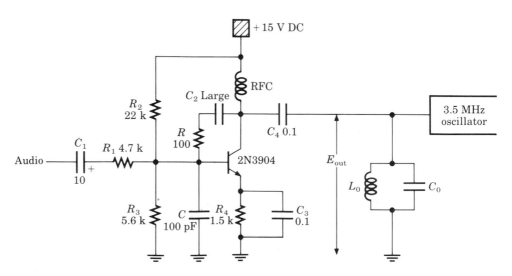

FIGURE 6.5
A reactance modulator as an equivalent value of inductance.

For field effect transistors

$$L_{\text{equiv}} = \frac{RC}{g_m} \quad \text{(in henrys)} \tag{6.5}$$

The advantages of this type of circuit compared to the varactor are stability and the alternate modes of capacitive and inductive behavior. The disadvantage is that the circuit is more complex than the varactor. Also, the values of R and C are exceptionally small, which makes the input impedance of the modulator very small. The small value of the series capacitor also means that the carrier oscillator cannot operate at a very high frequency. When the frequency goes above about 5 MHz, C becomes quite small, R becomes *very* small, and the input impedance becomes too small to be workable. This also prompts the need for frequency multipliers to raise the carrier frequency to the broadcast band frequencies.

6.3   FREQUENCY CHANGERS

The categories of frequency changers will be defined first, since we will encounter several different types and the type names have been abused over the years. The five categories are

1. Mixers
2. Modulators
3. Converters or translators
4. Dividers
5. Multipliers

6.3.1 Mixers

A mixer is a simple *linear* adder or summer. The number of input signals must be greater than one, and the output will be the undistorted sum of the input voltages at their respective shapes and frequencies.

6.3.2 Modulators

The modulator usually falls under the heading of a *nonlinear* wave shape or wave frequency changer. Two input signals are always required, where one signal effects a change in the size, frequency, or phase of the other signal. The output of the modulator will be the two original signal frequencies plus their sum and difference frequencies. The voltage level at each frequency is determined by the form of modulation. Frequency-selective filters may be required at the output.

6.3.3 Converters or Translators

Converters are nonlinear frequency shifters. Two input signals are required in all cases. The converter takes one of the input signals and shifts it, intact, to a different frequency range determined by the frequency of the second signal. The signal to be shifted may contain a group (or band) of frequencies. There is no loss in the meaning of the signal that has been shifted. The output is similar to that of the modulator: it contains the two original signal frequencies plus the sum and difference frequencies. Frequency-selective filters are required at the output to recover only the desired band. Converters are regularly (but mistakenly) called mixers, intended, of course, to mean RF mixers. Converters may up-convert or down-convert the desired signal frequency.

6.3.4 Dividers

Frequency dividers may be of the multivibrator type or of the relaxation oscillator variety, which depend on a trigger pulse to change states and, as such, are nonlinear down-converters. J-K flip-flop circuits are common digital down-converters. A divide-by-$(7 \times 5 \times 5 \times 3)$ relaxation oscillator circuit is shown in Figure 6.6. This relaxation divider is taken from the sync generator of a television transmitter, where twice the horizontal frequency is divided down to attain a synchronized vertical rate of 60 Hz. The divider is adjustable from divide-by-90 to divide-by-2200.

6.3.5 Multipliers

Multipliers are nonlinear frequency up-converters. They require only one input signal and multiply *all* frequency components of the input signal equally

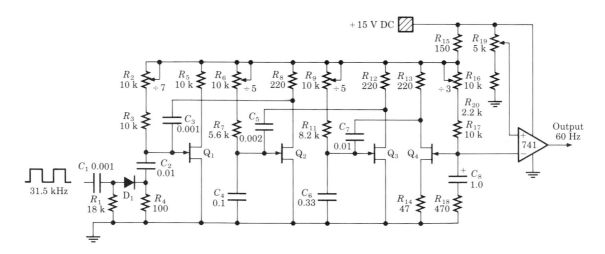

FIGURE 6.6
Frequency divider used in television broadcasting as the divide-by-525 circuit in the sync generator. Variable range: divide-by-90 to divide-by-2200.

by a known constant. The constant must be a whole number; it cannot be a decimal part of a number. Typical constants are 2, 3, and 4; on rare occasions, 5X and 7X multipliers are used. Frequency multipliers use a nonlinear device to change the signal to one rich in harmonics, such as a pulse wave; the wave is then filtered to recover the desired by-products. Such a system appears in Figure 6.7, which shows a tuned-input, tuned-output, class B or C RF amplifier.

This system departs from standard amplifier design in that the output resonant circuit is tuned to a multiple of the input signal and circuit frequencies. The multiplication constant is set by the resonant frequencies of the tuned circuits, the conduction angle of the active device, the amplitude of the input drive signal, and the external bias (if used). Table 6.1 is developed from typical multiplier measurements and indicates the range for the conduction angles of the amplifiers. The efficiencies are close approximations relative to the performance of the same amplifier operating at its fundamental frequency. Usually, the conduction angle is selected from a table similar to this one and used in Equation 6.6 to find the peak signal voltage.

$$V_{\mathrm{pk}} = \frac{V_{\mathrm{sat}} + V_{b\text{-}e}}{1 - \cos\left(\dfrac{\theta}{2}\right)} \tag{6.6}$$

where V_{sat} = collector-emitter voltage at $I_{c\,(\mathrm{max})}$
$V_{b\text{-}e}$ = base-emitter junction voltage (0.7 or 0.3 V)
θ = conduction angle from Table 6.1

FIGURE 6.7
A tuned-input, tuned-output RF frequency doubler with gain.

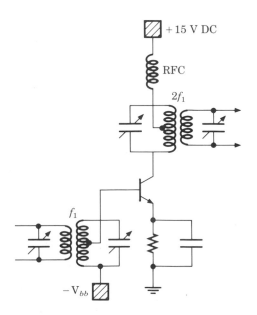

TABLE 6.1

Multiplier	Conduction Angle	Efficiency
Doubler	110–120	64%
Tripler	80–90	47%
Quadrupler	60–70	36%
Quintupler	45–55	27%

A transistor curve and load line are needed to find the value of V_{sat}. A close approximation sets V_{sat} at 5% of V_{cc}. When a fixed bias is used, then V_{sat} is replaced by the magnitude of the fixed bias voltage. Using this approximation, Equation 6.6 becomes

$$V_{\text{pk}} = \frac{0.05\ V_{cc} + V_{b\text{-}e}}{1 - \cos\left(\dfrac{\theta}{2}\right)} \tag{6.7}$$

For $\theta = 110°$ (frequency doubler), $V_{cc} = 15$ V, $V_{b\text{-}e} = 0.7$ V, and the peak signal input voltage is 3.4 V_{pk}. For a quintupler in the same circuit, the input signal would need to be 8.6 V_{pk}.

The circuit voltages may be known, in which case it becomes necessary to find the conduction angle. Equation 6.7 can be rearranged as follows:

$$\theta = 2\ \cos^{-1}\left[1 - \left(\frac{0.05\ V_{cc} + V_{b\text{-}e}}{V_{\text{pk}}}\right)\right] \tag{6.8}$$

For $V_{cc} = 12$ V DC, $V_{b\text{-}e} = 0.7$ V, and a peak signal of 10.75 V, the conduction angle would be 57°. This would constitute a frequency quintupler, where the output resonant circuit is tuned to five times the frequency of the input resonant circuit. The peak signal input could be reduced to lower the multiplication factor. These equations are based on the typical class C self-biasing circuits.

A *push-pull* class B amplifier can be used as a frequency multiplier at all odd-numbered multiples, since a push-pull amplifier cancels the even harmonics. Good efficiencies have been obtained to the fifth and seventh harmonics. The output resonant circuit is tuned to the desired multiple (see Figure 6.8).

A *push-push* amplifier that cancels all odd harmonics makes a very efficient frequency doubler. The push-push amplifier is a class B amplifier with out-of-phase input signals, like the push-pull amplifier, but with the amplifier output terminals connected in parallel. Such a circuit is shown in Figure 6.9.

Diodes, especially tunnel diodes, are often used as frequency multipliers at extremely high frequencies since they are nonlinear devices (see Figure 6.10). The output resonant circuit is tuned to a multiple of the input resonant signal and circuit. The disadvantage of diode multipliers is lack of gain. Combined with the low efficiencies of multipliers, this makes the diode multiplier a very lossy circuit. Diodes have very good power-handling capabilities, which accounts for their usage. The circuit of Figure 6.10 has 5 W output for 20 W input.

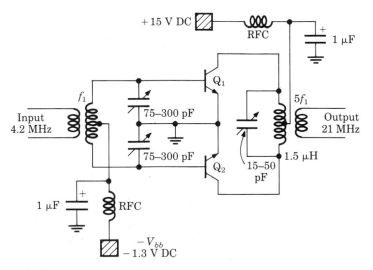

FIGURE 6.8
A class B push-pull frequency quintupler (5X) with adjustable capacitors at the input for balance.

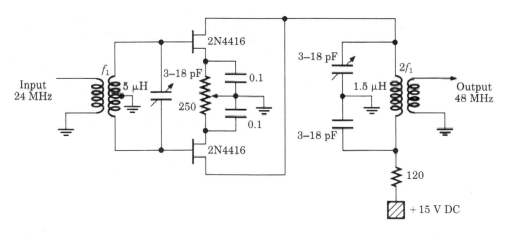

FIGURE 6.9
A class B push-push frequency doubler.

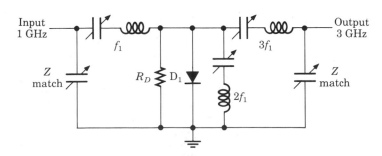

FIGURE 6.10
A diode frequency tripler in the GHz range. Input of 20 W provides 5 W output.

FIGURE 6.11

A phase-locked loop frequency synthe-sizer, used here as a frequency dou-bler. Any divide-by-*n* network could be used.

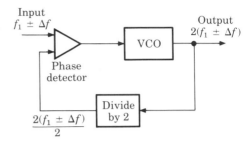

Phase-locked loops with a frequency divider in the feedback network will also multiply all of the input signal frequency components (see Figure 6.11). An FM signal of 5 MHz ±2 kHz deviation would be multiplied by 2 (in this case) to 10 MHz ±4 kHz deviation. The output signal of the VCO is supplied to the second input of the phase detector *after* it has been divided (by 2 in this case) to synchronize it with the original signal frequency. Note the deliberate omission of the low-pass filter from the error voltage line between the phase detector output and the VCO input. Any practical divider circuit will work well here: A divide-by-*n* network will multiply the output of the VCO to *n* times the input signal frequency.

Note the total absence of neutralization in any of the frequency multipli-ers circuits. Recall that three conditions must be met for an amplifier to go into parasitic oscillation. The one condition lacking in multipliers is that the tuned-input and tuned-output circuits are not at the same frequency. For this reason, multipliers do not need to be neutralized.

6.4   POWER AMPLIFIERS

The power amplifier in Figure 6.12 has been simplified in the area of the con-trol circuits for easy reading and left complete in the area of the signal circuits to identify the nature of the circuitry.

The two-stage RF class C power amplifiers are driven with a 300 mW signal at the 50 Ω input. Chebyshev wideband impedance-matching networks allow full 88–108 MHz coverage with no amplifier tuning. The input matching to the base of Q_3 is by capacitors C_{35} and C_{36} and a series section of microstrip transmission line (see Chapter 8). The RF driver Q_3 amplifies the signal to approximately 2 W. The matching network, consisting of capacitors C_{13} through C_{17} and the associated sections of microstrip line, match the collector impedance of Q_3 to the base impedance of Q_4 over the entire FM band. An output impedance of 50 Ω is developed by capacitor C_{22} and the output section of microstrip line, to deliver the 15 W maximum output into a 50 Ω load.

The RF output is through a directional coupler and a low-pass filter implemented with microstrip line techniques. C_{32} and its series section of

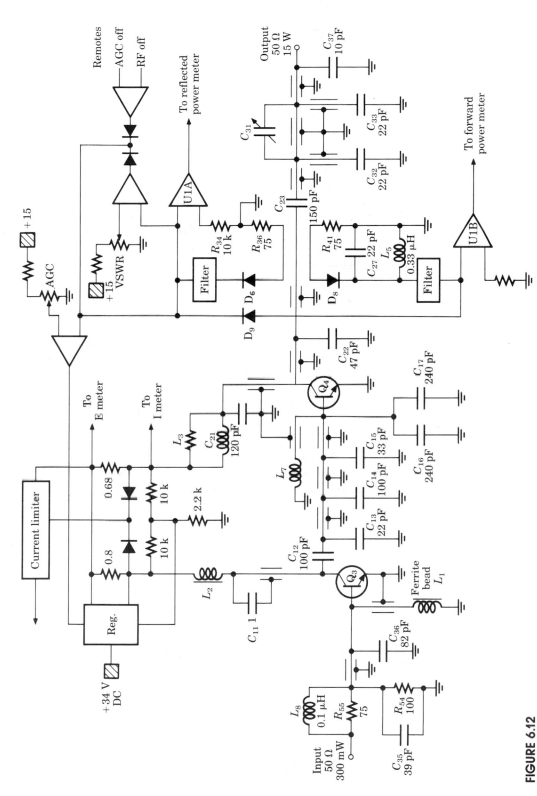

FIGURE 6.12

A 15 W power amplifier, for the frequency range of 88 MHz to 108 MHz, and control circuits. Note the extensive use of microstrip transmission line. (Courtesy of the Harris Corp. Broadcast Division, Quincy, Ill.)

microstrip trap the third harmonic, while C_{33} and its series section of microstrip trap the second harmonic, aided by C_{31}.

The directional coupler continuously monitors the RF forward and reflected output power. The forward power sensing is applied as one of the inputs to the AGC system and through amplifier U1B to the RF power-metering system. The reflected power sensing is coupled as the second input to the AGC system and through amplifier U1A to the RF power-metering system. The AGC system increases and decreases the DC supply voltage (not shown in this drawing) to the power amplifiers, as required to maintain constant power output, and will shut the system down should some self-destructive situation arise. No damage will occur under any conditions if the voltage standing wave ratio (VSWR) adjustment is correctly maintained. Monitoring of the forward and reflected power output, the level of VSWR, current limiting, DC voltage, and DC current drain, as well as temperature control, are functions built into the power amplifier as standard equipment. Due to basic microstrip principles, perfect directional coupling cannot be achieved, which limits the "zeroing" of the reflected power directional coupler.

A versatile alternative approach to power amplifiers is shown in Figure 6.13. Here the output from the exciter is applied to an RF amplifier such as the 15 W amplifier of Figure 6.12, which is then used to drive any number of identical final high-power amplifiers connected in parallel. In this manner, the one reliable low-power system could become a stand-alone low-power transmitter or the low-power section of a very high power transmitter. The design, inventory, field maintenance, product reliability monitoring, and parts stock could all be minimized with this system.

Review Section 2.3.1 on class C power amplifiers and see how much you can infer about the Q_4 class C amplifier of Figure 6.12.

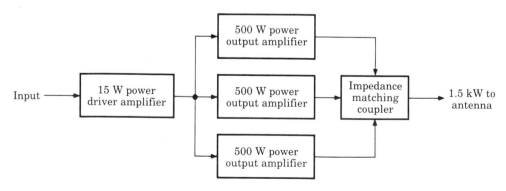

FIGURE 6.13
A low-power amplifier used to drive any number of higher-power final amplifiers.

6.5 FEEDBACK FREQUENCY CONTROL

Many direct-modulated FM transmitters are not sufficiently advanced to use the PLL carrier oscillator system. This does not negate the need for frequency control of the carrier oscillator.

One method used to crystal-control the carrier oscillator is shown in Figure 6.14 and is the forerunner to the PLL circuit. The carrier oscillator signal is picked up somewhere after the oscillator circuit and fed into a converter. This pickup point could have been (and often is) just before the final power amplifier stage (see Figure 6.14b). Any point is all right *except* right at the oscillator output. The sample signal is heterodyned with a signal from the crystal-controlled oscillator to result in the frequency difference signal. Figure 6.14a picks up the signal after the second frequency doubler at a frequency of 16.35 MHz, to be heterodyned with the 10 MHz signal from the crystal-controlled oscillator.

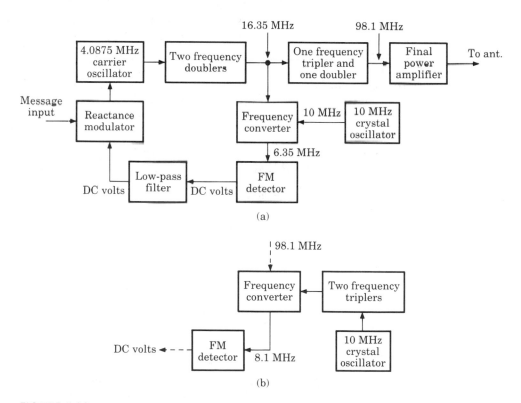

FIGURE 6.14

Feedback frequency control of a direct FM transmitter. (a) Fewer circuits and less sensitivity. (b) More sensitive system requires more circuits.

The difference frequency of 6.35 MHz will be found at the output of the frequency converter. The FM detector may be set so that a DC reference voltage (positive or negative) is established at the output. This same condition holds for the output of an AM detector, covered in Chapter 4, except that here the DC voltage will respond to the carrier frequency changes rather than to amplitude changes. The changing DC output voltage is filtered though a very low pass filter and used as a control voltage for the reactance modulator. The feedback voltage adjusts the reactance modulator to change the oscillator frequency in the proper direction to return it to the station's carrier rest frequency.

Because the feedback network senses changes in frequency, it will sense the deviation of the carrier. The very low pass filter is made to pass only those frequency changes that continue longer than a few seconds, as the carrier will do when it strays off the rest frequency. Even the lowest frequency deviation of 50 Hz would be too high to cause a noticeable change in the detector DC control voltage.

The advantage of the arrangement shown in Figure 6.14b is that with the carrier sampled at the output of the last multiplier, the frequency is six times higher than in the scheme of Figure 6.14a. Thus, the input to the detector will be six times more sensitive to carrier frequency changes. The disadvantage is that another frequency multiplier is needed to raise the crystal-controlled oscillator from 10 MHz to 90 MHz to put the converter output at a frequency low enough for the FM detector sensitivity.

6.6   THE PHASE MODULATOR

Phase modulation is defined as treatment of the carrier signal after it leaves the carrier oscillator. In Figure 6.15, transistor Q_1 presents an impedance to ground that forms a phase shift network with capacitor C_1. The DC collector current of Q_1 is

$$I_c = \frac{V_{cc} - 0.7}{R_c + R_e + \left(\frac{R_b}{\beta}\right)} = 4 \text{ mA}$$

The collector-to-ground DC voltage is then

$$V_c = V_{cc} - (I_c)\,(R_c) = 24 - (0.004)\,(5000) = 4 \text{ V DC}$$

A circuit that has 4 V DC across it and 4 mA of current through it looks like a 1000 Ω resistor. The reactance of C_1 at 250 kHz is 2122 Ω. A constant phase shift of the carrier oscillator frequency signal with no audio signal applied is valued at:

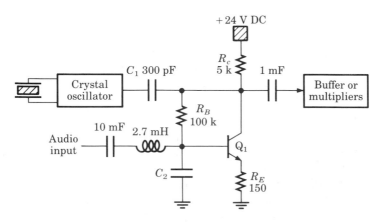

FIGURE 6.15
Transistor indirect frequency modulator.

$$\theta = \tan^{-1}\left(\frac{X_c}{R}\right) = \tan^{-1}\left(\frac{2122}{1000}\right) = 65°$$

As audio is applied, the collector current is increased by 0.5 mA (to 4.5 mA), and the voltage drop across the transistor decreases to 1.5 V DC, as found by an earlier equation. A device that has 1.5 V DC across it and 4.5 mA of current through it looks like a 333 Ω resistor. The capacitive reactance remains unchanged; therefore, the phase angle is found to be 81°, an increase of 16°.

When the audio swings in the other direction, the collector current drops to 3.5 mA (a decrease of 0.5 mA). The collector voltage is again calculated and found to be 6.5 V DC. A device with 6.5 V DC across it and 3.5 mA of current through it looks like an 1875 Ω resistor. Substituting this value into the tangent equation nets an angle of 49°, a decrease of 16°. The amount of frequency change that would result from a ±16° phase change is found by

$$f = \Delta\theta f_s \quad \text{(in Hz)} \tag{6.9}$$

where θ is expressed in radians and f_s is the lowest audio frequency used. A change of 16° is 0.279 radians, and the lowest audio frequency is 50 Hz. Therefore, the frequency change is ±13.96 Hz.

Equation 6.9 is valid for very low frequencies and is a reduced form of the standard equation:

$$f = \Delta\theta f_s \left(\frac{d\theta}{dt}\right) \tag{6.10}$$

The term $d\theta/dt$ is the time rate of change of the angle θ. This means that as the audio frequency increases, the amount of frequency modulation also increases at a linear rate. Increasing the audio frequency ten times would increase the carrier frequency change to ± 139.6 Hz. To compensate for this change, a low-pass filter called a "$1/f$ network" is added to the input of the phase modulator (in place of a deemphasis network) to reduce the audio frequency at a rate of 6 dB per octave. A reduction of 6 dB for each time the audio frequency is doubled amounts to about -49 dB attenuation at the highest audio frequency of 15 kHz.

In the circuit of Figure 6.15, the collector current was found to change by 0.5 mA. A β of 150 means a base current change of 0.5 mA/150, or 3.3 μA. The voltage required to change the current of 3.3 μA through the base resistor of 100 Ω is 330 mV peak voltage. There is no indication of whether the carrier frequency change resulted in 100% modulation at the antenna. The next step would be to make that determination by examining the number and type of frequency multipliers used and identifying the transmitted carrier rest frequency and permitted deviation for the type of service for which this transmitter is intended.

6.7   CARRIER NULL

Examine the trace for the carrier (J_0 term) on the modulation index graph of Figure 5.7. As the modulation index advances from zero, where the carrier is $+1.00$, note that the value of the carrier decreases. The carrier goes to zero at a modulation index of 2.4048. The carrier has a negative value for a period of time, again passing through zero at 5.5201. The zero crossings occur only for specific values of modulation index, called **Eigen values.** A list of the Eigen values appears in Table 6.2.

The Eigen values can be a useful tool for testing and measuring FM transmitter performance. Connect a spectrum analyzer across the output dummy load of the transmitter and an audio generator with a frequency counter at the monaural input of the transmitter. To set up a 30% modulated FM carrier, multiply the modulation factor by the maximum deviation for the system under test: 75 kHz $\times$ 0.3 = 22.5 kHz. From Equation 5.2, for the modulation index, select an Eigen value corresponding to an audio frequency below the -3 dB frequency on the preemphasis curve (2122 Hz). This example would require the fourth Eigen value, 11.7915. The maximum deviation (22.5 kHz) divided by the null value (11.7915) leads to an audio frequency of 1908.15 Hz. Set the audio generator to exactly 1908.15 Hz using the frequency counter. Then, while observing the spectrum analyzer display, slowly increase the audio voltage from zero to a level that shows the carrier voltage going to zero for the fourth time on the analyzer display. The transmitter is now at a reliable 30% modulation value and ready for other measurements and tests. A measure of true 100% modulation can be set up in a similar fashion.

TABLE 6.2

Order of carrier null	Eigen value of m_f
1	2.4048
2	5.5201
3	8.6531
4	11.7915
5	14.930
7	18.0711
8	21.2116
9	24.3525
10	27.4935

6.8   TRANSMITTER ADJUSTMENTS

Always start at the carrier oscillator and proceed toward the output at the antenna terminals. Make sure that the antenna has been replaced with a non-inductive dummy load resistor of equal resistance and equal or greater wattage to that presented by the antenna to the transmitter under normal load conditions. In series with the dummy load resistor, place a current meter that has ranging to cover the full antenna current.

Remove or greatly reduce the DC supply voltages to all stages. Be prepared to restore normal DC voltage to one stage at a time. When tubes are used, the heater voltage remains at normal rating at all times during these tests. See Figure 6.16 for a generalized approach.

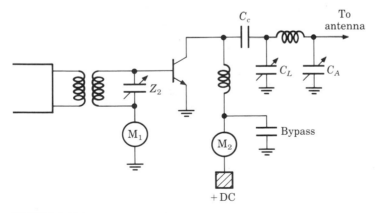

FIGURE 6.16
Tuning the final power amplifier of the transmitter.

1. With reduced DC voltage, adjust the oscillator circuit to its exact frequency using a calibrated frequency counter with NO modulation applied. Adjust the coupling tuned circuits between the oscillator and the following stage. Apply full DC voltage to the oscillator only, and repeat the above procedure.

2. Follow a similar procedure for all of the amplifiers up to, but not including, the final power amplifier(s). Where frequency multipliers are involved, ensure the correct frequency exchange. At reduced DC voltages, tune the output resonant circuits first, adjust the input resonant circuits second, apply full DC voltage, and repeat the steps. If there is a degree-of-coupling adjustment between stages, this should be done last and set for the signal level called for by the manufacturer.

3. The final stage is the most critical. If the final stage includes neutralization, refer to Chapter 2 for the adjustment procedure and do this first. Next, *badly detune* the output side of impedance Z_2 (at the input to the final) and *both* of the tuning elements between the final amplifier and the antenna terminals. Use a DC current meter in series with the DC supply to the final and a second meter to measure DC base current. (This assumes that the transmitter does not have permanent meters installed in the circuit.) Adjust C_L for minimum reading on the supply DC current meter. Adjust the output of Z_2 for maximum current on the base DC current meter, and readjust C_L for minimum DC current. Adjust the antenna matching capacitance C_A until the DC supply meter increases by about 25%, and adjust C_L again for the lowest DC current reading.

Apply full DC supply voltage. Repeat the three adjustments until the base current is at its maximum, the DC supply current cannot be reduced further by C_L, and adjustment of C_A causes no further increase in the DC supply current.

At this point, the transmitter is fully adjusted and ready for all other measurements. Further testing may cover percentage of modulation, modulation linearity, power output, harmonic distortion, and all of the audio tests associated with proof-of-performance testing.

QUESTIONS

1. What characteristics of the carrier oscillator signal are critically guarded? (a) frequency, (b) oscillator load, (c) symmetry, (d) amplitude, (e) all of the above.

2. A square wave signal will be restored to a fundamental sine wave frequency when it is (a) integrated, (b) passed through a tuned circuit, (c) differentiated, (d) modulated, (e) none of the above.

3. The programmable divider in Figure 6.2 would need to divide by _____ for the carrier oscillator to be set on channel 221 at 92.1 MHz. (a) 1842, (b) 29,472, (c) 3125, (d) 16.

4. What component determines the carrier oscillator frequency stability in Figure 6.2? (a) the crystal, (b) the VCO, (c) the programmable divider, (d) the error voltage, (e) the RC timing components of the VCO.

5. State the carrier rest frequency tolerance at the antenna of a station in the commercial FM radio band.

6. The automatic frequency control voltage of the transmitter VCO is a (a) sine wave voltage, (b) square wave voltage, (c) ramp voltage, (d) DC voltage, (e) none of the above.

7. List two conditions that must be observed at all times regarding the operation of varactor diodes.

8. Define *symmetry* and *linearity,* and identify the difference.

9. The capacitance of a varactor diode changes nonlinearly with frequency. The capacity of a varactor diode changes nonlinearly with the DC voltage across the diode. The frequency of the tuned circuit that uses a varactor diode changes _____ for a linear change in the DC control voltage. (a) linearly, (b) nonlinearly.

10. A discrete reactance modulator circuit presents _____ reactance to the oscillator circuit in which it is working. (a) an inductive, (b) a capacitive, (c) either inductive or capacitive, (d) both inductive and capacitive, (e) neither inductive nor capacitive.

11. To make a circuit look capacitive, the current through the circuit should _____ the voltage across the circuit by 90°. (a) lead, (b) lag, (c) be in phase.

12. What is the typical reactance to resistance ratio in a reactance modulator circuit? (a) $X_o = R$, (b) $X_c = 5R$, (c) $R = 5X_c$, (d) $C = R$, (e) $C = 5R$.

13. Why is the reactance to resistance ratio *not* equal to 1 in a reactance modulator circuit?

14. The equivalant reactance of the reactance modulator circuit is connected (a) in parallel with the frequency-determining elements of the oscillator, (b) as the biasing circuit for the oscillator, (c) as the load impedance for the oscillator, (d) in the feedback loop of the oscillator.

15. A reactance modulator that performs like a capacitive reactance can be made to appear as an inductive reactance by (a) interchanging the locations of X_c and R in the circuit, (b) replacing X_c with X_L, (c) inverting the X-to-R ratio, (d) The circuit cannot be changed.

16. What frequency is used to determine the reactance X_c or X_L?

17. List the frequency changers from Section 6.3 that are linear devices and state the minimum number of input signals required in each case.

18. List the frequency changers from Section 6.3 that are nonlinear devices and state the minimum number of input signals required in each case.

19. A frequency multiplier that conducts for 85° of the input cycle has the correct angle for a frequency (a) doubler, (b) tripler, (c) quadrupler, (d) quintupler.

20. What is the *peak* drive signal for a frequency tripler that uses a silicon transistor when operating from a 24 V DC supply?

21. Determine the conduction angle for a silicon frequency multiplier with 4.5 V peak input signal at 5 MHz when powered from an 18 V DC power supply.

22. To what frequency will the output circuit in Question 21 be tuned to be resonant at the correct multiple?

23. All frequency multipliers that use a single active device are operated in class (a) A, (b) B, (c) C, (d) D, (e) E.

24. A push-pull frequency multiplier is operated as a class _____ amplifier. (a) A, (b) B, (c) C, (d) D, (e) E.

25. A push-push frequency multiplier is a frequency (a) doubler, (b) tripler, (c) quadrupler.

26. A push-push frequency multiplier will cancel all odd frequency harmonics. (T) (F)

27. What characteristic of a diode makes it a candidate for frequency multiplication?

28. What is the advantage of using a diode as a frequency multiplier?

29. What is the disadvantage of using a diode as a frequency multiplier?

30. A $2\frac{1}{2}X$ frequency multiplier would have a conduction angle of (a) 100°, (b) 85°, (c) 65°, (d) 50°, (e) Multiplication is by integers only.

31. Why do frequency multipliers *not* need to be neutralized?

32. Crystal control of the carrier oscillator in a wideband direct FM transmitter is achieved by using a crystal in the feedback control loop. (T) (F)

CHAPTER SEVEN

FM RECEIVERS

7.1 INTRODUCTION

It is amazing how closely the block diagram of an FM receiver resembles the diagram of an AM receiver. The FM block diagram of Figure 7.1 is that of a typical superheterodyne receiver and represents practically all of the receivers in use today. If you are comfortable in your understanding of the block diagram for the AM receiver of Chapter 4, then the FM diagram should be an easy conquest.

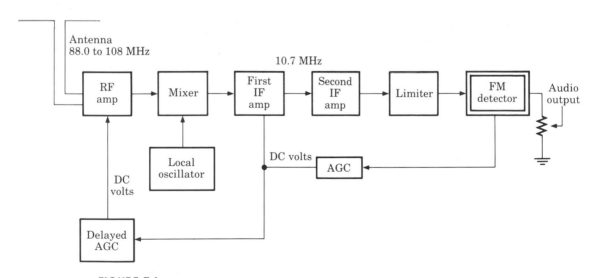

FIGURE 7.1
The block diagram of a commercial broadcast FM radio receiver.

The first circuit to examine in any receiver block diagram is the **detector.** This circuit (or block) will tell the most about the kind of receiver shown in the diagram. The second most revealing characteristic is the intermediate frequency. The most popular choice for the IF in high-frequency radio receivers is 10.7 MHz regardless of the band in which the receiver operates. This frequency was chosen to place the image frequency outside the broadcast band (see Chapter 4 for a review of image frequencies). It is extremely rare to find a receiver in the range of frequencies above 10 MHz without an RF amplifier. Of course, the FM radio broadcast band, at 88 to 108 MHz, and the public service and utility bands, at 150 MHz and 470 MHz, respectively, all fall above this limit. Note that the antenna symbol in the diagram is different than the one used for AM. This symbol is common for FM receivers but is not a standard. Noise is a prime consideration in RF circuits, although this is not a condition that can be studied from a block diagram or even a schematic.

The receiver circuit functions for FM are akin to those for AM. That is, the RF amplifier must still furnish gain at a time when the signal is relatively noise-free, it aids in the frequency selection process, and it isolates the antenna from the local oscillator. The local oscillator supplies a sine wave signal to be mixed with the RF signal to reduce the carrier frequency to the intermediate frequency. At these high frequencies, the local oscillator may operate at a frequency higher *or* lower than the incoming signal frequency. The mixer uses a nonlinear circuit to combine the RF and oscillator signals. The mixer output contains four frequencies: the two original signal input frequencies plus the sum and difference frequencies. The IF amplifiers select the difference frequency and introduce a gain equivalent to about 66% of the total system gain. The IF amplifiers also aid in the selectivity process. In FM receivers, a **limiter** may be found following the IF amplifiers, before the detector, and is a positive indication that the receiver is for FM only. The limiter clips excessive amplitude changes, such as large noise spikes. The detector that follows the IF amplifiers (and limiter) will bear the name *discriminator, ratio detector* or *phase-locked loop* to indicate that the signal to be detected is FM. The AGC circuit is exactly the same as for AM receivers, but with the variation of delayed AGC. The AGC delay is a *level* delay, not a time delay. The AGC voltage must exceed a specified value before the RF amplifier AGC voltage alters the RF amplifier gain. The audio circuits are similar to those of AM and will be covered in detail later in this chapter.

7.2   THE RF AMPLIFIER

Commercial FM transmissions are radiated as horizontally polarized signals. This is why the antenna is shown in a horizontal position in the diagram. These signals travel in a "line of sight" path, which limits the distance (on earth) over

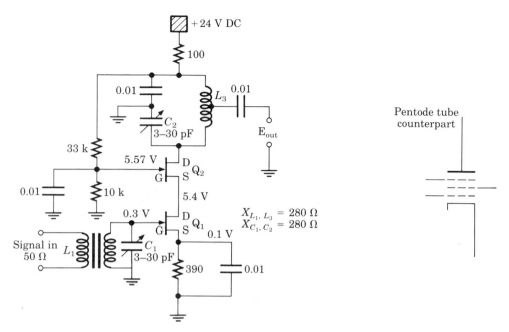

FIGURE 7.2
30 MHz cascode dual-gate MOSFET RF amplifier.

which the signal can be received, which in turn increases the need for an RF amplifier. Most high-frequency receivers employ RF amplifiers.

If an ideal RF amplifier could be made, it would have the *low noise* of a triode and the *high gain* of a pentode tube. One popular state-of-the-art version of such an amplifier is shown in Figure 7.2 and is called a **cascode** amplifier. The cascode amplifier features a common-source amplifier Q_1 (with all the traits of a triode tube) in series with a common-gate amplifier Q_2. Note how closely the cascode amplifier resembles a pentode tube circuit. The characteristics of the cascode amplifier using FETs are high input and high output impedances and Q_1 acting as a low-noise three-element amplifier. The effects of placing the gate of Q_2 at AC ground are to reduce the interelectrode capacitances and improve the high-frequency response of the amplifier. The grounded gate also acts as a shield to guard the amplifier from external noise pickup. When these two amplifiers are used in series, the circuit provides about 28 dB gain with good bandwidth and an excellent noise factor. A good dynamic range results in an RF amplifier that has the lowest DC current through the amplifier (called a starvation amplifier) with as high a load impedance as is practical yet consistent with good gain and low-distortion characteristics.

The dual-gate MOSFET of Figure 7.3 is another very popular RF amplifier configuration. Although tubes are almost obsolete, they feature many

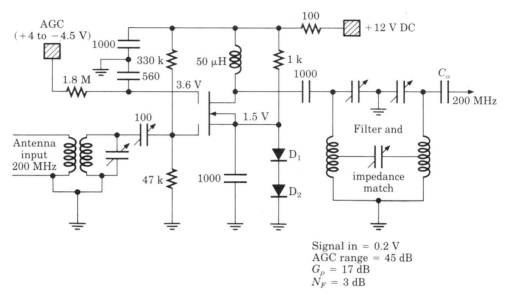

FIGURE 7.3
Dual-gate FET RF amplifier with AGC applications.

beneficial high-frequency characteristics that are imitated by the solid state replacements, such as high input impedance, low noise, and small interelectrode capacitances for high-frequency amplification. The dual-gate MOSFET in Figure 7.3 has the additional advantage that an AGC voltage can be applied to the second gate to control the gain without affecting amplifier linearity.

7.2.1 Noise in RF Amplifiers

The word *noise* has been used several times so far in this chapter to emphasize its importance in low-level amplifier circuitry. We know that every electronic circuit contaminates the signal with noise, but nowhere is noise more damaging to signal quality than in the RF amplifier. The reason, of course, is that the signal level from the antenna may be as low as 0.25 μV, and the input circuit noise will add directly to the output signal of the amplifier. Noise of 1–2 μV here could totally mask the signal. In circuits where the signal level is 100 μV or higher, 1–2 μV of noise will have very little effect. Noise originates from a variety of sources, and since a lengthy discussion of noise can fill an entire chapter, only the effects and measurements of noise will be covered here.

Noise is broken down into two basic categories: internal noise and external noise. Our major interest is the noises internal to the receiver, but both will be defined.

External Noise   Galactic noise originates outside the earth's atmosphere and is generated by all of the energy sources in our solar system. The sun is the most offensive because of its location relative to earth. Sunspots (massive atomic explosions on the sun's surface) cause electrical interference with radio transmissions and have a seven-year cycle between high and low interference levels. Galactic noise covers the frequency spectrum from about 15 MHz to about 500 MHz and decreases above this frequency range.

Atmospheric noises are those that originate within the earth's atmosphere. Electrical storms are the most offensive and can interfere with radio reception even if the lightning flashes are 1000 miles away. Electrical interference is limited to frequencies below 15 MHz and diminishes rapidly above that frequency. A given energy of interference will have greater effects closer to the equator. Man-made noises, such as electric motors, auto ignition systems, neon signs, and power lines, have their greatest effect below 15 MHz. Interference decreases rapidly above this frequency.

Internal Noise Internal noises are caused primarily by current flow through any of the electronic devices in use today. In basic electronics, DC current is considered to be a steady-state flow of electrons in the system. In reality, the exact number of electrons that pass through a resistor over a given time will differ depending on the number of collisions with other electrons and the directness of the path taken by any one electron. The difference in current flow may seem small, but it can be large compared to the signal current in some cases. This random flow of current was identified in 1928 by J.B. Johnson and is referred to as *Johnson noise*. This noise current is dependent on the temperature of the resistor and is therefore also called *thermal noise*. Thermal noise has equal energy at all frequencies of interest and was compared to the energy of light; thus, it also took on the name *white noise*. **Shot noise** is defined as the bombardment by electrons of the target element (plate, collector, drain). We would be hard-pressed to define the difference between shot noise and thermal noise. **Partition noise** results when a current path is divided into two or more different paths. The added fluctuations increase the already-present noise currents. A bipolar transistor is noiser than a diode for this reason. Thermal noise, shot noise, and partition noise are lumped together as internal noise. It is this noise we will contend with here.

Hum caused by 60 Hz/120 Hz power supply ripple also constitutes noise but is not part of this discussion because it may be eliminated by proper voltage supply filtering (decoupling) and good grounding techniques.

The noise that an amplifier contributes to the system is the **factor of noise** (*F*), which is a comparison of the input power S/N ratio to the output power S/N ratio:

$$F = \frac{P_{s(\text{in})}/P_{n(\text{in})}}{P_{s(\text{out})}/P_{n(\text{out})}} \qquad (7.1)$$

where $P_{s(in)}$ = input signal power
$P_{n(\text{in})}$ = input noise power
$P_{s(\text{out})}$ = output signal power
$P_{n(\text{out})}$ = output noise power

EXAMPLE 7.1 ━━━━━━━━━━━━━━━━━━━━━━━━━━━━━━━━━━━━━━

An amplifier has a signal input power of 5×10^{-16} W, a noise input power of 1×10^{-16} W, a signal output power of 5×10^{-12} W, and a noise output power of 4×10^{-12} W. The input-to-output S/N ratio factor F is thus

$$F = \frac{(5 \times 10^{-16})/(1 \times 10^{-16})}{(5 \times 10^{-12})/(4 \times 10^{-12})} = \frac{5}{1.25} = 4$$

━━

More often than not, the power ratio factor F is expressed in decibels, in which case it is converted to the **noise figure** (NF):

$$NF = 10 \log F \quad \text{(in decibels)} \tag{7.2}$$

In Example 7.1, NF = 10 log 4 = 6 dB.

The noise figure is a function of the random DC current fluctuations through the amplifier, and a different value of DC current reflects a different amplifier gain. This means that amplifier noise is also a function of the gain:

$$P_{n(\text{out})} = P_{n(\text{in})}GF \tag{7.3}$$

where G = power gain (not in dB). For Example 7.1, where the amplifier gain is 40 dB (10,000), the noise output power is

$$P_{n(\text{out})} = (1 \times 10^{-16})(10,000)(4) = 4 \times 10^{-12} \text{ W}$$

This is verified by the original conditions stated for the amplifier in Example 7.1. Due to the small value of $P_{n(\text{in})}$, it is often easier to calculate it through measuring the output noise power. Simple rearrangement of Equation 7.3 leads to:

$$P_{n(\text{in})} = \frac{P_{n(\text{out})}}{GF} \tag{7.4}$$

Calculation of NF may involve several cascaded amplifiers, as shown in Figure 7.4. The overall factor of noise F_{tot} in the system can be found for any number of amplifiers as follows:

$$F_{\text{tot}} = F_1 + \frac{F_2 - 1}{G_1} + \frac{F_3 - 1}{G_1 G_2} + \cdots \tag{7.5}$$

where G = power gain (not in dB)
F = factor of noise (not in dB)

EXAMPLE 7.2

Inserting the values indicated in Figure 7.4 into Equation 7.5 yields the total factor of noise F_{tot}:

$$F_{\text{tot}} = 20 + \frac{20-1}{20} + \frac{20-1}{(20)(50)} = 20 + 0.95 + 0.019 = 20.97$$

NF = 10 log 20.97 = 13.22 dB.

FIGURE 7.4

Three cascaded RF amplifiers with the same noise figure but different gains.

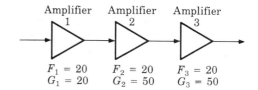

Amplifier 1   Amplifier 2   Amplifier 3

$F_1 = 20$ $F_2 = 20$ $F_3 = 20$
$G_1 = 20$ $G_2 = 50$ $G_3 = 50$

Notice that although each amplifier has the same factor of noise, it is the amplifier that processes the smallest signal that has the greatest effect on the system noise. This emphasizes again that the input RF amplifier must be the cleanest with regard to noise.

Using the noise figure to determine the performance of a high-frequency, low-noise, low-signal RF amplifier is fairly straightforward. The only precautions needed are to be careful about when to use the factor of noise F rather than the noise figure NF and when to express the gain in dB or not in dB.

Obtaining the values to be used in the calculations is a little more challenging. It has been observed that F is related to some readily obtainable information:

$$F = \frac{P_{n(\text{out})}}{KT_0BG} \tag{7.6}$$

where $P_{n(\text{out})}$ = output noise power with the input terminated
K = Boltzman's constant = 1.374×10^{-23} J/°K
T_0 = absolute temperature (°K)
B = amplifier bandwidth in Hz
G = amplifier power gain (not in dB)

An amplifier that contributes no noise would have $F = 1$. This is translated to NF = 10 log 1 = 0 dB, the *ideal* amplifier. Since such an amplifier is only theoretical, every amplifier has a noise factor greater than 1 and a noise figure greater than 0 dB.

Equation 7.6 is easily converted to a decibel noise figure:

$$\text{NF} = 10 \log \frac{P_{n(\text{out})}}{KT_0 BG} \tag{7.7}$$

From the definition of the log of a fraction, $\log A/B = \log A - \log B$. This approach may be taken with Equation 7.7 to separate the constants from the variables.

$$\text{NF} = 10 \log P_{n(\text{out})} - 10 \log GB - 10 \log KT_0 \tag{7.8}$$

The values of K and T_0 are constants, where K is taken from Equation 7.6 and T_0 is the absolute temperature in degrees Kelvin (273°) added to the ambient temperature in degrees Celsius. An ambient temperature of 75°F (24°C) added to 273°K equals 297°K. Substituting these values into the last part of Equation 7.8,

$$-10 \log KT_0 = -10 \log (1.374 \times 10^{-23})(297) = +204 \text{ dB}$$

Equation 7.8 then becomes,

$$\text{NF} = 10 \log P_{n(\text{out})} - 10 \log GB + 204 \text{ dB} \tag{7.9}$$

The values of noise output power, gain, and bandwidth depend on the statistics of the individual amplifier.

Table 7.1 sets some arbitrary boundaries for the noise quality of RF and IF amplifiers. These boundaries will change when tempered by the level of input signal in individual cases. A noise factor of 10 may be good for a 50 μV input signal but poor for a 2 μV input signal.

The input noise power sets the stage for determining the lower limit of the noise figure of the amplifier. When the two basic concepts of $P = E^2/R$ and $P = 4KT_0B'$ are combined, they provide the means to find the input noise voltage:

$$E_n = \sqrt{4KT_0 B' R_{\text{in}}} \tag{7.10}$$

This equation has been used with some success. B' represents one-half of the bandwidth; however, R_{in} is both evasive and theoretical. The bandwidth is estimated; therefore, the quantity B' is also an estimation. The value for R_{in} is the input terminating resistance of 75 Ω in parallel with the theoretical resistance R_{eq}, which is X_L/Q, where Q is the unloaded value of f_0/BW. Even under the worst conditions, R_{in} cannot exceed 75 Ω in this case. For the circuit of Figure 7.3, R_{in} is found to be less than 20 Ω. Substituting $R_{\text{in}} = 20$ Ω, $B' = 1.3125$ MHz, $T_0 = 295°$K, and $K = $ Boltzman's constant into Equation 7.10 leads to an input noise voltage of 0.64 μV.

A practical rule of thumb for determining bandwidth shows that when the interstage transformers are at critical coupling, the bandwidth (to a very close approximation) is 1% of the resonant frequency f_0. When the transformers are overcoupled, the bandwidth is 2.5% of f_0.

TABLE 7.1

Comments	Noise Factor	Noise Figure
Ideal	1.00	0 dB
Very good (RF)	1.259	1
	1.585	2
	2.000	3
Good to fair (RF)	2.51	4
	3.16	5
	3.98	6
	5.01	7
	6.30	8
	7.94	9
	10.0	10
Fair to poor (RF)	15.85	12
Good to fair (IF)	25.12	14
	39.80	16
	63.00	18
	100.0	20
Poor to very poor (RF)	158.0	22
Fair to poor (IF)	251.0	24
	316.0	25
	500.0	27
	1000.0	30

VHF and UHF amplifiers are better analyzed by using the noise figure to define amplifier performance, whereas at microwave frequencies, the term *noise temperature* identifies the noise effects better. The basics to remember are:

1. The noise factor is the ratio of the S/N power ratio at the input to the S/N power ratio at the output.
2. The noise factor will increase when *any* of the following parameters increase: gain, DC current, bandwidth, temperature, and the input resistance, which controls the input noise power.

7.3 THE LOCAL OSCILLATOR

There are very few differences in the local oscillators found in receivers that embrace the superheterodyne format of circuitry. The function of the local oscillator is to supply a sine wave signal to be used at the mixer to reduce the RF carrier frequency to the IF carrier frequency. Two limitations are placed on oscillators performing this function:

1. They must be spectrally pure, that is, low in distortion and free of harmonics. This condition helps determine the receiver quality and ease of alignment.

2. The oscillator must be frequency-stable. It should not drift off frequency due to changes in temperature or to component parts aging, and it should remain stable when component parts are changed during service.

The oscillator frequency may be higher or lower than the selected RF signal frequency by an amount equal to the IF (10.7 MHz in this case). The ratio of tuning capacitor change is about 1.5:1 to tune the oscillator from f_{max} to f_{min} in either instance, and the frequency change is small compared to the geometric center frequency of the FM band. For instance, a 0.1 μH coil will tune to 118.6 MHz with an 18 pF capacitor and to 98.8 MHz with a 26 pF capacitor. The tuning capacitor must have a range of 18–26 pF, a ratio of 1.444:1, for a frequency range of 20 MHz compared to the center frequency of 100 MHz, or 20% of the center frequency.

In FM receivers, a signal that is twice the IF (21.4 MHz) away from the selected RF signal places the image frequencies outside the FM band (88.1 + 21.4 MHz = 109.5 MHz and 107.9 − 21.4 MHz = 86.5 MHz). The local oscillator should have the ability to track the RF amplifier frequency well for a constant 10.7 MHz difference frequency.

7.4   THE MIXER AMPLIFIER

The mixer circuit of the receiver has all of the properties of a high-frequency modulator. It operates at or near cutoff on the nonlinear portion of the transconductance curve (I_c vs. V_{b-e} curve) and blends the RF carrier signal with the local oscillator signal to produce the sum and difference frequency signals at the output. The output is usually tuned to the difference frequency, but will work equally well at the sum frequency. Because of the nonlinear operating condition, the mixer is an efficient noise generator and can have a noise factor F as high as 500 with no RF input signal. As the RF is applied, the noise factor may drop to about 100 and reinforces the need for a low-noise RF amplifier. The RF carrier signal is almost always applied to the base of the mixer amplifier, whereas the oscillator signal may be applied to the base or the emitter of the mixer.

The oscillator signal voltage is typically ten times greater than the RF signal strength applied to the mixer for strong RF station signals. The ratio of the output signal voltage at the intermediate frequency to the input voltage at the RF signal frequency is called the **conversion gain:**

$$\text{Conversion gain} = 20 \log \frac{E_o(\text{IF})}{E_i(\text{RF})} \quad \text{(in dB)} \qquad \textbf{(7.11)}$$

The strong oscillator signal supports the mixer gain and reduces the mixer noise. An IF signal voltage of 10 mV out of the mixer for an RF signal input of 500 µV is a 20:1 ratio, or 26 dB. This does not account for the oscillator signal into the mixer of 5 mV (for 10 mV output), which indicates a gain of only 2. The difference stems from the fact that the oscillator signal is ten times larger than the RF signal. Conversion gain is thus a ratio of input signal strengths plus the amplifier gain rather than a measure of true amplifier gain.

7.5   IF AMPLIFIERS

Wide bandwidth, a sharp cutoff frequency response curve, and high gain with stability are the characteristics of the 10.7 MHz IF amplifiers.

One hundred percent deviation of the FM carrier signal is ±75 kHz; therefore, the bandwidth of the IF amplifiers should (after application of the shrinkage factor) be at least 150 kHz. Overcoupling in the interstage transformers helps to execute the steep response curve shown in Figure 7.5. The peaks of the response curve above and below the center frequency are located at

$$f_{\text{off-res}} = \frac{f_0}{1 \pm k} \tag{7.12}$$

where $f_{\text{off-res}}$ = frequency off resonance
f_0 = resonant (center) frequency
k = coefficient of coupling

The response curve of Figure 7.5 peaks at ±50 kHz off resonance when the coefficient is 0.005, and rolls off to the −3 dB points at ±75 kHz. The dip at the center frequency should not exceed 10% of the maximum signal level, or else the sound of the recovered audio will be distorted and mushy on loud passages of high-frequency tones.

The steep rolloff characteristics are equivalent to a circuit with a Q of about 500 when the dropoff rate is measured and the values substituted into

FIGURE 7.5
Typical overcoupled IF response curve for FM receivers.

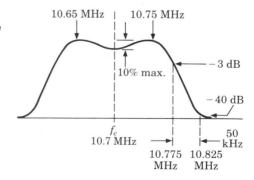

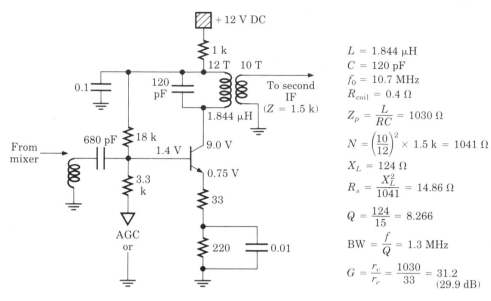

FIGURE 7.6
A typical 10.7 MHz IF amplifier for an FM radio receiver.

Equation 3.1 for relative Q, that is, a drop of 40 dB over a frequency change of 50 kHz at a center frequency of 10.7 MHz. The actual circuit Q is dramatically lower than 500 (closer to 10), but the steep response skirts make this circuit similar to one with a Q of 500. This severe rolloff will attenuate unwanted signals outside the band of interest and, in effect, enhance the receiver's selectivity.

The IF amplifier in Figure 7.6 has a voltage gain of 30 dB, $(E_o/E_i = 31.6)$, but because of the low coupling coefficient in the transformers, a 6 dB loss is realized at the output, making the gain per stage 24 dB or 15.85. Two identical stages would be used here for a gain of 48 dB (250), which will account for 60% of the total receiver gain.

Many of the circuit calculations have been worked out in Figure 7.6 and will not be repeated here (review Chapter 4, if necessary). Note the similarity of this circuit configuration and all of the other tuned RF and IF amplifiers shown thus far. If the circuit values were not included in the diagram, it would be nearly impossible to discern the differences.

7.6   LIMITERS

In frequency modulation, the signal amplitude is held constant while the carrier frequency is varied. Any noise that contaminates the signal will manifest itself as a change in amplitude. Two simple approaches used to remove these amplitude changes are shown in Figure 7.7.

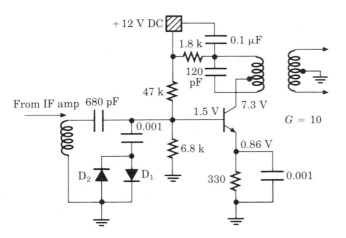

FIGURE 7.7
Two forms of limiting: D_1 and D_2 as low-level limiters and an amplifier driven into satuation and cutoff for high-level limiting.

The first limiter is a pair of back-to-back diodes D_1 and D_2. Any small-signal silicon diode will work well here. Diode D_1 will conduct when the input signal is greater than 0.7 V on the positive peak, and diode D_2 will conduct on the portion of the negative half cycle that exceeds -0.7 V_{pk} of the input signal. Even if a 20 V peak-to-peak signal were coupled through the input transform-er, the signal arriving at the base of the transistor would never exceed 1.4 V p-p.

The second form of limiting in the figure is the transistor amplifier itself, which has a gain of 10. When the base signal reaches 1.4 V p-p, the collector voltage is pressured to become ten times larger (14 V p-p). The half cycle that increases the collector voltage pushes the collector positive until it reaches the supply voltage of +12 V and then stops. The collector voltage changing from 7.3 V to 12 V is a positive change of only 4.7 V (rather than 7 V). When the collector is driven lower than 7.3 V DC, the collector and emitter currents increase, raising the emitter voltage at the same time that the collector is going lower. The collector voltage will not drop below 2.6 V DC, again a 4.7 V change. The total collector change is 9.4 V, limiting the output signal to 9.4 V p-p instead of the anticipated 14 V p-p. Simply stated, when a transistor amplifier is alternately driven into saturation and cutoff, it limits the signal ampli-tude.

7.7   **FM DETECTORS**

There are four versions of FM detectors that are noteworthy: the slope detector, the phase discriminator, the ratio detector, and the phase-locked loop.

7.7.1 The Slope Detector

When Major E.H. Armstrong developed the FM system, he needed a detector as well as a transmitter in order to verify and demonstrate his system. Although the **slope detector** is no longer commonly used, its simple and obvious operation allows an easier understanding of the detectors to follow. The slope detector makes use of a single tuned circuit that is tuned away from the carrier frequency by a small amount. Such a circuit is shown in Figure 7.8a, with the carrier frequency at 10.7 MHz and the resonant frequency of the tuned circuit at 10.8 MHz. The carrier signal at rest is located halfway up the side of the response curve, and the output voltage is only one-half of the maximum possible output voltage. Figure 7.8b shows that when the carrier deviation shifts the carrier *up* in frequency $(+\Delta f_c)$, the signal frequency moves up the response curve, causing a larger output voltage. When the carrier deviation shifts the carrier *down* in frequency $(-\Delta f_c)$, the output frequency moves down the response curve, causing a reduction in the output voltage. This part of the slope detector is a *frequency-to-amplitude* converter. Note that the output wave is still frequency-modulated, and that now it also has changes in amplitude. These amplitude changes follow the shape of the message signal that originally caused the modulation of the wave at the transmitter. All that is needed now to recover the modulation envelope is a simple diode detector and filter network, as shown in Figure 7.8c.

You might conclude that the slope detector of Figure 7.8c is identical to a simple AM diode detector, and you would be right. The only difference is that the transformer tuned circuit is tuned *off resonance*. Any AM detector can be made to detect FM in this fashion.

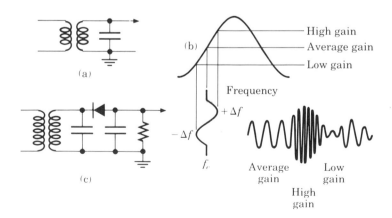

FIGURE 7.8
Output response of the slope detector. A change in frequency results in a change in amplitude.

The slope detector is simple, low-cost, and it works. The major disadvantage is nonlinearity. Only a small portion of the frequency response curve is linear, and this causes high distortion in the output signal.

7.7.2 Dual Slope Detector

Foster and Seeley recognized the poor linearity of the slope detector and reasoned that two slope detectors of opposite phase would cancel the nonlinearity of a single slope detector. This is shown in Figure 7.9, which has the two slope detectors tuned to different frequencies and connected to have the output voltages across R_1 and R_2 of opposite phase compared to ground. When the carrier signal is at its rest frequency of 10.7 MHz, the voltage appearing across *each* secondary is one-half of its maximum possible output voltage. Diodes D_1 and D_2 conduct equally, so that the voltage across R_1 is positive with respect to ground and equal to the voltage across R_2, which is negative with respect to ground. The voltages are equal and opposite and will thus cancel. The resulting output voltage is zero.

This condition is plotted in Figure 7.9b. As the discussion continues, follow the curve to check the detector's response for all other frequencies. When the carrier deviation shifts up in frequency $(+\Delta f_c)$, E_1 increases, and the signal frequency moves up its tuned response curve. Diode D_1's conduction increases, causing a larger positive voltage drop across R_1. The carrier frequency shift is away from the 10.6 MHz secondary, so E_2 decreases. The conduction of D_2 decreases, causing a smaller negative voltage drop across R_2. The output voltage, the sum of voltage drops across R_1 and R_2, is now some positive value.

As the carrier deviation shifts down in frequency $(-\Delta f_c)$, the opposite effect holds true. The conduction of D_1 is less, the positive voltage drop across R_1 decreases, the conduction of D_2 increases, the negative voltage drop across R_2 increases, and the sum of the voltages across R_1 and R_2 is now some

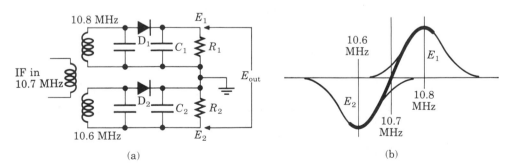

(a) (b)

FIGURE 7.9
Dual slope detectors. (a) The circuit. (b) The "S" curve that results from the inverted summing.

negative value. The effective sum of the response curves (the heavy line in Figure 7.9b) traces the "S" curve that is specified in all FM detector alignment instructions.

7.7.3 The Discriminator

Figure 7.10 shows the final circuit developed by Foster and Seeley for the phase discriminator. Note the addition of a third voltage supplied through coupling capacitor C_c to appear as a voltage across L_3. E_1 and E_2 are 180° out of phase, and both voltages are split by E_3, which lags E_1 by 90° and leads E_2 by 90°. The dual slope detector of the phase discriminator in Figure 7.10a resembles two low-voltage rectifier circuits, except they operate at a higher frequency, 10.7 MHz. At resonance, the AC voltage applied to D_1 is the vector sum of E_1 and E_3, and the AC voltage applied to D_2 is the vector sum of E_2 and E_3. The phase difference between E_{D1} and E_{D2} is of no consequence since the AC voltages will be rectified to result in a DC voltage across C_1, R_1 and C_2, R_2. If E_{R1} is +2 V DC compared to ground at the same time that E_{R2} is −2 V DC compared to ground, then it is easy to see that the resultant $E_o = 0$.

The phase diagram of Figure 7.10b shows that above the center frequency (say, 10.775 MHz), the resonant secondary becomes inductive, and voltage E_1 leads E_3 by an increased phase angle (say, 120° instead of 90°). E_2, representing the below-center frequency, looks capacitive, and voltage E_2 decreases in phase

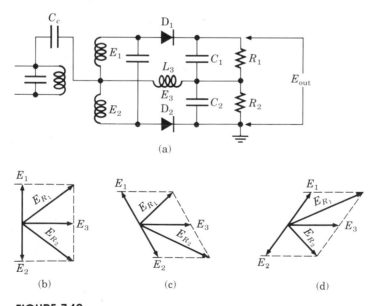

FIGURE 7.10
The phase discriminator as devised by Foster and Seeley.

compared to E_3 (say, 60°). It follows that E_1 and E_2 are always 180° out of phase. Note that the amplitude of E_1, E_2, and E_3 all remain constant, as expected of an FM wave, but that the phase-related vector sum of E_1 and E_3 becomes larger while the vector sum of E_2 and E_3 becomes smaller. With a larger AC voltage applied to D_1, the DC output voltage E_{R1} increases to, say, +2.5 V DC. With a decreased AC sum of E_2 and E_3 applied to D_2, the DC output voltage E_{R2} decreases to, say, −1.5 V DC. Thus, the sum of the DC output voltages is now +1 V DC.

Figure 7.10d represents the phase-related voltages for a below-resonance condition, which will shift the vector voltage in the opposite direction. The resulting effect, to produce −1 V DC at the output, will not be detailed here. The reader is encouraged to generate a step-by-step explanation.

The overall effect of the discriminator is to place a changing DC voltage at the output whose amplitude is proportional to the change in frequency of the deviated carrier. This voltage will, in a linear fashion, duplicate the message voltage that modulated the carrier at the transmitter.

7.7.4 The Ratio Detector

Comparing the ratio detector circuit of Figure 7.11 to the discriminator circuit of Figure 7.10 reveals several strong similarities. Both systems use the vector sum of the AC voltages at the transformer secondary to activate the rectifier diodes. That is, the vector sum of E_1 and E_3 is used to activate diode D_1, and the vector sum of E_2 and E_3 is used to activate diode D_2. Beyond this point, some differences become apparent:

1. the addition of C_3 across resistors R_1 and R_2,
2. the circuit separation between points A and B, where the output will be taken by connecting the load resistor to these two points, and
3. the reversed position of one of the diodes.

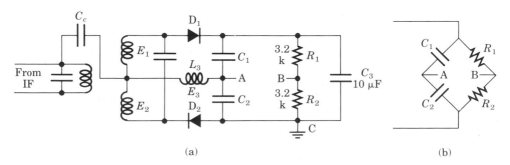

FIGURE 7.11
The Armstrong ratio detector. The output signal voltage is taken from points A and B.

After only a few cycles at the carrier signal frequency, capacitor C_3 will become fully charged to the peak value of the carrier voltage and will remain charged to the average peak value of the carrier voltage. This means that the *total* voltage across R_1 and R_2 will remain *constant* and negate any sudden changes of the carrier due to amplitude modulation or noise. This also indicates that the ratio detector does not require a limiter preceding the detector circuit. The RC time constant of $(R_1 + R_2) \times C_3$ should not be less than 0.025 s. The RC time of Figure 7.11 is 0.064 s.

The ability of the ratio detector to self-limit is termed *AM rejection* and is defined as the ability to eliminate the effects of amplitude modulation of the carrier signal while not degrading the frequency modulation of the carrier signal. The carrier is amplitude-modulated to 30%, and the output of the FM detector is a measure of the audio ouput in decibels below the carrier signal level, stated as dB of AM rejection.

Because diode D_2 is reversed relative to D_1, both diodes will conduct equally at the carrier frequency ($f_0 = 10.7$ MHz). A constant DC current will flow downward through the transformer secondary, through diode D_2 to point C, upward through R_2 and R_1, and then through diode D_1 to complete the circuit at the top side of the secondary. A charge-discharge AC current will flow down through the transformer secondary, through diode D_2, upward through capacitors C_2 and C_1 as well as through the parallel path of C_3, to complete the loop through diode D_1 to the top of the secondary.

The voltage drop across R_1-R_2 is held constant by C_3 at some level (say, +6 V DC), and $R_1 = R_2$; thus, the voltage at point B must be one-half of the total voltage, or +3 V DC. The network C_1-C_2 is in parallel with R_1-R_2; therefore, the total voltage across this network is +6 V DC, and when $C_1 = C_2$, the voltage at point A must be one-half of the total, or +3 V DC. The C_1-C_2-R_1-R_2 network forms a bridge circuit with the load between points A and B. When the voltage at A equals the voltage at B, the voltage drop across the load is 0 V. At f_0, the output voltage is zero. At this time, the current through the coil E_3 is also zero.

At a carrier frequency above $f_0 = 10.7$ MHz, diode D_1 conducts more than diode D_2 (because of the vector sum of the AC voltages E_1 and E_3), with the increased current flowing through coil E_3, through the load and up through R_1. The voltage across R_1 increases. The reduced D_2 current still flows through R_1, through the load resistance, and back through the E_3 coil to the cathode of D_2. With the voltage across R_1-R_2 held constant by C_3 and the drop across R_1 made larger, the voltage drop across R_2 is proportionally smaller. The voltage at B is lower than the voltage at A, and the output is the difference of these voltages.

At a carrier frequency below $f_0 = 10.7$ MHz, the conditions are reversed. D_2 conducts more than D_1, the voltage drop across R_2 increases, reducing the voltage drop across R_1, and the output signal goes up.

A ground connection may be made at point A, B, or C in Figure 7.11, with no change in circuit performance except output signal polarity. Figure 7.12 shows one of several variations for the output of the ratio detector.

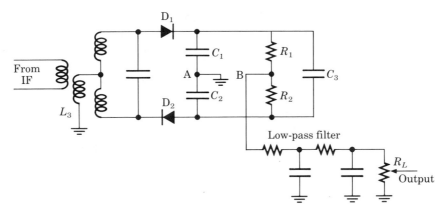

FIGURE 7.12
A variation for the output voltage using an Armstrong ratio detector.

Comparing the Ratio Detector and the Discriminator   In the ratio detector, the sum of the voltages remains constant while the ratio of the voltages changes with frequency. One of the diodes is reversed in the ratio detector, and the output voltage is typically one-half of the discriminator output voltage for equal signal levels input. The ratio detector has a large capacitor across the output, which limits amplitude changes.

In the discriminator, *both* the sum of the voltages and the ratio of the voltages change with frequency. The diodes face in the same direction, and the discriminator requires a limiter to reject amplitude changes of the carrier voltage.

7.7.5   The Phase-Locked Loop

The PLL in its most fundamental form is a feedback system made up of three basic blocks: a phase comparator, a low-pass filter, and a voltage-controlled oscillator. The low-pass filter is usually external to the chip circuit, as seen in Figure 7.13, while the phase comparator and VCO are integral parts of a single chip.

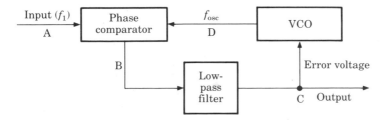

FIGURE 7.13
A phase-locked loop (PLL) used as an FM detector.

The VCO is a sine or square wave oscillator, having a free-running frequency that can be determined by an external *RC* time constant. Many charge-discharge circuits are dependent on both the *RC* time and the charge on the capacitor from the DC supply (V_{cc}), so a supply voltage different from that specified by the manufacturer may offset the free-running frequency.

The oscillator output signal (at D) is coupled to the phase comparator, where its frequency is compared to the frequency of the input signal (at A), which at this time is the IF signal of 10.7 MHz. When the VCO frequency is identical to the frequency of the IF signal, then the output of the phase detector is a DC voltage equal to $V_{cc}/2$. For a +12 V DC supply voltage, the phase detector output voltage will be +6 V DC. When a dual voltage supply is used (V_{cc} = +12 V and V_{ss} = −12 V), the phase detector output voltage will be 0 V DC.

The output from the phase detector is passed through a low-pass filter to remove the 10.7 MHz component of the IF carrier signal, and then used as a DC control voltage to stabilize the frequency of the VCO. The cutoff frequency of the low-pass filter is selected so that the highest message frequency of the system is reduced by only 3 dB. When the IF input carrier signal is in a modulated state, the output of the phase detector and low-pass filter will be the recovered modulation (audio) signal (at C). On the positive carrier deviation at A, the IF rises above the VCO frequency at D, and the two are compared. The DC output voltage from the comparator, at B, increases to make the VCO increase its signal frequency to match the higher IF signal. The DC voltage on the VCO control line, at C, rises above +6 V DC.

When the carrier deviation is below the 10.7 MHz rest frequency, it is lower than the VCO frequency at D, and the DC control voltage from the phase comparator drops below +6 V DC to lower the VCO frequency to match the new carrier input frequency. The changing DC voltage on the VCO control line is proportional to the IF carrier frequency change and is a true reproduction of the message signal.

The free-running frequency of the VCO is defined as the oscillating frequency that results when *no signal* is applied to the phase detector input (at A).

Numerically, frequency is the reciprocal of time, and time can be related to an *RC* constant. Therefore,

$$f = \frac{1}{T} = \frac{1}{R_o C_o} \quad \text{(in hertz)} \tag{7.13}$$

Some precautions are necessary when working with Equation 7.13. All integrated circuits have some internal wiring capacitances that will slightly offset the calculated frequency. The internal capacitance has its greatest effect when the VCO frequency is very high. At high frequencies, C_o becomes small, and the wiring capacitance becomes a large part of the total capacitance. The Motorola 4046 PLL has 32 pF internal capacitance, so for this circuit Equation 7.13 is changed to

$$f = \frac{1}{R_o(C_o + 32 \text{ pF})}$$

Other PLL circuits have a fixed resistor value built in, so that only a value of capacitance need be selected. The resistance and the internal capacitance are then both used to set a constant that is applied to the frequency calculations for a selected value of C_o. The XR2212 VCO has 5 pF internal capacitance and 5 kΩ fixed internal resistance, so the frequency selection becomes

$$f = \frac{2 \times 10^8}{C_o + 5} \quad (C \text{ in pF})$$

or

$$C = \frac{2 \times 10^8}{f} - 5$$

When $C_o = 0$, the upper frequency limit of 40 MHz is established for the VCO in the XR2212. In an FM radio receiver, a value of 18.7 pF would be needed for the VCO to oscillate at 10.7 MHz. Some manufacturers simply supply a graph of capacitance vs. frequency in place of equations.

When the values of R_o and C_o are both optional, there is an endless number of possible combinations. The frequency will be most stable for a large R-to-C ratio. The total capacitance is

$$C_{\text{tot}} = \frac{1}{f} \sqrt{\frac{4}{f}} \tag{7.14}$$

At 10.7 MHz,

$$C_{\text{tot}} = \frac{1}{10.7 \times 10^6} \sqrt{\frac{4}{10.7 \times 10^6}} = 57.14 \text{ pF}$$

Subtract from this the 32 pF internal capacitance for a value of 25.14 pF (use 20 pF). R_o may be found as

$$R_o = \frac{1}{C_{\text{tot}}f} \tag{7.15}$$

At 10.7 MHz,

$$R_o = \frac{1}{52 \text{ pF} \times 10.7 \text{ MHz}} = 1797 \ \Omega$$

Use $C_{\text{tot}} = 32 \text{ pF} + 20 \text{ pF} = 52 \text{ pF}$.

Using values of 52 pF and 1800 Ω in Equation 7.13 will result in a frequency of 10.684 MHz, which is close enough for the VCO to synchronize it with the 10.7 MHz IF carrier signal. Remember that 32 pF of the 52 pF are internal to the chip.

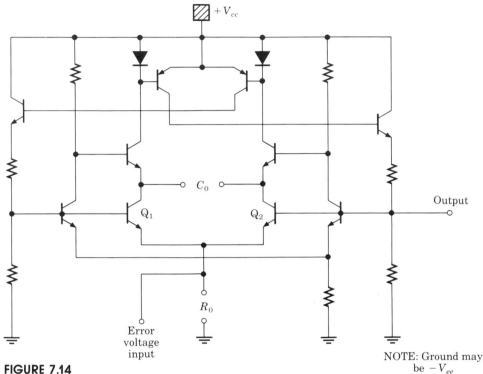

FIGURE 7.14
The circuit for the VCO.

A circuit representative of a VCO is shown in Figure 7.14. Note that this VCO is basically a differential multivibrator. The timing capacitor is found at the center of the diagram, where it will cause the on/off switching action of the multivibrator, Q_1-Q_2. The timing resistor, at the bottom of the diagram, is representative of the common-emitter resistor that acts as the constant-current resistor of a differential amplifier. Note also that the DC feedback voltage on the control line (error voltage) also regulates the emitter voltage across the constant-current resistor of Q_1-Q_2. Changing the DC error voltage will also change the frequency.

A typical phase detector circuit is shown in Figure 7.15. The differential input preamplifier is abbreviated, with the assumption that the reader understands the operation of this basic building block. However, it should be stated that whatever shape (sine or square wave) of input signal is applied to the preamplifier, the amplifier will saturate and output a square wave to the bases of Q_1 and Q_2. Q_1 and Q_2 will be alternately turned on and off (see wave A in Figure 7.16).

The VCO signal to the dual differential amplifier sets, Q_3-Q_4 and Q_5-Q_6, will also saturate these amplifiers because the VCO signal is very large and overdrives them. Again, the VCO signal may be a sine or square wave; either

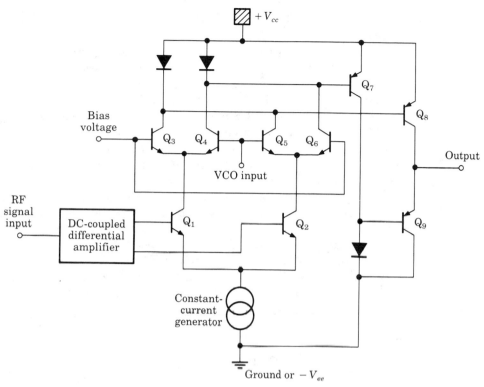

FIGURE 7.15
Phase detector circuit for the PLL.

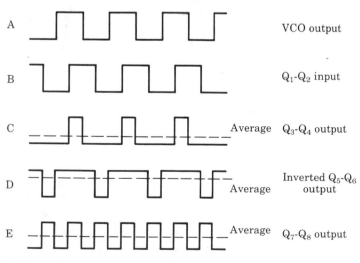

FIGURE 7.16
PLL waveshapes.

245

will saturate the amplifier. See wave B of Figure 7.16 and note that waves A and B are 90° out of phase.

The digitally oriented reader will recognize that the net effect of this circuit is that of an AND gate. For the analog-oriented reader, the net effect is identical to the function of a quadrature FM detector used in television receivers during the 1960s, 1970s, and early 1980s. Briefly review Figure 3.14 (balanced modulator) and you may recognize the extreme similarities between the chip circuitry of the balanced modulator and the phase detector.

Both Q_1 and Q_3-Q_5 must be turned on to turn on Q_8 in the push-pull output amplifier (see wave C of Figure 7.16). Note the reduction of the duty cycle and that the average voltage will be $V_{cc}/4$.

Ninety degrees away, Q_2 and Q_4-Q_6 must be turned on (see wave D), to turn on Q_7, which inverts the signal to turn on Q_9 and complete the second half cycle of the class B Q_8-Q_9 push-pull output amplifier. The end result is that wave C and inverted wave D are added together to form wave E as an output signal. Also note that the sum of waves C and D in wave E again form a 50% duty cycle wave with an average voltage of $V_{cc}/2$, stated earlier as the phase detector (error voltage) output. When the IF signal changes frequency, the comparative duty cycle changes, and the error output voltage changes to make the VCO frequency change to follow the IF change. Filtering out the carrier frequency from the error voltage will leave the average voltage change, which is equal to the message signal voltage.

The low-pass filter needed to pass the 15 kHz message signal and attenuate the 10.7 MHz carrier signal may again be any combination of values. A good RC ratio is achieved when the filter capacitance is

$$C_f = \left(\frac{1}{f_s}\right)^2 \quad \text{(in farads)} \tag{7.16}$$

where f_s = the highest message signal frequency. For a cutoff frequency of 15 kHz, the capacitor value is

$$C_f = \left(\frac{1}{15 \text{ kHz}}\right)^2 = 0.0044 \text{ }\mu\text{F} \quad \text{(use 0.0047 }\mu\text{F)}$$

For a -3 dB response, $R_f = X_c$ at 15 kHz, so that

$$R_f = X_c = \frac{1}{2\pi f_s C_f} = 2257 \text{ }\Omega \quad \text{(use 2200 }\Omega\text{)}$$

When the VCO self-adjusts its frequency to be identical to that of the input signal, the two signals are said to be *locked* (or synchronized or coincident). If there is a change in the frequency of the input signal, the VCO will self-adjust to be identical to the new frequency. The **lock range** is defined as the range of frequencies, in the vicinity of the VCO free-running frequency, over which the VCO will, once locked, remain synchronized to the input signal frequency. Sometimes called the *tracking bandwidth,* the lock range is dependent on the values of the VCO timing resistor, the filter resistor, and the supply voltage:

$$\begin{matrix} \text{Lock range} \\ \text{(tracking BW)} \end{matrix} = \pm f_L = \frac{2\pi f_0(R_0/R_f)}{V_{cc}} \qquad (7.17)$$

where f_0 = free-running VCO frequency
R_0 = VCO timing resistor
R_f = filter resistor
V_{cc} = total supply voltage

The **capture range** is always a narrower bandwidth than the lock range and is also called the *acquisition range*. Once the VCO loses synchronization with the input signal frequency, the capture range is a measure of how close to the VCO free-running frequency the input signal frequency must come for the VCO to resynchronize with it. An equation to approximate the capture range is:

$$\text{Capture range} = \pm f_c = \frac{1}{2\pi} \sqrt{\frac{2\pi f_L}{R_f C_f}} \qquad (7.18)$$

where f_L = lock range
R_f = filter resistance
C_f = filter capacitance

EXAMPLE 7.3

Find the lock range and capture range for a PLL using the values from Equation 7.16.

$$\text{Lock range} = \frac{(6.283)\,(10.7\ \text{MHz})\,(1800/2200)}{12} = \pm 4.584\ \text{MHz}$$

$$\text{Capture range} = .159155 \sqrt{\frac{(6.283)(4.584\ \text{MHz})}{(2200)(.0047\ \mu\text{F})}}$$

$$= \pm 265.6\ \text{kHz}$$

Equations 7.17 and 7.18 are reasonably reliable for very high levels of signal input voltages. However, as the signal input voltages get smaller, the input preamplifier may not be driven into saturation, and both the lock range and the capture range will become narrower. This effect is shown in Figure 7.17.

Use Equations 7.15 through 7.18 only as a starting point, with the understanding that some values may need to be altered during test. The equations are given here to assist the reader in value selection and to give a feel for the parameters that each component may affect. The overall circuit of the phase-locked loop as an FM detector is shown in Figure 7.18.

The PLL has one advantage over other FM detectors that will overshadow any disadvantage. That is, PLL never needs alignment. It will, within reason-

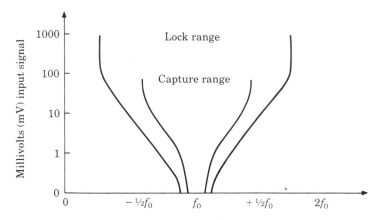

FIGURE 7.17
The lock and capture ranges for the 4046 PLL.

FIGURE 7.18
Working circuit for the 4046 PLL.

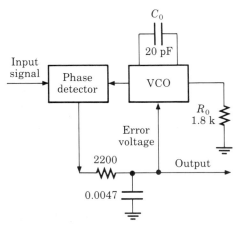

able limits, self-adjust to match the input frequency. Its chip circuitry, small size, and few component parts make the PLL the preferred detector circuit for commercial practice.

7.8   STEREO FM

The word *stereo* comes from the Greek word meaning "three-dimensional." In modern stereo, the three-dimensional effect is accomplished by a two-source sound system at a distance from the listener. The receiver must be equipped to separate a two-source signal, and the transmitter must be equipped to generate a two-source program. The treatment of stereo was omitted from the FM transmitter section because it is part of the audio (or message) circuitry and is developed before the modulator at the transmitter.

The two sources of sound in Figure 7.19 are indicated as microphones, but they could be any dual program source. They are labeled "L" for the left and "R" for the right channel. Each channel has an independent preamplifier, and there is a common gain control to balance the output levels of each channel. The FCC requires that any new system presented to the general public be compatible with existing systems. That is, the two sources of sound must be combined so that a single-channel receiver will receive both channels of sound through one speaker with no loss in quality. The listener who operates a dual-

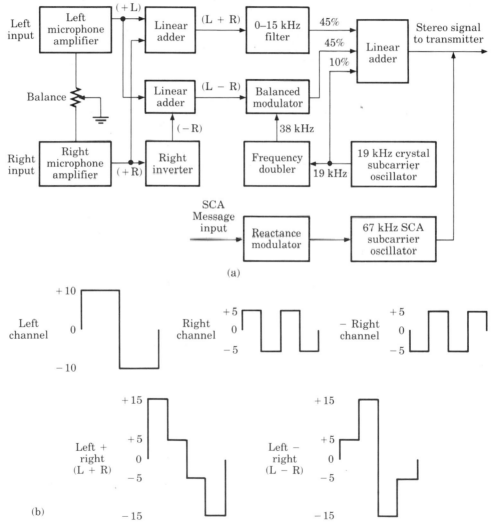

FIGURE 7.19

(a) The stereo generator (modulator) section of the audio portion of the FM transmitter, including the SCA message signal. (b) Stereo waveshapes.

channel receiver will hear the same program through two separate speakers simultaneously driven by the two sources of sound.

To accomplish the first part of this requirement, the left channel sound and right channel sound are simply added together over the frequency range of 50 Hz to 15 kHz to form the monaural program. This signal is called "L + R." It is shown in Figure 7.19 as the sum of the two square waves and in Figure 7.20 as the total frequency band between 50 Hz and 15 kHz. (Square waves are used here to simplify understanding of circuit performance; it is understood that real-life programming will be sinusoidal.)

The second part of the FCC requirement involves a few more steps. After the preamplifiers, the R signal is inverted relative to the L channel sound (180° phase shift). Any common-emitter amplifier or inverting op amp circuit will perform this task. The inverted R signal (−R) is then added to the L signal to become the L − R signal. This second set of signals will occupy the same frequency range, 50 Hz to 15 kHz, but the signals are *not* added directly to the first set. Instead, the L − R signal is first amplitude-modulated onto a 38 kHz carrier in a balanced modulator that suppresses the voltage at the carrier frequency. The double sidebands that are generated will thus occupy the frequency band from 15 kHz below 38 kHz (i.e., 23 kHz) to 15 kHz above 38 kHz (i.e., 53 kHz), as shown in Figure 7.20. The (balanced) modulated sidebands of the L − R signal can now be added to the unmodulated L + R signal with no chance of interference between the two sets because of the frequency separation and because one is modulated while the other is unmodulated.

Although the adder and inverter here are called "simple" circuits, keep in mind that they must be high-quality circuits that support the intent of FM broadcasting.

In order to recover, or demodulate, an $A3_a$ message signal, the carrier must first be reinserted and then demodulated as an AM signal. It would be asking too much to expect several million receivers to generate a 38 kHz carrier for reinsertion purposes and be *exactly* on frequency with the transmitter.

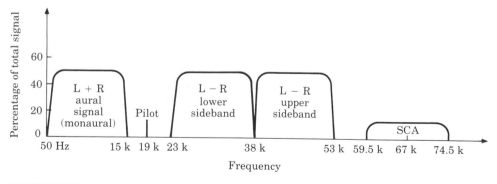

FIGURE 7.20
The frequency spectrum of a stereo signal before being frequency-modulated onto the station carrier signal frequency.

A frequency drift of 30 Hz (0.08%) would be heard as a rumble in the speakers. To avoid this problem, a 19 kHz **pilot** signal (one-half of 38 kHz) is transmitted along with the sound signals as a reference tone for the receivers. In Figure 7.19, the pilot tone is generated by a 19 kHz oscillator for transmission with the program and is shown having an amplitude not to exceed 10% of the total combined signal. The pilot signal is then doubled in frequency (to 38 kHz) to be used as the carrier input to the balanced modulator.

The total combined package of preemphasized L + R signal, pilot signal, the modulated upper and lower sidebands of the L − R signal, and the SCA signals (to be covered later) constitutes the message signal in Figure 7.20, which will be applied to the transmitter to frequency-modulate the station carrier.

Again, all of the stereo processing, although it involves some low-frequency modulation, is considered as falling within the audio section of the transmitter. Therefore, at the receiver, the stereo demodulation must also be considered as a process of the audio section and takes place after the FM detector.

The receiver processes the stereo signal in the reverse order of the transmitter. The last step at the transmitter is to combine all signals; therefore, the first step at the receiver is to separate all the signals, as shown in Figure 7.21. The low-frequency L + R signal is easily separated from the total signal with a low-pass filter. The only precaution here is to observe that the cutoff of the filter applies little or no attenuation to the 15 kHz signal, but reduces the 19 kHz pilot signal by a minimum of 20 dB. The L + R signal is deemphasized before further processing. The modulated L − R double-sideband signals are

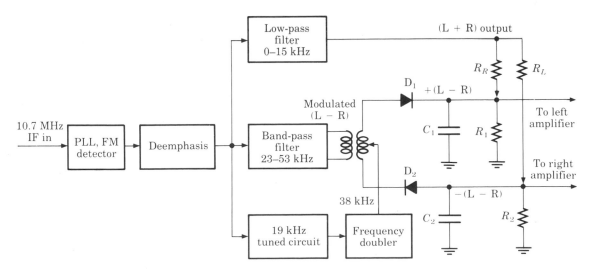

FIGURE 7.21
Demodulation of the stereo signal in the audio circuits. The output is applied to identical left and right audio amplifiers.

separated by a band-pass filter that allows only the frequencies between 23 kHz and 53 kHz to pass to the stereo demodulators. The pilot carrier signal is separated from the balance of the message by a high-Q parallel-resonant amplifier, possibly of the class C variety, with a tuned circuit of 0.5 μF and a 140 μH coil. A coil resistance of 0.8 Ω would present a Q of 20 and a bandwidth of less than 1 kHz.

The 19 kHz pilot carrier is then doubled in frequency to establish the 38 kHz subcarrier signal and to drive the stereo indicator. The double-sideband L − R signal is transferred through transformer T_1, and the 38 kHz carrier signal is coupled to the secondary center tap of T_1. The sidebands are recombined with the carrier in the secondary of the transformer and applied to rectifiers D_1 and D_2.

Diode D_1 is polarized to rectify only the positive half cycle of the recombined AM carrier wave. The filter circuit C_1-R_1 removes the carrier and passes the demodulated L − R signal to the *left* channel amplifier. At this point, the deemphasized L + R signal is fed back into the system, resulting in the *sum* of the two signals:

$$\begin{array}{r} L + R \\ + L - R \\ \hline 2L \end{array}$$

(7.19)

The positive right portion of one signal cancels the negative right portion of the other signal; the remainder is a signal of twice the amplitude of the left signal only (2L).

Diode D_2 is polarized to rectify only the negative half cycle of the recombined AM carrier wave. The filter C_2-R_2 removes the carrier and passes the demodulated −(L − R) signal to the right channel amplifier. At this point, the deemphasized L + R signal is fed back into the system, and the result is the sum of the two signals. −(L − R) converts to −L + R, so the resulting signal is:

$$\begin{array}{r} L + R \\ - L + R \\ \hline 2R \end{array}$$

(7.20)

The positive left portion of one signal cancels the negative left portion of the other signal, leaving a signal of twice the amplitude of the right signal only (2R). The separated R audio and L audio are then fed to individual (identical) amplifiers to drive separate sets of speakers. Tone controls, balance controls, and volume controls are all parts of the audio section.

A broadcast station may, in conjunction with standard programming, transmit control signals and privileged information on a subcarrier system. This transmitted information is labeled **SCA** (for *subsidiary communication authorization*). For monaural stations, the frequency range from 20 kHz to 75 kHz is available for this use with no restrictions on the number of subcarriers. The only limitation is that the SCA transmission shall not exceed 30% of the total modulation and that the "crosstalk" into the main program channel shall not exceed −60 dB. With stereo programming, the frequency range from 53 kHz to 75 kHz is open to subcarrier service. Modulation of the main carrier is limited to 10% and the crosstalk ratio again to −60 dB.

Commercial-free background music for offices and stores is the most familiar form of SCA. In large cities, the latest advances in medicine, treatment, and procedures are transmitted via SCA to the medical profession. This is usually by subscription from a service organization, which then maintains the equipment.

SCA is a frequency-modulated signal with 76 kHz as the most popular *fixed* carrier frequency. One hundred percent modulation is limited to ±7500 Hz deviation, with audio limits the same as for commercial audio, 50 Hz to 15 kHz. Each station combines the SCA signal with its regular FM monaural or stereo broadcast material, to become a part of the message signal modulating the station carrier. The double modulation process used here should be noted, as it becomes a useful tool in other transmission formats.

Normally, special receivers are made to decode only the SCA signal, which is then coupled to a sound distribution system in large installations. SCA is generally omitted from home entertainment units for two reasons:

1. SCA is monaural only, and the private sector trend is toward total stereo programming.
2. Although the audio range is the same as for the commercial signal, 50 Hz to 15 kHz, the modulation index is only 7.5 kHz/15 kHz = 0.5 rad. This is typical of narrowband FM with essentially only one set of sidebands, and, except for the noise immunity of FM, it has the characteristics of AM signal transmission.

A receiver block diagram for SCA reception is shown in Figure 7.22. Observe the double detection circuitry. The first detector removes the full range of frequencies from the 10.7 MHz IF carrier, and, following the band-pass filter, the second FM detector recovers the SCA signal from 67 kHz carrier. A squelch circuit is included to turn the audio amplifier off in the absence of a 67 kHz carrier.

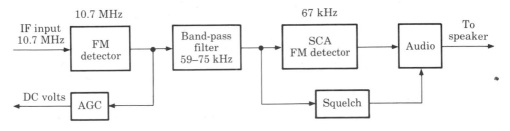

FIGURE 7.22
The demodulation circuit for an SCA receiver. Note the double detection due to double modulation.

Quadraphonic sound is four-channel sound. The proposal to transmit a system of sound distribution consisting of left rear, left front, right rear, and right front signals has been on trial since 1972. Of the many proposals, four configurations remain as contenders for the title of "The Acceptable System." Examine the transmitter block diagram of Figure 7.19a and visualize the left or right sound source as another complex set of circuits originating from a front and rear set of microphones. Or imagine the receiver of Figure 7.21 as splitting the left channel sound into a left front sound and a left rear sound. The electronics are not difficult, although some systems would include phase angles in the modulated process (L rear as $-45°$ and L front as $+45°$, for example). All systems must meet the letter of the law: total sound from one speaker for the monaural receiver, total stereo sound for the two-channel receiver, and four channels of sound for the quadraphonic receiver. The issue that is unsettled can be observed in Figure 7.20. Where in this frequency spectrum is the best place to add another modulated set of sidebands that will prove least harmful to existing services and that will not cause interference, by harmonic generation, with other signals? There is no system currently "acceptable" (by the FCC).

7.9   FM RECEIVER ALIGNMENT

Only two areas of receiver alignment need to be covered here. Except for the numerical values of the frequencies, the procedure to align the IF and RF of a receiver is almost universal, although the RF alignment for FM requires more care than for AM. The two areas open for discussion are (1) the dummy antenna and input signal connections, and (2) alignment of the FM detectors.

The input connections and the dummy antenna networks are shown in Figure 7.23. FM receivers have the advantage that there is always a set of antenna input terminals, and the input impedances have been standardized at the two levels of 300 Ω and 75 Ω. Even when the receiver has a built-in antenna, it is connected to these terminals and may be removed during tests. Remember to include a 101 dB attenuator between the RF signal generator and the input end of the chosen dummy antenna of Figure 7.23.

The second area of concern revolves around the detector. When a PLL is used as a detector, *no* alignment is required. When the detector is a discriminator or ratio detector, the vertical input of an oscilloscope is connected across the detector output load resistance. The signal generator must be a *sweep* generator, capable of modulating the FM carrier to two times 100% modulation for the service involved (± 150 kHz for standard FM receivers). The generator must also provide markers at 10.6, 10.7, and 10.8 MHz. To align the discriminator (or ratio detector), connect the high side of the RF signal generator output through a 0.001 μF capacitor to the collector of the RF amplifier, generator low side to ground. Set the sweep bandwidth for 300 kHz at a center frequency of 10.7 MHz. The sweep rate is 60 Hz. Set the RF output from the

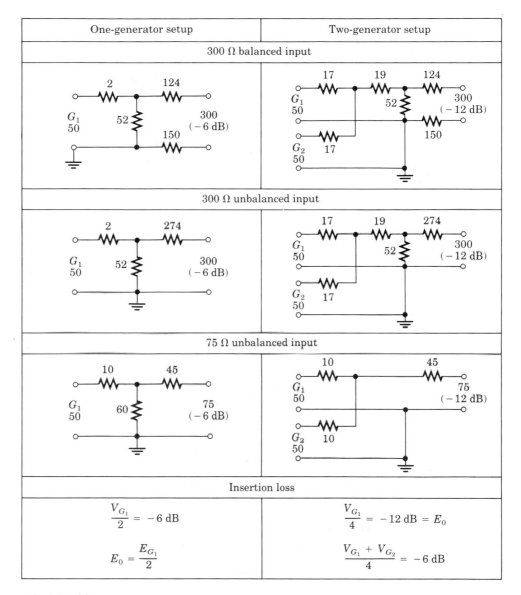

FIGURE 7.23
Impedance networks.

generator only high enough to display the "S" curve on the oscilloscope (see
Figure 7.24). Apply the markers at 10.6, 10.7, and 10.8 MHz with only enough
amplitude to show an indication on the scope display. Connect the sweep rate
output from the generator to the horizontal input terminal of the oscilloscope,
and set the scope sweep rate to "horizontal input."

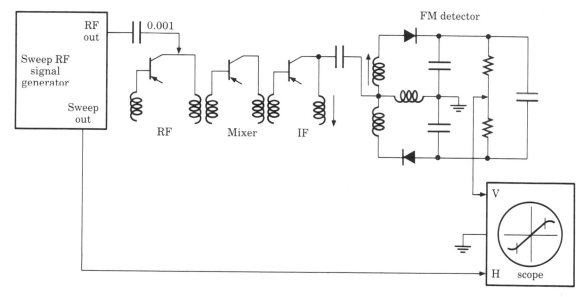

FIGURE 7.24
Test setup for sweep alignment of an FM receiver. Note the "S" curve display on the oscilloscope. External bias may be required for AGC.

Adjust the discriminator (or ratio detector) transformer secondary (top slug adjustment) to place the 10.7 MHz marker at the center crossover line on the scope "S" display. Adjust the detector primary (bottom slug adjustment) for the *most linear* plot on the "S" display with maximum amplitude.

7.9.1 FM Receiver Selectivity

The selectivity for FM receivers is measured in exactly the same manner as for AM receivers. The only difference is the dummy antenna system used for each type of receiver. Refer to Chapter 4 for the procedures, but use the dummy antenna selection from Figure 7.23.

7.9.2 FM Receiver Sensitivity

There are two methods prescribed for measuring the sensitivity of FM receivers: the 20 dB quieting method and the 12 dB SINAD method. Each has its advantages.

The *20 dB of quieting* method uses an unmodulated RF signal applied through a 101 dB attenuator and a selected dummy antenna to the antenna terminals of the receiver. The output is taken across the speaker terminals with the speaker replaced by an equivalent load resistance. Connect a true-rms-reading voltmeter in parallel with the resistive load. With the RF signal generator initially turned *off* (and the receiver *off of a station*), adjust the volume control for an output noise voltage of 1 V_{rms}. If 1 V cannot be obtained from maximum volume, use a convenient setting of some lower value (for example, 0.1 V). Turn on the signal generator, and ensure that the generator frequency matches the RF tuning. Now increase the generator output level until the noise at the output load *decreases* to one-tenth its original value (1 V down to 0.1 V or 0.1 V to 0.01 V). The receiver output is now 20 dB quieter than it was with no RF signal input because the AGC has reduced the IF gain for a noise reduction. Measure the RF signal at the generator and subtract the losses for the attenuator and −6 dB for the dummy antenna single input (or −12 dB for the two-generator connection). The RF at the antenna terminals is the 20 dB quieting threshold. This is the minimum required RF input signal that will reduce the receiver thermal noise to its −20 dB power level (one-tenth voltage level). Typical values are around 0.25 μV. This measurement would then be stated as 0.25 μV input for 20 dB of quieting.

The 12 dB SINAD (signal + noise + distortion) value is defined as the minimum value of the modulated RF generator voltage that will produce at least 50% of the receiver's rated audio output power with a (signal + noise + distortion)/(noise + distortion) ratio (SINAD ratio) of 12 dB.

The RF signal generator, frequency-modulated with a 1 kHz message frequency for ±3 kHz of deviation, is connected to the test setup at the antenna terminals. The output is terminated with a resistive load and a *distortion analyzer* to remove the 1 kHz tone signal. The RF signal generator is initially set for about 1000 μV, and the receiver volume control is adjusted for the receiver's rated output power, typically 5 W (4.5 V_{rms} across a 4 Ω load). The RF signal generator level is reduced until the measured output power of the receiver is one-half its rated power. The distortion analyzer is then adjusted to eliminate the 1 kHz tone and the difference in voltage (power) level is noted. The RF signal generator level is reduced further, until the audio output power is 12 dB lower than the half-power point, and the antenna input voltage here is noted as the 12 dB SINAD sensitivity.

For both tests, the RF signal generator frequency must be set for the discriminator zero crossing before the test tone is added.

The advantage of the 12 dB SINAD method is that a modulated signal is utilized and the receiver's thermal noise, harmonic, and intermodulation distortion are all accounted for in the tests; in the 20 dB quieting test, only the thermal noise is measured. The 20 dB test is quicker and easier, but the 12 dB SINAD test is more exact.

QUESTIONS

1. Which functional block of a receiver most readily identifies it as an FM receiver? (a) the RF amplifier, (b) the IF amplifiers, (c) the detector, (d) the mixer, (e) the audio.

2. What is the common standard intermediate frequency used in FM receivers? (a) 470 MHz, (b) 151 MHz, (c) 88 MHz, (d) 10.7 MHz, (e) 455 KHz.

3. In the commercial FM radio receiver, the local oscillator operates at a frequency that (a) is higher than the RF signal frequency, (b) is lower than the RF signal frequency, (c) may be either higher or lower than the RF signal frequency, (d) There is no pattern between the IF and RF.

4. What circuit is found in an FM receiver IF amplifier chain that is not found in an AM receiver? (a) AGC, (b) limiters, (c) RF amplifiers, (d) class D amplifiers, (e) a detector.

5. The function of the RF amplifier is to (a) render gain at the system's lowest noise point, (b) aid in the selection of a station carrier, (c) isolate the local oscillator from the antenna, (d) support the selectivity of the receiver, (e) none of the above.

6. An RF amplifier circuit that has a common-emitter amplifier in series with a common-base amplifier is called a (a) dual-input amplifier, (b) starvation amplifier, (c) cascade amplifier, (d) cascode amplifier, (e) none of the above.

7. Identify the characteristics of a starvation amplifier. (a) very low DC current, (b) high output load resistance, (c) low noise and distortion, (d) high dynamic range, (e) all of the above.

8. What advantage is obtained by applying the AGC voltage to one gate of a dual-gate FET and the RF signal to the second gate?

9. What is the single most important function of the RF amplifier?

10. What is the basic form of receiver found in Figure 7.1? (a) TRF, (b) super-regenerative, (c) double-conversion, (d) superheterodyne, (e) audiodyne.

11. What is the frequency range of galactic noise?

12. What is the most common source of galactic noise?

13. What is the frequency range of atmospheric noise?

14. What is the frequency range of man-made noise (excluding RFI)?

15. What does RFI stand for?

16. Internal noise caused by the random flow of DC current is called (a) Johnson noise, (b) thermal noise, (c) white noise, (d) all of the above, (e) none of the above.

17. The noise *factor* is a ratio of (a) the input signal-to-noise power, (b) the output signal-to-noise power, (c) the input signal-to-noise voltage, (d) the output signal-to-noise voltage, (e) a compared to b.

18. The noise *figure* is (a) equal to the noise factor, (b) the inverse of the noise factor, (c) the dB conversion of the noise factor, (d) the antilog of the noise factor, (e) not a function of the noise factor.

19. The noise figure is a function of the amplifier gain. (T) (F)

20. The noise factor of several amplifiers in series will add together directly. (T) (F)

21. The amplifier that has the greatest effect on the noise in a series string of amplifiers is the (a) input amplifier, (b) output amplifier, (c) All amplifiers have equal effect, (d) the amplifier with the highest gain, (e) the amplifier with the lowest gain.

22. Absolute zero, the temperature where all molecular motion stops, is considered to be $-273°K$. (T) (F)

23. What is the noise figure (NF) of an amplifier that has a gain of 18 over a frequency band of 15 kHz and an output noise power of 24 fW? (a) -18 dB, (b) -13.5 dB, (c) -46.2 dB.

24. The approximate bandwidth of a critically coupled transformer amplifier is: (a) 1% f_0, (b) 2.5% f_0, (c) 5% f_0, (d) 10% f_0.

25. The approximate bandwidth of an overcoupled transformer amplifier is (a) 1% f_0, (b) 2.5% f_0, (c) 5% f_0, (d) 10% f_0.

26. The gain of an RF amplifier should be measured at a peak-to-peak output level not to exceed (a) 10% V_{cc}, (b) 15% V_{cc}, (c) 20% V_{cc}, (d) 25% V_{cc}, (e) 33.33% V_{cc}.

27. What is the noise figure (NF) of three series amplifiers when $G_1 = 15$, $G_2 = 15$, $G_3 = 18$ and $F_1 = 8$, $F_2 = 12$, $F_3 = 14$?

CHAPTER EIGHT

TRANSMISSION LINES

8.1 INTRODUCTION

The word *lines,* associated with transmission, describes a hard-wire, point-to-point system. The energy put into the line travels the length of the conductors to the load and, except for small wire losses, does not lose energy through radiation. A *transmission line* is considered to be any electrically suitable confinement used to guide the flow of energy by physical contact between two points within a system.

8.1.2 Cable Types

Energy conductors fall within the following three basic classes of transmission lines:

1. two-wire parallel lines
2. coaxial lines
3. waveguide and fiber optics

Waveguide and optic fibers are covered in later chapters devoted to special high-frequency systems. Coaxial and two-wire parallel lines are illustrated in Figure 8.1. Choosing among the many available cables is done by comparing the properties of the cable to the requirements of the system at hand. Factors that may be taken into account are indoor versus outdoor use, operating frequency, power-handling needs, area environment, and electrical interference, as well as cost and size.

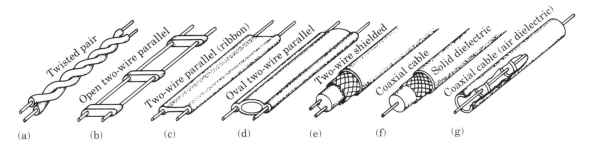

FIGURE 8.1
Types of cable.

8.2   CHARACTERISTIC IMPEDANCE

The impedance of the cable, termed the **characteristic** or **surge impedance** Z_0, is considered to be independent of the cable length and the operating frequency. This consideration is valid when the line is properly terminated and when the operating frequency is above a few tens of kilohertz, but below a few gigahertz.

The infinitely long line, (a) of Figure 8.2, has an input impedance equal to Z_0, measured as the ratio of E_i to I_i. Energy put into the line will travel the length of the line and never return. When the total length is shortened by removal of section (b), the remaining section, (c), is still infinitely long, by definition. Energy applied at the input of (c) will respond exactly as it did at the

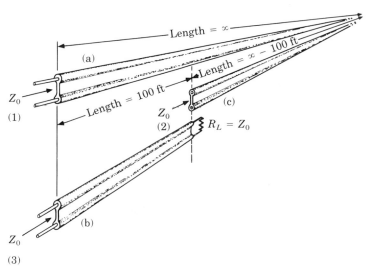

FIGURE 8.2
Impedance (Z_0) for an infinitely long line.

input of (a). The input impedance will still be Z_0, and the energy will travel the length of the line never to return. To make section (b) look like section (a), a resistor of value equal to Z_0 is attached to the output end of section (b). Section (b) now behaves as if section (c) were reconnected. As long as resistor Z_0 is attached to section (b), sections (a) and (b) will act identically, and section (b) can be of *any* length.

8.3

ELECTRICAL PROPERTIES OF THE LINE

Two-wire line and coaxial cables have some electrical similarities. Both are wire cables and have wire resistance (R). At high frequencies all wires exhibit inductive properties (L). Two conductors separated by an insulator will form a capacitance (C), while the insulating material (even air) contributes conductance (G). The line constants are evenly distributed in microscopic values over the entire length of the line. However, for easier understanding, these line constants are lumped together in Figure 8.3 to explore the principles of transmission lines.

Resistance R and inductance L represent the loop resistance and inductance, that is, the resistance and inductance down one line and back up the other line. Capacitance C and conductance G represent an endless number of small values in parallel across the line over its entire length. The T network shows that the series impedance $(R + j\omega L)$ is divided into two halves, while the shunt impedance $(G + j\omega C)$ is across the line at its center. We use the T network analogy so that the impedance into the line will look the same from either direction.

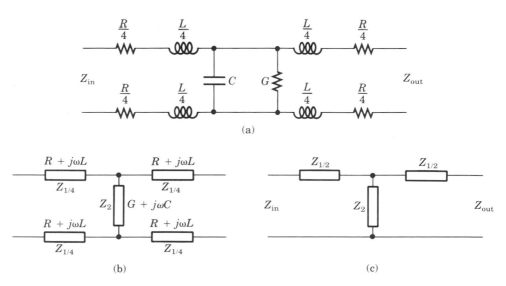

FIGURE 8.3
Equivalent impedance of a transmission line.

8.3.1 Constant Impedance

For the moment, the series and shunt impedances will be replaced with resistor values for easy calculation. If a very short section of line were selected and a second very short section of line were added on, then a third and a fourth, and so on, then the configuration of Figure 8.4 would materialize. Although the values of Figure 8.4 are out of proportion to an actual line, they are chosen here to simulate a 50 Ω cable.

Figure 8.4a shows a microscopic length of cable with 3.3 Ω in each series branch; the 330 Ω resistance represents the parallel branch of the network. These values can be substituted for the equivalent values of Z_1 and Z_2 in Figure 8.3. The input impedance of the cable would be the sum of the resistances, equal to 336.6 Ω. When the second very short section of cable is added on, as shown in Figure 8.4b, its 336.6 Ω will be in parallel with the 330 Ω resistor of the first section, forming a parallel equivalent of 166.63 Ω. To this the series value of 6.6 Ω is added, for a total input impedance equal to 173.23 Ω. Placing this 173.23 Ω in parallel with the 330 Ω of the third section, and adding to this the 6.6 Ω series resistance of the third section, equals a total line impedance of 113.6 + 6.6 = 120.2 Ω. Figure 8.4d shows what happens to the impedance of the cable as each new section is added on. With five sections, the impedance is reduced to 76.14 Ω; with ten sections, to 55.32 Ω; and with 20 sections, to

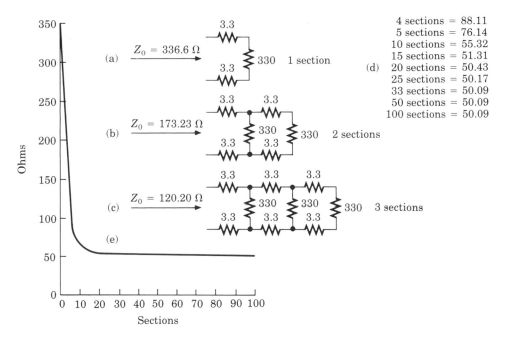

FIGURE 8.4
Cable impedance as a function of added sections.

50.43 Ω. With 33 sections added on, the impedance value reaches 50.09 Ω and remains at this value for any number of added sections. Therefore, this particular cable is classified as a 50 Ω cable and, since it is made to look resistive, it will remain as a 50 Ω cable regardless of its length or operating frequency. This **constant impedance** principle holds true for all types of transmission lines.

8.3.2 Cable Impedance from Line Constants

It has been shown that the line is made up of lumped constants R, L, C, and G, rather than just the resistances of the sample cable. A simple equation borrowed from filter theory is applicable here to determine the characteristic impedance of cables:

$$Z_0 = \sqrt{\frac{R + j\omega L}{G + j\omega C}}$$

where Z_0 = characteristic impedance
$Z_1 = R + j\omega L$ series impedance
$Z_2 = G + j\omega C$ parallel impedance

At high frequencies, where ωL is much larger than R and ωC is much larger than G, the values of R and G may be omitted from the impedance equation, which then reduces to:

$$Z_0 = \sqrt{\frac{L}{C}} \quad \text{(in ohms)} \tag{8.1}$$

Care need be taken only to state the values of inductance and capacitance for the *same length* of cable.

EXAMPLE 8.1
A cable has 2135 pF capacitance per 100 ft length and 11.38 µH inductance for the same 100 ft length:

$$Z_0 = \sqrt{\frac{11.38 \times 10^{-6}}{2135 \times 10^{-12}}} = 73 \ \Omega$$

Note that the measurements (L and C) are for the same length of cable and the impedance is stated in ohms. Had the measurements been stated for 1000 ft of the same cable, then both L and C would have been ten times greater. That is, the inductance would be 113.8 µH and the capacitance would have been 21,350 pF. However, the impedance would have remained the same, at 73 Ω. The impedance equation (8.1), then, is valid for all types of cable.

If, for any reason, the inductance and capacitance are given for different lengths, simply increase the shorter length to match the longer and change the component value by the same ratio. Either length and corresponding value may be altered to match the other length and its value.

EXAMPLE 8.2

Cable X is known to have 4.5 μH of inductance per 100 ft. The capacitance of a 500 ft section is measured as 250 pF. Therefore, the inductance needs to be increased by a factor of 5, to 22.5 μH, or the capacitance could be reduced by a factor of 5, to 50 pF. Try both alternatives, and you should arrive at the same impedance of 300 Ω. Again, length does not affect the characteristic impedance value.

8.3.3 Time Delay

It does not require a vivid imagination to see that some time will pass from the instant energy is applied to the input of the line to the instant the same energy arrives at the output end of the same line. This time difference is called **time delay** and uses the symbol T_d. Other references may call this period the *transit time*. The unit of measurement is seconds, although it is normally stated in microseconds or nanoseconds. This time delay really exists; it can be measured, and it has a bearing on the time relationships in circuits that are separated in distance by the line. The effects of time delay are more noticeable at higher frequencies and when electrically long distances separate the circuits. Delay in the line results from the time rate of change of the current through the inductance and on the charge-discharge characteristics of the shunt capacitance. Therefore, since inductance, capacitance, and length are factors, an adequate expression for time delay uses all these terms:

$$T_d = \sqrt{LC} \quad \text{(in seconds per unit length)} \tag{8.2}$$

The important notation in Equation 8.2 is *"per unit length."* Time delay is directly proportional to length. If the length is doubled, the time delay will be twice as long.

EXAMPLE 8.3

A 100 ft length of cable has an inductance of 2.5 μH per 100 ft, and 1000 pF capacitance per 100 ft. What is the time delay for this cable?

$$T_d = \sqrt{(2.5 \times 10^{-6})(1 \times 10^{-9})} = 50 \text{ ns/100 ft}$$

When the same type of cable is twice as long (200 ft), both the inductance and the capacitance double in values, and the time delay becomes twice as long, that is, 100 ns for 200 ft of cable.

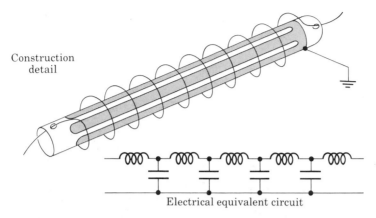

Construction
detail

Electrical equivalent circuit

FIGURE 8.5
Artificial delay line.

Artificial Delay Lines   Delay time can be desirable or undesirable, depending on the needs of the circuit at hand. One case of desirable time delay occurs in every color television receiver. The monochrome picture signal, called the luminance signal, must be delayed by 1 μs while the color signal is being processed. The two signals are then recombined and applied to the picture tube. If this were not so, the two signals would arrive at the picture tube at different times and cause picture misregistration.

If RG59/U cable is used to cause the 1 μs delay, the required length of cable can be found by knowing that RG59/U cable has 2100 pF capacitance and 10.92 μH inductance per 100 ft length (RG stands for radio grade cable). This calculates out to 0.1515 μs per 100 ft by the delay time formula. Therefore, 660 ft are required to delay the signal by 1 μs. However, the idea of 660 ft of cable coiled up inside every color TV set is too ridiculous to consider.

The **artificial delay line** is a practical solution. Shown in Figure 8.5 is an insulated tube about 4½ in long and ¾ in in diameter. Several strips of conducting material are bonded to the outside surface of the tube and joined at one end (see the shaded area in Figure 8.5). A single-layer coil is then wound around the coil form. Each turn of wire adds to the total inductance, and as each turn of wire passes over a conductive strip, it forms capacitance between the wire and the conductor. When 250 turns of #26 wire (.015 in diameter) are wound around the ¾ in diameter coil form, the coil will be approximately 4½ in long. The inductance (L) of a single-layer close-wound coil is

$$L = \frac{a^2 n^2}{9a + 10l} \quad \text{(in microhenrys)} \tag{8.3}$$

where a = coil form radius = .375 in
n = number of turns = 250
l = coil length = 4.5 in

The resulting inductance is 182 μH.

The capacitance formed by each turn of wire and the common conductive strip on the coil form is found by

$$C = \frac{A(8.85 \times 10^{-12})}{D} \quad \text{(in farads)} \tag{8.4}$$

where A = area of the smaller-capacitance plate
D = distance between the plates

The capacitance 8.85×10^{-12} F is the dielectric permittivity of air. The smaller-capacitance plate is the wire. It passes over the conductive material, which covers about 80% of the circumference of the coil form. Therefore, the area is 80% of the circumference times the wire diameter times the number of turns. The distance between the plates is the wire radius plus about 0.004 in for the varnish insulation. Using these approximations in Equation 8.4 gives the following results:

$$C = \frac{1.767(8.85 \times 10^{-12})}{2.84 \times 10^{-3}} = 5506 \text{ pF}$$

From Equation 8.2, the time delay is 1.001 μs. This time delay is typical for the average color TV receiver and is considerably more practical than using 660 ft of RG59/U cable.

8.3.4 Cable Connectors

There are around ten different types of cable connectors in common use today. Although they are similar in function, they differ in construction. The common feature is that all types of connectors must make separate contacts with both the cable center conductor and the outer conductor (shield or braid) of the cable. The connector must then mate with a like connector of the chassis or panel style without causing a shorted condition between the two contacts. Figure 8.6 shows a few considerations that are taken when a connector is put onto a cable.

After the assembly is completed, make sure to test for opens and shorts. There must be firm continuity between the center conductor at one end to the center conductor at the other end of the cable, likewise between the outer connector contact at one end to the outer connector contact at the other end of the cable. Always verify that the shield is *not* shorted to the center conductor. An ohmmeter using clip leads will confirm these conditions. Flex the cable connectors during testing to ensure a positive and firm contact void of shorts.

8.3.5 Determining Line Constants *R, L, C,* and *G*

The resistance, inductance, and capacitance of any given line are easily measured on a good laboratory impedance bridge. Keep in mind that R and L are

1. Cut end of cable even. Remove vinyl jacket 1-1/8"— don't nick braid.

2. Bare 3/4" of center conductor— don't nick conductor. Trim braided shield 1/16" and tin. Slide coupling ring on cable.

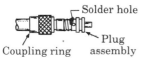

3. Screw the plug assembly on cable. Solder plug assembly to braid through solder holes. Solder conductor to contact sleeve.

4. Screw coupling ring on assembly.

UHF plugs

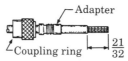

1. Cut end of cable even. Remove vinyl jacket 21/32"—don't nick braid. Slide coupling ring and adapter on cable.

2. Fan braid slightly and fold back over cable.

3. Compress braid around cable. Position adapter to dimension shown. Press braid down over body of adapter.

4. Bare 1/2" of center conductor—don't nick conductor. Pre-tin exposed center conductor.

5, 6. Same as 3 and 4 UHF plug.

UHF plugs with adapters

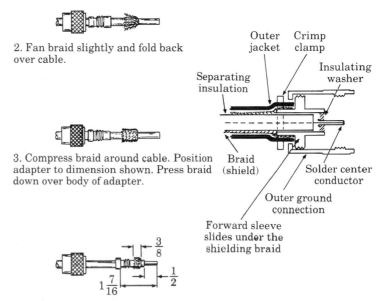

FIGURE 8.6
Types of connectors.

loop resistance and inductance whereas the capacitance is for a single length of line (see Figure 8.7).

The conductance G is a function of the molecular motion within the insulating material due to the rate of change of voltage (frequency) of the applied signal. Values of conductance are measured at room temperature and in relatively dry air. The conductance of solid insulation cables such as twin-lead are more subject to moisture absorption and thus have poorer conductance stability. Shielded cables and open-wire, air dielectric cables are better.

8.3.6 Dielectric Constants

The insulating materials used to separate the conductors of transmission lines introduce some losses in the cable. This property of these dielectrics, called

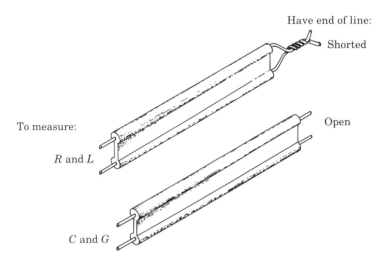

FIGURE 8.7
Treatment of cable to measure line constants.

permittivity, has a capacitive effect that introduces leakage. Air has the lowest leakage, is most stable, and is therefore the reference for all other dielectric materials. Air has a permittivity of 8.842×10^{-12} F per meter. The symbol for permittivity is epsilon (ϵ), referred to as the **dielectric constant.** Since air is the reference, it is given the symbol ϵ_0 and the dielectric constant of 1. The constants for other dielectric materials relative to air are provided in Table 8.1.

8.3.7 Wire Resistance

We will deal regularly with many types of standard copper wires during this discussion on transmission lines, and you will need a wire reference table. An American standards wire reference is found in Table 8.2.

8.4   **IMPEDANCE FROM CABLE GEOMETRY**

The electrical specifications for a given cable may not always be readily available. All is not lost. Two mechanical measurements taken as accurately as possible will enable you to find all of the electrical details you need to know about any standard cable. Of the two cable types (parallel lines and coaxial lines), the parallel lines will be treated first.

8.4.1 Two-Wire Parallel Lines

Figure 8.8a shows a typical twin-lead, or two-wire, ribbon cable. The impedance is

$$Z_0 = \frac{276}{\sqrt{\epsilon}} \log_{10}\!\left(\frac{2D}{d}\right) \qquad\qquad (8.5)$$

where ϵ is a dielectric constant from Table 8.1. If you feel more comfortable using the natural log,

$$Z_0 = \frac{120}{\sqrt{\epsilon}} \ln_e\!\left(\frac{2D}{d}\right) \qquad\qquad (8.5a)$$

TABLE 8.1
Dielectric constants of materials

Material	Dielectric Constant (Approx.)	Material	Dielectric Constant (Approx.)
Air	1.0	Nylon	3.4–22.4
Amber	2.6–2.7	Paper (dry)	1.5–3.0
Asbestos fiber	3.1–4.8	Paper (paraffin coated)	2.5–4.0
Bakelite (asbestos base)	5.0–22	Paraffin (solid)	2.0–3.0
Bakelite (mica filled)	4.5–4.8	Plexiglas	2.6–3.5
Barium titanate	100–1250	Polycarbonate	2.9–3.2
Beeswax	2.4–2.8	Polyethylene	2.5
Cambric (varnished)	4.0	Polyimide	3.4–3.5
Carbon tetrachloride	2.17	Polystyrene	2.4–3.0
Celluloid	4.0	Porcelain (dry process)	5.0–6.5
Cellulose acetate	2.9–4.5	Porcelain (wet process)	5.8–6.5
Durite	4.7–5.1	Quartz	5.0
Ebonite	2.7	Quartz (fused)	3.78
Epoxy resin	3.4–3.7	Rubber (hard)	2.0–4.0
Ethyl alcohol (absolute)	6.5–25	Ruby mica	5.4
Fiber	5.0	Selenium (amorphous)	6.0
Formica	3.6–6.0	Shellac (natural)	2.9–3.9
Glass (electrical)	3.8–14.5	Silicone (glass) (molding)	3.2–4.7
Glass (photographic)	7.5	Silicone (glass) (laminate)	3.7–4.3
Glass (Pyrex)	4.6–5.0	Slate	7.0
Glass (window)	7.6	Soil (dry)	2.4–2.9
Gutta percha	2.4–2.6	Steatite (ceramic)	5.2–6.3
Isolantite	6.1	Steatite (low loss)	4.4
Lucite	2.5	Styrofoam	1.03
Mica (electrical)	4.0–9.0	Teflon	2.1
Mica (clear India)	7.5	Titanium dioxide	100
Mica (filled phenolic)	4.2–5.2	Vaseline	2.16
Micaglass (titanium dioxide)	9.0–9.3	Vinylite	2.7–7.5
Micarto	3.2–5.5	Water (distilled)	34–78
Mycolex	7.3–9.3	Waxes, mineral	2.2–2.3
Neoprene	4.0–6.7	Wood (dry)	1.4–2.9

TABLE 8.2

Copper wire characteristics

AWG	Nom Bare Diameter (Inches)	Nom Circular Mils	Nom Feet Per Lb (Bare)	Nom Ohms Per 1000 Ft @20°C	Current Carrying Capacity @700 CM/Amp	Turns Per Linear Inch	
						Single Film Coated	Heavy Film Coated
0000	.4600	211600	1.561	.04901	302.3		
000	.4096	167800	1.969	.06182	239.7		
00	.3648	133100	2.482	.07793	190.1		
0	.3249	105600	3.130	.09825	150.9		
1	.2893	83690	3.947	.1239	119.6		
2	.2576	66360	4.978	.1563	94.8		
3	.2294	52620	6.278	.1971	75.2		
4	.2043	41740	7.915	.2485	59.6		4.80
5	.1819	33090	9.984	.3134	47.3		5.38
6	.1620	26240	12.59	.3952	37.5		6.03
7	.1443	20820	15.87	.4981	29.7		6.75
8	.1285	16510	20.01	.6281	23.6		7.57
9	.1144	13090	25.24	.7925	18.7		8.48
10	.1019	10380	31.82	.9988	14.8		9.50
11	.0907	8230	40.2	1.26	11.8		10.6
12	.0808	6530	50.6	1.59	9.33		11.9
13	.0720	5180	63.7	2.00	7.40		13.3
14	.0641	4110	80.4	2.52	5.87	15.2	14.8
15	.0571	3260	101	3.18	4.66	17.0	16.6
16	.0508	2580	128	4.02	3.69	19.0	18.5
17	.0453	2050	161	5.05	2.93	21.3	20.7
18	.0403	1620	203	6.39	2.31	23.9	23.1
19	.0359	1290	256	8.05	1.84	26.7	25.9
20	.0320	1020	323	10.1	1.46	29.9	28.9
21	.0285	812	407	12.8	1.16	33.4	32.3
22	.0253	640	516	16.2	.914	37.5	36.1
23	.0226	511	647	20.3	.730	41.8	40.2
24	.0201	404	818	25.7	.577	46.8	44.8
25	.0179	320	1030	32.4	.457	52.5	50.1
26	.0159	253	1310	41.0	.361	58.8	56.0
27	.0142	202	1640	51.4	.289	65.6	62.3
28	.0126	159	2080	65.3	.227	73.3	69.4

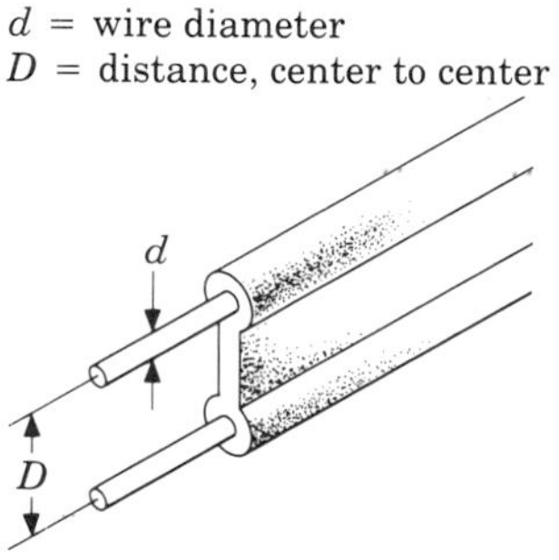

d = wire diameter
D = distance, center to center

Shielding braid (grounded)

Insulating jacket

Solid insulation

S_d = inside diameter of shield (outside diameter of solid insulation)

(a) Two-wire parallel (ribbon) cable

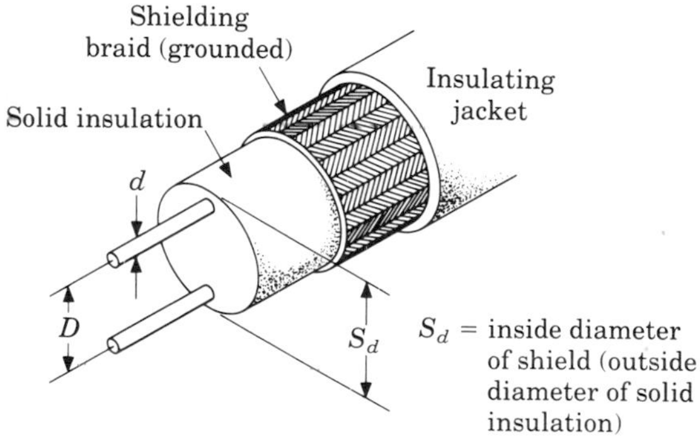

(b) Two-wire shielded cable

FIGURE 8.8

Cable construction details used to calculate line impedance of a two-wire line.

AWG	Nom Bare Diameter (Inches)	Nom Circular Mils	Nom Feet Per Lb (Bare)	Nom Ohms Per 1000 Ft @20°C	Current Carrying Capacity @700 CM/Amp	Turns Per Linear Inch	
						Single Film Coated	Heavy Film Coated
29	.0113	128	2590	81.2	.183	81.6	76.9
30	.0100	100	3300	104.0	.143	91.7	86.2
31	.0089	79.2	4170	131	.113	103	96
32	.0080	64.0	5160	162	.091	114	106
33	.0071	50.4	6550	206	.072	128	118
34	.0063	39.7	8320	261	.057	145	133
35	.0056	31.4	10500	331	.045	163	149
36	.0050	25.0	13200	415	.036	182	167
37	.0045	20.2	16300	512	.029	202	183
38	.0040	16.0	20600	648	.023	225	206
39	.0035	12.2	27000	847	.017	260	235
40	.0031	9.61	34400	1080	.014	290	263
41	.0028	7.84	42100	1320	.011	323	294
42	.0025	6.25	52900	1660	.0089	357	328
43	.0022	4.84	68300	2140	.0069	408	370
44	.0020	4.00	82600	2590	.0057	444	400
45	.00176	3.10	107000	3350	.0044	520	465
46	.00157	2.46	134000	4210	.0035	580	510
47	.00140	1.96	169000	5290	.0028	630	560
48	.00124	1.54	215000	6750	.0022	710	645
49	.00111	1.23	268000	8420	.0018	800	720
50	.00099	.980	337000	10600	.0014	880	780
51	.00088	.774	427000	13400	.0011	970	855
52	.00078	.608	543000	17000	.00087	1080	935
53	.00070	.490	674000	21200	.00070	1270	1110
54	.00062	.384	859000	27000	.00055	1430	1220
55	.00055	.302	1090000	34300	.00043	1560	1330
56	.00049	.240	1380000	43200	.00034	1690	1450
57	.000438	.192	1722000	54100	.00027	1960	
58	.000390	.152	2166000	68000	.00022	2160	
59	.000347	.121	2737000	85900	.00017	2450	
60	.000309	.090	3453000	108400	.00014	2740	

An example at this time will be helpful.

EXAMPLE 8.4

What is the impedance of a cable that uses #22 gauge wire separated by 9/16 in center to center and molded in polyethylene insulation? Table 8.2 shows #22 wire to have a diameter of 0.02536 in and Table 8.1 lists polyethylene with a dielectric constant of 2.5. Converting 9/16 to 0.5625, Equation 8.5 yields

$$Z_0 = \frac{276}{\sqrt{2.5}} \log_{10}\left(\frac{2(0.5625)}{0.02536}\right) \approx 300 \ \Omega$$

If a situation arises where the insulation type is not known, then familiariza-tion with Table 8.1 and some experience will lead you to discover that all common materials have a dielectric constant between 2.1 and 2.7. A good aver-age is 2.4, which can be used as a temporary substitute for the dielectric con-stant. When more accurate information is available, your solution will be more accurate.

Shielded Two-Wire Parallel Lines When the two-wire parallel cable is placed inside a shielded jacket and the jacket is grounded, then the capacitance between the wires and the jacket will increase the total capacitance of the cable. When the impedance is calculated from the L and C values, there is no alteration needed in the impedance formula, which uses the total capacitance. When the impedance is calculated from the mechanical dimensions, a third measurement is necessary. Figure 8.8b shows the inside diameter of the outer jacket as dimension S_d. The impedance equation is altered as follows:

$$Z_0 = \frac{276}{\sqrt{\epsilon}} \log_{10}\left(2D \frac{\left[1 - \left(\frac{D}{S_d}\right)^2\right]}{\left[1 + \left(\frac{D}{S_d}\right)^2\right]}\right) \tag{8.6}$$

Two-Wire Cable Impedance Limits Select any wire size, even as large as 10 in diameter. Place two such wires parallel to each other so that only the thick-ness of a human hair keeps them from shorting together. Figure 8.9 shows this setup. Now calculate the impedance. Regardless of the wire size selected or the tightness of their placement (without shorting), the impedance will calculate to 83.2 Ω. The *minimum impedance* of two-wire open-line cable is 83.2 Ω! Placing the two wires inside a jacket or using a high-conductance insulating material will lower the impedance of the cable. The dielectric material loses its effec-tiveness when the wires are closer than one-tenth of the wire diameter due to the *proximity* effect; that is, two wires carrying current that flows in opposite directions will have a cancellation effect. Therefore, it would be extremely dif-ficult to build a two-wire cable that has an impedance less than 75 Ω.

FIGURE 8.9
Minimum impedance of two-wire cable.

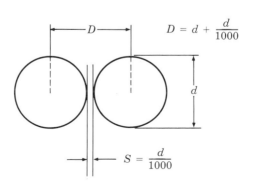

A similar statement could be made about the *maximum impedance* of two-wire parallel cables. Rearranging Equation 8.5 to solve for the distance D between the wires results in Equation 8.7. Using #22 wire, with a diameter of 0.02536 in, the distance between the two wires for a 2000 Ω cable is

$$D = \frac{d\left(\log_{10}^{-1} \frac{Z_0\sqrt{\epsilon}}{276}\right)}{2} \tag{8.7}$$

where D and d are in the same dimensions. The distance is found to be about 18,634.4 feet or 3.529 *miles*. Although this equation has a mathematical solution, you can see that it is impractical. Moreover, in Chapter 10 it will be shown that as the wires are separated by more than 0.2 wavelengths at the operating frequency, they act as antennas. It can be stated that there is *no absolute maximum* impedance for two-wire transmission lines. *Practical maximum limits* are found to be about 600 Ω for communication cables and about 800 Ω for power cables.

8.4.2 Coaxial Cables

The word *co-axial* describes a cable type in which the two wires share a center axis. The larger conductor must be a hollow tube so that the smaller conductor may fit on the inside along the center axis. To hold the conductors in a fixed relative position, spacers are positioned along the inside of the larger conductor. When solid insulation is used for separation, the dielectric constant will be that of the insulating material.

The outer conductor completely surrounds the center conductor and is normally grounded. At ground, the outer conductor acts as a shield to prevent signal radiation from the cable as well as to keep out interference and noise that could affect the desired signal. Because of this arrangement, coaxial cable is also called **shielded cable.** Figure 8.10 highlights the dimensions that are

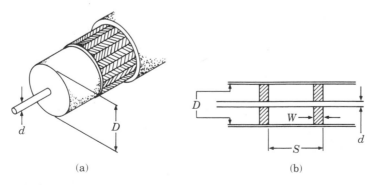

(a)   (b)

FIGURE 8.10
Construction details to be used when calculating line impedance of coaxial cable.

critical to finding the characteristic impedance of coaxial cable from the mechanical measurements. Dimension D is the inside diameter of the outer conductor, and d is the outside diameter of the smaller conductor. The impedance equation is

$$Z_0 = \frac{138}{\sqrt{\epsilon}} \log_{10}\left(\frac{D}{d}\right)$$ (8.8)

$$Z_0 = \frac{60}{\sqrt{\epsilon}} \ln\left(\frac{D}{d}\right)$$ (8.8a)

Rigid coaxial cables are often used for outdoor runs to antennas on top of transmitter towers. Air dielectric is best suited for outdoor use since air is least affected by moisture. The spacers used to hold the center conductor in place have a contributing effect, as shown in Figure 8.10b. The impedance equation is altered to:

$$Z_0 = \frac{138}{\left(1 + \frac{\epsilon - 1}{S}\right)W} \log_{10}\left(\frac{D}{d}\right)$$ (8.9)

where ϵ is the dielectric constant of the spacers.

Coaxial Cable Impedance Limits   Numerical values such as the ones used for two-wire lines could also be set up here to illustrate the minimum and maximum impedances of coaxial cables. Values can be selected to achieve a minimum impedance of 1 Ω, but such a cable would be extremely difficult to make. Values of $D = 0.1283$ in and $d = 0.1$ in with a solid dielectric of 2.23 would form a small cable with a *practical* impedance of 10 Ω. However, because of the cost and difficulty of manufacturing, coaxial cables with an impedance less than 50 Ω are rarely used.

The maximum impedance is just as difficult to define. The impedance is a function of the ratio between the dimensions, D/d. When this ratio is 42.5:1, the impedance is 225 Ω. For a ratio of 65:1, the impedance increases to 250 Ω. Since these are log functions, the ratio climbs faster than the impedance, up to a ratio of 150:1, where the impedance has only risen to 300 Ω. Minimum attenuation is obtained when the ratio is 3.6:1 and the dielectric is air, giving an impedance of 77 Ω. Maximum power-handling ability is reached when the ratio is 1.65:1; with air dielectric, the impedance is 30 Ω. A reasonable compromise between smallest attenuation and maximum power-handling ability is reached with a ratio of 2.3:1, an impedance of 50 Ω. Table 8.3 summarizes the *practical* impedance limits for communication cables, with the understanding that these values are not *absolute* limits.

TABLE 8.3
Summary of cable impedance limits

Cable Types	Maximum	Minimum
Two-wire	600 Ω	75 Ω
Coaxial	225 Ω	10 Ω

8.5 WAVELENGTHS

Every electrical signal has a repeating periodic time function. We have seen this on an oscilloscope and have learned to relate frequency to time per cycle. In the atmosphere that surrounds the earth, radiant energy travels at the speed of light. Therefore, by comparing how often the signal repeats its cycle to the velocity of light, the length of the wave can be determined. The symbol lambda (λ) is used to represent wavelength:

$$\lambda = \frac{V_c}{f \text{ (in hertz)}} \tag{8.10}$$

where V_c = velocity of light
$= 300 \times 10^6$ m/s (meters per second)
$= 984 \times 10^6$ ft/s
$= 1.1808 \times 10^{10}$ in/s
$= 186,400$ mi/s

One cycle of energy at 186.364 kHz would be a mile in length. If the signal could be frozen in time and a voltmeter placed across the line, the positive voltage peaks would be a mile apart. Because the wave is so physically long at the lower frequencies, the communications industry does not normally consider frequencies below 200 kHz as a practical range for transmission.

EXAMPLE 8.5

What is the length of the energy wave (in meters) at an operating frequency of 30 MHz?

$$\lambda = \frac{300 \times 10^6}{30 \times 10^6} = 10 \text{ m}$$

The length of one cycle of energy in air is 10 m long. If this signal were radiated from a transmitting antenna to a receiving antenna 10 m away, the radiated energy would be arriving at the receiving antenna at the same instant that the transmitting antenna had completed one full cycle. In feet, one wavelength at 328 MHz would measure 3 ft. Try some practice examples.

8.5.1 Velocity of Propagation along the Line

In Equation 8.10 the velocity of light was specified for the atmosphere (free air). When the signal is coupled from the atmosphere to a conductor, the velocity of the signal decreases. This is because the molecules are harder to set into motion in the solid dielectric; therefore, the energy moves at a slower rate through the wires. The velocity factor (F_v) is the ratio of the velocity of energy propagation down the line to the velocity of light in free air.

$$F_v = \frac{V_p}{V_c} = \frac{1}{\sqrt{\epsilon}} \qquad (8.11)$$

where ϵ = dielectric constant of the line
V_p = velocity of energy in the line

EXAMPLE 8.6
Find the velocity factor for a cable with a dielectric constant of 2.23.

$$F_v = \frac{1}{\sqrt{2.23}} = 0.6697 \text{ of the velocity in air}$$

The **propagation velocity** V_p is the speed at which the signal moves through the cable. From Equation 8.11,

$$V_p = F_v V_c = \frac{V_c}{\sqrt{\epsilon}} \qquad (8.12)$$

For Example 8.6, this velocity would be

$$V_p = (984 \times 10^6)\,(0.6697) = 658.985 \text{ ft/s}$$

The energy moves 67% as fast as the energy in free air, or 660 million ft/s. As a result, the *wavelength* of the signal as it moves through the line will also decrease. The new wavelength on the line, at 328 MHz, will be

$$\lambda = \frac{658.985 \times 10^6}{328 \times 10^6} = 2.0 \text{ ft} \qquad (8.13)$$

instead of the 3 ft previously determined for free air.

8.5.2 The Line Length in Wavelengths

Once the wavelength of the signal on the line is known and the linear length of the line is measured, we need a simple conversion to derive the electrical length of the line, measured in wavelengths, λ_0. This can be accomplished by using Equations 8.11, 8.12, and 8.13 separately or by using Equation 8.14, which combines the three:

$$\lambda_0 = \frac{\text{length } (f_0) \sqrt{\epsilon}}{V_c} \qquad (8.14)$$

where f_0 = operating frequency
ϵ = dielectric constant

The line length, V_c, and solution must all be in the same units. Again using Example 8.6, at 328 MHz, the length of a 176.4 ft cable with a dielectric constant of 2.23 would be

$$\frac{176.4(328 \times 10^6)\sqrt{2.23}}{984 \times 10^6} = 87.8 \text{ wavelengths}$$

A transmission line that is $\frac{1}{16}$ wavelength long or longer is considered to be a *long line*. One less than $\frac{1}{16}$ wavelength long is called a *short line*.

8.6   PROPAGATION COEFFICIENT

So far, the series resistance and the line losses have been ignored. This is perfectly acceptable when very short, low-loss lines are used. However, this may not always be the case.

When the line losses need to be evaluated, several approaches may be taken, depending on what is known about the line or what can be measured.

1. Using voltage or current information, when 100 mV input results in 10 mV output across the correct line termination, then the losses are

$$\text{Losses} = 20 \log \frac{E_o}{E_i} \quad \text{or} \quad 20 \log \frac{I_o}{I_i} \qquad (8.15)$$

$$= 20 \log \frac{10 \text{ mV}}{100 \text{ mV}} = -20 \text{ dB}$$

Stated in nepers,

$$\text{Losses} = -\ln \frac{E_o}{E_i} \quad \text{or} \quad -\ln \frac{I_o}{I_i} \qquad (8.16)$$

$$= -\ln \frac{10 \text{ mV}}{100 \text{ mV}} = -2.3026 \text{ nepers}$$

Since 1 neper = 8.686 dB, then -2.3026 nepers = -20 dB, which agrees with Equation 8.15.

2. In cases where the losses are stated in nepers on the spec sheet, for example, losses = −2.302585 nepers, these are the losses to the base e ($e = 2.71828$) of the natural log ($\ln = \log_e$), in this case raised to the power of −2.302585.

$$2.71828^{-2.302585} = 0.1 = \frac{E_o}{E_i} \tag{8.17}$$

Since 20 log 0.1 = −20 dB, this solution corresponds to that of Equation 8.15.

Attenuation is a linear function of length and an exponential function of frequency, so most cable manufacturers supply charts showing attenuation at selected frequencies per unit length. For RG59/U cable, the attenuation is:

3.4 dB per 100 ft at 100 MHz
4.9 dB per 100 ft at 200 MHz
7.1 dB per 100 ft at 400 MHz

To use these charts, the only information needed is the length of the cable and the highest operating frequency. The attenuation is linear and can be extrapolated for longer lines. A length of 500 ft at 200 MHz would have a total attenuation of 5 × 4.9 dB = 24.5 dB. With 100 mV input to the line, attenuation of 24.5 dB would result in 6 mV output.

When no specifications are available and immediate data are required, the mechanical measurements can be used in Equation 8.18 to furnish a very close approximation. The constant in this equation includes a power factor that may vary from cable to cable.

$$\text{Attenuation} = 3.948 \times 10^{-6} \, \frac{\left(\frac{f\sqrt{\epsilon}}{D}\right)\left(1 + \frac{D}{d}\right)}{\log \frac{D}{d}} \quad \text{(in dB/100 ft)} \tag{8.18}$$

where f is in hertz
ϵ is the dielectric constant
D is the inside diameter of the outer conductor
d is the outside diameter of the center conductor

As you work with any cable, you will *always* want to know the characteristic impedance. The attenuation must be known in approximately 80% of all cases. The cable capacitance and delay time are important factors in approximately one-half of the cases, while the need to know the inductance or the conductance range is important at higher frequencies in cases that occur from 25% of the time to never.

8.7 TOTAL LINE PARAMETERS

When a detailed accounting of a specific transmission line is required, the cable geometry will supply all of the necessary information to within about 3% of the measured values. We will be able to solve for Z_0, T_d, R, L, C, G, and attenuation per unit length.

It is helpful to first understand the behavior of transmission lines over the frequency spectrum. There is a low frequency, from 0 Hz to a few kilohertz, where R and L do not change. For the next few decades, both R and L obey a very complicated set of laws, which will not be discussed here. Approaching a frequency where $2\pi fL >> R$, called the **leveling frequency**, R starts to increase as a factor of the square root of frequency, and L becomes independent of frequency. The leveling frequency for two-wire lines is a few tens of kilohertz, whereas coaxial cable reaches a leveling frequency at a few hundred kilohertz. It is for this reason that the term *transmission line* applies only to frequencies above approximately 200 kHz. Below the leveling frequencies, the cables act as (and are called) cable pairs, for two wires, and shielded cable, for coaxial cables.

EXAMPLE 8.7

A coaxial cable has $d = 0.025$ in, $D = 0.15$ in, and the solid insulating material appears to be polyethylene; thus, a dielectric constant of 2.23 will be used. What can be determined about this particular cable?

1. By Equation 8.8, $Z_0 = 71.91\ \Omega$ (72 Ω).
2. Find the basis for determining leveling frequency, $L = 0.1091\ \mu\text{H/ft}$ $= 10.91\ \mu\text{H/100 ft}$.
3. By rearranging Equation 8.1, $C = 21.10\ \text{pF/ft} = 2110\ \text{pF/100 ft}$.
4. The leveling frequency where $2\pi fL = 10R$ is found as follows. From Table 8.2, $R = 16.14\ \Omega/1000\ \text{ft} = 1.614\ \Omega/100\ \text{ft}$. Therefore,

$$f = \frac{10R}{6.28L} = \frac{16.14}{6.28 \times 10.91 \times 10^{-6}} = 235\ \text{kHz}$$

5. As the frequency increases, the current density within the wire rises toward the surface of the wire; this is called *skin effect*. The DC wire resistance is then converted to AC resistance:

$$R_{\text{ac}} = 0.134d\sqrt{f}R_{\text{dc}} \tag{8.19}$$
$$= 0.134(0.025)\sqrt{235\ \text{kHz}}(1.614) = 2.452\ \Omega/100\ \text{ft}$$

The attenuation constant gives accurate results by the following formula:

$$\alpha = \frac{R_{\text{ac}}}{2Z_0}\sqrt{\frac{f}{f_1}} + \frac{GZ_0}{2}\left(\frac{f}{f_1}\right) \quad \text{(in nepers/100 ft)} \tag{8.20}$$

R_{ac} is from Equation 8.19, G is measured to be 0.5351×10^{-6} siemans, f is the frequency of the test condition, and f_1 is the leveling frequency from step 4. The attenuation can now be determined at any frequency by application of Equation 8.20.

At 100 MHz,

$$\alpha = \frac{2.452}{2 \times 72} \sqrt{\frac{100 \text{ MHz}}{200 \text{ kHz}}} + \frac{0.5351 \times 10^{-6}(72)}{2} \frac{100 \text{ MHz}}{200 \text{ kHz}}$$

$$= 0.01703(22.36) + 19.26 \times 10^{-6}(500)$$

$$= 0.3904 \text{ nepers} \times 8.686 = 3.39 \text{ dB/100 ft}$$

At 200 MHz,

$$\alpha = 0.01703 \sqrt{\frac{200 \text{ MHz}}{200 \text{ kHz}}} + 19.26 \times 10^{-6} \left(\frac{200 \text{ MHz}}{200 \text{ kHz}} \right)$$

$$= 0.5578 \text{ nepers} \times 8.686 = 4.845 \text{ dB/100 ft}$$

At 400 MHz,

$$\alpha = 0.01703 \sqrt{\frac{400 \text{ MHz}}{200 \text{ kHz}}} + 19.26 \times 10^{-6} \left(\frac{400 \text{ MHz}}{200 \text{ kHz}} \right)$$

$$= 0.80012 \text{ nepers} \times 8.686 = 6.945 \text{ dB/100 ft}$$

The cable in question here is RG59/U. Table 8.4 compares the manufacturer's specifications to the values calculated in this example for a 100 ft length of cable. It is interesting to note that the calculations in this example are within 3% of the published specifications. More care in the original measurements would have made the results even closer.

TABLE 8.4

Parameter	Calculated Values	Manufacturer's Values
Diameter (d)	0.025 in	0.02535 in
Diameter (D)	0.15 in	0.149 in
Dielectric constant	2.23	2.23
Inductance L	10.92 μH	not stated
Capacitance C	2110 pF	2100 pF
Impedance (Z_0)	72 Ω	73 Ω
Frequency (min)	235 kHz	not stated
Conductance (G)	0.54 μS	not stated
Resistance (R)	1.614 Ω	1.614 Ω
Attenuation (α)		
@ 100 MHz	3.39 dB	3.4 dB
@ 200 MHz	4.845 dB	4.9 dB
@ 400 MHz	6.945 dB	7.1 dB

8.7.1 Parallel Lines

Only a small modification to the approach used above for coaxial lines need be made for the examination of two-wire lines.

EXAMPLE 8.8 ═══

The diameter of each wire in a two-wire flat ribbon cable is 0.025 in, and they are separated by 0.53 in, center to center, by polyethylene insulation. Find the attenuation at 150 MHz.

1. Find the characteristic impedance by using Equation 8.5:

$$Z_0 = \frac{276}{\sqrt{2.23}} \log_{10}\left(\frac{2(0.53)}{0.25}\right) = 300.7 \ \Omega$$

2. The attenuation can be found by omitting the conductance factor from Equation 8.20 and solving for the losses from the resistive factor only. Below 500 MHz, the conductance has almost no effect on the losses, and above 500 MHz, the conductance varies so unpredictably with environment as to make any calculation meaningless. Therefore, only the AC resistance becomes an unknown. For 100 ft lengths, it is found to be

$$R_{ac} = \frac{2.01 \times 10^{-4}}{d}\sqrt{f} \tag{8.21}$$

$$= \frac{2.01 \times 10^{-4}}{0.025}\sqrt{1.5 \times 10^8} = 95.15 \ \Omega/100 \ \text{ft}$$

Because the *loop* resistance is needed, we must raise the resistance to $2 \times 95.15 = 190.3 \ \Omega$. The attenuation then becomes

$$\alpha = \frac{R_{ac}}{2Z_0} \tag{8.22}$$

$$= \frac{190.3}{2(300.7)} = 0.3164 \ \text{nepers} = 2.75 \ \text{dB}$$

The catalog lists this cable with an attenuation of 2.8 dB at 150 MHz, placing our calculations at 2% error.

8.8   **MAXIMUM POWER TRANSFER**

Up to this point, only transmission lines that are terminated in a resistive load equal to the characteristic of the line have been considered. This is the ideal case and certainly the target condition, but other load conditions are found that are less than ideal. **Maximum power transfer** occurs when the load is resistive and equal to the surge impedance of the line and, furthermore, when the line impedance is equal to the generator (transmitter) output impedance. Under these arrangements, the highest percentage of transmitter output power will appear across the line-terminating impedance.

8.8.1 Nonresonant Lines

The conditions for maximum power transfer describe a transmission line that is termed **nonresonant.** The load is resistive and matches the impedance of the line. All of the power put into the line by the transmitter will move down the line and appear across the load resistance (minus some small copper wire losses).

8.8.2 Resonant Lines

The transmission line is termed **resonant** when the load impedance *does not* match the line impedance or is *not* purely resistive. The load impedance may be either larger or smaller than the surge impedance of the line and may also be resistive *and* reactive. The energy sent down the line by the transmitter is *not all* absorbed by the load. The quantity of energy not absorbed is reflected back up the line to the transmitter. There are now two energy waves moving through the line at the same time, but in different directions. The **incident wave** is the wave moving from the generator to the load. The **reflected wave** moves in the opposite direction, along the line from the load back toward the generator. Both waves are continuous in nature; that is, they are two sine waves moving through the line in opposite directions at the same time. There will be moments in time when the peaks of the waves align and the voltages add together. There will be other moments when the two waves are totally out of phase and completely cancel each other. Figure 8.11 represents these two waves moving along the line.

FIGURE 8.11
Incident and reflected waves moving
along the line with a mismatched load.

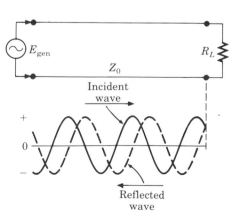

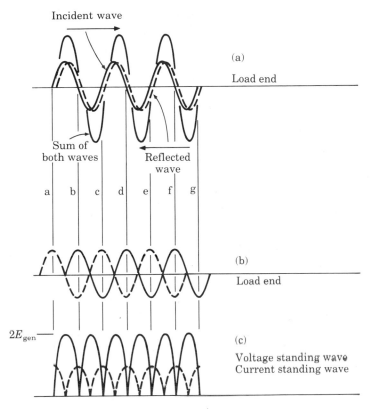

Incident wave

(a)

Load end

Sum of both waves

Reflected wave

a | b | c | d | e | f | g

(b)

Load end

$2E_{gen}$

(c)

Voltage standing wave
Current standing wave

FIGURE 8.12
The development of the standing wave on the line.

Figure 8.12a shows the same two voltage waves at the moment before they are exactly in phase. The sum of the voltage waves would crest at twice the peak voltage of either wave separately. In Figure 8.12b, each wave has traveled another quarter wavelength (90°) in its respective direction. The two waves are now almost exactly out of phase and will completely cancel.

Study Figure 8.12a and b very carefully. Also note that the sum of the voltages at each of the time intervals a, b, c, d, e, f, and g will be equal to zero. In Figure 8.12a, this is because both waves are passing through zero voltage at the same time. In Figure 8.12b, the voltage sums are zero because the two waves completely cancel each other.

Although the incident wave and the reflected wave are constantly in motion along the line, the positions along the line where the waves add together and cancel each other *do not move at all*. This creates a standing wave of voltage along the line that is *stationary*. The minimum value of standing wave voltage may go to 0 V, and the crest of the standing voltage wave will reach a maximum of two times the generator peak voltage ($2E_{gen}$). The magnitude of the standing voltage wave could be damaging to the insulation and cause a short circuit in the cable. Review Figures 8.11 and 8.12, and try to develop a mental image of the two sine waves moving in opposite directions, adding together at some points and canceling each other at other points. If you have difficulty visualizing this effect, then just accept it as fact and move on. The concept will be used many times in the forthcoming examples, and the cloud may lift then.

8.8.3 Standing Wave Ratio

Standing waves may be measured. The voltage pattern of Figure 8.12c presents another important principle of the standing wave, that is, the ratio of the *minimum* to *maximum* voltage standing on the line.

EXAMPLE 8.9

When 110 V of peak signal is applied to the line, and the load is mismatched to absorb only 80 V, the remainder of 30 V will be reflected back up the line toward the generator (transmitter). The standing wave pattern along the line (similar to Figure 8.12c) will have a peak value of the sum of the generator voltage plus the reflected voltage, 110 V + 30 V = 140 V. The minimum (trough) value will be the difference between the two voltages, 110 V − 30 V = 80 V. The voltage standing wave ratio (VSWR) will be

$$\text{VSWR} = \frac{E_{\max}}{E_{\min}} = \frac{140 \text{ V}}{80 \text{ V}} = 1.75:1 \qquad (8.23)$$

The numerical value of the voltage standing wave ratio has another application. It reveals the ratio of *load impedance* to *line impedance* and is always a number greater than 1. This expression is simply

$$\text{SWR} = \frac{Z_L}{Z_0} \quad \text{or} \quad \frac{Z_0}{Z_L} \qquad (8.24)$$

whichever arrangement generates a ratio greater than 1. The standing wave ratio is a dimensionless number and has the same relationship regardless of whether the load is larger or smaller than the line impedance. A load of 400 Ω on a 200 Ω line will have a ratio of 2:1. The same ratio will result when a 100 Ω load is connected to the 200 Ω line. The ratio does not tell you "what," only "how much." It is a statement of the degree of mismatch, but does not tell the direction or value of mismatch. If you read a specification of 3:1, all you would be able to conclude is that the load is three times larger than the line impedance *or* one-third the value of the line impedance. Knowing the value of line impedance will only help to establish two possibilities for the value of load impedance. The location of the voltage maxima and voltage minima along the line determine the direction.

Order of the Unknown   From a systems point of view, the line impedance is almost always known. The SWR is then the easiest unknown to measure and can be found by simple direct meter readings. The value of $E_{\max}$ and $E_{\min}$ are the next most easily measurable values, while the load impedance (because it is a complex variable) will require gathering more data. Knowledge of the first two values makes the remaining values easier to find. The procedure for finding these answers is outlined in Chapter 9 under the discussion of Smith charts.

Reflection Coefficient   The upper-case Greek letter gamma (Γ) is used as a symbol for the reflection coefficient. The value itself is a decimal number used to represent the percentage of energy coming down the line that *is not* absorbed by the load, in other words, the percentage of reflected energy. The coefficient can be determined by any of several standard ratios:

$$\Gamma = \frac{E_{\text{reflected}}}{E_{\text{applied}}} = \frac{E_{\max} - E_{\min}}{E_{\max} + E_{\min}} = \frac{\text{SWR} - 1}{\text{SWR} + 1} = \frac{Z_L - Z_0}{Z_L + Z_0} \qquad (8.25)$$

8.9   LIMITS OF THE LOAD IMPEDANCE

The load impedance is ideal when it is equal to the line impedance. In reality, the load impedance could be *any* value between zero and infinity. There are only a few occasions when the load would be left off ($Z_L = \infty$) or when the load would be short-circuited ($Z_L = 0$), but it is easy to see and understand that these are the extreme limits. In practice, it would be unusual to find the load impedance different from the line impedance by more than a factor of 5. These extreme limits are the easiest to explain and understand.

8.9.1   Open-Circuit Load Impedance

When the load impedance is removed, the output end of the transmission line becomes an open circuit and an infinite impedance. A train of waves moving through the transmission line from the generator will reach the open end of the line and encounter an impedance that will absorb no energy at all; the wave is completely reflected back up the line toward the generator. Figure 8.13 shows the effects when the incident wave is reflected by an open circuit.

Since there is no current through an open circuit, the voltage at the open-circuit terminals is at maximum, equal to two times the peak generator voltage. Looking at the voltage as it moves back up the line toward the generator, we can see that the incident wave and the reflected wave start to add and cancel at various points along the line. The first voltage minimum will occur one-quarter wavelength from the load terminals. Wave cancellation or wave summation is repeated every half wavelength, so that the second minimum will appear at one-half wavelength from the first minimum, or three-quarters of a

FIGURE 8.13
Voltage standing wave for an open load.

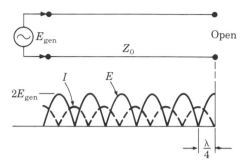

wavelength from the load terminals. The maximum voltage seen at the output terminals will be repeated at half wavelengths from the load terminals, and a standing wave pattern is thus formed that can be predicted for any point along the transmission line. A standing wave of current is also present along the line and is 90° out of phase from the voltage wave, that is, minimum current at the load terminals and maximum one-quarter wavelength away from the load terminals. The current standing wave is used less often than the voltage standing wave.

For the open-circuit load, E is maximum and I is zero at the load terminals. This is similar in effect to the conditions of a parallel-resonant circuit, where Z is high, E is large, and I is minimal. One-quarter wavelength from the load terminals, we find I to be maximum and E to be minimum. High currents indicate low impedance, and with low voltage at this location, the conditions match those of a series-resonant circuit. It is because of these extreme conditions that the line is termed a *resonant line,* which indicates a severely mismatched load condition. Furthermore, we have noted the line impedance change for any location along the line as a result of the change of E and I at this

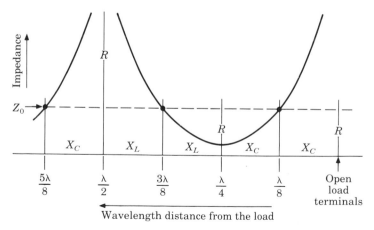

FIGURE 8.14
Impedance anywhere along the line for an open load condition.

FIGURE 8.15
Voltage standing wave for a shorted load.

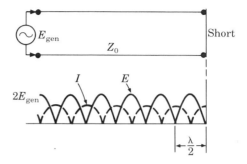

location. A graph of the line impedance as a function of the wavelength distance from the load is given in Figure 8.14. This graph is noteworthy here but will be detailed later.

8.9.2 Short-Circuit Load Impedance

The same general conditions apply to the short-circuit load. That is, the generator still feeds a train of waves down the transmission line, and the short-circuit load, which cannot absorb or dissipate energy, reflects the total energy back toward the generator. The reflected wave, however, is not the same as for the open-circuit load conditions, for here Z_L is zero. The current through zero resistance is maximum, while the voltage drop across the short circuit is minimum (or zero).

Compare the two reflected voltage waves in Figures 8.13 and 8.15. The electrical behavior of the waves in Figure 8.13 demonstrates true reflection; that is, a positive peak of the incident wave arriving at the open load will be reflected back as a positive peak. A positive peak of the incident wave arriving at the short-circuit load in Figure 8.15 is inverted as it is reflected. These conditions satisfy the basic AC voltage theory and the theory of waves and results in a standing wave of voltage equal to E_{max} across the open load terminals and 0 V across the shorted load terminals.

The impedance along the line having a shorted load differs from the open-load line impedance in that the maximum and minimum line impedance points are shifted by one-quarter wavelength (90°) along the line. This also shifts the locations of the voltage maximum and current maximum in the standing wave by 90° relative to the locations along the open-circuit line. Compare Figure 8.14 to Figure 8.16 to see these changes.

8.9.3 Input Impedance of the Line

Discounting any losses due to cable length, the load impedance at the output end of the line will affect the impedance at the input end. Maximum power transfer from the generator (transmitter) to the load takes place when the load impedance equals the line impedance and the line impedance equals the

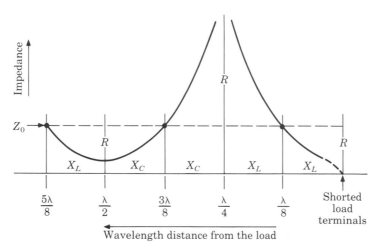

FIGURE 8.16
Impedance anywhere along the line for a shorted load condition.

generator impedance. In the case of transmission lines, the load reflects an impedance directly back to the transmitter output terminals.

Figure 8.17 shows the voltage standing waves (of Figures 8.13 and 8.15) compared to the impedance graphed in Figures 8.14 and 8.16. They show the line input impedance as seen by the generator at selected lengths of transmission line.

In the left column, when the shorted load is one-eighth wavelength away from the generator, the generator will see an inductive load equal in value to the characteristic cable impedance. When the short-circuit load is a distance of one-quarter wavelength from the generator, then the impedance seen by the generator makes the circuit equivalent to a parallel-resonant circuit. When the short-circuit load is at three-eighths of a wavelength from the generator terminals, the generator sees an impedance that looks capacitively reactive with a value equal to the cable impedance. When the short-circuit load is one-half wavelength away from the generator terminals, the generator sees a minimum impedance, which resembles a series-resonant circuit. At five-eighths of a wavelength from the generator, the shorted load looks exactly as it did at one-eighth wavelength. That is, the line input impedance looks inductively reactive and equal in value to the cable impedance.

The prime observation here is that the line impedance repeats itself *every half wavelength*. For instance, the line will look inductive when the shorted load is ⅛, ⅝, 1⅛, 1⅝, etc. wavelengths away from the generator terminals. Moreover, this means if we can understand and explain what happens within the first half wavelength from the shorted load, then whatever happens along the remaining length of line will be a repetition of what happened along the first half wavelength.

The open-load line impedance values, shown in the right column, repeat themselves like those of the short-circuit load. When the open load is one-eighth wavelength from the generator, the generator sees a capacitively reactive load impedance that is equal in value to the cable impedance. When the open end of the line is located one-quarter wavelength from the generator, the generator sees an input impedance that looks like a series-resonant circuit. With the open load at three-eighths of a wavelength away, the generator sees an inductive load with a reactance equal to the characteristic impedance of the transmission line. When the cable's open end is one-half wavelength from the generator, the line looks like a high-impedance circuit equivalent to a parallel-resonant tuned circuit. Beyond one-half wavelength, the performance of the line repeats itself continually for each half wavelength. Therefore, if you know

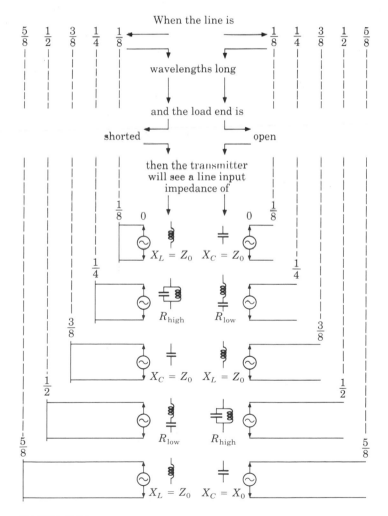

FIGURE 8.17
The input impedance to the line that the generator will see for varying line lengths under conditions of open and shorted loads.

the line length in wavelengths, you can examine the last half wavelength and disregard the remainder of the line (except for line losses). For example, the solution to Equation 8.14 gave that particular cable a line length of 87.8 wavelengths. Disregard 87.3 wavelengths, and examine only the half wavelength remaining to determine the impedance behavior along the entire line.

8.10   TRANSMISSION LINE AS CIRCUIT COMPONENTS

An open or shorted section of transmission line has been shown to exhibit electrical characteristics that may resemble capacitors, inductors, or resonant circuits. Indeed, transmission line is often used as such a circuit element.

The circuit shown in Figure 8.18 uses a quarter-wave shorted section of coaxial cable as a parallel-resonant circuit. From Figure 8.17, it is seen that a quarter-wave shorted section of line is equivalent to a parallel-resonant circuit. That is, it has one resonant frequency, it has Q, and it has bandwidth exactly like a discrete component circuit.

The shorted section of line was used here for two reasons. First, the circuit must be capable of passing DC current. Second, the open transmission line will radiate more signal into the surrounding system than will a shorted line.

A series-resonant circuit could have been constructed in a similar manner. That is, a quarter-wavelength open line or a half-wavelength shorted line will both look like a series-resonant circuit.

8.10.1 Capacitance

A short section of *open* transmission line can act like a capacitor. From a line of known impedance, select an operating frequency and value of capacitance and then solve for the length of cable that will serve as that capacitance value. First find the capacitive reactance, and then use Equation 8.26 to find the length.

$$\text{Length} = \lambda \; \frac{\arctan \left(\dfrac{Z_0}{X_C} \right)}{360} \tag{8.26}$$

EXAMPLE 8.10
Find the length of 75 Ω cable that will form a 0.5 pF capacitor at 500 MHz. The wavelength is 60 cm. The capacitive reactance is equal to 636.62 Ω at 500 MHz.

$$\text{Length} = 60 \text{ cm} \; \frac{\arctan \left(\dfrac{75}{636.6} \right)}{360} = 1.12 \text{ cm}$$

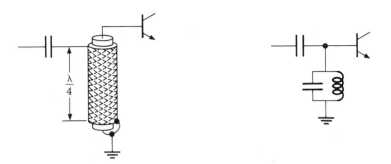

FIGURE 8.18
Cable used as a resonant circuit.

The cable length and wavelength use the same dimensions. It is more common to know capacitance, Z_0, frequency, and wavelength, and then solve for the cable length that will result in the desired capacitance value.

Sometimes the line length is known and it is the value of C that is the unknown. Use Equation 8.27 to find the reactance of the line from the line constants. Then use the value of X_C to find the capacitance.

$$X_C = \frac{Z_0}{\tan\left(\dfrac{360 \times \text{cable length}}{\lambda}\right)} \quad \text{(in ohms)} \tag{8.27}$$

It is interesting to note that the 1.12 cm cable will act as a 0.5 pF capacitor over a *very wide* range of frequencies.

It is really only necessary to know the frequency in order to set up Equation 8.26 and compare the cable length to the wavelength. This corresponds to Figures 8.14 and 8.16, where the impedance along the line is referenced to as a fraction of a wavelength. You will understand this if you set up and solve for selected values of C at widely separated frequencies. Calculate wavelength and reactance at each new frequency.

8.10.2 Inductance

A similar format is used to set up the conditions to find the inductance for a given length of transmission line. The notable difference is that the cable is *short-circuited* at the receiving end for the inductance condition. The equation is

$$X_L = Z_0 \tan\left(\frac{360 \times \text{length}}{\lambda}\right) \quad \text{(in ohms)} \tag{8.28}$$

The length is usually the most sought-after quantity, so Equation 8.28 is usually rearranged to solve for this value:

$$\text{Length} = \lambda \; \frac{\arctan\left(\dfrac{X_L}{Z_0}\right)}{360} \tag{8.29}$$

Compare Equations 8.26 and 8.29 and note that the only difference is the exchange of Z_0/X_C in Equation 8.26 for X_L/Z_0 in Equation 8.29.

These equations work well for all cases except one. This case is when the cable length is known and the search is for the value of C or L by solving to find X_C or X_L. X_C or X_L may be found to be a *negative* number, and our instincts tell us that reactance cannot be negative. The reverse polarity indicates that the cable is longer than one-quarter wavelength. The simplest remedy is to make the cable one-quarter wavelength shorter or longer and continue with the problem solving. A second method may be used when the cable length is fixed and a specified reactance is required. Change the load end termination from its present condition to the opposite condition, that is, shorted to open or open to shorted. The circuit must be examined to see if it will tolerate this change. The determining factor is that a shorted termination will pass DC current and an open load will not. Is there a DC current loop in the circuit, and will the circuit tolerate the DC current?

8.10.3 Impedance-Matching Transformers

A quarter-wavelength section of transmission line has the unique property of transforming an impedance. This is verified in Figures 8.14 and 8.16. The impedance one-quarter wavelength from a short circuit resembles an open circuit. It stands to reason that an impedance-matching transformer is a section of transmission line one-quarter wavelength long. The impedance change is due to the phase relationship between the incident wave and the reflected wave on the line. At frequencies above 100 MHz, the transmission line transformer is more efficient than a common wire-wound transformer of equal value. The disadvantage of the transmission line transformer is that it will be one-quarter wavelength long only over a very narrow range of frequencies. A value of $\pm 5\%$ of the center frequency seems to be the functional range for frequencies above 100 MHz, but this is sufficient to make these transformers a valuable tool.

The surge impedance of the matching section of transmission line placed between the load impedance and the line impedance in Figure 8.19a can be determined by a single equation:

$$Z_T = \sqrt{Z_0 Z_L} \tag{8.30}$$

All impedances are purely resistive and expressed in ohms. The transformer is connected directly at the load terminals for this case. In cases where the load impedance is both resistive and reactive, the matching transformer is placed in

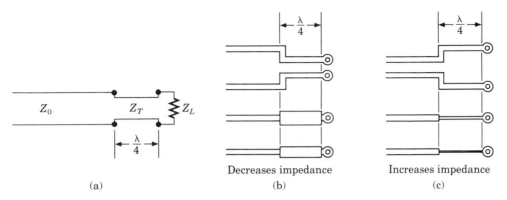

FIGURE 8.19
Two-wire line (quarter-wave transformer).

the line a short distance back from the load. These conditions are covered in Chapter 9 in the discussion of Smith charts.

EXAMPLE 8.11

What is the impedance of the transmission line transformer needed to match a 144 Ω load to a 225 Ω transmission line?

$$Z_T = \sqrt{(144)(225)} = 180 \ \Omega$$

Matching transformers for two-wire lines (open wire or insulated) are simple in construction. If there is not a commercially available line with the desired impedance, then use a fairly stiff wire and bend it into shape as needed to form the correct separation between conductors. Use Equation 8.5 to determine the correct separation for the wire size and impedance to fit the conditions. Wire diameter *and* separation affect impedance. Some practical examples are shown in Figure 8.19.

Coaxial cables require slightly more imagination when impedances are to be matched. Equation 8.30 still applies, but the mechanics of altering coaxial sizes are different. The changes are usually in the form of conductor size change. That is, the center wire size may be increased or decreased to change impedance, *or* the outer conductor size may be changed. Insert sleeving is commonly used in these cases. Usually, the matching transformer impedance is larger than the line impedance but smaller than the load impedance. Sleeves may be brazed onto the existing cable or sleeve inserts may be slid into place around the center conductor or inside of the outer conductor. In all cases, the impedance change is accomplished by a physical change in cable size. Several examples are shown in Figure 8.20, where the size is changed by both machining and sleeve addition.

To decrease impedance

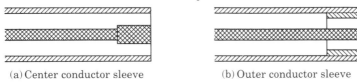

(a) Center conductor sleeve   (b) Outer conductor sleeve

To increase impedance

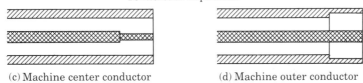

(c) Machine center conductor   (d) Machine outer conductor

FIGURE 8.20
Coaxial cable (quarter-wave transformer).

QUESTIONS

1. Define the term *transmission line.*
2. How many basic classifications are there in the transmission line family?
3. Of the basic types of transmission line, how many are metallic wire conductors?
4. The surge impedance of a transmission line terminated with a resistor equal to Z_0 is independent of length or frequency. (T)(F)
5. Find the value of characteristic impedance for a cable that has inductance of 0.25 μH per 100 ft and capacitance of 100 pF per 100 ft.
6. What is the time delay for 100 ft of the cable described in Question 5?
7. What is the time delay for 750 ft of the cable described in Question 5?
8. Find the impedance and delay time for 350 ft of cable that has $L = 0.281$ μH per 100 ft and $C = 50$ pF per 100 ft.
9. What is the delay time of a 150 Ω cable that is 500 ft long and has capacitance of 50 pF per 100 ft?
10. What is the surge impedance of a cable that has 4.5 μH per 100 ft and 500 pF per 1000 ft?
11. Two #25 wires are separated by ⅜ in center to center by a material with a dielectric constant of 2.23. What is the cable impedance?
12. What wire size would be needed to make a two-wire open transmission line of 150 Ω, when the center-to-center separation between the two wires is ⅜ in?
13. What center-to-center separation of a two-wire open line would be needed to make a 500 Ω transmission line using #20 wire?
14. What center-to-center separation of a two-wire open line would be needed to make a 70 Ω transmission line using #20 wire?

15. At an operating frequency of 236 MHz, 0.2 wavelengths is about 10 in. What is the open-wire cable impedance of two #22 wires separated by 10 in?

16. Below is a list of transmission line parameters. Identify which are frequency-sensitive for a stated length and which are dependent on length at a fixed frequency.
    a. Inductance (L)
    b. Capacitance (C)
    c. Impedance (Z_0)
    d. Resistance (R)
    e. Conductance (G)
    f. Attenuation (α)
    g. Propagation velocity (V_p)
    h. Propagation factor (F_p)

17. Of the values that are frequency-sensitive, which are linear with changes in the length of the cable?

18. Of the values that are sensitive to length at a fixed frequency, state if the change for each is linear or nonlinear with changes in length.

19. Of the values that are sensitive to length at a fixed frequency, state if the change for each is linear or nonlinear with changes in frequency.

20. A transmission line is considered to be any electrically suitable confinement used to guide the flow of energy, by physical contact between two points within a system. (T)(F)

21. The frequency of the applied signal and the length of a correctly terminated transmission line will have no effect on the surge impedance. (T)(F)

22. The length of the transmission line will affect the (a) input power, (b) line losses, (c) load impedance, (d) reflection coefficient.

23. The *practical minimum* impedance of a two-wire transmission line is considered to be _____ Ω.

24. The *practical maximum* impedance of a two-wire transmission line is considered to be _____ Ω.

25. The *practical minimum* impedance of coaxial transmission line is considered to be _____ Ω.

26. The *practical maximum* impedance of coaxial transmission line is considered to be _____ Ω.

27. Why is two-wire transmission line impractical below a specified impedance value?

28. Why is two-wire transmission line impractical above a specified impedance value?

29. Calculate the distance that would separate a 1 in diameter center conductor from the outer conductor for a coaxial line impedance of 1 Ω.

30. Calculate the distance that would separate a center conductor with a diameter of 0.005 in from the outer conductor for a coaxial transmission line of 3000 Ω impedance.

31. Which of the following values are needed to determine the characteristic impedance of a transmission line? (a) inductance, (b) capacitance, (c) inductance and

capacitance, (d) inductance, capacitance, and length for each value, (e) inductance, capacitance, length for each value, and load impedance.

32. Which of the following facts are needed to determine the time delay along a transmission line? (a) inductance, (b) capacitance, (c) inductance and capacitance, (d) inductance, capacitance, and length for each value, (e) inductance, capacitance, length for each value, and load impedance.

33. A transmission line is considered to be a *long* line when the cable length is equal to or longer than _____ wavelengths.

34. A correctly terminated transmission line has a load impedance that is (a) resistive, (b) inductive and equal to Z_0, (c) capacitive and equal to Z_0, (d) resistive and equal to Z_0, (e) open-circuited.

35. Which of the following is *not* one of the four principal electrical properties of transmission lines? (a) capacitance per unit length, (b) inductance per unit length, (c) series RF resistance, (d) series DC resistance, (e) shunt conductance.

36. For two-wire transmission lines, in which direction will the impedance change when the line capacitance increases? (a) increased impedance, (b) decreased impedance, (c) no change.

37. For two-wire transmission lines, in which direction will the time delay change when the line capacitance increases? (a) increased delay time, (b) decreased delay time, (c) no change.

38. For two-wire transmission line, in which direction will the impedance change when the center-to-center separation increases? (a) increased impedance, (b) decreased impedance, (c) no change.

39. For two-wire transmission line, in which direction will the line capacitance change when the center-to-center separation is increased? (a) increased capacitance, (b) decreased capacitance, (c) no change.

40. For coaxial cable, in which direction will the impedance change when the outer conductor is made smaller, with no change to the center conductor? (a) increased impedance, (b) decreased impedance, (c) no change.

41. For coaxial cable, in which direction will the impedance change when *both* the center conductor and the outer conductor are made smaller by one-half? (a) impedance increases, (b) impedance decreases, (c) no change.

42. When the delay time is 10 ns for 100 ft of cable, what will the delay time be for 2500 ft of the same cable? (a) 10 ns, (b) 250 ns, (c) 100 ns, (d) 2.5 ns, (e) 1000 ns.

43. What is the velocity of radiant energy in the atmosphere:
   a. in feet per second
   b. in meters per second
   c. in inches per second.

44. The velocity of energy in solid material will be (faster than) (slower than) (the same as) the velocity of energy in the atmosphere.

45. State the relationships that exist between frequency, velocity, and wavelength.

46. Wavelength and frequency are (directly proportional) (inversely proportional) to each other.

47. How many wavelengths are of prime concern when the cable is 17.8 wavelengths long? (a) 17.8, (b) 8.9, (c) 0.8, (d) 0.3, (e) none of the above.

48. What proportional length of transmission line is examined in order to understand the behavioral patterns of the line along its entire length?

49. Describe the relationships between the velocity of energy in free air to the velocity of energy through the transmission line.

50. Define the relationship between the neper and the decibel.

51. What frequency would be considered when selecting a cable for a wideband distribution system? (a) lowest frequency, (b) center frequency, (c) geometric center frequency, (d) highest frequency.

52. Which of the following conditions describes a resonant line? The load is: (a) capacitive and equal to Z_0, (b) resistive and equal to Z_0, (c) inductive and equal to Z_0, (d) larger than Z_0, (e) smaller than Z_0.

53. A resonant line (will) (will not) have standing waves.

54. Describe a nonresonant line.

55. A nonresonant line (will) (will not) have standing waves.

56. Standing waves on a transmission line are desirable. (T)(F)

57. Z_L/Z_0, Z_0/Z_L, and E_{max}/E_{min} are equations to describe what ratio?

58. The reflection coefficient is a measure of the percentage of power arriving at the load that is (a) absorbed, (b) reflected by the load back to the generator.

59. The standing wave ratio will always be a number greater than 1. (T)(F)

60. The reflection coefficient will always be a number greater than 1. (T)(F)

61. A shorted section of transmission line can be made to look like a circuit element representing capacitance, inductance, series tuned circuits, or parallel tuned circuits. (T)(F)

62. An open section of transmission line can be made to look like a circuit element representing capacitance, inductance, series tuned circuits, or parallel tuned circuits. (T)(F)

63. At the output terminals of an open transmission line, the voltage will be (a) maximum, (b) minimum.

64. During the first quarter wavelength of shorted transmission line from the load, the line impedance will look (a) inductive, (b) capacitive, (c) resistive, (d) none of the above, (e) all of the above.

65. What will the input impedance of an open transmission line look like at one-quarter wavelength from the load? (a) inductance, (b) capacitance, (c) resistance.

66. What is the characteristic impedance of a transmission line that has 6.64 μH of inductance per 100 ft and 295 pF of capacitance per 100 ft?

67. Find the characteristic impedance of a transmission line with 375 pF per 100 feet and 1.944 μH per 100 ft.

68. A 50 Ω cable has 50 pF per meter of capacitance. How much inductance per meter will this line have?

69. Find the capacitance of a 125 Ω cable, 3 meters long, that has 0.516 μH per meter.

70. How many nanoseconds of delay will 100 ft of cable introduce into the system when the cable has 100 pF capacitance and 9 μH inductance per 100 ft?

71. How many nanoseconds of delay will be introduced by 100 ft of cable that has 150 pF capacitance and 3.375 μH inductance per 100 ft?

72. What is the characteristic impedance of the cable described in the preceding question?

73. How many feet of 72 Ω cable with 19 pF per ft of capacitance is required to cause 100 ns of delay?

74. How many meters of 52 Ω cable, with 2200 pF per kilometer, is required to cause 500 ns of delay?

75. What is the wavelength (in feet) of a half wave of energy at 48 MHz in free air?

76. What is the signal frequency when one-half wavelength measures 20 cm in free air?

77. What is the frequency when one wavelength is 15 in long?

78. How many meters long is one wavelength (in air) for the frequency of 325.9 MHz?

79. How many wavelengths long is a 41.25 meter air dielectric cable operating at 120 MHz?

80. How many wavelengths long is a 15 meter air dielectric cable at 164 MHz?

81. What is the impedance of a two-wire parallel transmission line that uses two conductors of 0.142 in diameter separated by 0.25 in center to center?

82. Find the surge impedance of a two-wire parallel line that uses two conductors of 0.214 in diameter separated by 3/8 in center to center.

83. What is the center-to-center separation of the two conductors of a 250 Ω cable that uses 18 gauge (0.040 in) diameter wire?

84. What decimal size of wire would be used for a 210 Ω ribbon cable that had to have a wire separation of 0.092 in in air?

85. What size is the center conductor of a 50 Ω coaxial cable when the outer conductor has an inside diameter of 1.4375 in?

86. Find the impedance of a coaxial cable that has a center conductor made with 23/32 in copper tubing and an outer conductor made from copper tubing with a 2 5/8 in outside diameter and a wall thickness of 0.625 in.

87. What is the propagation velocity (in feet) of a cable that has a dielectric constant of 2.25?

88. What is the propagation wavelength (in feet) in a cable with a dielectric constant of 2.25 operating at 400 MHz?

89. One neper is how many decibels?

90. A cable with an attenuation of -0.5756 nepers has how many decibels of attenuation?

91. What is the dB attenuation of a 450 ft cable that has a loss factor of -0.3569 nepers per 100 ft?

92. An air dielectric coaxial cable with $D = 1.5$ in and $d = 0.65$ in will have how much dB loss per 100 ft when operating at 400 MHz?

93. How much input power is required for the cable in the preceding question when 1000 W is needed at the output of a 250 ft section of cable?

94. An air dielectric cable with $D = 3.5$ in and $d = 1$ in will have how much dB loss per 100 ft at 650 MHz?

95. How much input power is required to the cable in the preceding question for 5 kW output at the end of 350 ft of cable?
96. What is the standing wave ratio when a 150 Ω antenna is connected to a 60 Ω transmission line?
97. What is the standing wave ratio when a 24 Ω antenna is connected to a 60 Ω transmission line?
98. Find the SWR for the standing wave of voltage measured as E_{max} of 198 V_{pk} and E_{min} of 110 V_{pk}.
99. What is the reflection coefficient when a 50 Ω resistor is terminating a 75 Ω transmission line?

CHAPTER NINE

SMITH CHARTS

9.1 INTRODUCTION

There was a time when all transmission line matching problems were solved by rigorous mathematics. Around 1939, P.H. Smith developed a calculator, a circular slide rule to be exact, to reduce the complexity of solving these problems. The circular slide rule is still available today. However, a graph of the slide rule is also available, at a much lower cost, which will provide the user with a permanent record of the solutions to the problems. This graph is now in common use and is called the **Smith chart.** Since its inception, the Smith chart has expanded its role in complex electronic problem solving as new uses for it have been adopted.

9.2 CHART SCALES

There are nine scales on the chart that are of universal interest. They are

1. The pure resistance (or zero reactance) line
2. The resistance circle sets
3. Two circles sets for reactances X_L and X_C
4. Two wavelength scales
5. The reflection coefficent angle scale
6. The reflection coefficent magnitude scale
7. The dB of loss scale

Each will be identified in turn.

9.2.1 Pure Resistance Line

There is only one straight line on the Smith chart. It lies in a horizontal position across the chart, passing through the **prime center** of the graph and extending in both directions to the perimeter of the chart. This is the **pure resistance** or **zero reactance line.** It is calibrated from zero, on the left, to infinity, on the right. The value of 1 is at the graph's center.

9.2.2 Resistance Circle Sets

The **resistance circle sets** are shown in Figure 9.1. These are sets of eccentric circles, all tangent to the point of infinity at the right edge of the graph, with their centers all on the zero reactance line. The outer perimeter of the chart represents a value of zero resistance, and as the resistance value becomes larger, the circles become smaller. Of major concern to the user is the $R = 1$ circle. It is the circle that passes through the prime center of the chart as it crosses the zero reactance line. Any value of resistance smaller than 1 will be found as a circle crossing the zero reactance line at a point to the left of prime center, and any resistance value larger than 1 will be found as a circle crossing the zero reactance line at a point to the right of prime center. Infinite resistance is represented by a circle that has diminished to a point at the right end of the zero reactance line.

9.2.3 Reactance Circles (Arcs)

The area above the horizontal center line is reserved for *inductive* reactance values, and the lower half of the chart is reserved for *capacitive* reactance values. This explains why the horizontal line is called the zero reactance line. All values of resistance that fall on this line are pure resistances, with *no* reactance value.

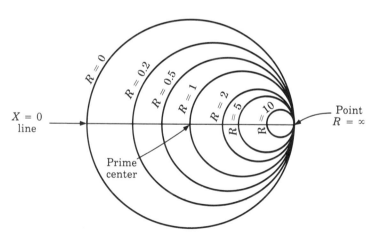

FIGURE 9.1
Concentric resistive circles.

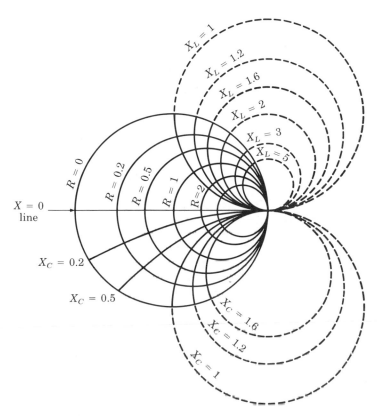

FIGURE 9.2

Concentric reactive circles. Only the arcs of these circles appear on the Smith chart.

The **reactance circle sets** are a series of eccentric circles above and below the horizonal center line and tangent to the point of infinity. Examine Figure 9.2 and note that although the reactances are full-circle values, only the portion of a circle that falls inside the resistance circle $R = 0$ is included on the Smith chart. An important point to note in each reactance hemisphere is that a reactance value equal to 1 crosses the zero resistance circle at infinity and again at a point $+90°$ from the zero reactance line, for inductance, or at $-90°$, for capacitive reactance. Reactance values greater than 1 ($+jX$ or $-jX$) can be located only in an area smaller than one-fourth of the total Smith chart area. This is in the upper right quadrant for values of $+jX$ and in the lower right quadrant for $-jX$ values. The area for reactance values less than 1 takes up the greater portion of the chart, including the area inside of the $R = 1$ resistance circle.

Figure 9.3 summarizes and illustrates the basic structure of the Smith chart. This figure combines the arcs and circles of Figures 9.1 and 9.2.

Plotting the location of an impedance or reading a value of impedance from the chart is a fairly large part of working with Smith charts. Several points have been plotted on the graph in Figure 9.4. They are points corresponding to $R \pm jX$, including A = 1.0 + j2.0, B = 2.0 − j0.5, C = 0.3 − j1.2, D = 0.3 − j0.2, and E = 0.3 + j1.0. Examine these values and make sure you understand why each point is located where it is on the graph. Note that the

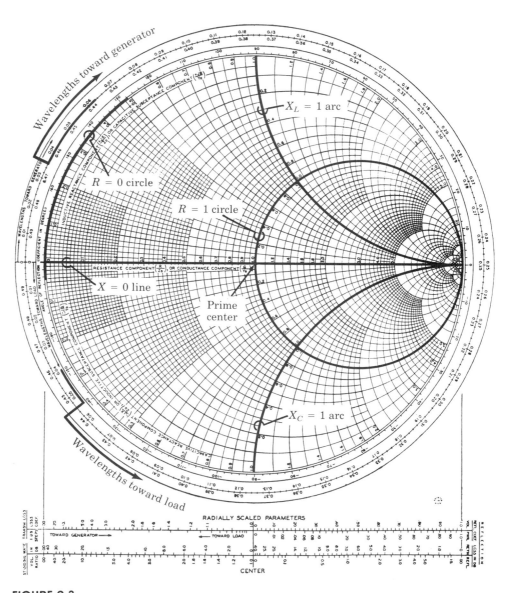

FIGURE 9.3
The reference line, circles, and arcs of the Smith chart.

values for resistance are printed on the chart along the zero reactance line and again along part of each $X = 1$ arc. The values for reactance are printed around the perimeter of the $R = 0$ circle and again along part of the $R = 1$ circle. The values of the plots are less important at this time than the ability to read or mark locations on the chart.

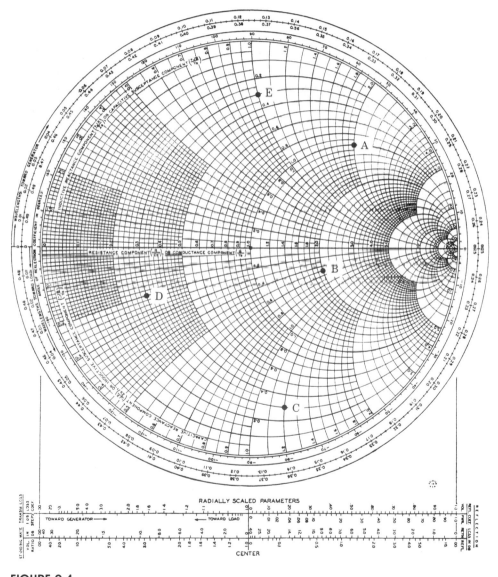

FIGURE 9.4
Locations of the points A, B, C, D, and E corresponding to values specified in the text.

9.2.4 Wavelength Scales

Around the outside edge of the chart are two wavelength scales. Both scales start at the left of the chart on the zero reactance line. The outer scale advances in a clockwise direction, making one full revolution around the chart in one-half wavelength, and is labeled "wavelengths toward the generator." The left center mark, identified as 0 wavelengths, represents both the starting point and end point. This measure of one-half wavelength is used because the transmission line patterns repeat themselves every half wavelength. The inner scale, labeled "wavelengths toward the load," advances in a counterclockwise direction. The selection of when to use which scale is dependent on what information is known and what information is to be determined. Scale selection will become second nature as the procedures are outlined in this chapter for each use of the chart. Note that the measure of one-quarter wavelength toward the generator and that of one-quarter wavelength toward the load are both at the same point (extreme right). On the Smith chart slide rule, the two wavelength scales are on separate slip rings and are freely positioned anywhere around the perimeter of the chart.

9.2.5 Reflection Coefficient

The third scale in from the edge is marked "angle of reflection coefficient in degrees." Zero degrees is found at the extreme right edge of the chart, and the scale advances to 180° at the left edge. Positive angles occupy the scale across the top half of the chart, and negative angles occupy the scale around the lower half of the chart.

9.2.6 Reflection Magnitude

Across the bottom of the chart are several scales identified as "radially scaled parameters." On the Smith circular slide rule, these scales are on a clear bar that is attached to the center of the calculator and is free to rotate around the chart. The two scales of interest for the **reflection coefficient magnitude,** at the upper right, and the **dB loss,** found on the upper left. Again, all of the scales will be treated in turn as the need arises and as the applications and uses of the chart are established.

9.3   NORMALIZED IMPEDANCES

The Smith chart that forms the foundation of this chapter is known as a **normalized impedance chart.** It is identified as a chart that has the $R = 1$ circle passing through the prime center and is made to process *all* values of impedance. There are other charts made for one impedance only. A 50 Ω chart, for example, would have the $R = 50$ circle passing through the prime center. All impedance values smaller than 50 Ω would be represented as circles larger than the $R = 50\ \Omega$ circle and would be found to the left of the center, while all impedance values larger than 50 Ω would be represented as circles smaller than the $R = 50\ \Omega$ circle and located to the right of center. The normalized reference impedance chart is the most common because one chart may be used to represent *all* values of load impedance. The conversion from a real-life load impedance to a normalized load impedance is executed by dividing the characteristic impedance of the transmission line into the load impedance:

$$Z_n = \frac{Z_L}{Z_0} \tag{9.1}$$

Even though the load impedance may be a complex number, such as $40 + j60\ \Omega$, the division by 50 Ω is done in two steps: $40/50 = 0.8$ and $60/50 = 1.2$. The normalized load impedance is $0.8 + j1.2\ \Omega$. This point is plotted on the chart of Figure 9.5 and noted by the symbol for the normalized load impedance, Z_n.

9.3.1 Standing Wave Ratio Circle

Draw a circle on the Smith chart using the prime center ($R = 1$, $X = 0$) as a pivot point and Z_n as a point on the circumference (see Figure 9.5). This circle is called the **standing wave ratio circle.** It tells us several things.

1. The point where this circle crosses the zero reactance line to the *right* of prime center is the numerical value of the standing wave ratio. In Figure 9.5, the SWR is 3.5, or 3.5:1.
2. The location and value of the normalized load impedance Z_n is located on this circle. Expanding this idea establishes that the standing wave circle is a plot of the impedance on the line at every point along the length of the line.
3. The same location at $X = 0$ on the SWR circle is the location of the maximum voltage of the standing wave (E_{max}), in wavelengths from the load (calculations to be shown later), and the value of the maximum impedance (Z_{max}) on the line, in wavelengths from the load.
4. Where the SWR circle crosses the zero reactance line to the *left* of center represents the location along the line where E_{min} and Z_{min} are found.

9.3.2 The Load Line

Using a straight edge, draw a line that starts at the prime center of the chart, passes through Z_n, and extends out past the wavelength scales. This is the **load line.** Note that the load line cuts the "wavelength toward the generator" scale at 0.159 wavelengths. As of now, this value has no significance except to establish a reference point. Move along the outer wavelength scale to the point where it intersects the zero reactance line, at 0.25 wavelengths. This wave-

length difference (0.25 − 0.159 = 0.091 wavelengths) is the distance from the
load to where the standing wave voltage is maximum and the impedance on the
line is maximum, as shown at point A in Figure 9.5. Following the standing
wave ratio circle in a clockwise direction, we see it crossing the zero reactance
line left of center where the "wavelength toward the generator" scale reads 0 or

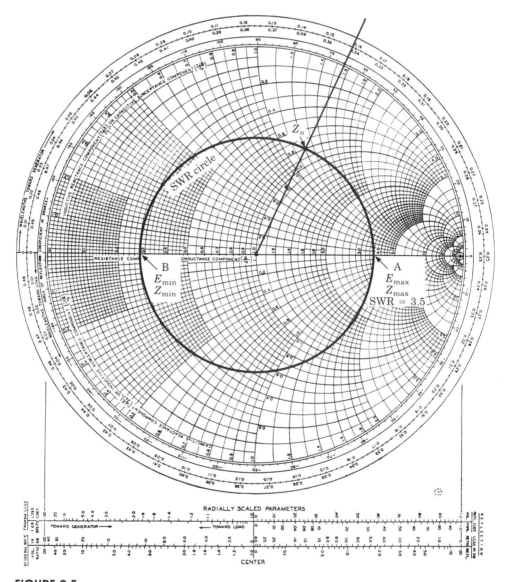

FIGURE 9.5
Drawing the "standing wave ratio" circle, plotting the load impedance and load
line, and locating the value of SWR, E_{max}, Z_{max}, E_{min}, and Z_{min}.

FIGURE 9.6
A standing wave of voltage along a
mismatched transmission line.

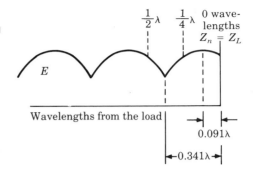

Wavelengths from the load

0.091λ

0.341λ

0.5 wavelengths (point B). The minimum impedance on the line and the minimum voltage of the standing wave are at $X = 0$ and $R = 0.28$ as shown in
Figure 9.5.

Figure 9.6 is a graph of the standing wave voltages and corresponds to the
chart of Figure 9.5. In particular, note the distance of E_{max} and E_{min} in wavelengths from the load. Correlate the values in Figure 9.5 to those of Figure 9.6
before reading on. As the standing wave ratio circle is followed still further, it
would end up at Z_n, completing one revolution of one-half wavelength.

9.4   MATCHING LOAD IMPEDANCES

When the load impedance does not match the impedance of the line, several
plans of attack can be used to reverse this condition. Plan (a) in Figure 9.7
shows R_L larger than Z_0. Here a resistor placed in parallel with the load would
match the total load impedance to the line impedance, but R_1 would siphon off
33% of the power intended for the load. In plan (b), the series resistance would
make the total load impedance equal to the line impedance but again would
dissipate about 30% of the load power. Plans (c) and (d) are highly efficient
methods of removing the standing waves along the line by matching the load
impedance to the line impedance. In plan (c), a capacitor is placed in parallel
with the line at a point away from the load where the impedance on the line is
purely resistive. The capacitor is adjusted in value until the line impedance is
matched. Reactances can neither absorb nor dissipate power, so no power is lost
due to the reactance. However, a small section of the line between the load
terminals and the place where the capacitor is connected will still have standing waves present and will dissipate a small amount of power, causing some
losses. Plan (d) is almost the same as plan (c) except an inductor is used in place
of the capacitor. Note also that the location of the inductor is at a different place
than the capacitor. The equivalent L or C, and the placement of each, will be
determined by the amount and nature of mismatch in each case.

In practical situations, a section of the same transmission line to be
matched is used to form a capacitor or inductor (see Chapter 8) and is connected

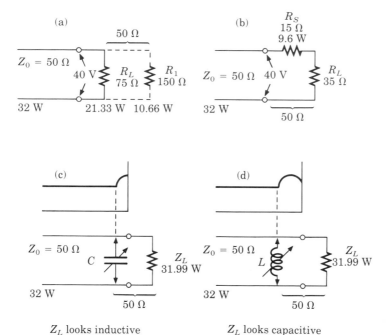

FIGURE 9.7
Correct impedance matching techniques (c) and (d) compared to incorrect techniques (a) and (b).

in parallel with the line being matched. This section is called a *matching stub*. It is the function of the Smith chart to determine the size and placement of the matching stub that will cause an irregular load impedance to match the line impedance.

9.5   PROBLEM SOLVING PROCEDURES

Two problem solving procedures that use the Smith chart are presented here. The first is used when the line impedance and the mismatched load impedance are known quantities. The second procedure will be covered in Section 9.7.1 and is used when the load impedance is unknown. Refer to Figure 9.8 as you go through this outline.

Method 1   A load impedance (antenna) of $Z_L = 150 + j60\ \Omega$ is connected to a $100\ \Omega$ transmission line. Find the size and placement of the matching stub that will remove all standing waves and match the antenna to the line.

1. Normalize the load impedance by dividing it by the characteristic impedance of the line, using Equation 9.1.

$$Z_n = \frac{150 + j60}{100} = 1.5 + j0.6 \ \Omega$$

2. Locate and plot a point on the Smith chart that represents the value of the normalized load impedance Z_n (see Figure 9.8).

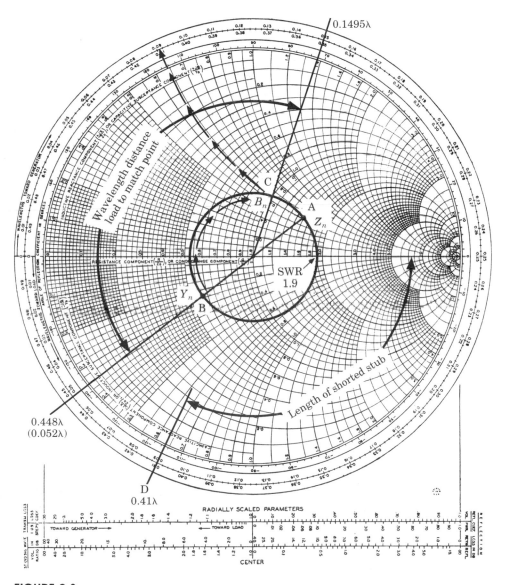

FIGURE 9.8

Using the Smith chart to find the length and location of a shorted impedance-matching stub (see text).

3. Draw the SWR circle, using the prime center of the chart as a pivot point and Z_n as a point on the circumference of the circle. Record this value as SWR = 1.9:1 (see Section 9.3.1).
4. With a straight edge, draw a line from the normalized load impedance, through the prime center of the chart, out to the wavelength scale opposite from Z_n. Record this reading on the "wavelengths toward the generator" scale. This example uses a reading of 0.448 wavelengths. Furthermore, note where this line crosses the SWR circle opposite from the normalized load; this is the normalized **admittance** Y_n, labeled point B on this chart.
5. From the normalized admittance point, move clockwise around the SWR circle to where it crosses the $R = 1$ circle *for the first time*. This is point C on the chart and denotes the normalized **susceptance** (B_n). This value is $1 + j0.64$.
6. Draw a line from the prime center of the chart, through the susceptance point (c), to the wavelength scale. Note the reading on the "wavelengths toward the generator" scale (0.1495 wavelengths in this case). Record this value.
7. The difference in wavelengths between the reading of step 4 and that recorded in step 6 (moving clockwise) is the *distance* from the load terminals to the point where the matching stub will be connected— for this example, 0.2015 wavelengths.
8. The reactive portion recorded in step 5 (normalized susceptance of $+j0.64$) must now be canceled. Find the opposite value, $-j0.64$, on the $R = 0$ circle on the bottom of the chart, noted as point D. Record the "wavelengths toward the generator" reading at point D (0.41 wavelengths) and measure the difference in wavelengths from the point of infinity (0.25 wavelengths) to point D. In this example, $0.41 - 0.25 = 0.16$ wavelengths. The *length* of the *shorted* stub connected to the match point will be 0.16 wavelengths. The line will now be correctly terminated and there will be no standing waves of voltage along the line.

9.5.1 Explanation of the Procedures

Now that the procedure for method 1 has been set forth and the steps listed, an explanation of each step is in order. The following outline will include what is done, how it is done, and why it is done this way.

Step 1 The load impedance is changed to a value that may be plotted on a universal graph made to accommodate any combination of load impedance and line impedance. The conversion is made by dividing the characteristic line impedance into the load impedance. Later, a value may be read from the graph and *denormalized* to a real-life impedance value by multiplying the chart value by the line impedance.

Steps 2 and 3   Plotting the normalized values onto the graph and drawing the SWR circle require no further explanation. If there is still some doubt, reread Section 9.3.

Step 4   The load line here is extended in the inverse direction to find the admittance of the load and a reference wavelength associated with the admittance. The final correcting impedance will be placed in parallel with the load at a determined distance from the load. When the correcting circuit is to be placed in parallel with the load, the starting point is the admittance of the load. When a correcting circuit is selected to be in series with the load (rarely used), then, the starting point is the load impedance.

Any admittance can be found from a given impedance by using the Smith chart. Normalize the impedance by dividing it by any convenient Z value, plot Z_n on the chart, draw a circle and a load line, find the normalized admittance, and then multiply the normalized admittance by the selected Z value.

Step 5   The object of the Smith chart matching exercise is to find a location along the transmission line where the impedance on the line has a resistive component equal to the line impedance, and to cancel the reactance at that location. Moving around the SWR circle establishes the $R = 1$ location, so that when 1 is multiplied by Z_0 in the denormalizing process, the result will be Z_0.

Steps 6 and 7   The distance in wavelengths is determined from the admittance point to the point where $R = 1$, which tells us the distance from the load to where the matching stub is to be connected. Always start at the normalized admittance, and move in a clockwise direction to the normalized susceptance point (C). In this example, from 0.448 wavelengths, move clockwise through 0.5 wavelengths, and from 0 wavelengths to 0.1495 wavelengths, for a total of 0.2015 wavelengths. This instruction directs a movement from the load (actually the admittance of the load) *toward* the generator to find the point where the impedance on the line is equal to the line impedance itself.

Step 8   This step simply involves recording a reactance, finding an equal value of the opposite polarity, and using one to cancel the other. In our example, when $R = 1$ and $X = +j0.64$, an impedance of $R = 0$ and $X = -j0.64$ is needed to cancel the reactance without changing the resistive component. At the $R = 0$ circle, the reverse polarity $X = -j0.64$ is found at the bottom of the chart at the wavelength scale reading of 0.41 wavelengths toward the generator. The matching stub is a shorted stub to reduce radiation of energy. An open stub could have been used (0.25 wavelength longer), but the RF interference to surrounding equipment would be more noticeable. Because the Smith chart is designed to work with the inverse function for shunt elements, the matching stub takes on the inverse of the desired condition. A shorted matching stub is desired, so its length is determined by a starting point for that which is the opposite of a short, an open circuit. The open circuit has infinite impedance at the open terminals, so the start of the stub length determination must be from the infinite impedance point on the Smith chart, at 0.25 wavelengths. The stub length is then calculated as the difference between 0.25 and 0.41 wavelengths, or 0.16 wavelengths.

9.6   DIRECTIONAL COUPLERS

To sense the direction of power travel as well as the amount of power, a sensing device must have a reversible diode as one of the circuit elements. The circuit of Figure 9.9 uses a reasonably long line placed along side of, and close to the center conductor of, a rigid coaxial cable. Energy traveling from input to output through the cable will cause the sampled voltage to build across the sensing device in an additive fashion. The reflected voltage from the load to the transmitter will build as it moves in the sampling line, but in a subtractive mode due to the direction of the diode. The reflected signal will be absorbed in the load resistor R_1, and the wanted signal will be sensed across the meter circuit. The reflected wave will be rejected when the circuit is measuring the forward power, and the forward power will be rejected when the reflected power is sampled. The **directivity** of a directional coupler is a measure of the coupler's ability to isolate the forward wave from the reflected wave and is stated in decibels.

9.6.1 VSWR Meter

The meter connected to the directional coupler of Figure 9.9 is calibrated to display percentage of power on one scale and SWR directly on the other scale. Figure 9.10 is typical of the scales used for SWR meters.

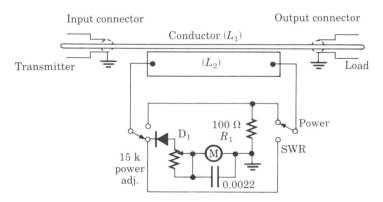

FIGURE 9.9
A meter circuit and transmission line used as a directional coupler to measure SWR.

FIGURE 9.10
The dial scale to indicate values of SWR.

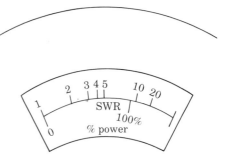

The function switch is set to the power position, and the "power adjust" control is varied until the meter reads 100% on the power scale. The function switch is then turned to the SWR position to reverse the diode. The meter will now read SWR directly on the top scale with no further adjustment. Zero deflection is an SWR of 1, the ideal condition, where all of the power is absorbed by the antenna. A 2:1 ratio would read 2 on the meter and so on. It may not always be possible to obtain a 1:1 SWR, so a decision must be made as to what limits of SWR are acceptable. Table 9.1 is included here to aid in making that decision. The standard acceptable limit is SWR = 1.5:1 = 3.5% power loss.

TABLE 9.1

SWR	Power Loss
1:1	0
1.3:1	1.9
1.5:1	3.5
1.7:1	5.9
2:1	11.0
3:1	24.0
4:1	38.0
5:1	48.0
6:1	56.0
10:1	70.0

9.7 SLOTTED LINES

Several other useful measurements involve the Smith chart and include the use of an instrument called a **slotted line.** A section of rigid coaxial transmission line has a slot cut into the side wall along its length. A measuring probe is inserted into the slot and moved along the length of the line to record the voltage at any location over the length of the slot (see Figure 9.11). The position of the probe is noted by the pointer on a centimeter scale along the length of the

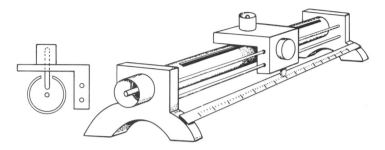

FIGURE 9.11

A "slotted line" used to measure voltage standing waves along the line.

assembly. The location of the voltage minimum and maximum of the standing wave can be measured with great accuracy using a slotted line. The voltage minimum is usually preferred as the indicator of the standing wave voltage because the very sharp voltage change is easier to read than the slower changing crest of the voltage wave. The equipment setup follows the arrangement of Figure 9.12, in which the slotted line is connected in series with the transmission line between the transmitter and the antenna. The detected signal may then be heterodyned down to a convenient frequency as needed.

A typical set of voltage readings is graphed in Figure 9.13. The minimum voltages are noted in terms of distance from the load in centimeters, and the values of $E_{\max}$ and $E_{\min}$ are used to find the SWR by Equation 8.23:

$$\text{SWR} = \frac{E_{\max}}{E_{\min}} = \frac{228}{80} = 2.85 : 1$$

Depending on the type of equipment used, these reading may be relative voltage ratios rather than actual voltage levels.

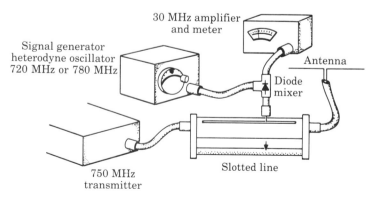

FIGURE 9.12

A typical "standing wave" measurement test setup.

FIGURE 9.13
A graphic display of the standing wave
voltage for the text example.

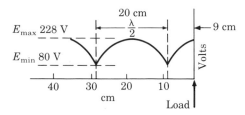

The distance between two successive voltage minima is, in this example,
29 cm − 9 cm = 20 cm, which represents one-half wavelength. One wavelength
is therefore 40 cm or 0.4 m. The frequency of the signal is the velocity of light
divided by the wavelength:

$$f = \frac{300 \times 10^6}{0.4} = 750 \text{ MHz}$$

Of still greater significance is that now the normalized load impedance
can be found. Figure 9.14 shows the construction of an SWR circle with a value
of 2.85:1, as determined earlier. The location of a voltage minimum on the
Smith chart is *always* found at the junction of the SWR circle and the zero
reactance line to the *left* of prime center, point A in Figure 9.14. The load is
9 cm away from this voltage minimum. A distance of 9 cm converts to wavelengths according the ratio:

$$\lambda = \frac{9 \text{ cm}}{40 \text{ cm}} = 0.225$$

Starting from the voltage minimum, move in a direction toward the load (counterclockwise) 0.225 wavelengths (to point B) and construct a load line to the
prime center of the graph. Z_n, at point C, has a value of $2.4 - j1.0$ Ω. When the
transmission line impedance is 50 Ω, the antenna impedance (load) would be
equal to $50 \times (2.4 - j1.0) = 120 - j50$ Ω.

9.7.1 Matching with a Slotted Line

The procedure demonstrated in Figure 9.14 shows the initial steps for finding
the location and length of a shorted, impedance-matching stub in the cases
where the load impedance is the unknown quantity. The value of the standing
wave of voltage and the wavelength distance from the load to the first voltage
minimum are the important criteria to be measured. Method 2 can be stated in
procedural step form as follows:

1. Graph the pattern of standing wave voltages along the line by using a
    slotted line. Record the values of E_{max}, E_{min}, and the distance from the
    load to the first voltage minimum.
2. Resolve and plot the value of SWR from E_{max} and E_{min}. Draw the
    standing wave ratio circle.

3. Convert the scale distance into wavelengths for the distance from the load to the first E_{min}. Locate this distance on the "wavelengths toward the load" scale, starting at E_{min} and rotating around the graph in a counterclockwise direction. Construct a load line joining the wavelength scale to the prime center of the graph. Mark and record the point where the load line crosses the SWR circle as the normalized load impedance Z_n.

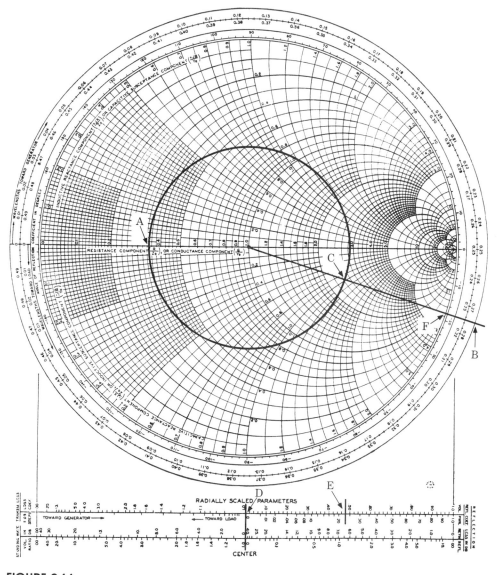

FIGURE 9.14
Using the Smith chart to find the magnitude and angle of the reflection coefficient.

The conclusion of method 2, steps 4 through 8, are the same as in the procedures outlined in Section 9.5 to find the length and location of the shorted stub along the line.

9.7.2 Unknown Z_L with a Slotted Line

It would be nice if the distance from the load terminals to the first null were always a known quantity. In ship installations, for example, the transmission line may go through several bulkheads and then up to the top of the mast. The exact length from the slotted line would be impossible to measure with the degree of accuracy needed here. An alternate method may be adopted to eliminate any guesswork or trial-and-error approaches.

1. Disconnect the load (antenna) from the output end of the transmission line.
2. Using a slotted line placed as close to the transmitter as possible, plot a graph of standing wave voltages. (*Note:* Make sure that the transmitter will not be damaged when operated at full carrier into an open-circuit load. If this presents a problem, either reduce the carrier power to a safe level or use a signal generator in place of the transmitter at the same frequency as the transmitted carrier center frequency.) Measure and record at least one *full* wavelength of voltage showing two E_{max} and two E_{min} positions. Record these locations on a graph similar to Figure 9.15.
3. Referring to Figure 9.15, the voltage minima at A and B are recorded with the open load. According to theory, the voltage is maximum at the open load terminals, and a minimum will occur one-quarter wavelength in the direction of the generator. This behavior repeats itself every half wavelength, so location C is representative of the load,

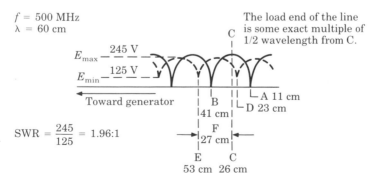

FIGURE 9.15
The plot of standing wave voltages for an open transmission line and a poorly matched loaded line.

which may be any number of half wavelengths away. Record the position of $E_{\max}$ on the centimeter scale (or graph) as $Z_L = \infty$. Thus, the exact distance to the load is no longer a factor for matching purposes, because a replica of the load is now a location on the slotted line scale, and this distance can be accurately measured.

4. Reconnect the load, apply power, and record the standing wave voltages under these conditions (shown dotted in Figure 9.15). The load was established at C = 26 cm on the scale. The first $E_{\min}$ from the load is found at 53 cm on the scale. The distance from the load to the first $E_{\min}$ is 53 − 26 = 27 cm.
5. Convert 27 cm into wavelengths as 27 cm/60 cm = 0.45 wavelengths.
6. Calculate the value of the standing wave voltage as 245/125 = 1.96. Draw the standing wave circle, and starting from $E_{\min}$ on the SWR circle and $X = 0$, move 0.45 wavelengths toward the load, construct a load line, and determine the value of the normalized load impedance Z_n.
7. Once the SWR circle is drawn and a value for Z_n is plotted, the standard procedure may be followed to find the position and length of the shorted impedance-matching stub.

9.7.3 Matching Transformers for Complex Load Impedances

A quarter-wavelength section of transmission line has been defined as an impedance-matching transformer for resistive loads. The previous few sections have shown that where the SWR circle crosses the $X = 0$ line, the impedance on the line is purely resistive, but *not* equal to the line impedance Z_0. To the right of prime center, the resistive equivalent on the line is

$$R_L = Z_0(\text{SWR}) \tag{9.2}$$

and to the left of prime center, the resistive equivalent is

$$R_L = \frac{Z_0}{\text{SWR}} \tag{9.3}$$

The distance from the load clockwise to the $X = 0$ line is the location from the load terminals to where the quarter-wave matching transformer will be inserted into the line (see Figure 9.16). The conversion equation is Equation 8.29. The transformer impedance is

$$Z_t = \sqrt{Z_0 R_L}$$

and may follow any of the mechanical forms expressed in Chapter 8.

FIGURE 9.16
Location of a quarter-wave impedance matching transformer when the mismatched load is not purely resistive.

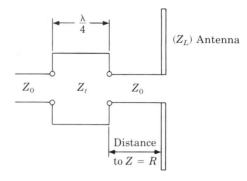

9.7.4 Reflection Coefficient, Magnitude, and Angle

Return to Figure 9.14. Using the same compass setting used to draw the SWR circle, at 2.85, move down to the right top radial scales. Place the pivot point at the center line position of the chart (D in Figure 9.14), and draw an arc as shown at location E. The value of reflection coefficient is marked as 0.48 and will correspond to the calculation from Equation 8.25:

$$\Gamma = \frac{\text{SWR} - 1}{\text{SWR} + 1} = \frac{2.85 - 1}{2.85 + 1} = 0.48$$

The Smith chart is used to find the value of the reflection coefficient at the same time that the SWR circle is being drawn.

The load line of Figure 9.14 from B to the prime center also passes through a third scale on the graph. This scale indicates the angle of the reflection coefficient, as stated earlier. Here the load line passes through $-18°$, at point F, which means that the load is capacitive with a lagging angle of $-18°$.

9.7.5 Line Input Impedance

Ideally the input impedance to the line should always be equal to Z_0, which should, in turn, match the output impedance of the source. This condition is met when the line is correctly and exactly terminated. A small degree of mismatch can be tolerated with negligible power loss.

As before, set up the slotted line close to the transmitter output terminals, connect the load (antenna), apply the carrier power, and set the carrier center frequency.

1. Measure the standing wave voltage pattern, making note of $E_{\max}$, $E_{\min}$, and the location on the slotted line for the value of $E_{\min}$.
2. Compute the value of SWR and construct the SWR circle.

3. *Accurately* measure the distance from the transmitter output terminals to the position of E_{min} on the slotted line. Convert this distance into wavelengths.
4. Starting from E_{min} on the SWR circle of the Smith chart, move clockwise (toward the generator) around the SWR circle by the number of wavelengths determined in step 3. The stopping point on the SWR circle is the value of the normalized input impedance. Multiply the normalized impedance by Z_0 to give the true input impedance of the line.

9.8   LOSSY LINES

To this point, lines have been assumed to be lossless in order to limit the number of variables. Now that a base of understanding has been laid down, the enemy of all electronics (attenuation) may be covered. We know that there will be I^2R copper wire losses, leakage losses through the dielectric, and some radiation losses over the length of the cable. The signal arriving at the end of the cable will be smaller than the signal put into the line. Cables with solid dielectrics are the worst offenders.

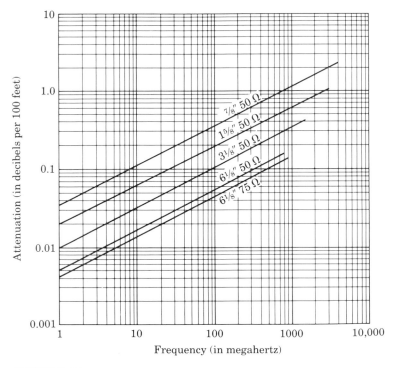

FIGURE 9.17
A plot of cable losses compared to frequency.

A not-so-obvious effect is that the standing wave ratio at the load is larger than the SWR at the input to the line. We have deceptively drawn the SWR circle as a circle, when in truth it is a spiral. This problem is overcome by simply drawing two standing wave circles, one to represent the SWR at the input to the line and a larger SWR circle for the load end of the line. The means of compensation for this condition depends on which end of the line the measurements are taken at and what the total line losses are in decibels.

For our example, a 3⅛ in diameter 50 Ω cable is used. The slotted line measurements are taken at the 500 MHz transmitter output, and the antenna is at the far end of 350 ft of rigid coaxial cable. From Figure 9.17, the cable is found to have 0.23 dB attenuation per 100 ft at 500 MHz. The total line losses are $0.23 \times 3.5 = 0.805$ dB. From Figure 9.18 it is noted that this cable will handle 22 kW at 500 MHz with peak pulsed power up to 920 kW. This goes well with this 15 kW_{rms} transmitter.

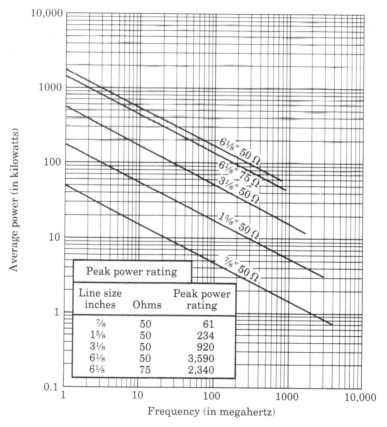

FIGURE 9.18
Cable power-handling ability compared to frequency.

FIGURE 9.19
The standing wave voltage measurements for an unknown antenna load
impedance.

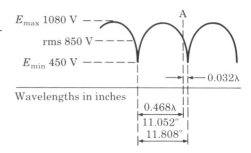

The measurements taken on the slotted line at the transmitter end appear in Figure 9.19. The dashed line at A shows the position of the voltage E_{max} at 28 in on the scale when the antenna was disconnected. With the antenna back in place, $E_{max} = 1080$ V and the two E_{min} points of 450 V are measured at 16.95 in and 28.75 in from the generator. This means that the first E_{min} from the load impedance is now 11.052 in from the antenna load. Use Figure 9.19 to calculate the input SWR:

$$\text{SWR} = \frac{1080}{450} = 2.4{:}1$$

Plot the SWR on the Smith chart, shown in Figure 9.20 at A, and construct the SWR circle.

The length of one wave at 500 MHz is 23.616 in, so start on the SWR circle at E_{min} (B in Figure 9.20), and move around the SWR circle toward the load for a distance of:

$$\lambda = \frac{11.052}{23.616} = 0.468 \text{ wavelengths}$$

This is marked as point C. Construct a load line from point C to the prime center of the chart. Do not plot the load impedance at this time.

Reset the compass to draw the SWR circle of 2.4, and move the compass down to the lower half of the upper left set of scales at the bottom of the Smith chart. Mark this reading as point D (see Figure 9.20). This scale reads in 1 dB steps of transmission loss along the line. The divisions are not numbered because the same scale is used when going toward the load and toward the generator with no preset starting point. Point D appears at about 20% of a division left of a mark. The 0.805 dB of transmission line losses are then added to the value at D in the direction toward the load, or about to point E. Expand the compass to stretch from the zero center mark to E, and use this setting to construct the SWR circle at the load. Now you may label the normalized load as $0.34 + j0.18$ Ω and SWR = 3:1. The only procedural change that may occur is when the measurements are made at the antenna end of the line and the input impedance is sought. Then simply construct the load SWR circle first, drop a line down to the 1 dB step scale (at E), add the cable losses back in (shift to D), move back up to construct the input SWR circle, and continue in the conventional manner.

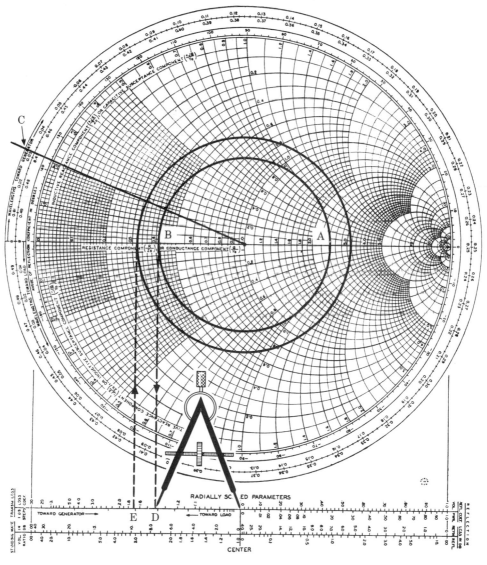

FIGURE 9.20
Compensating the Smith chart when the line losses are known.

9.9   MULTIPLE-STUB TUNING

The major disadvantage to single-stub tuning is that the matching conditions are correct for only one frequency. It is possible that at a slightly different frequency the load impedance and SWR could be worse than with no matching at all. The multiple stub allows for matched conditions over a wider frequency range. More stubs means wider frequency coverage. The calculations, however, are all done at the geometric center frequency of the band to be used.

The user has 2 selections to make. Referring to Figure 9.21, select the space between the two stubs to be *any* convenient part of a wavelength. A popular choice for two-stub matching is three-eighths of a wavelength. This is simply because common drafting tools make it easy to draw a three-eighths wavelength spacing circle. It has nothing to do with electronics. A popular choice for three-stub tuning is spacing of one-eighth of a wavelength, for similar reasons. The second choice to be made is the distance between the antenna terminals and the point where the first stub is connected. Here the limits are clearly defined by the value of SWR. These limits will become apparant as the sample problem unfolds.

The spacing circle is a unity-conductance circle that is rotated on the chart toward the generator by an amount equal to the chosen wavelength distance that separates the stubs. The choice of spacing used in this text will always be three-eighths of one wavelength, and the characteristic impedance of the matching stubs will be the same as the characteristic impedance of the feed lines, in order to simplify analysis. Other choices are available.

The reciprocal of resistance (R) is conductance (G). Therefore, the reciprocal of the $R = 1$ circle is the $G = 1$ circle (see Figure 9.22), called the *unity-conductance circle*. The unity-conductance circle is advanced toward the generator by the amount of spacing between the stubs, three-eighths of a wave-

FIGURE 9.21
Double stub matching.

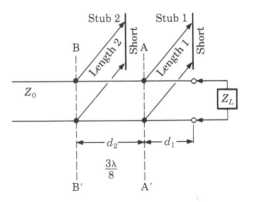

length, as shown in Figure 9.22, and is now called the *spacing circle*. The spacing circle will always be located as shown in Figure 9.22 when three-eighths wavelength stub separation is used.

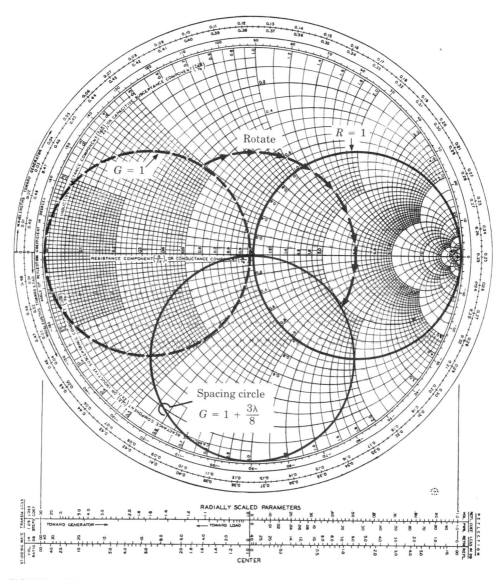

FIGURE 9.22
Rotating the "unity-conductance" circle for double-stub matching.

EXAMPLE 9.1

An antenna load of $100 + j100\ \Omega$ is connected to a $50\ \Omega$ transmission line. Find the length and spacing for a two-stub impedance-matching system with three-eighths wavelength separation between stubs (see Figures 9.21 and 9.23).

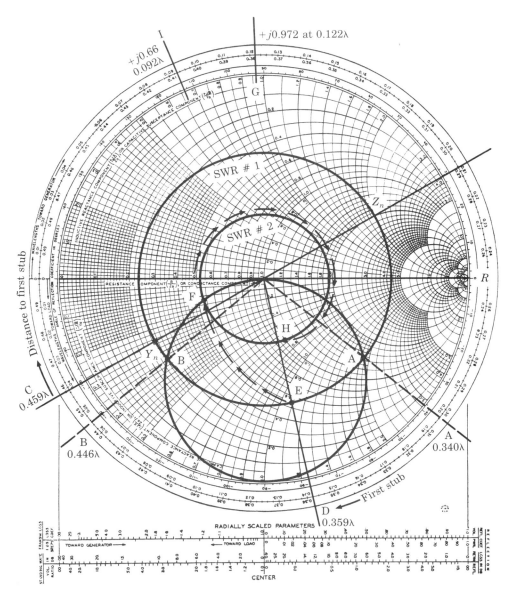

FIGURE 9.23
Finding the length of a double-stub matching system.

1. Normalize the load impedance, plot Z_n, construct the SWR circle, draw a load line, and record the wavelength value at Y_n. The resulting values are ($Z_n = 2 + j2$, SWR = 4.25, $Y_n = 0.245 - j0.245$ at 0.459 wavelengths (see Figure 9.23).
2. Construct a three-eighths wavelength spacing circle.

Now it is time to make the second choice. How far from the antenna terminals should the first matching stub be connected? Starting at the wavelength reading at Y_n, move clockwise around the wavelength scale so that you end up *anywhere* between the dashed lines A and B in Figure 9.23. Lines A and B describe an arc between two radii that defines the portion of the SWR circle *inside of* the spacing circle.

3. This example uses a distance of 0.4 wavelengths to the first matching stub (from 0.459 clockwise to 0.359 wavelengths), which places the distance reading at line D. Line D crosses the SWR circle at $0.53 - j1.08$ Ω, at point E.
4. Follow the resistance circle through point E (0.53) in the direction of a smaller reactance. Move to the left, in this case, to point F at the edge of the spacing circle, where the coordinates are $0.53 - j0.13$.
5. Find the *difference* in reactance between points E and F ($X_E - X_F$ = amount to be canceled). In this example, $-j1.08 - (-j0.108) = -j0.972$, and $0 + j0.972$ is found at location G on the $R = 0$ circle, at 0.122 on the "wavelengths toward the generator" scale.
6. The stub length is found in the same way as for single-stub tuning. Start at infinite impedance ($\lambda = 0.25$) and move clockwise to the 0.122 wavelength position, at point G. The stub length is $0.25 + 0.122 = 0.372$ wavelengths.

So far we know the distance to the first stub, 0.4 wavelengths (partly selected), the length of the first stub, 0.372 wavelengths, and the separation between the two stubs, 0.375 wavelengths (totally selected). All we need now is the length of the second stub.

7. Using point F as a circumference location and the prime center of the chart as a pivot point, construct a second SWR circle (SWR = 1.82:1).
8. From point F, move around the second SWR circle in a clockwise direction until you reach the $R = 1$ circle *on the inside of the spacing circle*. This is point H, at $1 - j0.66$, and indicates the susceptance to be canceled, $-j0.66$. The canceling value, $0 + j0.66$, is found at I, at 0.092 wavelengths. Therefore, the wavelength distance from infinity clockwise to 0.092 wavelengths is $0.25 + 0.092 = 0.342$ wavelengths. Stub 2 is 0.342 wavelengths long and shorted (see Figure 9.24).

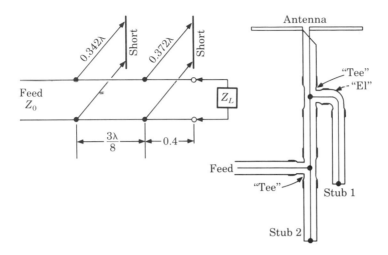

FIGURE 9.24
The electrical diagram of a double stub and the physical connections of the stubs.

9.10   OTHER USES OF THE SMITH CHART

The Smith chart is finding other uses today for plotting the impedances of high-frequency amplifiers. This is most evident in systems that use microstrip circuitry, which takes its theory from basic transmission line principles. The radiation patterns of circularly polarized antennas also lend themselves to the Smith chart format.

9.11   THE CARTER CHART

At about the same time the Smith chart was presented to the public, P.S. Carter of RCA was developing a chart of very similar format. The Smith chart was developed to use the rectangular notation of the complex number system, $R \pm jx$, while the Carter chart was designed to treat the polar coordinate system of numbering, $Z \angle \theta$.

The $R = 1$ circle on the Smith chart becomes the $R = 1$ line on the Carter chart and represents the Z values. The arcs across the top and bottom of the Carter chart represent the positive and negative angles of the impedances ($\pm \theta$). Although the Carter chart does not enjoy the popularity of the Smith chart, you should be familiar with it for its use in special cases.

QUESTIONS

1. How many straight lines are there on the Smith chart?
2. What is the designation of the horizontal straight line that passes through the prime center of the chart?

3. The Smith chart represents what portion (multiple) of a wavelength? (a) ⅛, (b) ¼, (c) ½, (d) 1, (e) 2.

4. The load impedance is normalized so that (a) larger and easier numbers can be used, (b) a single common line impedance can be used, (c) the SWR can be determined, (d) any set of impedance values may be plotted on one chart, (e) the distance is shorter.

5. The normalized load will always be found (a) to the right of prime center, (b) above the horizontal center line, (c) below the horizontal center line, (d) to the left of the prime center, (e) somewhere on the SWR circle.

6. The value of load impedance is normalized when it is (a) added to Z_0, (b) divided by Z_0, (c) multiplied by Z_0, (d) subtracted from Z_0, (e) not affected by Z_0.

7. Inductive reactances would be plotted on the Smith chart (a) only to the right of prime center, (b) only below prime center, (c) only above prime center, (d) only to the left of prime center, (e) as close to prime center as possible.

8. All resistance circles are joined at the point of infinity. (T)(F)

9. The reactance arc $X_L = 1$ passes through the perimeter of the chart at infinity and again at _____ .

10. The outer perimeter of the Smith chart represents what resistance value? (a) 0, (b) 0.25, (c) 0.5, (d) 1.0, (e) larger than 1.0.

11. Smith charts are made to accommodate only one value of impedance. (T)(F)

12. Load impedances are represented on the Smith chart in (a) rectangular form, (b) polar form, (c) triangular form, (d) trapezoidal form, (e) parallel form.

13. The SWR circle is a trace of (a) the change in load impedance, (b) the impedance at different frequencies, (c) the impedance at all points along the line, (d) none of the above.

14. The point where the SWR circle crosses the zero reactance line to the right of prime center also marks the location of (a) E_{min}, (b) E_{max}, (c) Z_{min}, (d) Z_{max}, (e) both (b) and (d).

15. Why are stubs used to match impedances on transmission lines rather than resistances?

16. A matching stub can resemble a (a) capacitive reactance, (b) inductive reactance, (c) either an inductive or a capacitive reactance.

17. The voltage standing ratio will always be a numerical value on the Smith chart where the SWR circle (a) goes to zero resistance, (b) crosses the zero reactance line, (c) crosses the zero reactance line to the left of prime center, (d) crosses the zero reactance line to the right of prime center, (e) intercepts infinity.

18. The load line joins (a) the SWR circle and E_{max}, (b) Z_n, Y_n, and the prime center, (c) the admittance and susceptance points, (d) E_{min} and the zero wavelength scale, (e) $R = 1$ and $X = 0$ points.

19. E_{max} and E_{min} will always be found at the zero reactance line. (T)(F)

20. The intent of the Smith chart is to (a) raise the line input impedance, (b) lower the line input impedance, (c) increase the SWR, (d) change the load impedance to equal Z_0, (e) none of the above.

21. Which resistance circle passes through the prime center of the Smith chart? (a) $R = 0$, (b) $R = 1$, (c) $R = 2$, (d) $R = 10$, (e) $R = \infty$.

22. All reactance circles are joined at the point of infinity. (T)(F)

23. Resistance values less than 1 are found to the (right) (left) of prime center.

24. Why is one full revolution on the Smith chart calibrated in half wavelengths?

25. To move along the transmission line from the transmitter to the load would be a (clockwise) (counterclockwise) rotation on the Smith chart.

26. An open-ended stub *or* a shorted stub can be used as the matching device for transmission line impedance matching. (T)(F)

27. Which stub type will radiate less energy into the surrounding atmosphere? (a) shorted, (b) open-ended, (c) either.

28. When the antenna load is removed from the line, the SWR will change to (a) zero, (b) no change, (c) maximum, (d) high at the load terminals only, (e) low at the load terminals only.

29. What is the value of SWR when E_{max} of 48.1 V and E_{min} of 26 V are found at 7.752 in and 17.952 in from the load?

30. What is the frequency of the signal in question 29?

31. What is the value of load impedance in question 29?

32. At 0.198 wavelengths from the load, E_{min} is 5 V. At what wavelength distance from the load would E_{max} of 9 V be found? (a) 0.250 wavelengths, (b) 0.198 wavelengths, (c) 0.357 wavelengths, (d) 0.302 wavelengths, (e) 0.448 wavelengths.

33. The normalized load impedance seen by the transmission line at the point where the correct matching stub is connected is (a) $0 \pm j0$, (b) $0 \pm j1$, (c) $1 \pm j0$, (d) none of the above.

34. The magnitude of the reflection coefficient identifies the percentage of signal that is *not* absorbed by the load. (T)(F)

35. If you have the choice between a shorted matching stub and an open matching stub, which would be better?

36. Why is one choice better than the other in the preceding question?

37. With $Z_0 = 100$ Ω and $Z_L = 150 + j60$ Ω, determine the following values at 738 MHz.
   a. Z_n
   b. SWR
   c. the distance from the load to the stub
   d. the length of the shorted stub

38. With $Z_0 = 80$ Ω and $Z_L = 48 - j28$ Ω, determine the following values at 590.4 MHz.
   a. Z_n
   b. SWR
   c. the distance from the load to the stub
   d. the length of the shorted stub
   e. the reflection coefficient

39. The separation between the stubs in a double-tuned matching system is how many wavelengths? (a) ⅛, (b) ¼, (c) ½, (d) not fixed.

40. Using a double-tuned matching network with three-eighths wavelength spacing, what is the minimum and maximum distance (in wavelengths) from the load terminals to the first stub when the SWR is 2:1?

CHAPTER TEN

ANTENNAS

10.1 **INTRODUCTION**

Whole books have been written on antennas. The topic appears to be boundless, and it may be. However, certain basic concepts appear in every antenna design. The most outstanding of these is the one-half wavelength measurement. Other antenna sizes are sometimes referred to in theoretical explanations, but when it comes down to building a practical transmitting element, the half-wave dipole antenna is the basis for *all* antenna systems.

There are three fundamental forms used to describe how an antenna launches energy into the surrounding atmosphere:

1. The isotropic radiator (point source)
2. The elementary doublet
3. The half-wave dipole (All other antennas are modified forms of the half-wave dipole antenna.)

The **isotropic radiator** is impossible to build. However, if one could be made, its behavior would be useful in analyzing (and explaining) the radiation patterns and power densities from a distant source that are dispensed into the atmosphere. The radiation pattern of the point source antenna is similar to that of the sun; it gives off energy in every conceivable direction. Further discussion of this radiator appears in Chapter 12. The **elementary doublet** antenna is defined as being one-tenth of a wavelength long. The current in the antenna is considered to be constant at every point along this length and is therefore useful for analyzing the power in the transmitted wave relative to direction. The elementary doublet antenna is easily constructed, but has limited applications. Its service to science is as a theoretical tool with which to build theories.

10.1.1 The Reciprocity Theorem

Memorize this statement: Every antenna will work equally well for transmitting *and* receiving. Every antenna to be discussed in this text will have the
same properties when transmitting and when receiving. The gain is the same,
the radiation patterns are the same, and the frequency and beam width are the
same. The only property of the antenna that will differ is its power-handling
ability. Transmitting antennas need to emit high power levels and are made of
thick-gauge tubing. Receiving antennas operate on low power, often in microwatts, and are made from small-diameter conductors.

10.2   THE HALF-WAVE DIPOLE ANTENNA

The most useful concept of antenna theory is taken from transmission line
theory. Figure 10.1a reviews principles of open-ended transmission line and
shows the standing waves of voltage and current along the line. Current is zero
(or minimum) through the infinite impedance of the open load, and the voltage
is maximum across the load terminals A and D. Assume that, at this instant,
the voltage at point A is positive with respect to the voltage at point D. One-
quarter wavelength from the open terminals, at B and C, the voltage is minimum and the current is maximum. Now bend the wire at B so that A swings
away from center, and bend the other wire at C so that D swings away from
center, as shown in Figure 10.1b. The potential difference between A and D is
still maximum, with A positive compared to D, and the current at B-C is still
maximum. The impedance of the antenna at B-C is typically 73 Ω, so that a
73 Ω transmission line would see a matched load impedance and the standing
waves from B-C back to the line input would be canceled. However, in the

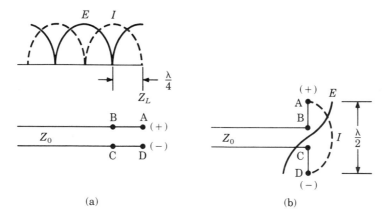

(a)   (b)

FIGURE 10.1
Open transmission line and standing waves of current and voltage form the principle
of the half-wave dipole antenna.

extended half-wavelength section (A to D), there are still standing waves of voltage and current that are required to release electromagnetic waves into the surrounding atmosphere. The wave of voltage across the antenna is standing, in that it is always maximum at the ends and zero at the center, although the ends may alternate between positive and negative at the frequency of the carrier signal. The current wave also alternates between positive and negative but remains maximum at the center and zero at the ends. The section of line from A to D is one-half wave in length and is a two-pole (dipole) radiating element (antenna). The separation at the center between the quarter-wave sections is necessary for cable attachment. The size of the gap is not important above 14 MHz, but it is required and is included in the half-wave length measurement. This *is* the Hertz half-wave dipole antenna. As other antennas are treated in this chapter, they will be compared to the performance of this reference antenna.

10.2.1 The Radiated Wave

The changing voltage and current in the antenna system will set up an electric field and a magnetic field around the dipole antenna, as shown in Figure 10.2. The electric field (E) surrounds the dipole with lines of force from end to end of the antenna. The magnetic field (H) surrounds the antenna like orbiting rings around the rod, strongest at the center and weaker toward the ends of the antenna. The electromagnetic waves that leave the antenna do so in a plane that is perpendicular to *both* the electric and the magnetic fields. The energy will leave the antenna in a pattern that encompasses 360° of the field, at 90° to the plane of the antenna. This plane would include the A, A', B, and B'

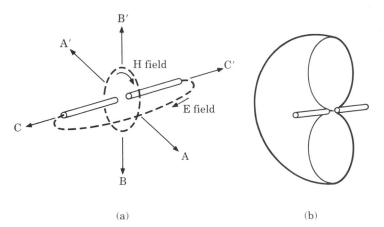

(a)   (b)

FIGURE 10.2
(a) The electric and magnetic fields of a half-wave dipole antenna. Energy propagates toward A, A', B, and B' but *not* toward C or C'. (b) The half-field radiation pattern.

directions in Figure 10.2; *no* signal will be radiated from the ends of the antenna, in the C or C' directions. The energy will create a doughnut-shaped radiation pattern, with the dipole antenna through the center of the hole. This pattern has been sliced in half to show the cross section in Figure 10.2b.

10.2.2 Polarization

The antenna is polarized in the direction of the plane of the electric field. The antenna in Figure 10.2 is lying in a horizontal plane, so the electric field is in a horizontal plane; therefore, this antenna is said to be *horizontally* polarized. A half-wave dipole antenna that stands on end (such as a CB or AM car radio antenna) has the electric field in the same plane as the length of the antenna and is therefore *vertically* polarized. For best results, the transmitter and receiver antennas should be positioned in the same polarity plane.

10.2.3 Radiation Patterns

Figure 10.3 shows the comparative radiation patterns for the point source, elementary doublet, and the Hertz half-wave dipole antennas for the same relative power to the antennas. Because the point source dispenses energy in every direction, its field intensity takes on the shape of a globe and covers a limited range. The elementary doublet restricts energy at the ends of the antenna and therefore has a greater intensity in the perpendicular plane. The pattern is a sine wave function and appears as a true circle in both the horizontal and vertical planes of the radiation fields. The half-wave dipole has a pattern such that the energy at one tip of the antenna is 180° out of phase from the energy at the other tip; thus, it will have a canceling effect in the near field for angles close to the axis of the antenna (see Figure 10.4)

The equation for the relative field strength e at any angle away from the axis of the antenna is

$$e = A_{vn}\left[\frac{\cos\,(90\,\cos\,\theta)}{\sin\,\theta}\right]^{n} \tag{10.1}$$

where A_{vn} = antenna gain
n = number of elements
θ = degrees away from the antenna axis

For the basic dipole antenna, $A_v = 1$ and $n = 1$. The radiation pattern of Figure 10.5 results from Equation 10.1 for a single element Hertz half-wave dipole antenna for $\theta = 0°$ through $\theta = 90°$.

This is not to say that any area outside of the described pattern will receive no signal, or that all points inside of the pattern will receive equal signal. It merely defines a graph of equal-energy points within the radiated field.

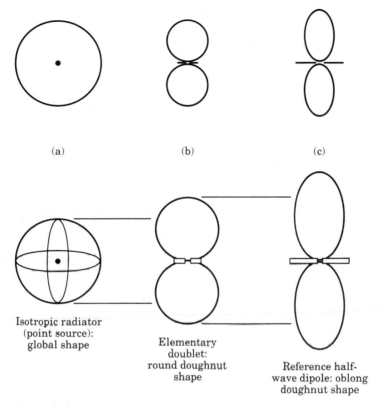

(a)   (b)   (c)

Isotropic radiator
(point source):
global shape

Elementary
doublet:
round doughnut
shape

Reference half-
wave dipole: oblong
doughnut shape

FIGURE 10.3
Comparative fields of the three basic forms of antenna.

FIGURE 10.4
Partial cancellation of the energy in the
direction P-P′ by an amount equal to
λ/2 cos θ due to out-of-phase voltage
from A and B.

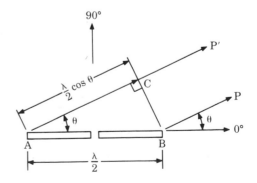

339

FIGURE 10.5
The radiation pattern of a half-wave dipole antenna, showing the determination of the beamwidth angle.

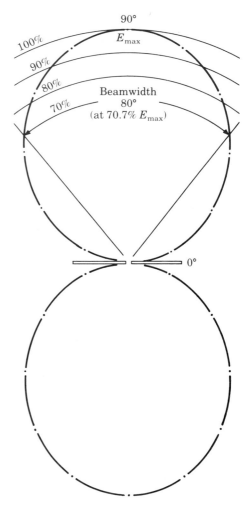

A fictitious but realistic field strength pattern is mapped in Figure 10.6 for a town having the horizontally polarized antenna at the center of the town, radiating primarily to the north and south. Field strength measurements have been made at all 1-mile coordinates in the northeast quadrant and are assumed to be the same for all four quadrants. At 5 miles from the antenna on the center line of radiation, the signal strength is measured at 250 μV. Join all points on the map that have a 250 μV field strength reading. The points on the center line closer to the transmitter have a signal strength *much* stronger than 250 μV, and locations outside of the 250 μV pattern will have signal strengths acceptable for normal to fringe-area reception. The further from the antenna, or the further off center, that the receiver is located, the weaker the signal to the receiver.

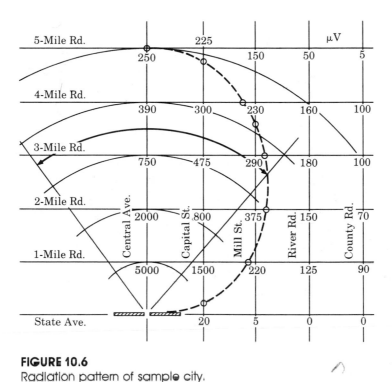

FIGURE 10.6
Radiation pattern of sample city.

Beamwidth On the center line of Figure 10.6 and 5 miles away from the transmitter, the field strength is 250 μV. Maintaining the 5-mile distance, the degree to which you can move off center before the signal decreases to −3 dB determines the **beamwidth** angle. During measurement procedures, the transmitter and receiver are at fixed locations and the antenna is rotated. The beamwidth is the *total* of the −3 dB points on either side of the center line. When a radiation pattern is given, find the −3 dB level on the center line, then with a compass, find the intercepts at either side of the pattern to determine the beamwidth angle. The beamwidth for Figure 10.6 is ±39° = 78° total.

10.2.4 Antenna Impedance

The input resistance at the center of a half-wave dipole antenna is independent of the gap size for frequencies above 14 MHz but is closely related to the antenna length-to-diameter ratio and the current distribution throughout the length of the antenna. The difference in current at various parts of the antenna is compensated for by the variable exponent in the input resistance equation. The

FIGURE 10.7
Exponent values for n of the input resistance.

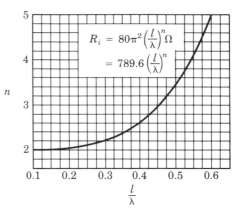

exponent n in Figure 10.7 is valid for antenna lengths from 0.1 to 0.6375 wavelengths. Beyond this size, a new breed of antenna called the *collinear antenna* is used (to be covered later). An antenna 95% of one-half wavelength long, center-fed, has an input resistance of

$$R_i = 80\pi^2\left(\frac{l}{\lambda}\right)^n \qquad (10.2)$$

The length is $0.95(\lambda/2) = 0.475$ wavelength and $n = 3.2$; thus, the input impedance is

$$R_i = 789.6(0.475)^{3.2} = 72.92 \ \Omega \quad \text{(use 73 } \Omega\text{)}$$

An end-fed half-wave dipole antenna (no center gap) will have an input impedance of approximately 2500 Ω.

10.2.5 Antenna Size

Because energy moves more slowly in a solid than in air and because the current near the ends of the rod does not have the space to move as freely as current near the center, the ends of the antenna exhibit a capacitive effect. This effect is countered by making the physical size of the antenna slightly smaller than the calculated half-wavelength size. The calculations are based partly on the conductivity of the antenna material, partly on skin effect at the ends of the antenna, and partly on the length-to-diameter ratio of the radiating element. For the self-supporting horizontally polarized antenna, the length to diameter ratio ranges from about 50:1 to about 150:1. Thin wire antennas that are supported at both ends may have a length-to-diameter ratio as high as 1000:1. Adjustment values of 0.94 to 0.97 of the calculated length are common. This text will use the average of 0.95 or 95% of a half-wavelength as the physical size of the Hertz half-wave dipole antenna.

To fabricate an antenna for operation at 75 MHz, first use Equation 8.10 to find the size of one-half wavelength, then multiply it by 0.95:

$$\frac{300 \times 10^6}{2 \times 75 \times 10^6} = 200 \text{ cm} \times 0.95 = 190 \text{ cm}$$

(Without further mention, all future size calculations will include the 5% reduction factor.)

10.3   THE FOLDED DIPOLE

Due to the distributed inductance and capacitance, the half-wave dipole antenna is a resonant circuit at all frequencies for which the length is equal to a whole number of half wavelengths.

The tuning of the antenna is proportional to its Q. The Q of the antenna may be lowered by increasing the diameter, causing a decrease in the inductance. From the relationship X_L/R, Q decreases in order to respond to a wider band of frequencies. The antenna of Figure 10.8 is called a **folded dipole** antenna and is a popular way to effectively increase the antenna diameter. A half-wave dipole antenna is also shown as a reference. The folded dipole is one-half wavelength long, end to end, and has one driven rod and one parasitic rod. The driven rod is center-fed at the separation. To the electromagnetic wave, the folded dipole antenna looks like a large-diameter half-wave dipole antenna. The closely spaced parasitic conductor increases the antenna

FIGURE 10.8
Comparison of the Hertz half-wave dipole (a) and the folded dipole (b) antennas. The diameter determines the antenna Q.

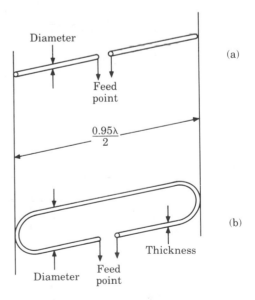

impedance to about 300 Ω, which remains constant over a fairly wide range of frequencies. The impedance of the folded dipole antenna is

$$Z_{fd} = 73N^2 \qquad (10.3)$$

where N is the number of parallel radiators. In Figure 10.8, there are two parallel radiators for the folded dipole antenna (one driven, one parasitic) so that the impedance is $73 \times 4 = 292$ Ω, usually referred to as 300 Ω. Typical ratios for rod spacing to rod diameter range between 5:1 and 10:1.

The diagrams and equations in Figure 10.9 represent the exact calculations for the impedance of the folded dipole antenna. As can be seen, the values depend largely on the differences in the rod diameters. When all of the rods are of the same diameter, then Equation 10.3 applies. When the diameters differ from one rod to another, then the equation for the associated folded dipole antenna configuration in Figure 10.9 is used. Antenna impedances much over

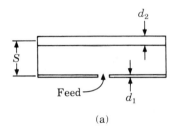

(a)

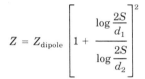

$$Z = Z_{dipole} \left[1 + \frac{\log \dfrac{2S}{d_1}}{\log \dfrac{2S}{d_2}} \right]^2$$

d_1 = driven diameter
d_2 = parasitic diameter 1
d_3 = parasitic diameter 2
S = center-to-center separation

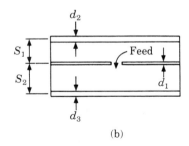

(b)

$$Z = Z_{dipole} \left[1 + \frac{\log \dfrac{2S_1}{d_1}}{\log \dfrac{2S_1}{d_2}} + \frac{\log \dfrac{2S_2}{d_1}}{\log \dfrac{2S_2}{d_3}} \right]^2$$

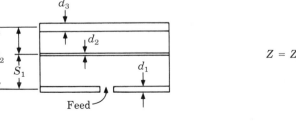

(c)

$$Z = Z_{dipole} \left[1 + \frac{\log \dfrac{2S_1}{d_1}}{\log \dfrac{2S_1}{d_2}} + \frac{\log \dfrac{2S_2}{d_1}}{\log \dfrac{2S_2}{d_3}} \right]^2$$

FIGURE 10.9
Variations of the folded dipole antennas and the associated impedance equations.

650 Ω are inappropriate in view of the maximum values of cable impedances from Chapter 8. The Q of the antenna is roughly estimated at

$$Q_{\text{ant}} = \frac{\text{rod length}}{2 \times \text{rod diameter}} \qquad (10.4)$$

For the basic dipole antenna, use the rod diameter. For the folded dipole antenna, instead of the rod diameter, use the center-to-center spacing of the parallel rods.

The gain, beamwidth, and radiation pattern of the dipole and folded dipole are the same. Only Q, the bandwidth, and the center-fed impedance change. The flat surface of the folded dipole antenna faces in the direction of transmission and reception.

10.4   THE CONICAL ANTENNA

A differential-diameter half-wave dipole called a **conical antenna** is shown in Figure 10.10, which illustrates the cone-shaped dipole from which the antenna takes its name. The conical antenna introduces another popular method of increasing the diameter to lower the Q and increase the bandwidth.

The shaded area in Figure 10.10b is equal to the area that is unshaded. The distance between the lines X and X' is made to be 95% of one-half wave-

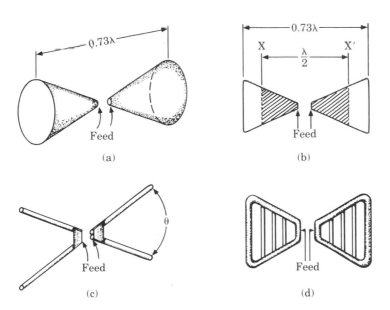

FIGURE 10.10
The conical antenna. The evolution from a cone (a) to the rod (c) to the bow-tie (b) to the modified bow tie (d).

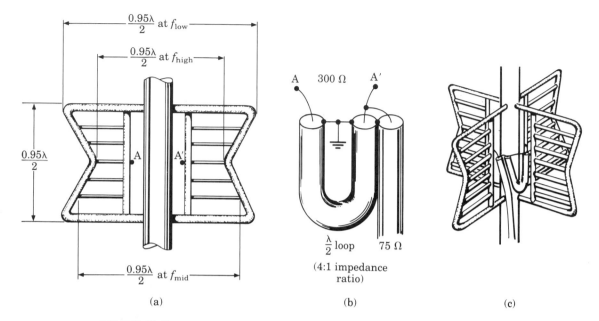

FIGURE 10.11
The batwing antenna, a modified bow tie. Part (c) shows two batwings configured for all-direction radiation.

length at the geometric center frequency of the band of frequencies to be covered. This works out so that the overall length of the conical antenna is 0.73 wavelengths. By definition, the conical antenna is still a half-wavelength dipole, center-fed, 300 Ω, low-Q, wide-bandwidth antenna. The major advantage of the conical antenna is its constant impedance and gain over a 4:1 frequency spread. It was developed for the VHF television band of 54 MHz to 216 MHz, which is exactly a 4:1 frequency spread. Its major disadvantages of bulk and poor resistance to high winds were overcome by the "fake" conical antenna of Figure 10.10c. Here two rods are spread open to resemble a cone shape. The performance is slightly diminished, but the tradeoffs were considered to be worthwhile.

The next transition was a conical antenna made from a flat sheet of steel, called the "bow tie" antenna, shown in Figure 10.10b. The bow tie antenna works as well as the conical and is easier to make. Although the bow tie antenna still has a large cross section, which presents resistance to high winds, it finds popular use at very high frequencies, where the total area can be kept small. A bow tie antenna made in the form of a grill, shown in Figure 10.10d, will greatly reduce the wind resistance and yield high performance when the space between the grillwork remains less than 0.1 wavelength at the highest operating frequency. The impedance of the conical antenna is determined by

$$Z = 120 \ln\left(\frac{\theta}{2}\right) \tag{10.5}$$

where θ is the angle shown in Figure 10.10c and Z is the impedence seen at the center space, where the signal is applied. The batwing antenna of Figure 10.11 is a modified bow tie (or conical) and has the grill rods in the horizontal plane. The shortest rod resembles a half-wave dipole at the highest frequency, and the longest rod favors the lowest frquency of interst. The feed point is at the center terminals, marked A and A'. The half-wave dimension is also identified in the figure. The batwing antenna also has characteristics that resemble a folded dipole with five parallel conductors.

The conical, bow tie, and batwing antennas all have the same gain, beamwidth, and radiation pattern as the reference half-wave dipole antenna.

10.5 THE TURNSTILE ANTENNA

All of the previously mentioned half-wave, horizontally polarized dipole antennas have the same ability to transmit front to back, but not side to side. When a second horizontally polarized antenna of like construction is added in parallel, but rotated 90° so that its front-to-back radiation pattern fills in the nonsignal areas of the first dipole (see Figure 10.11c), then the combined radiation patterns will cover 360° while remaining horizontally polarized.

Figure 10.12 illustrates such an antenna arrangement, called a **turnstile** antenna. The advantage is omnidirectional coverage with horizontal polarization. The disadvantages are a -1.5 dB reduction in gain compared to the reference dipole antenna and a decrease in antenna impedance. The dipole antennas are connected in parallel, as shown in Figure 10.12, where $Z = 38\ \Omega$.

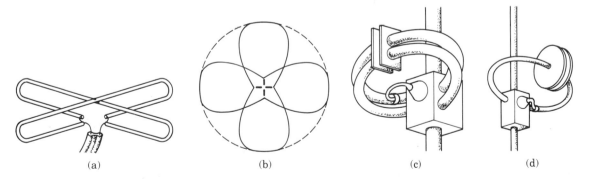

(a)　　　　　(b)　　　　　(c)　　　　　(d)

FIGURE 10.12
The turnstile antenna (a) and its radiation pattern (b). The Gates "cycloid" (c) and the "Collins ring" (d). All are horizontally polarized with 360° compass radiation patterns.

10.6   PARASITIC ARRAY ANTENNAS

An array is an orderly grouping of objects, in this case antenna elements. There are three basic parts to the parasitic array antenna, the *driver* or *driven element,* the *director,* and the *reflector.* The driven element is a half-wave dipole antenna and is the only element to directly receive power from the transmitter or deliver power to the receiver. Placing a rod in the near field, parallel to the dipole antenna but not electrically connected to it, will increase the far-field radiation signal strength in one direction and decrease the far-field signal strength in the opposite direction. Such rods or elements, including the director and reflector, are called parasites. They feed on the main energy source in that they receive their excitation by near-field coupling from the driven element (see Figure 10.13).

A director is shorter than the driver and is located in the direction of desired transmission or reception (in front of the driver). A reflector is longer than the driver and is placed in a plane away from the direction of desired transmission or reception (in back of the driver). The directors and reflectors are grounded to the support boom at their centers, while the driver, split at the center feed point, must remain insulated from ground.

A wire or rod cannot have gain, but focusing the energy pattern of Figure 10.2 into a specific direction through the addition of directors and reflectors makes the far-field signal strength greater in that direction than would have been measured from the driver alone. This increased concentration of energy is noted as *directivity gain*. It is only in this sense that an antenna is said to have gain. The gain in one direction is achieved by the sacrifice of signal strength in other directions. The driven element may be any of the half-wave dipole antennas covered so far, that is, a single straight half-wave dipole, a folded dipole of any number of conductors, a conical, a bow tie, or a batwing, but *not* a turnstile antenna. Adding on only one reflector of length 5% longer than the driven element and placed 0.15 wavelengths in back of the driven element will increase the forward signal strength by 5 dB. At the same time, the signal

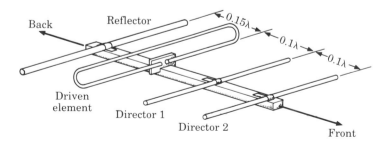

FIGURE 10.13
Assembly of a parasitic array, with a folded-dipole driven element, one reflector, and two directors.

pickup from the back end of the antenna will decrease by −15 dB, giving a front-to-back ratio of 20 dB.

The change in the radiation pattern is noted in Figure 10.14 as an increase to a balloon-shaped major lobe in the direction of desired radiation, a sharp reduction to a small lobe at the back of the antenna, and the appearance of several minor side lobes. The horizontal and vertical plane radiation patterns are plotted in the form that would appear in the technical antenna specifications. Removing the reflector and adding only one director 5% shorter than the driver, placed 0.1 wavelengths in front of the driver, will result in a radiation pattern almost identical to Figure 10.14, except for a smaller reduction in the front-to-back ratio. The directivity gain is still 5 dB, but the back lobe is reduced by only 10 dB, for a front-to-back ratio of 15 dB.

Two elements, one director and one reflector, with lengths and spacings as stated before, will develop a radiation pattern of the general shape of Figure 10.14 but will have a directivity gain of +8 dB, a front-to-back ratio of −23 dB, and a narrower beamwidth angle.

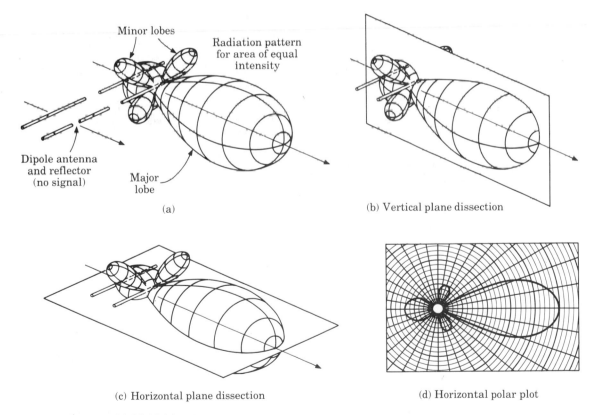

(a)

(b) Vertical plane dissection

(c) Horizontal plane dissection

(d) Horizontal polar plot

FIGURE 10.14
Dissection of radiation patterns for purposes of polar plotting.

Adding more directors, each 5% shorter in the direction away from the driven element, will increase the forward signal strength of the antenna array. The back lobe will decrease for each added element, but at a much smaller rate (see Table 10.1 for the details of gain, beamwidth, and front-to-back ratios as each element is added). The spacing for directors is constant at 10% of a full wavelength. Adding more than one reflector has such a small effect that it is seldom done. However, if more than one reflector is to be used, the length of each element increases by 5% over the length of the closer element to the driven dipole. The spacing of reflectors is constant at 15% of a full wavelength.

The dB gain column of Table 10.1 is referenced to the half-wave dipole antenna but includes the gain (−dB) of the isotropic point source and the elementary doublet (in parentheses). These values are given here only for reasons of comparison and are not to be used as a reference. The voltage gain values in Table 10.1 are for use in Equation 10.6.

Table 10.1 is converted to graph form in Figure 10.15 to show the dB gain compared to the number of elements. Here it is apparent that the gain

TABLE 10.1

Type	Voltage Gain	dB Gain (Iso.)	Beamwidth Angle	Front-to-back Ratio (dB)
Isotropic source	0.780	−2.16 (0)	360	0
Elementary doublet	0.942	−0.52 (1.64)	90	0
Half-wave dipole	1.0	0 (2.16)	80	0
+1 element (1 director or 1 reflector)	1.778	5 (7.16)	52	−15 or −20
+2 elements (1 refl. + 1 dir. or 2 dir.)	2.512	8 (10.16)	40	−23
+3 elements (1 refl. + 2 dir.)	3.126	9.9 (12.06)	36	−24
+4 elements (1 refl. + 3 dir.)	3.589	11.1 (13.26)	34	−25
+5 elements (1 refl. + 4 dir.)	3.936	11.9 (14.06)	32	−26
+6 elements (1 refl. + 5 dir.)	4.169	12.4 (14.56)	30	−27
+7 elements (1 refl. + 6 dir.)	4.365	12.8 (14.96)	28	−28
+8 elements (1 refl. + 7 dir.)	4.571	13.2 (15.36)	27	−28.5
+9 elements (1 refl. + 8 dir.)	4.677	13.4 (15.56)	26	−29
+10 elements (1 refl. + 9 dir.)	4.732	13.5 (15.66)	24	−29.5

approaches a plateau, so that adding on more elements will provide so little gain as to make it impractical. Each added element increases the total weight of the array. The cost of a sturdier structure to support the added weight under high wind conditions could offset the gain advantage of adding another element.

Table 10.1 and Figure 10.15 are taken from an array with 10% spacing for directors, 15% reflector spacing, and a length change of 5% per element. Antenna arrays with different constants and ratios will perform in generally the same manner but may have altered beamwidth and gain characteristics. Each variation will need to be built and field-tested for its own set of specifications. The exact length of directors and reflectors and their distance from the driven element, and each other, is the little bit of magic that remains in electronics.

The radiation patterns for antennas are plotted on graph paper as a result of field tests, and the beamwidths are then determined from the graphs. The field strength is recorded at a right angle to the antenna axis as the maximum strength, and the antenna is then rotated to a direction away from the right angle to record other values of field strength.

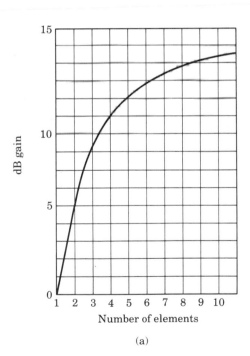

(a)

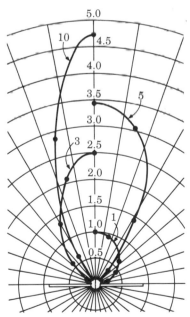

Relative radiation patterns in quadrants
for one, three, five, and ten elements

(b)

FIGURE 10.15
Gain vs. radiation patterns for up to ten elements.

Figure 10.15 shows the field strength radiation patterns that will result from an increase in the number of elements. Using the data from Table 10.1, the same patterns could be approximated with the use of Equation 10.1.

Much of the original work on parasitic array antennas was conducted by Professor H. Yagi and his protégé, Shintaro Uda. Their array has become known as the *Yagi-Uda antenna* or simply the *Yagi antenna*. This array had one half-wave driven dipole, one reflector that was 5% longer than the driven element, and up to 13 directors all of the same length (5% shorter than the driven element): however, the spacing between directors was made progressively smaller as the distance from the driven element increased. This laid the foundation for the many variations of parasitic array antennas in use today.

In any array construction, there is a point where the radiation pattern begins to self-destruct. This condition is called *lobe splitting*. At some number of elements, the major lobe begins to pull back on itself and multiply, as shown in Figure 10.16a and b. As the number of elements further increases, the condition progresses to a point where the major lobe splits into a full cloverleaf, as seen in Figure 10.16c, with almost no signal emitted in the desired direction. The main points to remember about the parasitic array antenna family is that

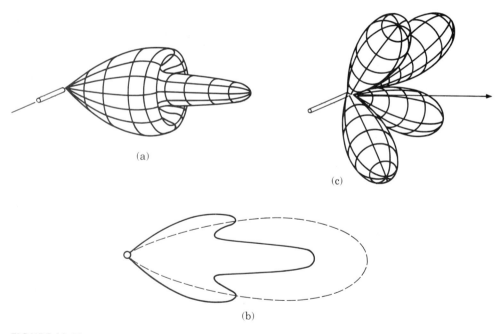

(a)

(c)

(b)

FIGURE 10.16
Contraction of the field pattern for excessive addition of elements. (a) Start of contractions. (b) Field pattern plot. (c) Lobe splitting.

only one element is driven and the rest are parasites and change size with distance from the driven element. The spacing between elements is generally constant but differs for the directors and reflectors. The number of elements will generally not exceed 15. The driven element is split at the center and is insulated from ground, while the parasitic elements are grounded to the boom at the center. The parasitic array is most often used in the horizontally polarized position but is not restricted to this mode.

10.6.1 The Driven Arrays

The **driven array** is a family of half-wave dipole antennas in which signal is applied to all elements and the length and spacing may differ from one array member to another. This family includes the end-fire antenna, the broadside antenna, the collinear antenna, and the log-periodic array antenna. It is not uncommon to find combinations of driven array members in a single antenna assembly.

The End-Fire Antenna The two-element *end-fire* antenna, shown in Figure 10.17a, has the characteristics of two half-wave dipole antennas connected in parallel. Because the separation is exactly one-half wavelength, the current in D_1 leads the current in D_2 by 180°, and the fields to be radiated will aid each other to form equal lobes to the front (F) and to the rear (R) of the array, completely canceling any radiation to the sides.

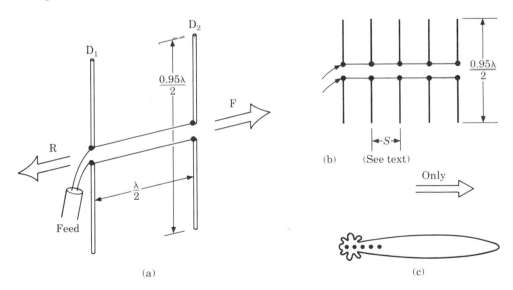

FIGURE 10.17
Vertically polarized end-fire antenna (may be horizontally polarized by 90° rotation).

Decreasing the separation (S) between elements to

$$S = \frac{\lambda}{2}\left(1 - \frac{1}{2N}\right) \qquad (10.6)$$

where N equals the number of driven elements, will cause the rear lobe to completely disappear and will reinforce the forward lobe to form a pencil-shaped radiation pattern in one direction. The radiation will be in the direction from the element leading in current phase to the element lagging in current phase. In Figure 10.17 the signal is applied to D_1; therefore, current will flow in D_1 ahead of the current in D_2 because of the delay encountered in the harness between the two elements. Current through D_1 leads current through D_2, so the radiation pattern to the rear (R) will be canceled while the forward radiation (F) will be strengthened. When S is less than one-half wavelength, the element-to-element current phase α is

$$\alpha = \left(\frac{2\pi}{\lambda}\right)S \quad \text{(in radians)} \qquad (10.7)$$

where S is a fraction of a wavelength.

Increasing the number of driven elements will, of course, increase the array gain. Gain measurements of 20 dB for a five-element end-fire array are typical. The end-fire antenna will operate equally well in the horizontal and vertical planes by a simple rotation around an axis through the center of the harness.

Collinear Arrays   It should be remembered that a half-wave dipole antenna will have a high input impedance when driven at either end. When two such devices are joined end to end, as shown in Figure 10.18, the assembly is called a **collinear array.** The radiation pattern of the collinear is similar to the reference half-wave dipole antenna. That is, it is doughnut-shaped in all directions

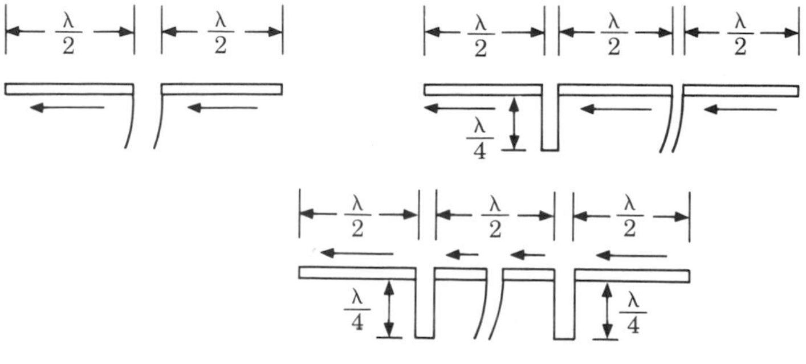

FIGURE 10.18
Collinear array antenna. Sections are one-half wavelength long and separated by a one-quarter wavelength shorted stub of infinite impedance.

that are perpendicular to the axis of the antenna, with no signal radiated off either end. However, the collinear pattern has a narrower beamwidth angle.

The connection between the half-wave sections is basically a quarter-wave length of shorted transmission line. It exhibits a high impedance one-quarter wavelength away from the short and is used to maintain the in-phase current through all of the sections. When multiple sections are used, as in Figure 10.18b and c, the antenna may be fed either at one of the interconnection points or at the mechanical center of the antenna, with no noticeable difference in the energy distribution of the transmitted signal. The gain will increase as more elements are added on, and the beamwidth will decrease with more elements. Generally, no more than four elements are used because of the large size. The size also influences the polarization, making the horizontal position less cumbersome than the vertical, although the antenna will perform equally well in either plane.

The Broadside Array The *broadside array* antenna is an extension of both the end-fire array and the collinear array (see Figure 10.19). All elements are driven and are one-half wavelength long on each side of the feed point. All elements are separated by one-half wavelength. This arrangement behaves like the end-fire antenna with collinear-sized elements. To cause the radiation to leave the antenna from the flat surface to the sides of the array, the feed lines are crisscrossed to change the phase of the current in every other element. The radiation pattern is a fan-shaped lobe on both sides of the array, as shown in Figure 10.19b, which displays a short vertical field and a wide horizontal field. When a wire mesh screen is placed 0.15 wavelengths from the array, an electrical reflector is formed; one lobe will be canceled and all of the energy will radiate from only one surface of the antenna. Note the mechanical assembly detail in Figure 10.19d and the change from the vertical polarization in Figure 10.19a to the horizontal polarization in Figure 10.19c.

Combinations One of the most confusing aspects of antenna theory to the newcomer is that the standard antenna is rarely found. Most antennas in use are either modifications of the basic antenna form or combinations of several different forms. Two of the many possible combinations are shown in Figure 10.20.

Figure 10.20a is a two-element collinear array, horizontally polarized. Because the input signal is applied to the exact electrical center of the array, it arrives at each element at the same time to develop in-phase currents in the two halves. When the currents are in phase, the antenna becomes a broadside radiator. The array is then called a "horizontally polarized, collinear broadside array." In shoptalk, it is called a "lazy H." The gain varies with the element spacing, as shown in the drawing. Figure 10.20b is also a horizontally polarized two-element collinear array. The signal is again applied to the exact electrical

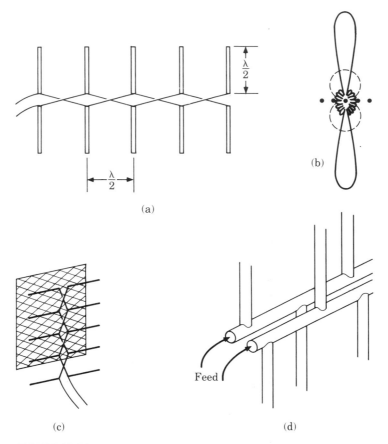

FIGURE 10.19
The broadside array. (a) The basic array configuration. (b) Radiation pattern (solid) compared to the reference dipole (dotted). (c) A reflector screen added to eliminate one lobe and provide a unidirectional pattern. (d) One structural approach.

center, but the phasing harness is cross-fed to result in out-of-phase currents in the two halves. The array thus becomes an end-fire array. Again, the spacing varies the gain, but note that the spacing never approaches one-half wavelength. Each antenna in Figure 10.20 has an input impedance of thousands of ohms and requires some type of impedance-matching network; also, each antenna has about a 2:1 frequency ratio. The collinear end-fire shows a shorted matching stub of 3/16 wavelength to help offset the high impedance.

10.6.2 Log-Periodic Array

The **log-periodic array** is unique and has earned its own place in the lineup of antennas. It is driven from the *short* end into a cross-fed series of half-wave dipole antennas that progressively increase in size and distance apart. The

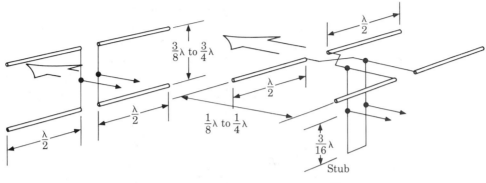

Broadside Collinear			End-fire Collinear	
Space	Gain		Space	Gain
$(3/8)\lambda$	4.4 dB		$(1/8)\lambda$	6.2 dB
$(1/2)\lambda$	5.9 dB		$(1/4)\lambda$	5.7 dB
$(5/8)\lambda$	6.7 dB		45° beamwidth	
$(3/4)\lambda$	6.6 dB			
(a)			(b)	

FIGURE 10.20
Combination collinear antenna arrays. Each is horizontally polarized and has a 2:1 frequency ratio.

changing length and space ratios are determined from the chart in Figure 10.21 and will always be a number less than 1. Typically the scale factor r ranges between 0.78 to 0.98 for the individual antenna array. The scale factor is

$$r = \frac{L_2}{L_1} = \frac{L_3}{L_2} = \frac{L_4}{L_3} = \frac{S_2}{S_1} = \frac{S_3}{S_2} = \frac{S_4}{S_3} \qquad (10.8)$$

The array takes its name from the constant input impedance, which is a periodic function of the logarithm of the frequency over which the antenna operates. The input impedance is a function of the length-to-diameter ratio of the dipole:

$$Z_{\text{in}} = 120\left[\ln\left(\frac{\lambda}{2d}\right) - 2.55\right] \qquad (10.9)$$

where d = rod diameter. The gain, SWR, beamwidth, and front-to-back ratio all remain fairly constant over the same frequency range, with the input impedance ranging from about 200 to 800 Ω for any given array. Therefore, once the basic features of the antenna are selected and the input impedance calculated or measured, an impedance-matching transformer is almost always required.

The input signal is applied to the shortest dipole. A current is induced in this element, which then couples energy to the second element through

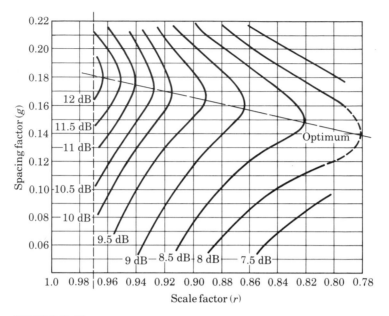

FIGURE 10.21
Log-periodic gain vs. spacing and length factors.

radiation. The conducted current arrives at the second dipole later in time than the current to the first dipole. Because of the cross-feed arrangement, the phase delay caused by the spacing and the separate phase delay caused by the coupled energy between elements will result in a signal radiation from the small end of the array only. Different spacing between elements could be selected to make the antenna radiate from both ends, similar to the end-fire antenna.

The log-periodic antenna is capable of operating over a 4:1 frequency ratio. The longest element length is determined as 95% of one-half wavelength at the lowest frequency in the operating band, and the shortest element is 95% of one-half wavelength at the highest frequency. The length and spacing ratios are determined from the chart in Figure 10.21 only after the antenna gain has been decided.

One may wonder why any gain factor other than maximum gain would be considered. The reason is found in Figure 10.21, which shows that at the highest gain, the rate of reduction in element length is the slowest and the spacing ratio is the widest. Thus, the antenna may be extremely long at the highest gain. Over the TV band (54 to 216 MHz), a 4:1 frequency ratio, the gain of 12 dB makes the length and spacing factor 0.93. The resulting antenna would have 37 elements and would be 8.5 ft wide by 64 ft long. On the other hand, at a gain of 9 dB, the antenna would have 12 elements and be 8.5 ft wide by only 13 ft long.

EXAMPLE 10.1 ══════════════════════════════════

A log-periodic antenna is to cover the frequency range of 200 MHz to 600 MHz and have a gain of 10 dB.

SOLUTION: The length of longest element (L_1) is

$$\frac{(V_c)(12)(0.95)}{(2)200 \text{ MHz}} = 28.044 \text{ in}$$

The length of shortest element is

$$\frac{(V_c)(12)(0.95)}{(2)600 \text{ MHz}} = 9.348 \text{ in}$$

From the 10 dB curve in Figure 10.21, a scale factor (r) of 0.92 is selected, resulting in a spacing factor (g) of 0.158. The space S_1 between the longest element and the second element is

$$S_1 = 2gL_1 = (2)(0.158)(28.044) = 8.862 \text{ in} \qquad \textbf{(10.10)}$$

The other spaces are found to be:

$$S_2 = rS_1 = (0.92)(8.862) = 8.153 \text{ in}$$
$$S_3 = 0.92S_2 = 7.5 \text{ in}$$
$$S_4 = 0.92S_3 = 6.9 \text{ in}$$
$$S_5 = 0.92S_4 = 6.348 \text{ in}$$

$$\cdot$$
$$\cdot$$
$$\cdot$$

$$S_{12} = 0.92S_{11} = 3.645 \text{ in}$$
$$S_{13} = 0.92S_{12} = 3.353 \text{ in}$$

The lengths are found to be:

$$L_1 = 28.044 \text{ in}$$
$$L_2 = 0.92L_1 = 25.8 \text{ in}$$

$$\cdot$$
$$\cdot$$
$$\cdot$$

$$L_{13} = 0.92L_{12} = 10.31 \text{ in}$$
$$L_{14} = 0.92L_{13} = 9.48 \text{ in}$$

L_{14} is close enough to 9.348 in to stop the calculations here. The antenna will have 14 elements and will be 2.5 ft wide and 6.1 ft long with a gain of 10 dB. The reader is encouraged to complete the intermediate calculations in this example and verify the conclusions.

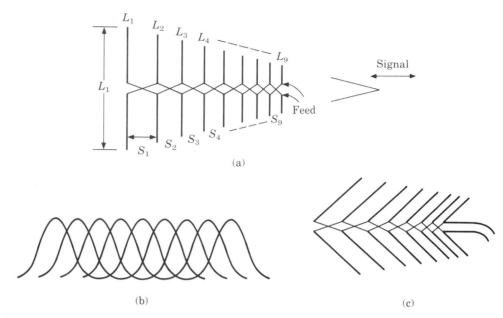

FIGURE 10.22
The log-periodic phased array antenna. (a) Spacing and length identification. (b) Gain response, likened to a family of resonant tuned circuits over a band of frequencies. (c) A log-periodic "V."

Figure 10.22c shows the log-periodic "V." Any half-wave dipole antenna may be bent into the V shape to minimize the side lobes and to improve the front-to-back signal ratio. This illustrates the frequency *foldover* principle, which has found application particularly in the TV band, where a center gap in the frequency range exists:

Channels 2–6: 54 MHz to 88 MHz
Channels 7–13: 174 MHz to 216 MHz

To cover this band, a log-periodic antenna was made with the longest element at 95% of λ/2 at 54 MHz = 104 in. The shortest element was made to be 95% of λ/2 at 88 MHz = 64 in. The 104 in element was also resonant as a 3λ/2 dipole at 162 MHz, and the 64 in element was resonant at 264 MHz as a 3λ/2 dipole. This antenna worked well over the two separate bands. The lower effective cross-sectional area at the higher frequencies tended to lower the gain, but this was compensated for by the additional gain of the 3λ/2 elements, resulting in flat gain over the full frequency range.

The log-periodic antenna can be viewed as a group of stagger-tuned circuits that cover a range of frequencies, as indicated in Figure 10.22b.

10.7   ANTENNA STACKING

Increased antenna gain can be achieved by simply adding on more antennas. The addition, however, is not random. The philosophy is based on a 3 dB gain for each time the antenna area is doubled. *Any* of the antenna types covered thus far can be stacked for greater gain as long as a few basic rules are observed.

1.   The polarization of each section must be the same.
2.   The signal applied to each section must be in phase.
3.   All sections must be of the same family.

This is not to say that a parasitic array could not be teamed up with a logperiodic antenna. It merely says that this combination would have unpredictable results and require extensive field tests. Antennas of the same type would yield predictable results.

This discussion will be limited to horizontally polarized antennas; the treatment of multiple vertical antennas will be held until later. Mounting one antenna above the other, as shown in Figure 10.23a, effectively doubles the antenna area and yields a 3 dB gain. The beamwidth remains the same as for a single dipole antenna. The separation between dipoles is normally held to onehalf wavelength and is not critical; however, the signal current phase to each

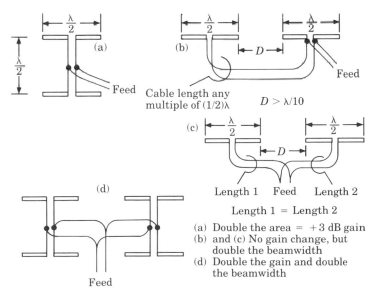

FIGURE 10.23
Stacking antennas of like construction.

dipole *is* critical. The normal connection is to supply the signal to the center of the harness so that it will be delivered to both antennas at the same time.

Figure 10.23b and c shows two horizontally polarized dipoles placed side by side. The physical distance between the antenna ends may be any measurement greater than one-tenth of a wavelength. The interconnecting harness should have a length equal to any multiple of one-half wavelength. When the harness length equals an even number of half wavelengths, then the antennas are simply connected in parallel. When the harness length equals an odd number of half wavelengths, then the connections must be transposed to keep the two antennas in phase. The result of this stacking arrangement is to maintain the single antenna gain while doubling the beamwidth. Expanded arrays such as the one in Figure 10.23d will double the gain (+3 dB) *and* double the beamwidth.

Figure 10.24 is an eight-bay stacked antenna assembly with identical signal phase to all dipoles. The antenna gain is +9 dB (the area of 1 doubled is 2, 2 doubled is 4, and 4 doubled is 8, with +3 dB per doubling over the gain of one section by itself). A signal of 1000 W fed into the antenna of Figure 10.24 would have an ERP of 8000 W.

The antenna gain is the power increase due to the alteration of the directivity pattern. With the stacking principle, the gain is due to the increase of the effective antenna area:

$$\text{Gain (dB)} = 10 \log (\text{number of bays}) \qquad \textbf{(10.11)}$$

This equation can be used where the number of bays is not an even doubling of dipoles, that is, is not a power of 2. For instance, a 12-bay stacked antenna has a

FIGURE 10.24
Eight-bay stacking for three times the antenna area for a 9 dB gain.

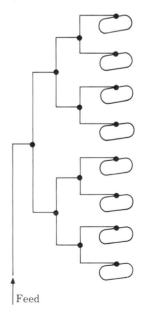

Feed

FIGURE 10.25
Cancellation of strong unwanted signals by placing a second antenna as
an out-of-phase input.

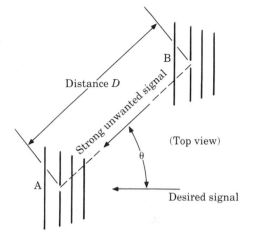

gain of 10.8 dB. When two or more array antennas are stacked, the principle of
stacking determines the dB gain to be added to the gain of the single array. For
example, say a parasitic array with a gain of 8 dB is stacked with an identical
parasitic array. The gain due to stacking is 3 dB. Added to the 8 dB gain of the
one antenna, this yields a total system gain of 11 dB. This could be an expensive way to achieve an additional 3 dB of gain but is a worthwhile approach in
some instances.

Horizontal stacking can also be used to cancel strong unwanted side signals. In Figure 10.25, antenna A can be rotated to find the direction of the
renegade signal and then repositioned as before to receive the desired signal. A
second antenna is placed in line with the unwanted signal, at a distance D from
the main antenna but aimed at the desired signal. The distance D, found from
Equation 10.12, is the interconnecting harness length (in the same dimensions
as the wavelength). The physical distance between the two antennas should be
as great as possible without putting undue tension on the cable.

$$D = \frac{3\lambda}{2 \sin \theta} \quad \text{(in same units as } \lambda) \tag{10.12}$$

10.8   MULTIBAND ANTENNAS

A dipole antenna can be modified to operate at several distinct frequencies. A
common antenna for amateur radio, shown in Figure 10.26, covers the 2 m,
6 m, and 10 m bands and is referenced in terms of wavelengths. The 2 m (150
MHz) center dipole is isolated from the remainder of the antenna by high-
impedance parallel-resonant circuits, labeled Z_1 in the diagram. Z_1 is resonant
at 150 MHz. The second band, of 6 m (50 MHz), will see the 150 MHz resonant
circuits as a low impedance and almost completely ignore them as an interrupt
in the 6 m dipole antenna. Z_2 is resonant at 50 MHz and isolates the 6 m dipole

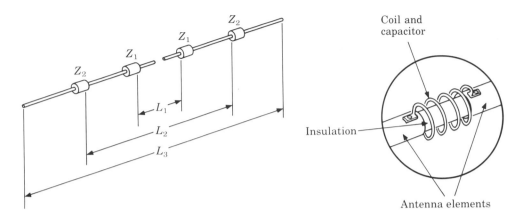

FIGURE 10.26
Multiband antenna.

from the ends of the antenna. Of course, the 10 m (30 MHz) signal will not see either tuned circuit but will view the whole antenna as simply a 10 m half-wave dipole antenna.

Reflectors and directors may be used in conjunction with the driven element by placing tuned circuits at the appropriate positions within the overall length of the parasitic elements. All of the basic guidelines apply to the reflectors and directors as covered earlier.

10.9   QUAD ANTENNAS

The **quad antenna** takes its name from the geometric shape of a quadrangle (having four sides and four angles). The unusual features about this antenna are that the sides are 3% longer than one-quarter wavelength and that the quad will operate at very low heights compared to most antennas. Generally, when an antenna is mounted close to the ground, the input impedance and the radiation patterns are both affected. Not so for the quad. This is due mainly to its diamond shape, where the sides are not parallel to the earth.

Figure 10.27b and c illustrates how this antenna can be made to operate in a horizontally or vertically polarized position by simple rotation of the feed terminals. A single-element quad has 2 dB more gain than the reference dipole antenna. A quad antenna used in the parasitic array configuration of Figure 10.27a will have a gain of 7 dB for two elements, 10 dB for three elements, and 12 dB for a four-element array, compared to the standard half-wave reference dipole antenna.

The antenna of Figure 10.27a is a multiband parasitic array for the 10 m, 15 m, and 20 m bands. Element 1 consists of the reflectors for all three bands.

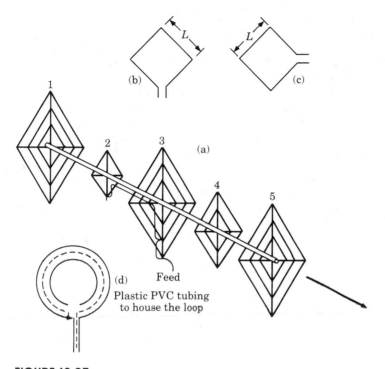

FIGURE 10.27
The quad antenna, a five-element, multiband parasitic array, (a), (b) Vertically polarized. (c) Horizontally polarized. (d) The loop, a cousin to the quad antenna.

Element 2 is the driven element for the 10 m band. Element 3 holds the driven elements for the 15 m and 20 m bands (two outside loops) and the first director for the 10 m band (inside loop). Element 4 is the first director for the 15 m band and the second director for the 10 m band. Element 5 serves as directors for all three bands in turn. The overall length is about 25 ft, which is about 40% shorter than comparable parasitic arrays. The input impedance is about 100 Ω for the driven element only, and drops to about 50 Ω for the four-element array shown. The main support boom is generally aluminum tubing, and the cross arms for the antenna wires are fiberglass, with holes drilled to string the wires at the appropriate dimensions. The quad antenna offers excellent gain in a compact size that will perform well at low heights.

The *loop* antenna, shown in Figure 10.27d, has a circumference of one wavelength and a radiation pattern that extends to the left and right of the diagram. The loop antenna is considered part of the quad family and is often used as a direction-finding device. However, it does not have a sharp directivity pattern, so its performance is only fair.

10.10 HELICAL BEAM ANTENNAS

Because the **helical beam antenna** has good characteristics over a 2:1 frequency ratio and can be made to be nearly circularly polarized with a modest beamwidth, it finds popular use as a satellite-tracking antenna.

The wire diameter d of the helical antenna in Figure 10.28 must be sufficient to support the weight of the coil. Each turn of length L is separated by a distance S, which sets the pitch α and determines the power gain G_p as a function of NS, where N is the number of turns.

$$G_p = 15\left(\frac{\pi D}{\lambda}\right)^2 N\left(\frac{S}{\lambda}\right) \tag{10.13}$$

$$\text{dB gain} = 10 \log G_p$$

The pitch of the coils is typically 12° to 15°, L is generally between 0.75 and 1.4 wavelengths, and the number of turns must be greater than 3. The upper limit of D is

$$D_{\max} = \frac{\sqrt{\dfrac{2S}{\lambda} + 1}}{\pi} \quad \text{(in same units as } S \text{ and } \lambda\text{)} \tag{10.14}$$

The reflector length must be greater than $\lambda/2D$.

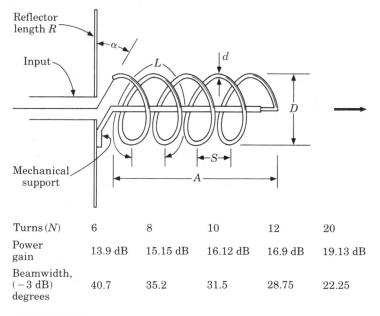

Turns (N)	6	8	10	12	20
Power gain	13.9 dB	15.15 dB	16.12 dB	16.9 dB	19.13 dB
Beamwidth, (-3 dB) degrees	40.7	35.2	31.5	28.75	22.25

FIGURE 10.28
The helical beam antenna.

Other definable characteristics of the helical antenna are given here as close approximations. The input impedance is purely resistive over the frequency range and is found to be between 100 and 200 Ω. The input resistance is

$$R_i = \frac{140\pi D}{\lambda} \tag{10.15}$$

The polar plot for the relative field strength is found by

$$e = \sin\left(\frac{\pi}{2N}\right) \frac{\sin\left(\frac{N\Phi}{2}\right)}{\sin\left(\frac{\Phi}{2}\right)} \tag{10.16}$$

where N is the number of turns and

$$\Phi = 2\pi\left[\frac{S}{\lambda}\cos\theta - \frac{LV_c}{\lambda V_h}\right] \qquad \theta = \tan^{-1}\frac{S}{D} \tag{10.17}$$

where V_c is the velocity of light in free air and V_h is the velocity of energy along the helix:

$$V_h = \frac{V_c}{\left[\frac{S}{\lambda} + \left(\frac{2N+1}{2N}\right)\right]\left(\frac{\lambda}{L}\right)} \tag{10.18}$$

From these equations, the half-power beamwidth is

$$\text{BW } (-3 \text{ dB}) = \frac{52}{\frac{\pi D}{\lambda}\sqrt{\frac{NS}{\lambda}}} \tag{10.19}$$

and the first null on either side of the bore sight center line is

$$\text{First null} = \frac{115}{\frac{\pi D}{\lambda}\sqrt{\frac{NS}{\lambda}}} \tag{10.20}$$

The *axial ratio* (AR) fashions the polarization and is found by

$$\text{AR} = \frac{2N+1}{2N} \tag{10.21}$$

When N is large, the ratio approaches unity and the polarization of the antenna approaches a true circular form. When N is small, the antenna can be either vertically or horizontally polarized, but it then loses one of its major advantages. Circular polarization means that the satellite antenna may align to either position for good reception or transmission so long as it is not axially aligned.

10.11   ## THE MARCONI ANTENNA

Among his many contributions to the science of physics, the 1909 Nobel prize winner, Guglielmo Marconi, opened up a whole new area of experimentation with the introduction of a vertically polarized *quarter-wave* dipole antenna, which bears his name. In the conceptualization of the quarter-wave antenna (see Figure 10.29), Marconi theorized that the earth would act as the second quarter-wave antenna to make up the half-wave dipole antenna. An electromagnetic wave leaving the antenna from point A, striking the ground, and bouncing back into the air in direction B would be identical to a wave leaving the "image" antenna from point C. This basic theory has become the industry standard for all AM transmissions below about 5 MHz.

EXAMPLE 10.2

An AM broadcast station operating at 720 kHz would have a wavelength of 1366.67 ft (416.67 m). Ninety-five percent of a half wavelength is 649.16 ft (197.9 m); such a structure would be far too large to operate as a horizontally polarized half-wave dipole antenna. However, the same station could transmit from a vertically polarized quarter-wave Marconi (tower) antenna of 324.58 ft (98.958 m), that is, 95% of a quarter wave. In these cases, the whole tower *is* the antenna. The structure should not be mistaken for an antenna mounted on top of a 324 ft tower.

Many field tests have been conducted since the introduction of the original Marconi antenna, with very interesting results. The standard tests require

FIGURE 10.29
The basic Marconi quarter-wave antenna.

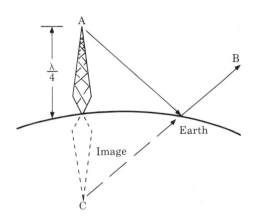

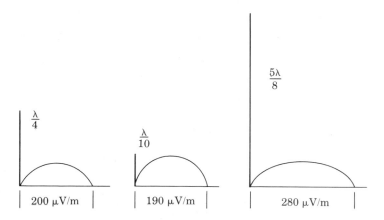

FIGURE 10.30
Relative field strength for various antenna heights (f = 720 kHz, λ = 1366.67 ft or 416.67 m).

that 1000 W be applied to the antenna and that the field strength be measured on a 1 m receiving antenna placed 1 mi from the test antenna. The measured signal strength is then stated in microvolts per meter, the standard unit of measurement for comparison purposes. The quarter-wave antenna used here as the reference will yield about 200 μV/m under the conditions specified in Figure 10.30a.

Surprisingly, reducing the test antenna length to about 0.1 wavelength (plus 0.1 wavelength for the image) will reduce the received signal strength to only about 5% below that of the quarter wave (plus the quarter-wave image) antenna. Maximum signal strength is recorded at the receiver when the length of the test antenna is 95% of five-eighths of a wavelength (plus the same-sized image), as shown in Figure 10.30c.

For maximum-distance transmission, the antenna for the 720 kHz carrier would need to be a vertical tower 1366.67 × 0.95 × 0.625 = 811.46 ft (247.4 m) tall, or 2.5 times as tall as Marconi's original predictions. The distance covered will be about 40% greater than that covered by the original quarter-wave antenna. A 40% increase in distance is not overly impressive compared to a 250% increase in tower height, but it must be remembered that the vertical antenna is omnidirectional, and that a 40% increase in the radius of a circle almost doubles the area coverage. For commercial transmissions, this may mean twice the number of customers. For the police and fire departments, this means twice the protection area.

The Marconi antenna radiation pattern can be shaped in much the same way as that for the driven array antennas. It would be a waste of energy for a coast guard station to transmit half of its power over dry land when 99% of its communications are to ships at sea. To accomplish this, several towers are erected and the signal applied to each is phase-delayed by a predetermined

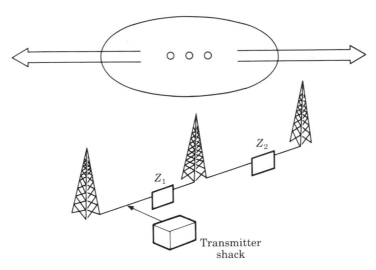

FIGURE 10.31
Phase array of Marconi antenna towers.

amount to create the radiation pattern desired (see Figure 10.31). The pattern can range from a full circle to a semicircle to a cigar shape.

The installation of a Marconi antenna requires only one provision, and that is that the ground on which the antenna stands (and transmits over) has good conductivity. Conductivity is directly related to the moisture content of the earth that the transmission is to cover. Desert is the poorest, and salt water is the best. In poor-conductivity areas, radial rods, as shown in Figure 10.32, are required to boost the conductivity at the antenna tower area.

FIGURE 10.32
Radial rods (only one quadrant shown).
Four rods minimum up to any number,
even one per degree.

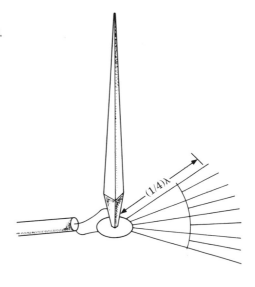

The radial rods are each a quarter wavelength long, and the number of rods used will depend on how poor the conductivity is. The rods are buried just below the surface of the ground and are often connected to heavy-gauge copper ground rods driven into the earth. The number of radial rods should not be less than four and has been known to range as high as 360 (one for each degree). Extremely arid soil may also require a sprinkler system to improve conductivity. The common connection of the coaxial feed line to the radial rod system is shown in Figure 10.32. For wireless microphones, the antenna is usually a coil on the inside of the case, and a short length of wire called a *counterpoise* acts as the radial rod.

The Marconi antenna is ideally suited for mobile communications and may be used at high frequencies for AM or FM systems that use vertical polarization. The whip, or quarter-wave antenna, is joined to the metal body of the vehicle, which serves as the counterpoise (or radial rods). For best performance, the antenna is mounted at the center of the metallic surface area so that the counterpoise area is equal in all directions. This will form a radiation pattern that is nearly equal in all directions, as shown in Figure 10.33. When the antenna is mounted to the bumper or trunk, at point B or C in the figure, the strongest signal will be in the direction of the greatest metallic mass, causing signals in all other directions to suffer in intensity. The preferred antenna length is one-quarter wavelength, but this size is awkward for some frequencies. The one-tenth wavelength antenna, with its 5% signal loss, is most common for mobile installations. Such a device is the 42 in whip antenna used for CB transmissions in place of the 104 in whip.

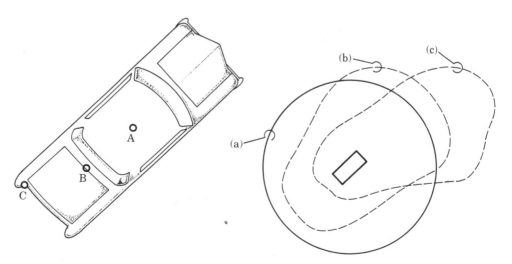

FIGURE 10.33
Mobile antenna installation and typical radiation patterns.

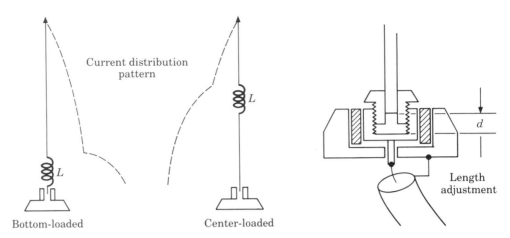

FIGURE 10.34
Coil loading of whip antennas to shorten the height.

Modifications to the Marconi antenna that improve performance without increasing the height include *coil loading,* shown in Figure 10.34. The current distribution curves are shown as dashed lines to indicate increased current at the input terminals of the antenna, whose height is unchanged since the effective length increase is in the coils. The center-loaded antenna has a slight edge over the bottom-loaded antenna. The antenna is increased in effective length, in the same units as the coil diameter, by approximately

$$\text{Length increase} = \sqrt{\pi N D} \qquad \textbf{(10.22)}$$

where N is the number of turns and D is the diameter of each turn. Also shown in Figure 10.34 is the mechanical adjustment to the length of the antenna, which will control the impedance matching to the coaxial line that affects the SWR. With an SWR meter connected in series with the antenna line and in the transmit mode, move your hand close to the antenna. If the SWR reading increases, the antenna is to long. If the SWR reading goes down, the antenna is too short. Loosen the hold-down nut on the whip and adjust the length accordingly.

10.12   IMPEDANCE MATCHING TO ANTENNAS

The easiest and simplest way to match the antenna impedance is to use a cable with an impedance identical to that of the antenna, that is, a 75 Ω cable for a 75 Ω antenna. However, it is beneficial to know and understand some alternate methods of matching impedances.

10.12.1 Transformer Matching

The quarter-wavelength transmission line impedance-matching transformer was covered in Chapter 8. A good application of this information is included here (refer to Figure 10.35a).

A physical change in the transmission line that extends for a quarter wavelength *exactly* can be used to match *any* two unequal impedances. Again, at very high frequencies, a transmission line transformer will perform better (fewer losses and better frequency response) than a wire-wound transformer. Review Chapter 8 for these particulars.

When an impedance transformation of $4:1$ is required, the alternative of Figure 10.35b is the most popular. Here we see a half-wavelength section (L) connected to provide the $4:1$ impedance ratio and also to convert a single-ended input feed line to a balanced line connection at the antenna terminals. This is a step-up transformer, so a 75 Ω single-ended feed line (Z_0) will match a 300 Ω balanced antenna (Z_{ant}) through the transformer (Z_t). This system works best when the antenna impedance is nearly a pure resistance and works only when the impedance ratio is 4 to 1. Changing the length of the transformer (L) to some other dimension (shorter or longer than one-half wavelength) will result in conditions that are worse than what the transformer was designed to overcome.

10.12.2 Matching Stubs

Chapter 9 covered the use of the Smith chart to evaluate the length of a shorted matching stub (L) and its position (d) along the feed line that will result in a perfect match to an antenna that is both resistive and reactive ($R \pm jX$).

It is good practice to calculate the length of the shorted stub (L) and its distance (d) from the antenna as a first step (see Figure 10.36). Use a cable section that is of the same type as the feed line, short one end (crimp and

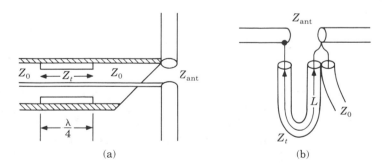

(a)   (b)

FIGURE 10.35
Transformer matching to antenna. (a) Impedance transformation 1:1. (b) Impedance transformation 4:1.

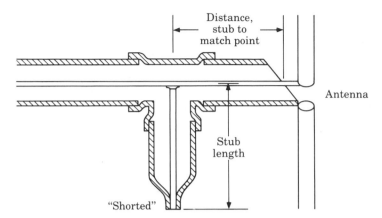

FIGURE 10.36
Stub matching to antenna.

solder), and then cut off a section of line that is slightly longer than the calculated stub length. Attach this stub to the open end of the feed line. Cut off a second section of the same cable of a length equal to the calculated distance (d) and attach this section to serve as the length (d) between the stub and the antenna. Connect the antenna to the unused end of this section. Better accuracy is achieved by using this procedure, which also requires fewer steps. In the post-installation testing stage, you can shorten the stub, in small steps, to the correct length by minor crimping at the shorted end while measuring the SWR. If flexible cable is used, the stub can be folded back and clamped along the feed cable to present a neater appearance as well as to prevent damage to the stub from winds or other environmental conditions.

10.12.3 The "T" Match and Delta Match

With the "T" and delta matching transformations, the driven element is *not* split at the center, as has been the case in all diagrams used so far. Although the antenna rod in Figure 10.37 looks like a short-circuit connection across the input feed line, it is in fact an AC impedance to the input signal at high frequencies. The driven element is attached to the support boom for parasitic arrays and is at ground potential. The "T" and delta matching systems are used for balanced input feed lines only, and it is assumed that points X and Y in Figure 10.37 are equidistant from the center of the dipole. A low impedance is established when points X and Y are close to the center of the dipole, and the impedance increases as the points are moved a greater distance from the center. The antenna impedance seen by the transmission line will be purely resistive when the antenna is at resonance. The behavior of an antenna at resonance is identical to that of a parallel-resonant tuned circuit.

The rod assembly connected between points X and Y has the general appearance of a folded dipole antenna that is center-fed at the input T_1-T_2. The

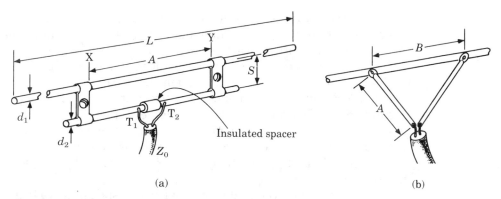

FIGURE 10.37
(a) "T" matching to the antenna. Insulated space is in the matching network rather than the main antenna rod. (b) "Delta" match.

calculations to follow will treat the antenna impedance as a folded dipole with length A, separation S, and rod diameters d_1 and d_2, as noted in the figure. The impedance of the matching section (Z_m) is determined by

$$Z_m = 276 \log \frac{2S}{\sqrt{d_1 d_2}} \qquad (10.23)$$

With the antenna impedance and the feed line impedance known, the length A can be found as the turns-to-impedance ratio of a transformer. Dipole length L is known, so a simple equation can be set up:

$$A = L \sqrt{\frac{Z_0}{Z_m}} \quad \text{(in the same dimensions as } L\text{)} \qquad (10.24)$$

EXAMPLE 10.3
The length L of a dipole operating at 150 MHz is 37.392 in. The antenna has a diameter d_1 of 0.5 in and is separated by 3 in from a matching element with a diameter d_2 of 0.25 in. The impedance to be matched, from Equation 10.23, is

$$Z_m = 276 \log \frac{2(3'')}{\sqrt{(0.25)\,(0.5)}} = 339.4 \ \Omega$$

When the antenna is connected to a 150 Ω transmission line, length A is

$$A = 37.392 \sqrt{\frac{150}{339.4}} = 24.86 \text{ in}$$

The antenna impedance of 340 Ω will be transformed down to match the 150 Ω line as a pure resistive load.

The **delta matching** network, or "Y" match, as it is sometimes called, is less popular than the "T" or gamma matching system because the metal rods that form the network also act as part of the radiating system. The slanting of the tie bars causes coupling between the feeder and the dipole element, which may upset the line balance and efficiency. Unbalanced currents will lead to uncontrolled radiation in an undesired direction and of the wrong polarization. Despite this unfavorable reputation, the delta match system continues to be used with reasonable success. More pruning of the end product is required at installation than with other matching systems, but approximate dimensions can be formulated by the following basic equations. Use Figure 10.37b as a visual reference.

It is known that the center impedance of a split half-wave dipole antenna is approximately 73 Ω and that the end-fed half-wave dipole has approximately 2500 Ω of impedance. Using this concept as the basis for impedance matching, it can be seen that dimension A in Figure 10.37b will be large when the impedance to be matched is large and that it will be short for low impedances. A theoretical impedance must first be determined that is based on the dipole length-to-diameter ratio:

$$Z_{\text{ant}} = 138 \log \left(\frac{L}{\pi \sqrt{d_1 d_2}} \right) \tag{10.25}$$

where L, d_1, and d_2 are in inches.

Next we need to establish a ratio of antenna impedance (Z_{ant}) to line impedance (Z_0) as a function of antenna length (L) to tap length (A):

$$A = L \left(\frac{Z_0 - Z_{\text{ant}}}{2500} \right) \quad \text{(in the same units as } L\text{)} \tag{10.26}$$

Now the tie rod length B can be found:

$$B = 2 + \sqrt{\left(\frac{A}{L} \right)^3} \quad \text{(in the same units as } A \text{ and } L\text{)} \tag{10.27}$$

The antenna is now ready to be assembled, tested, and adjusted for minimum SWR. It is recommended that adjustable clamps be provided at the rod and dipole junction for easy modification at the field site.

10.12.4 The Gamma Match

The most popular single-ended antenna impedance-matching method used for medium- and low-power antennas is the gamma system. It is straightforward, allows easy adjustment, and will present a purely resistive load impedance to the transmission line at the antenna design frequency. The gamma match resembles half of the "T" matching network but includes a reactive series element in the form of a shorted transmission line section.

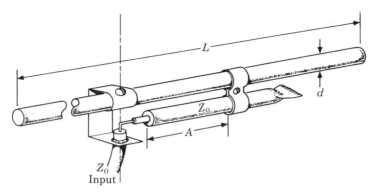

FIGURE 10.38
Gamma matching to antenna.

The dipole length is determined in the usual fashion. When made slightly shorter than one-half wavelength, it has a capacitively reactive term in the antenna impedance. The capacitance is canceled by a small length of transmission line shorted at the far end and open at the input end. A sliding clamp is included in the assembly to permit fine-tuning for minimum SWR at the time of installation. The dipole capacitance is first determined:

$$C_{\text{ant}} = \frac{17L}{\left(\ln \dfrac{24L}{d}\right) - 1} \quad \text{(in picofarads)} \qquad \textbf{(10.28)}$$

where L is in feet and d is in inches. After converting the capacitance into capacitive reactance at the center frequency, the length of the shorted stub (l) can be found:

$$l = \lambda \left[\frac{\tan^{-1} \dfrac{Z_0}{X_c}}{360} \right] \quad \text{(in the same units as λ)} \qquad \textbf{(10.29)}$$

Assemble the matching section as shown in Figure 10.38, grounding the shield (outer conductor) of the feed line to the center of the dipole element (and boom, if used). The matching stub is open at the input end but should be protected from the weather with heat-shrinkable tubing or the like. The far end of the stub is crimped to short the center conductor to the outer conductor. Seal this end with wax, if necessary.

10.13   PARABOLIC AND CORNER REFLECTORS

The corner reflector and parabolic reflector are introduced here only to establish a general idea of how these devices are used. Several additional concepts

concerning the parabolic reflector need to be introduced before a full understanding is possible. These are presented in Chapter 11 under the discussion of
microwaves, where the parabolic reflector achieves its greatest potential. Both
the corner reflector and the parabola are parasitic reflecting elements. They
are *not* antennas. The corner reflector acts as a reflecting mirror, and the parabolic reflector acts as a focusing lens.

The corner reflector in Figure 10.39a is a 90° formed flat metallic sheet (or
screen) placed one-quarter wavelength behind the driven element. In addition
to acting as an efficient reflector, with 0.5 dB more directivity gain than the rod
reflector, it reduces the front-to-back ratio by an additional 5 dB and reduces
the side lobes. The corner reflector finds its greatest application where interference from heavy signal traffic is a problem.

The gain of the parabolic reflector in Figure 10.39b is a function of the
ratio of dish diameter to the wavelength of the signal being received or transmitted and can be extremely high. The larger the diameter, the higher the
gain. Thus, the parabolic reflector finds its greatest use at higher frequencies.
It finds some use at lower frequencies because of its narrow beamwidth angle
and complete absence of side lobes. The parabolic reflector of Figure 10.39b is
shown in a modified form in Figure 10.39c. Here we see the half-wave dipole
antenna at the focal point of the parabola, but since the dipole has little radiation away from the center-line directivity path, the surface of the parabola is
not useful and has been omitted in the fabrication of the dish. The center-line
gain and beamwidth do not change. Also note that the parabola shell is made of
a grillwork to reduce the wind effects on its large surface area. The separations
in the grillwork must be less than one-tenth of a wavelength at the operating
frequency. See Chapter 11 for a detailed evaluation of the parabolic reflector.

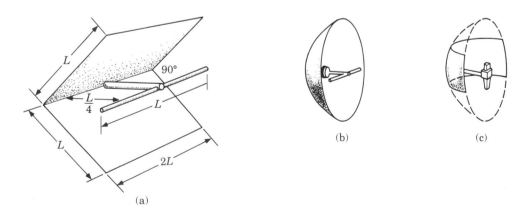

(a)

(b)

(c)

FIGURE 10.39
Antenna reflectors. (a) Corner reflector. (b) Parabolic reflector (lens), horizontally polarized. (c) Cutaway parabola, vertically polarized.

10.14 CIRCULAR POLARIZATION

The **polarization** of an antenna identifies the polarity of the electric field wave radiated by the antenna measured in the far field (a distance greater than two wavelengths). When the electric field builds and collapses along the same straight-line path, it is called *vertical linear polarization* when the path is vertical, and *horizontal linear polarization* when the path line is parallel to the surface of the earth. This description matches the definition of polarization for all the antennas covered so far, except for the helical antenna.

There are, however, some special cases. One such case is **circular polarization.** When the electric field builds in a horizontal direction at the start of the cycle, smoothly progresses through an angular change, and ends the collapsing segment of the cycle as a vertically polarized wave, it has added rotation to the vector. In the next cycle, the electric field would begin to build in the vertical direction and would end its collapsing segment as a horizontally polarized wave. Note that each cycle of the electric field wave experiences only a 90° phase (or polarity) change.

This wave motion may advance in either of two directions, referred to as the *sense of rotation*. A wave traveling toward the observer with a counterclockwise rotation has *right*-hand polarization. A clockwise-rotating vector has *left*-hand polarization. This identification is established by pointing the thumb in the direction of the wave propagation with the fingers curled in the direction of the electric field rotation. The hand that matches the conditions for the particular wave determines its rotational characteristic. A helical antenna with a counterclockwise spiraled coil will radiate a right-hand circularly polarized electric field.

A dipole antenna that is either vertically or horizontally polarized will receive a signal equally well from a right- or left-hand circularly polarized transmitting antenna. However, when *both* the receiving and transmitting antennas are circularly polarized, they must have the same sense of rotation. Circular polarization finds favor where one station is fixed and the other is mobile and can assume any attitude, as in satellite tracking and airport communications. A left-hand rotation is characterized as a +90° rotation and a right-hand spiral as a −90° rotation.

A 3 dB reduction in signal strength is the penalty paid for circular polarization (as compared to a linearly polarized antenna of like construction), but where circular polarization is a requirement, this is a small, token sacrifice.

Three practical systems are used to excite an antenna for circular polarization. One, of course, is the helical antenna, which has been covered in detail. The second method, shown in Figure 10.40, calls for a 90° mechanical rotation as well as a 90° electrical delay in the driven signal to the two-dipole array system. Driven element A and its parasitic elements are horizontally polarized and receive an *in-phase* signal from the transmitter. Driven element B and its parasitic element are vertically polarized and receive a signal that is delayed

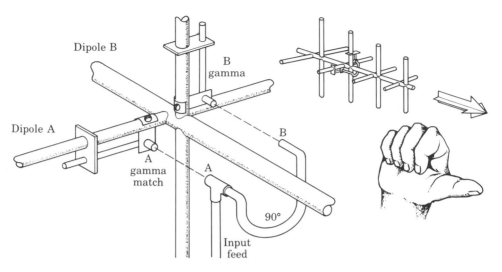

FIGURE 10.40
Matching connections for circular polarization of a parasitic array.

by 90° compared to the A signal. Both arrays in Figure 10.40, either singularly or collectively, will propagate in the same direction as any parasitic array. Since element A receives a signal 90° before element B, the sense of rotation is counterclockwise or right-hand rotation. Figure 10.40 shows the gamma match and the connecting harness, where the signal to B is delayed by 90° from the A signal. The propagated wave has unidirectional, right-hand circular polarization with the gain of a Yagi antenna, less 3 dB.

The antenna of Figure 10.41 is an omnidirectional, circularly polarized antenna with all four elements driven. Element A receives a signal at zero phase shift, the C signal is delayed −90°, dipole D is at −180°, and element B is at −270°, returning the signal back to A at 360° (or 0°). Sections A and C form a north-south dual-dipole assembly, in this case, each radiating in its own direction. Sections D and B form an east-west assembly of like nature. The back of each element also encounters the signal from its opposing element at −90°. The combination of the 90° forward electrical delay and the back reflected space delay requires that the mechanical shift be only 30° for each element (a total of 60° between opposing elements). This angle aids in the effects caused by the turnstile combination of the other two elements. The 360° coverage of this antenna is enhanced by the wide vertical beamwidth and circularly polarized nature of the propagated wave. This antenna finds use in ground-to-air communications at a large number of airports.

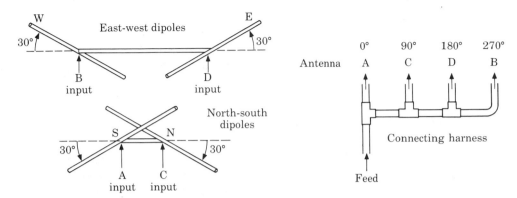

FIGURE 10.41
A four-bay circularly polarized dipole antenna.

10.15   MODEL ANTENNAS

Field testing of antennas requires that a physical product be made and taken to an open area for realistic gain and electric field strength measurements. This is practical for small to medium-sized antennas but not for large antennas. The situation is further complicated when multiple antennas are to be used in a confined area and in the presence of massive metallic structures, such as on board ship.

The problem is simplified by scaling down the antenna to one-eighth or one-tenth of full size and testing the scale model. The measurements will be true to life in all respects and will be applicable to the finished product. The only point that must be remembered is that if the size is reduced, the frequency must be increased by the same ratio. That is, one-eighth the size means the frequency must be eight times higher.

In building multibillion dollar ships, which may have as many as 50 antenna systems, it is common to include the cost of a scale model ship for field testing. The scale model is usually constructed of wood and then covered with a layer of metal to simulate the effects of the finished product. The antennas are then mounted and tested in the presence of the metal mass to ensure correct field patterns and freedom from interference from one another. The models are fabricated on an oversized turntable so that they may be rotated to avoid the need to move the test station for each compass change. This same practice is utilized for commercial antennas and saves the cost and time of disassembly and reassembly of the full-size product at the user location. In addition, the same model may require only slight modifications to be useful for other applications.

QUESTIONS

1. What is the size of the basic practical (reference) antenna? (a) zero wavelengths, (b) one-tenth wavelength, (c) one-fourth wavelength, (d) one-half wavelength, (e) one full wavelength.

2. Why is the basic practical (reference) antenna size used in the preceding question? (a) It is a natural law, (b) Maximum voltage appears across the minimum size, (c) The current is constant throughout its length, (d) It results in the highest gain, (e) It matches the transmission line impedance.

3. Can a metallic rod have gain? (a) yes, (b) no, (c) sometimes.

4. What is the gain of the basic practical (reference) antenna? (a) 3 dB, (b) 6 dB, (c) 1 dB, (d) 0 dB, (e) It differs with length.

5. What is the input impedance of a center-fed Hertz antenna? (a) 0 Ω, (b) 50 Ω, (c) 73 Ω, (d) 300 Ω, (e) It differs with length.

6. What would be the physical size of a Hertz half-wave dipole antenna (in inches) operating at 74.784 MHz? (include the size reduction factor) (a) 71 in, (b) 75 in, (c) 79 in, (d) 150 in, (e) 158 in.

7. What is the input impedance of an end-fed Hertz antenna? (a) 0 Ω, (b) 50 Ω, (c) 300 Ω, (d) 2500 Ω, (e) It differs with length.

8. How is the gain of an antenna increased? (a) by making the antenna size larger, (b) adding several antennas in parallel, (c) focusing the radiated energy in one desired direction, (d) applying more power, (e) making the antenna rods thicker.

9. Reflectors are placed (a) above the driven element, (b) below the driven element, (c) in front of the driven element, (d) to the rear of the driven element, (e) alongside of the driven element.

10. Directors are placed (a) above the driven element, (b) below the driven element, (c) in front of the driven element, (d) to the rear of the driven element, (e) alongside of the driven element.

11. What is the "reciprocity theorem"?

12. A center-fed folded half-wave dipole antenna of constant rod diameter has what input impedance? (a) 50 Ω, (b) 73 Ω, (c) 150 Ω, (d) 300 Ω, (e) 450 Ω.

13. The center space of the antenna (at the terminal connections) has no effect above what frequency?

14. The polarization of a half-wave dipole antenna is in the plane of the (a) far field, (b) electric field, (c) near field, (d) magnetic field, (e) none of the above.

15. The beamwidth of the antenna polar pattern is a measure of the antenna's (a) gain, (b) sensitivity, (c) directivity, (d) frequency response, (e) bandwidth.

16. The beamwidth is measured to either side of center at an angle where the relative field strength has dropped to 0.7071 of the maximum strength. (T)(F)

17. As the gain increases, the beamwidth (a) increases, (b) stays the same, (c) decreases.

18. An end-fed half-wave dipole antenna will have an impedance of approximately (a) 73 Ω, (b) 150 Ω, (c) 300 Ω, (d) 1000 Ω, (e) 2500 Ω.

19. Because of the "capacitive" end effect of an antenna, the dipole is made _____ the calculated half wavelength. (a) exactly the same as, (b) 5% longer than, (c) 10% longer than, (d) 5% shorter than, (e) 10% shorter than.

20. When the rod diameter of the half-wave dipole increases, the Q of the antenna (a) increases, (b) decreases, (c) stays the same.

21. When the Q of an antenna increases, the bandwidth (a) increases, (b) decreases, (c) stays the same.

22. Describe the diameter of the conical antenna.

23. The mechanical length of a conical antenna is 5% shorter than (a) 0.25 wavelengths, (b) 0.5 wavelengths, (c) 0.73 wavelengths, (d) one full wavelength, (e) can be any length.

24. The electrical length of a conical antenna is (a) 0.25 wavelengths, (b) 0.5 wavelengths, (c) 0.73 wavelengths, (d) one full wavelength, (e) can be any length.

25. A bow tie antenna is a flat-surface conical antenna. (T)(F)

26. The batwing antenna has the properties of a multirod folded dipole antenna. (T)(F)

27. The turnstile antenna may only be polarized (a) vertically, (b) horizontally, (c) circularly.

28. The gain of a turnstile antenna compared to the reference half-wave dipole antenna is (a) the same, (b) +3 dB, (c) −3 dB, (d) −1.5 dB, (e) +4.5 dB.

29. The input impedance of a turnstile antenna is (a) 73 Ω, (b) 73/2 Ω, (c) 2 × 73 Ω, (d) 300 Ω, (e) 2500 Ω.

30. The parasitic array antenna has how many driven elements? (a) one, (b) two, (c) three, (d) four, (e) any number.

31. The directors of a parasitic array antenna are placed in parallel to the driven element in a plane (a) above the driven element, (b) below the driven element, (c) in front of the driven element, (d) to the rear of the driven element, (e) all of the above.

32. The reflectors of a parasitic array antenna are placed in parallel to the driven element in a plane (a) above the driven element, (b) below the driven element, (c) in front of the driven element, (d) to the rear of the driven element, (e) all of the above.

33. Directors are _____ the driven element. (a) longer than, (b) shorter than, (c) the same length as.

34. Reflectors are _____ the driven element. (a) longer than, (b) shorter than, (c) the same length as.

35. The typical spacing between directors is (a) 10% of a half wavelength, (b) 10% of a full wavelength, (c) 15% of a half wavelength, (d) 15% of a full wavelength.

36. The typical spacing between reflectors is (a) 10% of a half wavelength, (b) 10% of a full wavelength, (c) 15% of a half wavelength, (d) 15% of a full wavelength.

37. Parasitic array antennas may be polarized either vertically or horizontally. (T)(F)

38. Antenna gain in one direction (directivity gain) is achieved through the sacrifice of signal strength from another direction. (T)(F)

39. Comparing the signal strength arriving at the driven element from the desired direction to the signal strength reaching the driver from the opposite direction is called the (a) gain, (b) sensitivity, (c) beamwidth, (d) front-to-back ratio, (e) frequency response.

40. Increasing the number of parasitic elements will increase the gain (a) linearly, (b) endlessly, (c) logarithmically, (d) by the square law, (e) by the inverse square law.

41. The list of antenna gains in Table 10.1 fits (a) all antennas, (b) all parasitic array antennas, (c) antennas of specific spacing and length factors, (d) all horizontally polarized antennas, (e) all vertically polarized antennas.

42. A driven array antenna is one that has (a) a driver at both ends, (b) all elements driven, (c) all half wavelength elements, (d) no elements driven, (e) no transmitting characteristics.

43. What separation between a four-element end-fire driven array is required to cancel the back lobe and reinforce the forward lobe?

44. The forward lobe of an end-fire driven array is toward the end (a) where the signal is applied, (b) that last received signal, (c) either, depending on spacing, (d) either, depending on element length, (e) either, depending on gain.

45. The broadside array antenna radiates signal from (a) the tips of the dipole rods, (b) the ends of the antenna, (c) the flat surface of the array.

46. The broadside array antenna has elements that are (a) one-half wavelength end to end, (b) one-half wavelength each side of center, (c) one full wavelength each side of center.

47. The broadside array antenna has a (a) teardrop-shaped radiation pattern, (b) high vertical and narrow horizontal pattern, (c) short vertical and wide horizontal pattern, (d) wide horizontal and wide vertical pattern.

48. A log-periodic array antenna has (a) increasing element length, (b) increasing element spacing, (c) all elements driven, (d) alternate phasing of elements, (e) all of the above.

49. A log-periodic array antenna has its input signal applied to (a) the large-element end, (b) the small-element end, (c) either end.

50. A log-periodic array antenna will radiate (a) toward the small-element end, (b) toward the large-element end, (c) toward whichever end the signal is input, (d) away from the signal input end.

51. The log-periodic array antenna has a constant spacing factor and a constant length reduction factor. (T)(F)

52. The log-periodic array antenna has a frequency ratio of (a) 1:1, (b) 2:1, (c) 3:1, (d) 4:1, (e) 5:1.

53. An input-impedance-matching transformer (is) (is not) generally required with a log-periodic array antenna.

54. The gain of four identical horizontally polarized antennas stacked one above the other and fed in phase will be _____ .

55. The beamwidth of the stacked antennas in the preceding question will be (a) the same as one antenna, (b) 2 × the beamwidth of one antenna, (c) 3 × the beamwidth of one antenna, (d) 4 × the beamwidth of one antenna, (e) 8 × the beamwidth of one antenna.

56. What effects will stacking two identical horizontally polarized antennas side by side have on gain and beamwidth?

57. What is the dB gain of 29 stacked bays of identical antenna sections?

58. List one feature of the quad antenna radiation pattern that is different from most other antennas.

59. The impedances that separate the elements of a multiband antenna should be (a) matched impedances, (b) high impedances, (c) low impedances, (d) a short circuit.

60. What is the minimum number of turns that will work as a helix antenna?

61. To achieve circular polarization with a helix antenna, should the coil turns be few or many?

62. The input impedance of the helix antenna (in ohms) is from (a) 100 to 200, (b) 200 to 300, (c) 300 to 400, (d) 400 to 500, (e) over 500.

63. What is the dB gain of a helix antenna that has 12 turns of 6 in diameter coils, spaced 5 in apart, when operating at 1.476 GHz? (a) 9 dB, (b) 12 dB, (c) 18 dB, (d) 28 dB, (e) 33 dB.

64. Find the −3 dB beamwidth for the helix antenna in the preceding question.

65. How many degrees away from the bore sight center line will the first null appear for the helix antenna in Question 63?

66. The real part of the basic Marconi antenna is 95% of _____ wavelengths long. (a) one-eighth, (b) one-fourth, (c) three-eighths, (d) one-half, (e) five-eighths.

67. The real part of the basic Marconi antenna will add to the imaginary part to look like a _____ wavelength antenna. (a) one-eighth, (b) one-fourth, (c) three-eighths, (d) one-half, (e) five-eighths.

68. All Marconi antennas are vertically polarized. (T)(F)

69. How much less field strength will a one-tenth wavelength Marconi antenna have than the basic Marconi antenna? (a) 6%, (b) 14%, (c) 22.5%, (d) 41%, (e) 57%.

70. How much more field strength will a five-eighths wavelength Marconi antenna have than the basic Marconi antenna? (a) 6%, (b) 14%, (c) 22.5%, (d) 41%, (e) 57%.

71. Because of their polarization, Marconi antennas should not be phase-arrayed. (T)(F)

72. For the basic Marconi antenna, the whole tower *is* the antenna. (T)(F)

73. The parabola is a focusing lens reflector. (T)(F)

74. The corner reflector is an antenna of the parasitic array family. (T)(F)

75. To work with a right-hand circularly polarized transmitting antenna, the receiving antenna must be (a) vertically polarized, (b) horizontally polarized, (c) right-hand circularly polarized, (d) left-hand circularly polarized, (e) may have any polarization.

76. At what frequency would a one-sixteenth scale model half-wave dipole antenna that measures 3.5 in be tested? (a) 100.156 MHz, (b) 478.667 MHz, (c) 982.133 MHz, (d) 1.6025 GHz, (e) 3.1644 GHz.

77. What is the operating frequency of the full-scale model antenna from the preceding question? (a) 100.156 MHz, (b) 478.667 MHz, (c) 982.133 MHz, (d) 1.6025 GHz, (e) 3.1644 GHz.

CHAPTER ELEVEN

MICROWAVE SYSTEMS AND DEVICES

11.1 INTRODUCTION

A **microwave** is a signal that has a wavelength of one foot (30.5 cm) or less. This converts to a frequency of 984 MHz, so all frequencies above 1000 MHz (1 GHz) are considered microwaves. The frequencies immediately below this border are considered ultra-high frequencies.

The upper end of the microwave range contains the light frequencies, about 10^{15} Hz. However, because electronic transmission is so closely geared to half-wavelength devices, the practical upper limit is about 300 GHz, where one wavelength is about 0.04 in (0.1 cm). Smaller devices are being made, but their power-handling abilities are also micro.

11.2 THE MICROWAVE SYSTEM

A fairly detailed radar system is shown in Figure 11.1. The system is intended to transmit a very high power, very short duration pulse. This pulse will reflect off of the surface of some distant object and bounce back to the antenna from which it was launched. By measuring the time span from launch to return, the distance between the two objects can be determined from the known velocity of wave travel in our atmosphere. A time lapse of 22 μs from launch to return at a velocity of 984,000,000 ft/s, for example, indicates a total distance traveled of 21,648 ft. A one-way trip to the target is 10,824 ft, or 2.05 mi.

A rotating antenna synchronized to a rotating display monitor adds the dimension of *direction* to the data already collected on distance. The word **radar** is a conjunction of the phrase "RAdio Direction And Ranging."

Referring to the numbered blocks in Figure 11.1, the pulse generator (1) sets the *pulsewidth* (pulse "on" time) and the *pulse repetition rate* (time between pulses) of the signal to be transmitted. The modulator (2) turns on the power

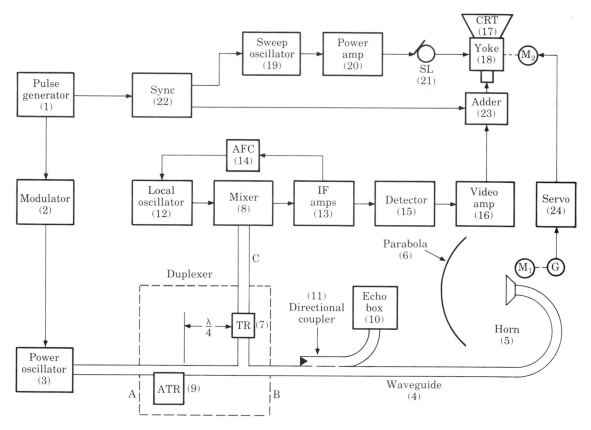

FIGURE 11.1
Simplified block diagram of a typical radar transmitter/receiver system.

oscillator (3) for the period of time and at the rate determined by the pulse generator. The power oscillator oscillates at the RF carrier frequency to send out short bursts of high-frequency energy.

The waveguide (4) carries the high-frequency, high-power energy to the horn antenna (5) and the parabolic reflector (6) to concentrate the energy into a narrow pencil beam to the target. The TR (transmit-receive) tube (7) is a special gas tube that short-circuits the waveguide going into the mixer (8) so that the high-energy pulses from the power oscillator do not burn out the sensitive diode mixer during transmit time. The ATR (anti-transmit-receive) box (9) is a gas-filled tube exactly like the TR tube. The difference is that the ATR tube is in series with the waveguide that connects the power oscillator to the antenna. In the presence of a high-energy pulse, the ATR tube short-circuits and forms a low-impedance path for the power oscillator signal going to the antenna. In the receive mode, the ATR tube does not short and is a high impedance for the receive signal. It reflects all of the signal back into the mixer. The distance between the TR and ATR tubes is critical and must be exactly one-quarter wavelength at the operating frequency.

The directional coupler (10) and echo box (11) are tuned far off of the carrier frequency during normal operation. The echo box is used for testing the system when no targets are within range. A pulse is transmitted while the echo box is motor-tuned through resonance to indicate a return signal at 1 mi distance.

The receiver functions as a standard superheterodyne radio receiver minus the RF amplifier. The mixer (8) is fed a signal from the antenna (5 and 6) as well as a local oscillator signal. The local oscillator (12) operates at a frequency higher than the RF signal and requires a special high-frequency tube circuit. The output from the mixer is selected as the difference frequency at about 30 MHz. The IF amplifier section (13) usually consists of five or six amplifiers to ensure high gain and approximately 10 MHz bandwidth.

The AFC (14) is an automatic frequency control circuit that is equivalent to a phase-locked loop. The mixer acts as a comparator whose output is amplified by the IF circuit. The signal is then converted to DC in the AFC circuit relative to frequency change and is used to control the frequency of a VCO. Since this is an amplitude-modulated system, the detector (15) need be only a simple diode rectifier and filter system.

The video amplifiers (16) must have wideband frequency response to accommodate the multiple odd harmonics of the fundamental square wave modulating signal. A 1 μs pulse will contain at least the ninth harmonic of the fundamental frequency, requiring a 9 MHz bandwidth amplifier. The output of the video amplifiers is used to intensity-modulate the CRT monitor. The CRT (cathode ray tube) (17) is a display device similar to an oscilloscope with two distinct differences.

1. Deflection only in one direction is needed.
2. The trace is DC-offset so that it starts at the center of the CRT and scans to one edge.

The deflection oscillator (19) is generally a multivibrator circuit that generates a trapezoidal wave signal that is amplified by a power amplifier (20) for application to the deflection coil (18), which is wrapped around the outside of the neck of the CRT. The oscillator must have the ability to work uniformly well over the frequency range from about 2 kHz (40 mi range) to about 90 kHz (<1 mi range).

The synchronizing circuit (22) notes each pulse from the pulse generator (1) and starts the CRT sweep at the center of the display monitor as each pulse leaves the transmitter. Range markers may be added to the display by the adder (23) for a full time-distance indication.

A motor servo system (24) controls the speed and position of the rotating antenna assembly (5 and 6) and the CRT deflection coil (18) to maintain a correspondence between the antenna direction and the monitor display position.

11.3   WAVEGUIDE

The part of the microwave system that establishes the theory of operation for all of the other devices is the interconnecting hardware called **waveguide.** The conductors of microwave energy constitute a departure from conventional cables in that they resemble a coaxial cable with the center conductor removed. Microwave energy is carried through the waveguide by reflection along its inside walls. This is possible only if the guide is larger than one-half the length of the applied voltage wave. Thus, the waveguide size is directly related to frequency.

Once again, transmission line theory will be used to help solidify the basic principles of waveguides. A quarter-wavelength section of transmission line shorted at the load end will resemble a high-impedance resonant circuit at the open end, one-quarter wavelength away from the short circuit. When two such sections are joined at the open ends, they form a rectangular shape, as shown in Figure 11.2a, that has high impedance at its center. When many such sections are joined side by side, as seen in Figure 11.2b, they form the equivalent of a waveguide section.

The inside dimensions of the guide are used in many of the waveguide calculations and are designated by a for the wide side and b for the narrow side.

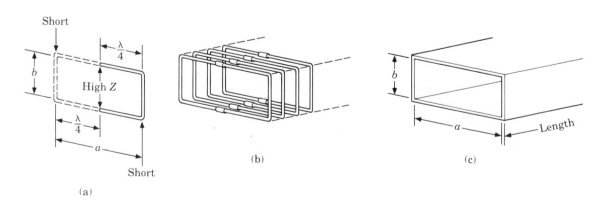

FIGURE 11.2
The evolution of waveguide. (a) Two quarter-wave shorted sections of transmission line. (b) Multiples of a. (c) Standard waveguide with designators (no flange).

The b dimension is always smaller than a by a factor of 0.4 to 0.5, for example, $a = 10$ cm and $b = 5$ cm. The a dimension is equal to one-half wavelength at the lowest frequency that will propagate through the guide, called the **cutoff frequency** (f_{co}). Since a and b are stated in wavelengths and the size is a measure of frequency, the term *cutoff wavelengths* is sometimes used in place of *cutoff frequency*. In essence, a waveguide acts as a high-pass filter and severely attenuates all frequencies that fall below the value where $a = \lambda/2$.

At a frequency of $2f_{co}$, a is a full wavelength long and the waveguide changes operating modes. Therefore, a given waveguide would be operated between the frequency limits of f_{co} and $2f_{co}$, with minimum attenuation occurring at approximately $1.33f_{co}$. The cutoff frequency is found by:

$$f_{co} = \frac{V_c}{2a} \quad \text{(in hertz)} \tag{11.1}$$

where V_c is the velocity of light in our atmosphere.

The standard waveguide frequency bands are listed in Table 11.1. The need for microwave systems was so urgent in the early years of World War II that the care generally given to band assignments was abandoned in lieu of more pressing matters. The letter assignments used to denote the bands became industry standards. As the space program took on more significance, the Department of Defense (DOD) redefined the frequency bands; the new designations are listed in Table 11.2. Newer systems use the DOD listings, while the established programs may still be using the WWII designations.

Waveguide sizes are standard according to the JAN (joint Army-Navy) specifications of WWII. For this reason, examples and problems in this text will use WWII designations.

TABLE 11.1
Microwave frequency bands (World War II)

Band	JAN Number	Frequency (GHz)	Size (inches) (a × b)	Cutoff Frequency (GHz)
L	WR650	1.12–2.70	6.50 × 3.250	0.9083
S	WR284	2.60–3.95	3.00 × 1.500	1.9680
G	WR187	3.95–5.85	1.87 × 0.870	3.1572
C	WR159	4.90–7.05	1.59 × 0.795	3.7132
J	WR137	5.85–8.20	1.37 × 0.620	4.3095
H	WR112	7.05–10.0	1.12 × 0.497	5.2714
X	WR 90	8.20–12.4	0.90 × 0.400	6.5600
M	WR 75	10.0–15.0	0.75 × 0.375	7.8720
P	WR 62	12.4–18.0	0.62 × 0.310	9.5226
N	WR 51	15.0–22.0	0.51 × 0.255	11.5764
K	WR 42	18.0–26.5	0.42 × 0.170	14.0570
R	WR 28	26.5–40.0	0.28 × 0.140	21.0857

TABLE 11.2
Microwave bands redefined by the Department of Defense May 24, 1970

Band	Frequency (GHz)
A	0.10–0.250
B	0.25–0.500
C	0.50–1.000
D	1.00–2.000
E	2.00–3.000
F	3.00–4.000
G	4.00–6.000
H	6.00–8.000
I	8.00–10.00
J	10.0–20.00
K	20.0–40.00
L	40.0–60.00
M	60.0–100.0

The waveguide factors that deserve attention are

1. attenuation (losses per unit length)
2. size selection,
3. coupling methods,
4. guide impedance, and
5. power-handling ability.

Other important factors will become apparent in the process of studying these five major conditions.

11.3.1 Attenuation

At frequencies below cutoff, one wavelength of signal is longer than twice the a dimension of the guide, and the energy wave will see a short circuit at the side walls. For still lower frequencies (and longer wavelengths), the effects of the short become more severe and the losses increase. The equation for power loss (in decibels per unit length) is

$$\text{Power loss} = \frac{54.6}{\lambda_{co}} \quad \text{(in dB/ft)} \tag{11.2}$$

for λ_{co} in inches, or

$$\text{Power loss} = \frac{455}{\lambda_{co}} \quad \text{(in dB/m)}$$

for λ_{co} in centimeters. This indicates that attenuation must be less at frequencies above cutoff. A generalized format for finding the attenuation α in rectangular waveguide is given in Equation 11.3.

$$\alpha = \frac{0.01107}{\sqrt{a^3}} \, R \left[\frac{\frac{a}{2b}\sqrt{\left(\frac{2a}{\lambda_0}\right)^3} + \frac{1}{\sqrt{\frac{2a}{\lambda_0}}}}{\sqrt{\left(\frac{2a}{\lambda_0}\right)^2 - 1}} \right] \quad \text{(in dB/ft)} \quad \textbf{(11.3)}$$

$$= \frac{0.3883}{\sqrt{a^3}} \, R \left[\frac{\frac{a}{2b}\sqrt{\left(\frac{2a}{\lambda_0}\right)^3} + \frac{1}{\sqrt{\frac{2a}{\lambda_0}}}}{\sqrt{\left(\frac{2a}{\lambda_0}\right)^2 - 1}} \right] \quad \text{(in dB/m)} \quad \textbf{(11.3a)}$$

The constants 0.01107 and 0.3883 include the velocity of light, 2π, and the permeability of the space inside the guide ($4\pi \times 10^{-12}$ F/m). The term R is the resistivity of the material on the inside wall compared to copper waveguide and air dielectric. Copper with air is equal to 1. Table 11.3 lists other common materials and compares them to copper with air. Aluminum and copper are the most common materials used for waveguide because they are easy to extrude. The most common plating materials for the inside walls (and their resistivity) are listed in Table 11.3.

The ratio $a/2b$ accounts for the *modes* (to be covered later) that can exist inside the guide, and the ratio $2a/\lambda_0$ accounts for the difference between the cutoff wavelength and the operating wavelength. The attenuation may vary by as much as 1 dB/100 ft from the results of Equation 11.3 due to inside surface roughness, even with well-machined waveguide. This is particularly true for the smaller guide sizes. In these cases, silver plating is applied to the inside surfaces of the guide. Silver has a higher resistivity than copper but yields a

TABLE 11.3
Resistivity of common plating metals

Silver	1.041
Copper	1.000
Gold	0.841
Aluminum	0.781
Brass	0.495
Platinum	0.415
Steel	0.186

smoother surface and reduces the losses. Attenuation for air-filled copper waveguide in the J band at 6.0 GHz is calculated at -2 dB per 100 ft. The transmitter would need to supply 1585 W to radiate 1000 W into an antenna 100 ft away.

11.3.2 Waveguide Selection

In a given system, say an operating frequency of 4.95 GHz is assigned. The required waveguide could be in either the G band, the C band, or the J band, from Table 11.1. In the G band, the cutoff frequency is 3.1572 GHz and twice the cutoff frequency is 6.3144 GHz. Since 4.95 GHz is between these two limits, the wave will propagate in the guide. The cutoff frequency for the C band is 3.7132 GHz and twice the cutoff frequency is 7.4264 GHz. Again, 4.95 GHz is within these limits. The J band cuts off at 4.3095 GHz, with 8.619 GHz as $2f_{co}$, and will also propagate the 4.95 GHz signal. All three bands would have to be examined for attenuation to see which has the lowest losses. Using Equation 11.3 (with copper walls) results in 0.01048 dB/ft for band G, 0.01506 dB/ft for the C band, and 0.02523 dB/ft for the J band. From these results, the G band has the lowest attenuation and would be selected. Other factors are also taken into account at this time for final choice of waveguide size. If the length of the guide is only 2 or 3 ft, as may be the case in some satellite installations, then other factors such as size, weight, and cost may have a greater effect on the final selection of guide size. When the length of the guide is 20 ft or more, then losses alone could be the deciding factor. When great lengths are coupled with the requirement for low losses, then plating is a possible alternative, with higher costs as the tradeoff.

11.3.3 Coupling Methods

Coupling energy to the waveguide is accomplished by one of three means. A *very* short section of coaxial line joins the transmitter to the input of the wave-guide. Here the input termination may be a **probe,** a **loop,** or a **slot.** The probe and loop are discussed in depth here, but a deeper understanding of waveguide fields is needed to explore slot coupling, so this will be covered in a later section in this chapter.

The probe input connector of Figure 11.3a is seen from the b side of the guide. The placement is typically one-quarter wavelength from the shorted back wall at the operating frequency. This wall is sometimes made adjustable as a means of impedance matching because the physical size of one-quarter wavelength will change when the operating frequency changes. The probe depth into the guide is also a means of varying the impedance, and the probe diameter can be varied in size and shape to accommodate different levels of

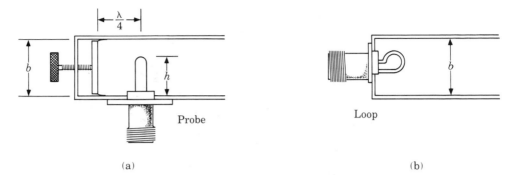

FIGURE 11.3
Waveguide input terminations. (a) Probe. (b) Loop.

power. The probe is associated with "electrostatic" or "capacitive" coupling, while the loop is designated "inductive" or "transformer" coupling.

Loop coupling simply involves a round wire loop placed on the inside of the waveguide along the wall surface, either at the closed back of the guide or along one side wall of the guide, but *never* along the center line of the *a* wall, which is where the probe is placed.

The style of input coupling determines the modes of propagation of the wave inside the guide, which sets the polarization of the transmitted signal at the antenna.

11.3.4 Modes

The theory of modes is based on the *electric* field set up about a conductor and the *magnetic* field that surrounds the conductor as a result of an electric current flowing through the wire. These are the same principles presented to explain the fields that surround the antennas in Chapter 10. There it was said that the energy will propagate in a direction that is perpendicular to *both* the electric and magnetic fields. The same is true for waveguide.

Figure 11.4a shows the electric field (E) set up by a probe inserted into the waveguide along the center line. The electric field lines of force are parallel to the probe and are strongest at the probe, becoming weaker with distance from the probe, as explained for antennas in Chapter 10. The weakest electric field is found one-quarter wavelength from the probe (at the side walls and the back wall) in Figure 11.4a. Here it is shown that the *a* wall of the guide must be at least a half wavelength long to prevent the electric field from shorting out at the *b* wall. The *b* wall here is shown as one-half of *a,* or a quarter wavelength long.

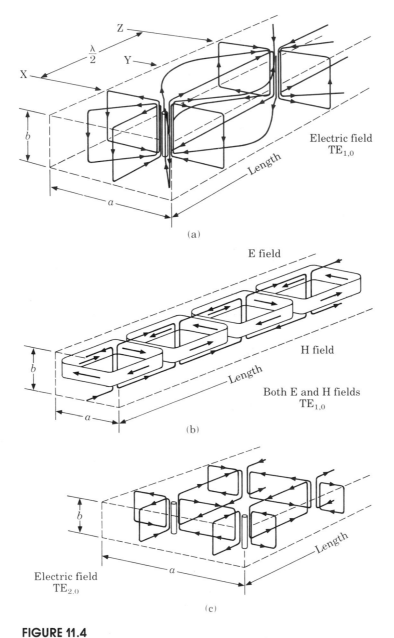

FIGURE 11.4
Electric and magnetic fields inside of the guide. (a) The electric field due to probe input. (b) The chain link.

A heavy concentration of the electric field current will again flow across the guide center at a distance of one-half wavelength (and each multiple of one-half wavelength) from the probe, at location Z. The heaviest electric fields are located at X and Z in Figure 11.4a, while the weakest electric field is found at location Y on the inside of the guide.

The magnetic field (H) surrounds the probe and is shown as wide bands interlinking the electric fields in Figure 11.4b. The magnetic field is strongest where the electric field is weakest, and the magnetic field is weakest where the electric field is the strongest. The overall combination of electric and magnetic fields resembles the links of a chain. Although Figure 11.4b is an oversimplification of the coexistence of the two fields, it can be used as a model for understanding other, more complicated field patterns. The guides in figure 11.4a and b have one half wavelength along the a wall and less than one half wavelength along the b wall, so the guide designation, stated in half-wavelength measurements, is 1,0. These are the number of **modes** in each plane. Figure 11.4c has two half wavelengths along the a length and less than one half wavelength along the b plane, so it carries the mode designation of 2,0. The length of a is a full wavelength, while the length of b is still less than one-quarter wavelength.

The diagrams of Figure 11.4 are all probe input diagrams, and the electric field is predominant with the probe. Because of the way the electric field is propagated through the waveguide, the waveguide is termed **transverse electric** or TE waveguide. The full designation of the guide in Figure 11.4a and b is $TE_{1,0}$. The guide in Figure 11.4c is termed $TE_{2,0}$. When a loop is used as the input connection, the magnetic field is generally predominant, and the guide is termed **transverse magnetic,** or TM waveguide. Examples are shown in Figure 11.5.

Remember these two statements: Electric lines are parallel to the probe, and magnetic lines move around the probe. The magnetic lines of a loop coupling go through the center of the loop, and the electric lines are in the same plane as the loop wire. Once these two conditions are understood, the variations of input coupling shown in Figure 11.5 can be recognized on sight.

Look for the probe or loop in each diagram. For *probe* input connections, visualize the electric field. The magnetic field will be rotated 90° from the electric field. Decide which field is in the b plane of the guide to identify an electric or magnetic field of propagation. For *loop* input connections, visualize the magnetic lines going through the loop and the electric field lines in the same plane as the loop wire. Decide which field is in the b plane of the guide to identify an electric or magnetic field of propagation. Study Figure 11.5, and review the definitions for TE and TM guide until the basic designations become easily identifiable.

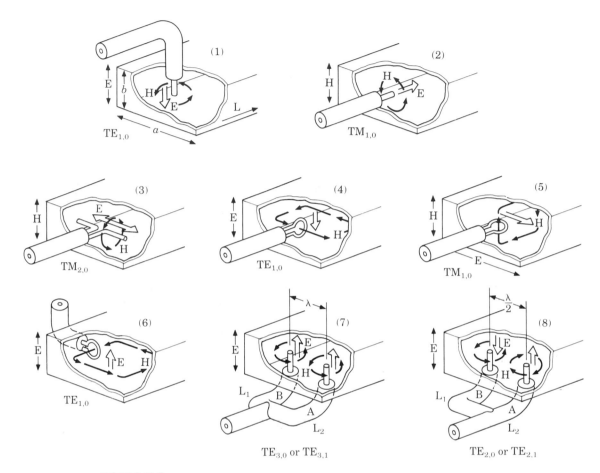

FIGURE 11.5
Variations of the probe and loop input termination used to establish a TE or TM mode of propagation.

EXAMPLE 11.1

Refer to diagram (2) in Figure 11.5. The probe enters the guide from the back wall. The electric field is parallel to the probe. The magnetic field moves around the probe. The magnetic field is aligned with the b wall of the guide. Therefore, the mode is transverse magnetic (TM). The size of the guide is 3 in $\times$ 1.5 in (S band), and the operating frequency is 2.624 GHz. The a dimension of the guide (3 in) is larger than one-half wavelength at the operating frequency of 2.624 GHz (4.5/2 = 2.25 in), so the guide *will* propagate this signal frequency. There is more than one half wavelength, but less than one full wavelength, in the a direction and less than one half wavelength in the b direction, so the waveguide designation is $TM_{1,0}$.

11.3.5 Slots in Waveguides

The principle of electric and magnetic waves existing along the walls of a waveguide is restated in Figure 11.6, which also shows several slots placed in the walls of the guide for the purpose of transferring energy either into or out of the waveguide. The position of a strong electric or magnetic wave will be stationary relative to the input coupling device and therefore will dictate where a slot should be placed to best transfer each form of energy. Again, a strong electric field will exist where the magnetic field is weakest, and a strong magnetic field will be found where the electric field is weakest.

Slots 1 and 2 are placed so as to interrupt the flow of a strong electric field current across the narrow dimension of the slot. This will induce an electric field of RF energy to be coupled through the slot when the slot length is $\lambda_0/2$ at the operating frequency. The slot width controls the frequency bandwidth of the signal. A narrow slot will pass a narrow band of frequencies centered around the carrier frequency (f_0). The maximum slot width, $\lambda_0/4$, will pass a wide band of frequencies also centered around the carrier frequency.

The impedance of the slot will remain constant at 487 Ω and is independent of the slot width. The slot impedance is determined by

$$Z_{\text{slot}} = \frac{(120\pi)^2}{4Z_d} = \frac{(377)^2}{292} = 486.74 \ \Omega$$

where 120π is the impedance of free space and Z_d is the impedance of a complementary half-wave dipole antenna, 73 Ω. The slot impedance will depart slightly from this value, depending on the thickness of the guide wall. For general purposes, the slot impedance is considered to be approximately 500 Ω.

Slot 3 in Figure 11.6 is located where the magnetic field is strongest across the narrow dimension of the slot. Slots 1 and 2 are designated as E field couplers, and slot 3 is designated as an H field coupler.

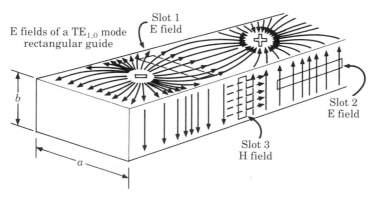

FIGURE 11.6
Slot coupling.

11.3.6 Working Factors of Waveguide

There is a group of quantities that must be determined when working with waveguide, which, when solved for, can yield information that is valuable to the user. One such factor is the propagation constant k, which is found by

$$k = \sqrt{1 - \left(\frac{\lambda_0}{2a}\right)^2} \tag{11.4}$$

which compares the operating wavelength (λ_0) to the cutoff wavelength ($\lambda_{co} = 2a$). The propagation constant will always be a number less than 1. When the dielectric inside of the guide is something other than air, the number 1 inside the radical sign is replaced by the dielectric constant ϵ. See Chapter 8 for a table of dielectric constants for various materials.

The propagation constant is used to identify the wave's angle of reflection θ off of the walls as it moves through the guide.

$$\theta = \sin^{-1} \sqrt{1 - \left(\frac{\lambda_0}{2a}\right)^2} \tag{11.5}$$
$$= \sin^{-1} k \quad \text{(in degrees)}$$

The propagation constant also shows the proportionality of the velocity of energy in the guide V_g to the velocity of energy in free space:

$$V_g = V_c \sqrt{1 - \left(\frac{\lambda_0}{2a}\right)^2} \tag{11.6}$$
$$= V_c k$$

The wavelength of the energy inside the guide λ_g is

$$\lambda_g = \frac{\lambda_0}{\sqrt{1 - \left(\frac{\lambda_0}{2a}\right)^2}} \tag{11.7}$$
$$= \frac{\lambda_0}{k}$$

The wave impedance for the energy in the guide Z_w is also a function of the propagation constant:

$$Z_w(\text{TE}) = \frac{377}{k} \quad \text{(in ohms)} \tag{11.8}$$

$$Z_w(\text{TM}) = 377k \tag{11.9}$$

where 377 Ω is the impedance of free space. Notice that the impedance of the transverse electric wave will always have a value greater than 377 Ω, while the impedance of the transverse magnetic wave will always be less than 377 Ω.

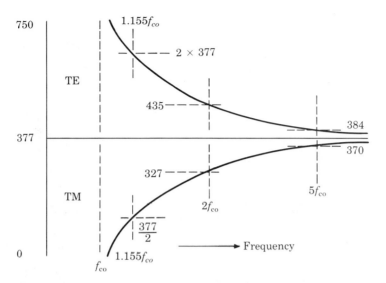

FIGURE 11.7
A graph of the wave impedance variance with respect to frequency for the TE and TM guide.

Figure 11.7 examines the impedances of the two waves at various frequencies and shows that at the cutoff frequency, both impedances are far from 377 Ω. As the frequency is increased, both values become asymptotic, that is, they approach the value of 377 Ω but never equal 377 Ω. An important fact to recognize here is that wave impedance means just what it says, the impedance of the fields inside of the waveguide. It is *not* the input impedance of the guide.

11.3.7 Input Impedance to the Guide

For probe input coupling to the guide, it was noted that the input impedance is a function of the position of the probe relative to the side walls and the closed wall at the back of the guide, as well as to the depth of the probe insertion length. Equation 11.10 compares the physical location of the probe and the depth of the probe to the wave impedance in order to establish the guide input impedance, Z_{in}

$$Z_{in} = \frac{Z_w(\text{mode})}{\left[\left(1 + \frac{a\lambda_0}{2c}\right) - \left(\frac{da}{2}\right)^2\right] + \left[\frac{ab}{c} - (2hb)\right]^2} \tag{11.10}$$

where Z_w is the wave impedance for the mode of propagation and Z_{in} is the cable impedance to be used. The other factors, $a, b, c, d,$ and h, are all taken from Figure 11.8. Z_{in} will yield a Gaussian response curve, reaching a maximum value of about 150 Ω and decreasing rapidly as the control values move away

FIGURE 11.8
Probe position considerations used to match the input impedance of the guide to the wave impedance.

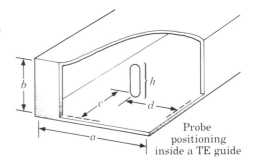

Probe positioning inside a TE guide

from the critical quantities: Z_w large (at f_0 close to f_{co}), distance $c = \lambda_0/4$, and insert length $h = b/2$. When $f_0 = 1.33f_{co}$, Z_{in} will equal about 50 Ω with correctly selected control values. Thus, it is safe to say that the input impedance will be nearly 50 Ω when the probe is correctly positioned and adjusted.

11.3.8 Power Limits of Waveguide

One of the major advantages of waveguide is its ability to handle two times the power of coaxial cable of the same size. The physics are based on the breakdown voltage of free air at sea level. With no center conductor, the distance between the conducting surfaces in a waveguide is twice the distance from the center conductor to one wall of the coaxial cable. A value of 15,000 V/cm breakdown voltage of air has been obtained experimentally and has proven valid for several decades. The shortest distance across the guide is generally in the b direction, and the maximum field strength is found at the center of the a dimension. The power at this point must be evaluated as peak power and used to determine the worst-case condition. Waveguide specification sheets will list the power limits for a given guide size. However, when the spec sheets are not available or when there is any question regarding a particular microwave device or operating frequency, then the maximum power P may be determined by

$$P = 7 \times 10^5 \; ab\left(\frac{\lambda_0}{2a}\right) \quad \text{(in watts)} \tag{11.11}$$

for a waveguide size in inches, and

$$P = 1.5 \times 10^5 \; ab\left(\frac{\lambda_0}{2a}\right)$$

for a waveguide size specified in centimeters.

The constants of Equation 11.11 are lumped together and include the breakdown voltage, the permeability of air, and the size relationships of the guide. The power range will be from a few kilowatts to several megawatts.

Altitude Effects The primary effect of high altitudes is power loss. The breakdown voltage of air decreases as the altitude increases. The power loss is related to the altitude, A, in thousands of feet, as

$$\% \text{ of max power} = \frac{1}{\ln A} \times 100 \qquad (11.12)$$

At 25,000 ft the power rating would be reduced to 31% of its sea level rating. At 40,000 ft, the power would be reduced to 27.1% of maximum. Below 2718 ft altitude, there is no change in the power rating of the guide. Moreover, this equation provides good correlation even at satellite elevations of 22,400 mi. At 22,400 mi (1.18272×10^5 thousands of feet), the power is reduced to 8.56% of maximum rated power.

To relieve the condition of power loss with altitude change, the waveguides are made airtight and held at a positive pressure relative to the pressure at sea level so that the power loss due to altitude is minimized and cold, humid air cannot enter the waveguide to reduce power through frost buildup. (The effects of humidity are included in the constant of Equation 11.11, but were not identified separately.) An inert gas is usually the element chosen to fill the waveguide and maintain the positive pressure.

11.3.9 Corners, Bends, and Tees

A slang name for waveguide is "plumbing" because waveguide consists of rigid piping between two or more pieces of equipment. Because waveguide is rarely installed in one straight line, the limitations associated with bends and corners are highlighted in this section.

The energy is reflected off of the walls as it propagates through the waveguide. When the guide changes direction too rapidly, one component of the wave may travel a different path length than the next part, and the phase of the signal will change, causing some cancellation. Also, the mode of propagation could be modified to one that is totally unacceptable. The waveguide could suddenly look like a short circuit or an open circuit to the energy, with the associated result of complete elimination of the signal.

Figure 11.9 shows some simple rules to follow when selecting or working with waveguide fittings for changing direction. Figure 11.9a is an E-plane bend whose radius is limited (at the outside of the guide) to no less than 1.5 times the b dimension of the guide. The H-plane bend of Figure 11.9b should never have an outside radius smaller than 1.5 times the a dimension of the guide. The 90° corners in Figure 11.9c and d have a mean length not less than $3a/2$. All bend fittings may be larger by an amount that will increase the length in half-wavelength increments. This interval will minimize phase distortion and hold down the losses. The exact relationship for mean length L is

$$L = \left(2n + 1\right)\frac{\lambda_g}{4} \qquad (11.13)$$

specified in the same units as the guide wavelength, where n is number of half-wavelength increments. Twists are somewhat more critical to phase changes, and lengths are therefore limited to two wavelengths minimum for

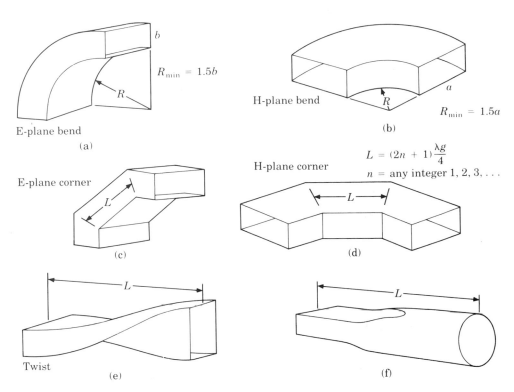

FIGURE 11.9
Bends, corners, twists, and tapers (minimum dimensions).

each 90° of spiral rotation. Twists may, of course, be longer if the same rule is followed as for turns and bends, that is, increase the length in half-wavelength increments beyond the two-wavelength minimum.

The conversion from rectangular to round waveguide may be encountered at the joints of rotating antennas. Good continuity is again achieved when the length of the converter is held to two wavelengths minimum and increased in length by half-wavelength steps.

The tee couplers shown in Figure 11.10 are also categorized by the E-plane and H-plane units. The E-plane tee of Figure 11.10a shows that when port 1 is the input, the electric field will hinge at the radius of the turn (at A) and move to output port 2, maintaining the phase of the E field. However, the same E field input to port 1 will encounter a 180° phase change as it goes past the channel of port 2 on its way to output port 3. Visualize this progression by means of the arrows in Figure 11.10. The electric field, shown with the arrowhead pointing up to indicate positive polarity, reaches the corner (at A) and "hinges" to make the turn and leave the guide at port 2. The field rotates at A until the tail of the arrow touches the corner at B. The tail of the arrow then comes to represent the hinge at B, swinging the head of the arrow downward as

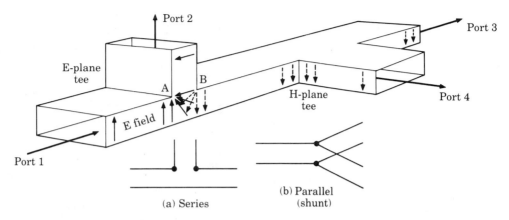

FIGURE 11.10
E- and H-plane tees, the relative electric fields at each port, and the tee equivalent circuit.

the wave moves toward output port 3. Thus, the output at port 3 is 180° out of phase from the input, while the output of port 2 is in phase with the input at port 1. When port 2 is the input, the polarity of the signals at all three ports will remain the same. The E-plane tee is termed a *series tee* and would be used as the ATR connection (9) in Figure 11.1.

The H plane tee of Figure 11.10b is termed a *shunt tee* and has the transmission line equivalent of a parallel connection, as shown in the figure. The H-plane tee has *no* phase change to the signal at any of its ports. A positive-polarity signal will be the same at all ports in the connector. The H-plane tee is used as the TR termination (7) in Figure 11.1.

11.3.10 Directional Couplers

Although directional couplers were discussed in the context of AM transmitters, they are reexamined here because the mode of coupling is different although the principle of operation is the same, that is, to siphon off a trickle amount of energy for measurement purposes without restricting the flow of energy in the main channel of the sensing device. In Figure 11.11, the input port is labeled port 1, the main channel output is port 2, and the sensing terminal is port 3.

In Figure 11.11b, the input signal (S_1) passes through hole X and enters the auxiliary channel as signal S_2 heading toward auxiliary port. The same signal through hole X is seen as signal S_3 heading in the other direction, only to be absorbed by the load resistance, a ferrite wedge labeled R. Signal S_1 continues to hole Y and enters the auxiliary channel as signal S_4 to cancel signal S_2. Signal S_4 travels one-half wavelength further than signal S_2, so that it is exactly 180° out of phase from signal S_2. The signal through hole Y leaves by

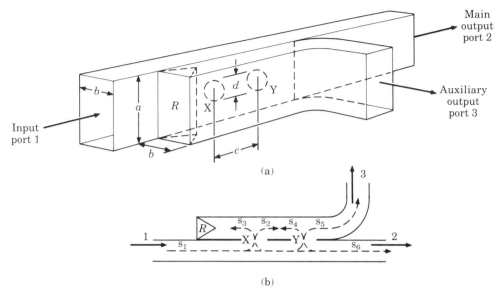

FIGURE 11.11
A two-hole E-plane directional coupler.

auxiliary port 3 as signal S_5, greatly reduced from the signal strength of input S_1. The main-channel signal S_1 continues through the main channel, leaving by port 2 as signal S_6, almost unaffected by the minor reduction in power.

The coefficient of coupling C_c (in dB) is taken to be

$$C_c = 10 \log \frac{\text{Power at auxiliary 3}}{\text{Power at main channel 1}} \qquad \textbf{(11.14)}$$

and is controlled by the diameter of the hole relative to the wall size of the guide. The hole diameter ratio is

$$\text{Ratio} = \frac{2D\sqrt{h}}{a} \quad (a \text{ side coupler})$$

$$= \frac{D\sqrt{h}}{b} \quad (b \text{ side coupler}) \qquad \textbf{(11.15)}$$

where D is the hole diameter and h is the number of holes, with a or b representing the guide wall to which the coupler is to be attached. Rearranging Equation 11.15 to solve for the hole diameter,

$$D = \frac{(a)\text{Ratio}}{2\sqrt{h}} \quad (a \text{ side coupler})$$

$$= \frac{(b)\text{Ratio}}{\sqrt{h}} \quad (b \text{ side coupler}) \qquad \textbf{(11.16)}$$

The distance between the holes is any whole number of quarter wavelengths of the guide. Select a number of quarter wavelengths (n) and substitute it into Equation 11.17,

$$\text{Center to center} = (2n - 1)\left(\frac{\lambda 2a}{4}\right) \qquad \textbf{(11.17)}$$

which is specified in the same units as the guide wavelength.

An example will show the use of the equations to select the right coupler for the job.

EXAMPLE 11.2

A 10 mW full-scale deflection meter is used to measure 10 W of power in the main channel. A two-hole coupler in the J band is to be used. The J band guide has a size ratio of $b/a = 0.62/1.37 = 0.45$. The coefficient of coupling is 10 log 10 mW/10 W = -30 dB. Using the graph of Figure 11.12, find where C_c of -30 dB intersects the size ratio curve of 0.45. The result indicates a diameter ratio of 0.7. Using Equation 11.16 to find the hole diameter size,

$$D = \frac{(a)\text{Ratio}}{2\sqrt{h}} = \frac{1.37 \times 0.7}{2 \times \sqrt{2}} = 0.339 \text{ in (2 holes)}$$

The hole centering is (for three quarter wavelengths)

$$\text{Center to center} = [(2 \times 3) - 1]\left(\frac{1.24}{4}\right) = 1.55 \text{ in}$$

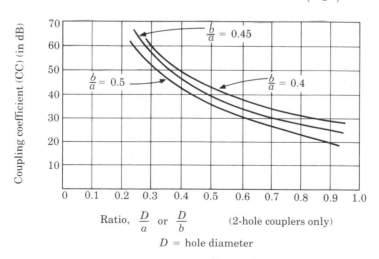

FIGURE 11.12
Coupling coefficient graph for a two-hole coupler.

The *directivity* of the coupler describes the isolation at auxiliary output port 3 when the same amount of power is applied in the reverse direction (input to port 2) of the main channel. Both measurements are taken at the auxiliary port and compared.

$$\text{Directivity} = 10 \log \left(\frac{\text{Power aux 3, forward}}{\text{Power aux 3, reversed}} \right) \quad \text{(in dB)} \qquad \textbf{(11.18)}$$

Minimum values of directivity should exceed 40 dB and typically reach 60–70 dB isolation.

Adding more holes to the coupler will increase the bandwidth of the coupled signal and improve the directivity as well. However, both the directivity and the coefficient of coupling are *very* frequency-sensitive and should be calculated, bored, and positioned very accurately. The holes should be lined up carefully and should be free of burrs or other irregularities.

11.3.11 Planned Waveguide Obstructions

A thin metallic plate that fills the guide in the a direction and partially blocks the transmission in the b direction (E plane) will have the effect of a capacitor shunting the waveguide across the short (b) dimension at the location of the window (see Figure 11.13a). The degree of obstruction and the corresponding value of capacitance is a ratio of the obstruction height (h) compared to the guide dimension (b) and the obstruction thickness. As the percentage of filling increases, the capacitance also increases.

An obstruction across the b dimension (H plane) would be the equivalent of an inductance across the b dimension of the guide at the window location (see Figure 11.13b). The ratio of the obstruction width (W) to the guide's a size and obstruction thickness all contribute to the value of inductance. The larger the obstruction, the higher the value of inductance.

Combining the obstacle effects of the E plane and the H plane would form the equivalent of a parallel-resonant circuit placed across the short (E plane) dimension of the waveguide, as seen in Figure 11.13c. Larger values of h and W will increase the Q of the circuit to better control the slope of the resonance response curve. The circuit looks resistive, and there are no reflections at resonance.

A metallic post inserted at the center of the a dimension, as seen in Figure 11.13d, will look inductive when the post extends fully across the waveguide. The inductance is proportional to the diameter of the post. When the post extends only part of the distance across the b dimension, the post looks capacitive, with a value proportional to the post length and diameter. However, as the length of the post increases, at some frequency the post will have equal inductive and capacitive reactances, and the post will look like a series-resonant circuit placed across the b dimension (E field) of the guide.

A ferrite material placed in the magnetic field of the guide, as shown in Figure 11.13e, will look like a pure resistance shunting the guide. The size of

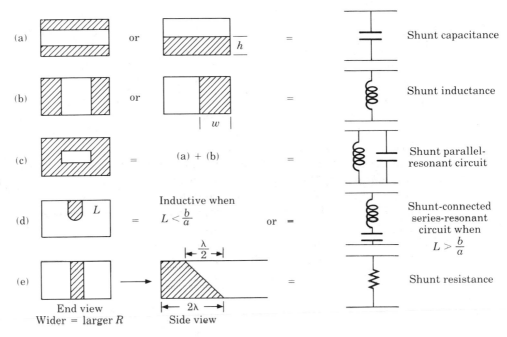

FIGURE 11.13

Reactive and resistive vanes placed inside of the waveguide to act as circuit components.

the obstruction, as well as the shape and location of the ferrite material inside of the guide, will determine the resistive value and the power rating of the equivalent resistance. These ferrite materials are sensitive to frequency and will change in resistive value as the frequency changes. Only one form of ferrite resistive vane is shown in Figure 11.13.

11.4   HORN ANTENNAS

Rectangular waveguide by itself could be used as a radiating antenna simply by letting the energy spill out from the open end of the waveguide. About 4 dB of gain over an isotropic radiator could be realized, depending on the relationship between the guide size and the wavelength at the operating frequency. This gain results from concentrating the energy in one direction. The radiation pattern would be a balloon-shaped pattern like that of the half-wave dipole antenna, except the vertical and horizontal patterns radiated from the guide would be identical. This practice, however, is not followed because of the serious mismatch between the waveguide impedance and the impedance of free space. The resulting high standing wave of voltage rules it out as a possible consideration.

The impedance mismatch is overcome by flaring the open end of the wave-guide in either one or both directions to form a horn. In addition to matching the impedance and reducing the SWR to an acceptable level, the flare offers a bonus package of good directivity gain, a narrower beamwidth angle, small size, light weight, and simple construction. In addition, because there are no resonant parts, the horn antenna will cover a wide frequency range.

Note the correlation of the size designators for the horns in Figure 11.14. Dimension a for the width of the guide at the throat corresponds to A for the larger width of the horn at the mouth. The same is true for the b dimension of the guide and the B dimension of the horn. A standard $TE_{1,0}$ mode of excitation is assumed for the rectangular waveguide in Figure 11.14. This places the E field across the b dimension of the guide and makes the horn in Figure 11.14a an H-plane sectional horn, that of Figure 11.14b an E-plane sectional horn, and the horn in Figure 11.14c, which flares in *both* directions, a **pyramidal horn.** The pyramidal horn is most common in microwave communications. The direction of the E field in the horn establishes the polarization of the transmitted signal. Figure 11.14 shows vertical polarization only.

The factors that control gain and beamwidth are the A/a and B/b size ratios, the length of the horn (L) from throat to mouth, and the operating

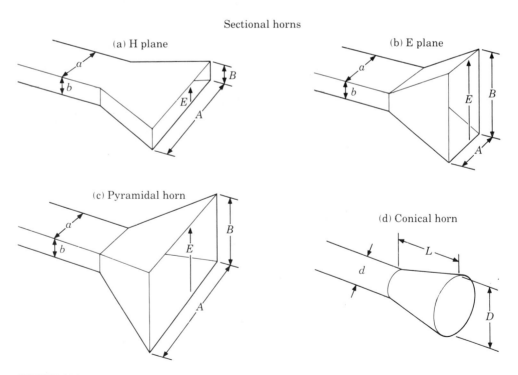

FIGURE 11.14
Horn antennas.

frequency. These values are as given in Figure 11.15. Before trying a sample case to find the gain and beamwidth angles, it is necessary to review the factors that control their limits.

The total mechanical flare angle of the horn antenna controls both the directivity gain and the beamwidth of the energy to be emitted from the horn. However, the flare angle is a function of the bore sight length and the size of the mouth of the horn. Therefore, gain and beamwidth are usually stated in terms of horn length compared to the mouth area $(A \times B)$ in wavelengths.

As noted in Figure 11.15, when the waveguide flare angle is expanded from 0° through ±15°, the directivity gain increases and the beamwidth narrows. Also during this time, however, the wave front leaving the horn is spherical in form, which is undesirable for this application. As the flare angle increases beyond 15°, the wave front becomes planar (desirable) and the directivity gain increases to the optimum value and then starts to decrease. The beamwidth continues to narrow during this phase.

An upper limit is reached when the slant length $(R + \Delta R)$ becomes 0.375 times longer than the center line length (R), that is, when $\alpha = $ arc cos $R/1.375R = \pm 43.4°$. This is noted as approximately ±45°, equal to a total flare angle of 90°.

The extremes are expressed as a minimum length and an optimum length resulting from these two mechanical angle limits for a specific mouth area. The horn antenna may be longer than the optimum length, but the gain decreases with extra length, which is why the best length value is termed the optimum

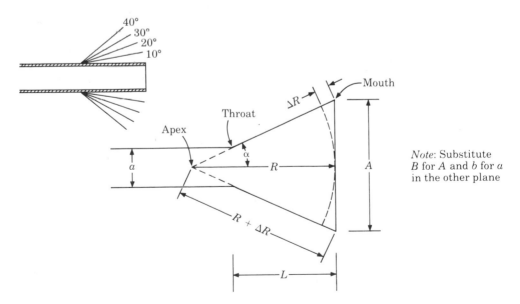

FIGURE 11.15
Dimensions and angles that affect the gain and beamwidth of a horn antenna.

length rather than the maximum length. Also, for lengths shorter than minimum, the main radiation lobe will begin to separate into two lobes, then three lobes, and so on, reducing the directivity gain and broadening the beamwidth. The minimum horn length (L_{min}) becomes a function of the maximum flare angle and can be determined by comparing the mouth area to the throat area:

$$L_{min} = \frac{(A - a) + (B - b)}{4} \tag{11.19}$$

The optimum length of the horn is

$$L_{opt} = \sqrt{\frac{AB}{\log 2}} \tag{11.20}$$

The most commonly used expression for gain [10 log ($7.5AB/\lambda^2$)] and the associated beamwidth equations are assumed for optimum conditions only. In practice, the optimum horn length is regularly sacrificed in favor of the greater gain and narrower beamwidths offered by other components in the microwave antenna system. In all instances, the best system may be assembled by knowing the exact gain and beamwidth angles for each component, evaluated in advance of the actual field testing. Equation 11.21 can be used to calculate the gain in decibels for any horn length, mouth area, throat area, and operating frequency.

$$\text{Gain dB} = 10 \log \frac{2\pi AB}{\lambda^2} + 10 \log \frac{1}{\sqrt{1 + (L_{opt} - L)^2}} \tag{11.21}$$

where A and B are the mouth measurements, L_{opt} is from Equation 11.20, L is the length from throat to mouth along the center line, and λ is the wavelength. All measurements are in the same units, centimeters, inches, feet, or meters. The first half of Equation 11.21 is the standard equation for optimum gain. The second half compares the actual horn length to the optimum length and converts them to decibels.

The -3 dB beamwidth angles may differ for the A and B directions and are found by

$$\theta_{A(-3dB)} = \text{arc tan} \left(\frac{A - a}{L}\right)^2 \quad \text{(in degrees)}$$

$$\theta_{B(-3dB)} = \text{arc tan} \left(\frac{B - b}{L}\right)^2 \tag{11.22}$$

The -10 dB beamwidth comes into play when the horn antenna is used with a reflector, and may be determined by

$$\theta_{A(-10dB)} = \text{arc tan} \left(\frac{A-a}{0.7L}\right)^2 \quad \text{(in degrees)}$$

$$\theta_{B(-10dB)} = \text{arc tan} \left(\frac{B-b}{0.7L}\right)^2$$

(11.23)

EXAMPLE 11.3

An X band waveguide system operating at 9.3 GHz is to be used with a 25.4 cm horn having a mouth that measures 18.46 cm in the A direction (H plane) and 14.55 cm in the B direction (E plane). Find the directivity gain and the −3 dB beamwidth angles for the E and H planes in this system. Also establish the minimum and optimum horn lengths.

SOLUTION: The X band waveguide inner dimensions are $a = 0.9$ in = 2.2855 cm, $b = 0.4$ in = 1.015783 cm, $2a = 4.571$ cm = $1.417(\lambda_{co})$, $\lambda_0 = V_c/f =$ 3.2258 cm, $A - a = 16.1745$ cm, and $B - b = 13.5342$ cm. Therefore,

$$L_{min} = \frac{16.1745 + 13.5342}{4} = 7.427 \text{ cm}$$

$$L_{opt} = \sqrt{\frac{(18.46)(14.55)}{0.30103}} = 29.87 \text{ cm}$$

$$\text{Gain dB} = 10 \log \frac{(6.2832)(18.46)(14.55)}{10.406}$$

$$+ 10 \log \frac{1}{\sqrt{1 + (29.87 - 25.4)^2}}$$

$$= 22.1 \text{ dB} + (-6.609 \text{ dB}) = 15.49 \text{ dB}$$

Note that a maximum gain of 22.1 dB could have been achieved for an optimum length of 29.87 cm.

$$\theta_{A(-3dB)} = \text{arc tan} \left(\frac{16.1745}{25.4}\right)^2 = 22.073°$$

$$\theta_{B(-3dB)} = \text{arc tan} \left(\frac{13.5342}{25.4}\right)^2 = 15.85°$$

Note that these equations for horn gain and beamwidth angles are accurate to within 0.25 dB under all conditions tested but are *not* recommended for design equations. Horn antennas are often used as gain standards and need to be accurate to within less than 1%. However, these equations are excellent for very close approximations and have proven far more accurate than the optimum equations under test conditions for a wide variety of sample horn antennas.

FIGURE 11.16
Conical horn antenna.

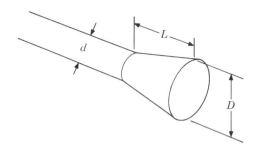

Round (or cone-shaped) horns, called **conical horn** antennas, are some-times used in rotating antenna systems that sweep across 360° of the horizon. The waveguide must be converted from a rectangular to a round guide for the rotary joint and remains round when the distance from the rotary joint to the horn antenna is short (see Figure 11.16). Conical horn antennas behave in much the same way as pyramidal horns except that conical horns have high cross-polarization and lack symmetry for large mouth diameter (D) values. The E- and H-plane field radiation patterns are equal when $D = 1.23\lambda$. The best results are obtained when the flare angle is between 15° and 50°. The gain and beamwidth angle equations for the cone horn are

$$\text{Gain dB} = 10 \log \frac{\pi L}{(D - d)}\left(\frac{D^2}{\lambda_0}\right) \tag{11.24}$$

$$\theta_{-3\text{dB}} = 2 \text{ arc tan} \left(\frac{dD}{L\lambda_0}\right) \quad \text{(in degrees)} \tag{11.25}$$

There are so many varieties of horn antennas that only the most popular types have been mentioned here.

11.5   PARABOLIC REFLECTORS

A geometric shape that satisfies the equation

$$Y = \sqrt{4FX} \tag{11.26}$$

will result in the type of graph shown in Figure 11.17. A value of focal length F is selected, and then a value for Y can be found for each value of X that is substituted in the equation. Different values of focal length will cause the curve to open or close more rapidly, as seen in Figure 11.17, but the curve maintains the same basic shape. A larger value of F causes the curve to open more rapidly. Y can also be given negative values.

All line lengths from the focal point F, at any angle, to the inside surface of the curve (where it will reflect along a path parallel to the X axis), and then to a vertical line through any value of X (called the *directrix*) will be of equal length. In Figure 11.17, the length of line FAB will be exactly the same as that of line FCD.

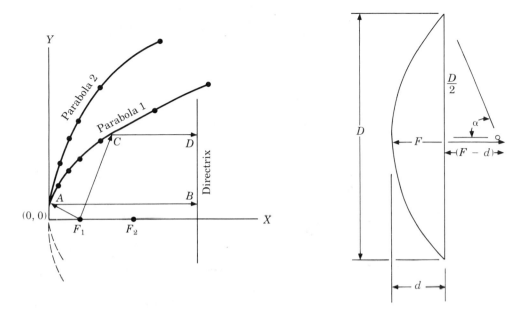

FIGURE 11.17
The curve of the parabola equation for two values of focal length.

The **parabolic reflector,** so common in microwave communications, is formed by rotating the graph of Figure 11.17 around the X axis to form a saucer- or bowl-shaped three-dimensional surface called a *paraboloid*. The reflector **dishes,** as they are called, are made in graduated sizes based on mouth diameter (minimum $D = 2\lambda$) and on the ratio of focal length to mouth diameter. The range of F/D ratios is from 0.15 to 2.5, with the most common ratio at $F/D = 0.5$.

A circular wave front emanating from an energy source at the focal point (F) will strike the inside surface of the parabola, and a time-related sequence will cause the wave to leave the parabola mouth as a straight-line wave front. Such a wave front is seen in Figure 11.18 as the directrix, which will move away from the reflector to the right in the figure.

Substituting one-half the mouth diameter ($D/2$) for Y in Equation 11.26 and the dish depth (d) for values of X will allow us to solve for all parabolic reflector measurements:

$$F = \frac{D^2}{16d}$$

$$d = \frac{D^2}{16F} \tag{11.27}$$

$$D = 2\sqrt{4dF} = 4\sqrt{dF}$$

For any fixed diameter, the dish depth and focal length are dependent variables. A shallow dish will result in the greatest focal length, the widest 3 dB

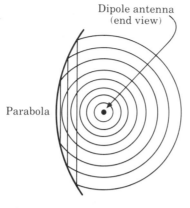
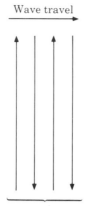

Parabola

Reflected wave   Emitted wave   E-field planar waves
is planar   is spherical

FIGURE 11.18
A spherical wave is converted to a planar wave by the parabolic reflector.

beamwidth, and the lowest reflector directivity gain. The deepest dish will result in the smallest focal length, the sharpest 3 dB beamwidth, and the highest reflector directivity gain.

With the energy radiator placed at the focal point, momentarily ignoring the source-to-reflector efficiency, we can calculate the dB directivity gain:

$$\text{Gain dB} = 10 \log 2 + \left[\frac{d}{D}\left(\frac{\pi dD}{3\lambda}\right)^2\right] \tag{11.28}$$

The -3 dB beamwidth has the relationship

$$\theta_{-3\text{dB}} = 2 \text{ arc tan } \frac{\lambda F}{d^2 D} \tag{11.29}$$

When a short dipole antenna, which radiates energy perpendicularly around the dipole, is placed at the focal point, the radiated energy will strike the inside surface of the parabola and be reflected out of the mouth as parallel rays of energy. There will be considerable spillover, as shown in Figure 11.19, but it may be greatly reduced by adding a parasitic director 0.1 wavelength away from the driven element, in front of the focal point. With a dipole antenna as the driven element, a parasitic rod is used as director 1 and the parabolic reflector becomes parasitic reflector 1. Note that the parabola is not an antenna, but simply a focusing reflector or lens. When a horn antenna is the energy source at the focal point, all of the energy is directed toward the parabola. Very little spillover takes place, and the system efficiency greatly improves.

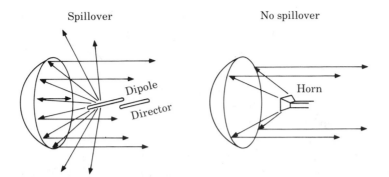

Spillover   No spillover

Dipole
Director

Horn

FIGURE 11.19
Two examples of a radiating source using a parabolic reflector and the resulting
spillover.

In the receive mode, parallel rays of energy that enter the mouth will
strike the parabola surface and be deflected toward the focal point.

Rather than discuss the wide number of choices and decisions that must
be made during the design phase, a known combination of horn antenna and
parabolic reflector will be examined as an example.

EXAMPLE 11.4

An M band system at 11.808 GHz has a wavelength of 1 in and horn throat
dimensions of 0.75 in × 0.375 in, placing the carrier at 1.5 times the cutoff
frequency of the guide. A 4 in long horn was selected, having mouth measurements of 8.75 in in the H plane by 7 in in the E plane. From Equations
11.19 and 11.20,

$$L_{min} = \frac{(8.75 - 0.75) + (7 - 0.375)}{4} = 3.65626 \text{ in}$$

$$L_{opt} = \frac{(8.75)\,(7)}{0.30103} = 14.264224 \text{ in}$$

The 4 in horn was selected with the knowledge that less-than-optimum gain
will result; however, the wide beam angles of the horn will better match the
full illumination of the parabolic reflector. The horn gain is found as

$$\text{Gain dB} = 10 \log \left(\frac{(6.283185)\,(8.75)\,(7)}{1} \right)$$

$$+ 10 \log \frac{1}{\sqrt{1 + (14.264224 - 4)^2}}$$

$$= 25.835 + (-10.134) = 15.7 \text{ dB}$$

The -10 dB beamwidth angles are found using Equation 11.23:

$$\theta_A = \text{arc tan} \left(\frac{8.75 - 0.75}{0.7 \times 4} \right)^2 = 83°$$

$$\theta_B = \text{arc tan} \left(\frac{7 - 0.375}{0.7 \times 4} \right)^2 = 79.87°$$

The result is an average angle of 81.4°.

A 60 in diameter parabolic reflector with an F/D ratio of 0.6667 was selected. This represents a focal length of $60 \times 0.6667 = 40$ in. The dish depth was calculated from Equation 11.25 to be 5.625 in. Note from Figure 11.17 that the mechanical angle of the focal point is the arc tan of 30 in divided by the focal length minus the dish depth, resulting in an angle of $41.1° \times 2 = 82.224°$. This closely matches the -10 dB average angle of 81.4° and confirms that the two components are mechanically well suited for full illumination of the reflector. The reflector gain is then calculated:

$$\text{Gain dB} = 10 \log 2 + \left[\frac{5.625}{60} \left(\frac{(3.141592)\,(5.625)\,(60)}{(3)\,(1)} \right)^2 \right] = 40.6865 \text{ dB}$$

The -3 dB beamwidth of the parabola is

$$\theta_{-3\text{dB}} = 2 \text{ arc tan} \left(\frac{(1)\,(40)}{(3.141593)\,(5.625)\,(60)} \right) = 4.32°$$

The total power gain of the horn and parabola is 15.7 dB + 40.6865 dB = 56.4 dB at a 4.32° beamwidth angle.

A double-lens reflecting system, called a **Cassegrain feed** (shown in Figure 11.20), has several advantages over the single-paraboloid system:

A. The paraboloid surface is more uniformly illuminated by the second reflector, which increases the efficiency to about 75%.
B. There is less spillover to cause nuisance interference.
C. A greater apparent focal length ($F/D = 1$ or more) allows the use of a longer horn with higher gain.
D. Shorter waveguide feed results in lower feed line losses.
E. There is a greater reduction in side lobe power, although the side lobes are closer to the main lobe.
F. The noise level is lower.
G. There is less cross-polarization.

The disadvantages are the more complex structure and design and the very critical mechanical alignment. Besides the size, shape, and placement of the second reflector, and the multiple choices for the new horn, there are now three

FIGURE 11.20
The dimensioning of the Cassegrain double reflector.

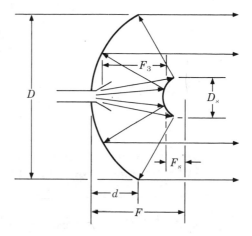

focal lengths to control: the original parabola focal length, the new hyberbola focal length, and the horn focal length, which, in all probability, will differ from the parabola focal length.

The second (smaller) reflector has the shape of a hyperbola, which is compared to the parabola in Figure 11.21. The hyperbola decreases in curvature as it extends further away from the focal point and complements the parabola. The parabola has high gain and narrow beamwidth when the focal length is short, which opposes the advantages of the horn antenna. With the Cassegrain feed, the longer effective focal length can take advantage of the higher gain possible from a longer horn.

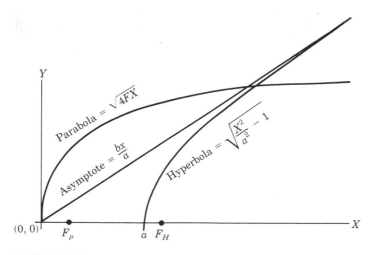

FIGURE 11.21
Comparing the curve of a parabola to the curve of a hyperbola.

There are two basic equations for the hyperbola that define the curve:

$$Y = \sqrt{\left[\frac{x^2}{a^2}\right] + 1} \qquad (11.30)$$

for the curve and

$$Y = \frac{bX}{a} \qquad (11.31)$$

for the asymptote of the curve. An asymptote is a value which Y will approach but never reach. The origin of the asymptote is the apex of the parabola ($X = 0$, $Y = 0$).

EXAMPLE 11.5

At 12 GHz, one wavelength measures 0.984 in, and a parabola with a mouth diameter of 96 in and a depth of 10 in will have a focal length of 57.6 in. Minimum signal blockage will occur when the hyperbola diameter D_s is

$$D_s = \sqrt{2\lambda F} = 10.65 \text{ in} \qquad (11.32)$$

where F is the focal length of the parabola. The focal length and diameter of the hyperbola have the same relationship as the focal length and diameter of the parabola. Therefore, the focal length of the hyperbola F_s is

$$F_s = \frac{D_s F}{D} = 6.4 \text{ in} \qquad (11.33)$$

where D is the parabola diameter. The focal length (F_3) of the horn (to hyperbola) will be a function of the -10 dB beamwidth of the chosen horn. A conical horn with a gain of 15 dB and a -10 dB beamwidth of 26.6°, designated as angle a, will yield a horn focal length of

$$F_3 = \frac{D_s}{2 \tan (a/2)} = 22.52 \text{ in} \qquad (11.34)$$

The relationships of the Cassegrain feed system show a magnification factor M of

$$M = \frac{D}{4F}\left(\frac{1}{\tan (a/2)}\right) = 1.763 \qquad (11.35)$$

The system would have a total gain of 57.64 dB and a -3 dB beamwidth angle of 0.64°.

The reader is encouraged to work through the individual steps of this example to arrive at the solution.

Some blockage will occur when any form of hardware is placed in the beam path of the propagated wave. A horn antenna, a hyperbolic reflector, or even the support struts for such devices will have a shadow effect on the transmitted pattern. The size of the hyperbola is just one factor to be observed. The shadow effect may be reduced by making the focal length of the parabola equal to any exact number of quarter wavelengths at the operating frequency. However, the side lobes in the radiated pattern are minimized most effectively when the focal length is one-quarter of the size of the parabola mouth diameter ($F = D/4$). Therefore, some form of compromise may be in order.

The *hoghorn* parabolic reflector of Figure 11.22 is a common microwave antenna that combines a horn and parabola into one assembly. It offers the advantages of straight-line waveguide routing and zero shadow effects. The mouth opening is usually covered with a protective screen that does not interfere with the signal, yet keeps out dust and debris and prevents birds from nesting inside of the structure. Hoghorns are used for stationary transmitting and receiving systems.

The *Cutler feed* of Figure 11.23 is favored when low shadow effects are a high priority on a mobile parabolic reflector. The waveguide leading into the split reflector is tapered in the *b* plane to improve the impedance match as well as to clear the path for the return wave to the lens reflector. The center-to-center separation of the two slots is exactly one-half wavelength at the operation frequency, and the typical F/D ratio of the parabola ranges from 0.33 to 0.45 because of the wide dispersion angle of the device. The set screw is a factory-adjusted impedance-matching device.

FIGURE 11.22
Hoghorn antenna.

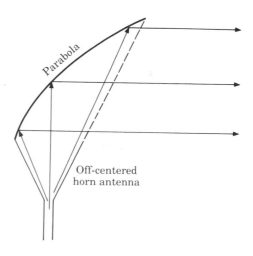

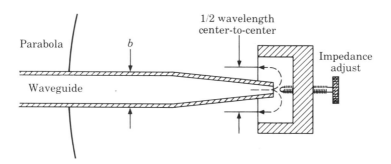

FIGURE 11.23
Cutler feed to a parabolic reflector, resulting in minimal shadow effect.

11.6   THE DIELECTRIC ROD ANTENNA

It is unreasonable to assume that an antenna mounted on the outside airfoil of a high-speed aircraft would have any life expectancy at all. Yet communication to these aircraft is essential, not to mention radar surveillance systems. On large aircraft, the parabolic reflector is mounted inside a protective nose cone assembly. On sleek, high-speed aircraft, the **dielectric rod antenna** is a practical radar transmitting device and is detailed in Figure 11.24.

The dielectric rod may be any electrically porous, high-strength material having a dielectric constant between 2.5 and 32. The energy is supplied through a short interconnecting coaxial cable to a dielectric-filled waveguide section, where the energy is propagated through the dielectric. When the cross section of the dielectric rod is large compared to a wavelength, the energy propagates through the dielectric at the velocity of V_g/V_c. For smaller rod cross sections, the energy moves to the outside of the dielectric rod and propagates at the velocity of light in free air.

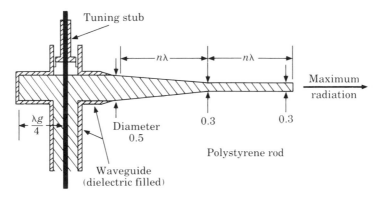

FIGURE 11.24
Dielectric rod antenna.

Tapering the rod transfers the energy from the inside to the outside of the dielectric rod and launches it in a forward direction along the length of the rod. Typical dimensions are detailed in Figure 11.24 and are used to calculate gain and the −3 dB beamwidth angle:

$$\text{Gain dB} = 10 \log 8 \left(\frac{L}{\lambda_0}\right) \tag{11.36}$$

The −3 dB beamwidth angle is

$$\theta_{-3\text{dB}} = \frac{60}{\sqrt{L/\lambda_0}} \tag{11.37}$$

As the length L increases, the gain increases and the beamwidth decreases.

On occasion, large parabolic reflectors of stationary ground radar systems are used to scan only a few degrees of arc. It is generally more practical and less expensive to switch signals to several horns, as shown in Figure 11.25, than to mechanically rotate a very large assembly. The use of waveguide switches requires great care and should not be taken lightly, but they are sometimes favored over complicated mechanical devices.

Early-warning radar systems use very narrow scanning angles in both vertical and horizontal directions, with extremely high power for long range. Figure 11.26 illustrates the concept of an entire wall of a five- or six-story building acting as the radiating antenna surface. In this arrangement, several thousand small dipole antennas are supplied with signal power, where the sum of the powers equals the total power. Four advantages are achieved with this scheme.

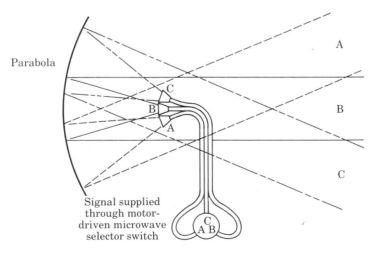

FIGURE 11.25
Multifeed horns to a single parabolic reflector has the effect of direction change.

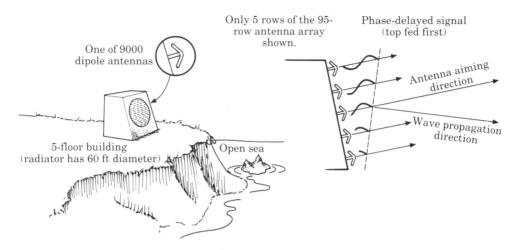

FIGURE 11.26
Typical installation of the phased array antenna system of an early-warning detection plant.

1. The use of many small antennas reduces the stringent requirements of one huge antenna.
2. Power may be selectively applied to antennas as a form of radiation pattern control.
3. Feeding a phase-delayed signal to an individual vertical or horizontal row of antennas will have the same effect as tilting the antenna up and down or sideways. Thus a scanning pattern can be arranged without moving the antenna.
4. The scanning repetition rate can be changed from the usual one scan per second to hundreds of scans per second.

The major disadvantage is the complexity of switching at microwave frequencies and controlling the microsecond phase-related signal to hundreds of antennas, using microwave hardware.

11.7   RESONANT CAVITIES

When a shorted quarter wavelength of transmission line is pivoted around the open terminals, as shown in Figure 11.27a, a chamber is formed that has all the electrical properties of a high-Q parallel-resonant circuit. Loaded circuit Qs of 20,000 have been reported. The enclosed chamber is called a **resonant cavity.** The height of the cavity represents the capacitive branch of the resonant circuit (see Figure 11.13); therefore, applying external pressure to the chamber will change the capacitance, changing the resonant frequency, as shown in Figure 11.27f.

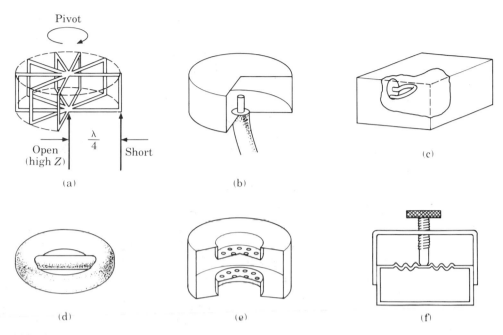

FIGURE 11.27
Resonant cavities. (a) Theory of construction. (b) Cylinder type. (c) Cubical type. (d) Ring type. (e) H type. (f) Tunable cavity.

The closed end of the waveguide, as described earlier, sets up a standing wave at that end of the guide. When the other end of the guide also has a shorting end plate, placed $n/2$ wavelengths away from the first shorting wall, the waveguide section will oscillate upon proper excitation. Cavities with this simple geometry could have many harmonically related resonant frequencies because the chamber will be some multiple of a half wavelength at a variety of frequencies. This departure from the discrete-component resonant circuit makes it unsuitable for pulse transmission. Odd-shaped cavities like those of Figure 11.27d and e are more practical for pulse circuits. However, these irregular shapes make it difficult to calculate the resonant frequency of the chamber, so laboratory scale models are often constructed until the desired performance is achieved. Unlike scale model antennas, here the chamber is larger than the finished product, so the frequency must be proportionately lower than the final operating frequency.

11.7.1 Transit Time

Every electronic product with two or more terminals will have interelectrode capacitance across the terminals and inductance in series with the terminals. Both effects limit the usefulness of the device at high frequencies. Values of

2 pF and 0.02 μH are typical interelectrode reactances that erode the high-frequency signals.

A second limitation is the time it takes for the electron to travel from one electrode to another electrode, called the **transit time.** At microwave frequencies, the time for one cycle of RF energy is often shorter than the transit time of the device. The measures required to improve one effect are the opposite of what is needed to correct the other, and are therefore counterproductive. For this reason, the principle of using the transit time, instead of fighting it, has become basic to many of today's microwave devices.

11.7.2 Electron Bunching

A highly focused beam of electrons moving past a series of alternately polarized magnets will be acted upon separately by each pole. Each positive magnetic pole will attract the negatively charged electrons as they approach the pole and will retard their movement as they move away from the pole. The negative magnetic pole will oppose the movement of electrons as they approach the pole and repel them as they move away from the pole. Neither pole is sufficiently strong to stop the flow of DC current in the beam, but the alternate action of increasing and decreasing the electron velocity will result in the current arriving at the anode in varying waves of dense and sparse electrons, as shown in Figure 11.28. The magnetic poles, as will be seen, are the poles of an electromagnet (rather than a permanent magnet), so changing the coil current in the electromagnet is equivalent to modulating the DC beam current of the device. *Intensity modulation* is another name for electron bunching.

11.7.3 The Klystron

The klystron uses a resonant cavity as its principal mode of operation. The reflex klystron is a single-cavity oscillator (see Figure 11.29) and is basically a diode, as most microwave tubes are. Among its many advantages is the ability to oscillate with *no* external frequency-selective components. Figure 11.29 identifies the cathode at a high negative potential compared to the focusing

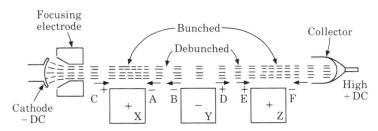

FIGURE 11.28
Effective bunching and debunching of the electrons in the DC beam current.

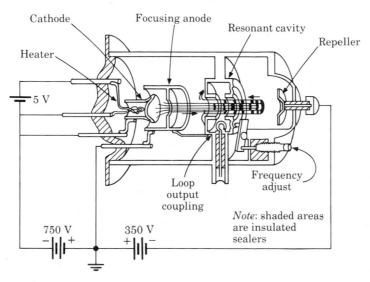

FIGURE 11.29
The reflex klystron.

anode and cavity. The high emission from the cathode is focused by the anode into a controlled beam of electrons, which passes through the cavity gap. The oscillatory energy in the cavity (which can be initiated by any small disturbance in the flow of electrons) bunches the electrons in rapid order as they move through the input chimney, past the gap, and out the exit chimney.

At the exit chimney, the electrons are at their maximum compression and encounter a high negative electrostatic potential on the repeller. The repeller voltage is set to stop the movement of the electron, turn it around, and send it back through the cavity gap. As the beam reenters the gap, the oscillatory cycle of the cavity, if correctly timed, will be at its zero value and moving toward the positive half. The returned electrons will give up their energy to sustain oscillations and then pass through the cavity wall as DC current. Thus, the repeller voltage is set so that the electron completes its travel through the gap, around, and back through the gap in three-quarters of one cycle of the RF signal frequency. The travel time can be any number of whole cycles plus three-quarters of a cycle to sustain oscillation:

$$T = \left(n + \frac{3}{4}\right)\lambda$$

It should be noted that the electrons never reach the repeller; there is *no* repeller current. If the electrons did reach the repeller, the tube could be destroyed. For this reason, most instructions direct that the repeller voltage be applied *before* the cathode voltage is turned on. Repeller voltages are somewhere between -250 to -750 V DC, the anode and cavity are at ground potential (for safety reasons), and the cathode can be -1250 to -1500 V DC.

X-rays are produced when there is a fast change in energy levels of a high-velocity electron. This is common with many of the microwave tubes, including klystrons, and is also common with every color television receiver. Precautions are taken to protect the personnel who work with such devices, and, of course, the general public, from exposure.

Reflex klystrons are capable of about 10 W output at frequencies to about 25 GHz and find common use as

1. microwave signal generators,
2. local oscillators in high-frequency receivers (see device 12 in Figure 11.1),
3. power oscillators in microwave ovens,
4. pump oscillators in parametric amplifiers (to be covered later), and
5. frequency-modulated oscillators in microwave links.

Frequency modulation is accomplished by adding an AC signal in series with the repeller supply voltage. It should be noted here that a change in repeller voltage will change the return time of the electron beam in the reflex oscillator and thereby change the frequency. The frequency may also be changed by the compression set screw in Figure 11.29.

Figure 11.30 shows the klystron as an amplifier. Although the reflex klystron is the most commonly used low-power microwave oscillator, the multi-cavity klystron finds use as a power amplifier. Continuous power ratings of 12 kW to 20 kW are the approximate upper range for power, but pulsed power klystrons have been successfully installed for use at 2 MW and 10 GHz. The two-cavity klystron in Figure 11.30 has a typical power gain of 30–35 dB, while a four-cavity klystron amplifier has a gain of close to 65 dB.

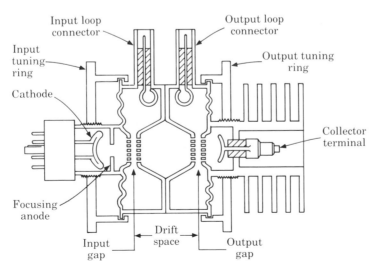

FIGURE 11.30
Klystron amplifier.

To reach these higher power levels, the power supply must deliver 17.5 kV DC at about 2 A (35 kW input power) for 15 kW approximate output power at about 40% efficiency. (Review input power, output power, and efficiency from Chapter 2, if necessary.) The theory of operation follows that of the reflex klystron through the first cavity. That is, an RF signal coupled into the first cavity will intensity-modulate the focused electron beam as it moves through the gap. The bunching process is started here, so the input cavity is also called the *buncher* cavity. The section between the two cavities is called the *drift space*. In this region, the electrons that had been accelerated will continue to accelerate, and the electrons that had been retarded will continue to slow as they move through. The applied voltage, the input frequency, and the length of the drift space are all selected so that the bunching action is maximized as the modulated beam passes through the gap of the second cavity (the last cavity for multicavity tubes), called the *catcher* cavity. The AC signal in the wave, due to bunching, excites the output cavity at resonance, and the amplified signal leaves the tube through the connection of the last cavity. The two cavities are tunable through a reasonable frequency range so that they can be matched to the center frequency. Adjusting the tuning rings will deform the bellows and change the cavity size and the frequency.

11.7.4 Magnetrons

Very high power radar, as we know it, would be impossible without the **magnetron.** Pulsed power levels of 10 MW (10,000,000 W) are practical using the magnetron, and continuous power levels of 25,000 W at 100 GHz mark the upper reaches of the magnetron's range of operation.

The name *magnetron* is a conjunction of the words *magnet* and *electron* and identifies one of the major components, a very powerful magnet. The second major component is a cylindrical copper block, drilled and channeled as shown in Figure 11.31. The center opening is called the *interaction chamber*.

FIGURE 11.31
The anode cylinder block of a multi-cavity magnetron.

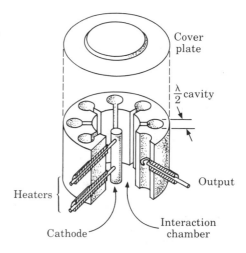

Cover plate

$\frac{\lambda}{2}$ cavity

Output

Heaters

Cathode

Interaction chamber

FIGURE 11.32
A "C" core magnet completes the magnetron assembly with the anode of Figure 11.31.

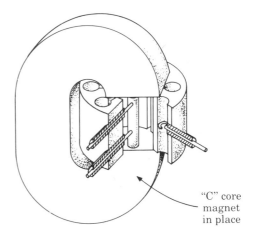

"C" core magnet in place

The holes drilled around the outer edge have a diameter equal to one-half wavelength at the operating frequency and are called *resonant chambers*. There will always be an even number of resonant chambers, usually no less than 6 and no more than 16.

With the magnetron used as a diode, the copper block becomes the anode, and a directly heated cathode is placed at the center of the interaction space. The chamber is sealed with top and bottom cover plates and the air is drawn out to form a vacuum. The output connection is a wire loop in one of the chambers that feeds to a coaxial cable fitting on the side wall of the block anode. Because the anode is exposed to the user, it is placed at ground potential and the cathode is at a high negative potential. The magnetron will only operate as an oscillator (never as an amplifier) and finds its greatest use as a power oscillator, such as device 3 in Figure 11.1.

The exploded view in Figure 11.32 has the uncovered cylinder block in position to be slid into the opening of a "C" shaped magnet. The magnetic lines of force between the poles of the magnet will pass through the interaction chamber of the anode block to react with the electron flow of DC current that passes from the cathode to the anode. The reaction of the magnetic lines of force upon the electrons in the interaction chamber is detailed in Figure 11.33.

When an electron travels through free space, it has a magnetic field around it that follows the left-hand rule just like the current through a wire. The electron is shown moving from left to right in Figure 11.33. As it passes through the magnetic field of the permanent magnet, with lines of force that press from the top toward the bottom in Figure 11.33, the magnetic field is lessened at the front of the electron path in the diagram because the directions of the two forces oppose each other. They are strengthened at the rear side because the directions of the two forces aid each other. The stronger force field toward the rear in the diagram and the weaker force at the front will move the electron in the direction of the weaker field, toward the front of the diagram. *The electron is deflected along a path that is perpendicular to both of the original fields.*

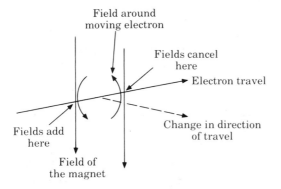

FIGURE 11.33
The effects on the path of an electron in the presence of an external magnetic field.

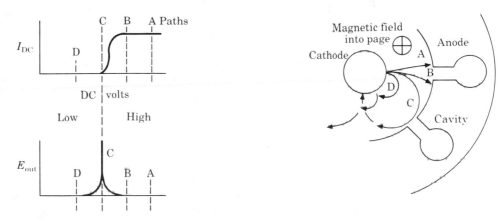

FIGURE 11.34
The change in the electron path in the presence of a magnetic field under different amounts of applied DC anode voltage. Condition C is best for oscillation.

The basic theory of operation, shown in Figure 11.34, is as follows. The magnetic lines of force move in the direction from the reader to the page. Path A is the path of the electron under a very high DC potential force. The electrons strike the anode with full force, and a heavy DC current flows. Observe each electron path while comparing I_{DC} and E_{out} from the graphs of Figure 11.34. Path B is the route taken by the electron with a lesser amount of DC voltage. The electron still strikes the anode with full force, and the value of DC anode current is unchanged. The B path length has a longer transit time, however. Path C, with a still lower DC potential, shows the electron grazing past the channel to a cavity, not striking the anode at all. The DC current drops to zero at this time, but the energy in the electron, being reactive, will be transferred into the channel, causing the cavity to oscillate at its resonant frequency. In the top graph of Figure 11.34, it is seen that the DC anode current is almost zero and the output level is maximum when the electron just misses colliding with the anode. Any lesser DC voltage, such as at D, will result in no DC current and no output signal due to the distance of the electron from the cavities of the anode. The returned electrons of paths C and D will be reenergized at the cathode and recycled. The near collision of the electron with the anode as it passes the channel to a cavity is the critical setting for maximum power output of the magnetron.

Figure 11.34 is a simplistic diagram of electron travel in the magnetron tube. The whole process is enhanced by the presence of an electric field across each channel in the anode block that changes polarity with each half cycle of the cavity's resonant frequency. The electron could be held in a suspended orbit inside the interaction chamber, making several trips around the cathode before being returned to the cathode for recycling. The many electrons in orbit at one time, combined with the alternating electric fields across each channel gap, cause an effect identical to electron bunching. The net result, of course, is sustained oscillation in the cavities, with power supplied by the in-phase energy from the orbiting electrons. The exact distance from the electron path to the cavity channel is so critical that even the consistency of a permanent magnet and a well-regulated power supply voltage may not be sufficient to prevent mode switching. That is, if the phase of the energy at one channel is not related to the phase of the energy at the adjacent channel, the magnetron could change resonant frequencies.

The frequency of the magnetron will remain most stable when any one channel differs in phase from its immediate neighboring channels by an exact multiple of $\pi/4$ radians. Best results are obtained at $4\pi/4$ radians = π radians = 180°. This is called "the π mode" of operation. To ensure this phase shift of 180°, alternate channels are strapped together as shown in Figure 11.35. This also requires that there be an even number of cavities. Use caution when wearing a watch near the powerful magnets of these tubes.

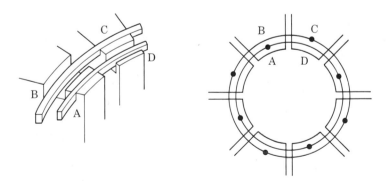

FIGURE 11.35
Strapping of alternate anode channel pole pieces to ensure 180° phase shift, π mode operation. A and C are strapping contacts; B and D are not.

11.7.5 The Traveling Wave Tube (TWT)

There is no question that the magnetron is capable of the highest power output, but there are times when the size and weight of the magnetron are not acceptable for a given application, such as airborn equipment. In these instances, the **traveling wave tube** is a viable substitute. The low noise figure of the TWT gives it advantages over other microwave tubes.

The traveling wave tube does exactly the same thing as the multicavity klystron: it sets up the electron bunching process to intensity-modulate the DC beam current. However, it bunches the beam over the entire length of the helix rather than just within the cavity. A helically wound center conductor, insulated at both ends, is placed in the drift space of the TWT so that the DC current (electron beam) passes through the longitudinal center of the helix. Because the helix is long (rarely less than 12 in or 30.5 cm), the electron beam tends to become unfocused toward the collector end of the tube. An electromagnetic sleeve is fitted around the body of the tube to maintain focus along the full length of the tube. Permanent magnets are sometimes used. The collector is typically at a high positive potential, the focusing anode is at ground potential, and the cathode is at a high negative potential.

Although the gains are lower than for other tubes, rarely exceeding 50 dB, the wide frequency band from 1 GHz to 100 GHz and the low noise characteristics make the TWT a common tube in RF amplifiers. The power limits of 3 kW in continuous wave transmitters and 10 kW pulsed power are suitable for general usage.

The velocity of the electron in the beam V_b will be less than that of the RF electric field in air because the beam velocity depends on the mass and charge of the electron and the potential of the applied voltage. The length of the tube is not a factor. The charge on the electron and its mass are constants; therefore, the controlling factor is the applied voltage:

$$V_b = 5.927 \times 10^5 \sqrt{E} \quad \text{(in m/s)} \tag{11.38}$$

We know that the velocity of the RF electric field in air (3×10^8 m/s) will be higher than the velocity of the beam electron from Equation 11.38. Furthermore, the electron with the lower velocity will take energy from the faster electron; therefore, the RF electric field velocity must be reduced to a value that is slightly below the velocity of the beam electron for there to be gain. This is accomplished by making the RF energy travel in a circular path around the electron beam current. Consequently, its velocity will be reduced by a factor equivalent to the circumference of the helix times the tangent of the pitch angle: The helix velocity V_h is

$$V_h = \frac{3 \times 10^8}{\pi D \tan \theta} \quad \text{(in m/s)} \tag{11.39}$$

where D is the helix diameter and θ is the pitch angle from Figure 11.36. With the RF electric field moving at an axial velocity that is slightly less than the velocity of the beam current, the DC voltage of the collector can be adjusted to achieve maximum gain. For this reason, the TWT is also called the *slow wave tube*.

An attenuator of some form is placed along the length of the tube body to retard the progression of a feedback voltage due to load mismatch, which would support oscillations. Although some gain is sacrificed, the attenuator is a practical necessity.

There are four generally accepted means to couple energy into or out of the traveling wave tube. The coupled helix is shown in Figure 11.36 and provides the best impedance match and, therefore, the lowest VSWR. This method is equivalent in performance to transformer coupling and is useful over a one-octave frequency spread (2 to 1 ratio) but is not the best choice for high-power applications. A second method uses a cavity at both ends of the tube for input and output coupling, with a coaxial loop as the pickup device. The efficiency of the cavity is high, which gives it good bandwidth characteristics even at high power levels. The third method also uses a cavity, but the coupling is directly to the waveguide. Again, the efficiency is good, but this method is more frequency-restrictive than the coaxial coupling because of the high Q of the guide. The last and simplest method has the helix connected through the glass body of the tube to a direct connection to the coaxial cable. The impedance match is poorest, yielding a high VSWR that will cause some overheating at high power levels.

The *backward wave oscillator* (BWO) is an extension of the traveling wave tube. The construction is modified to connect the helix to the collector, which causes a severe mismatch and generates a high VSWR. The attenuator at the center of the tube is removed, and oscillations are allowed to form as a result of the reflected wave. The only connection to the oscillator TWT is at the cathode end of the tube and may take any of the aforementioned forms. The cathode

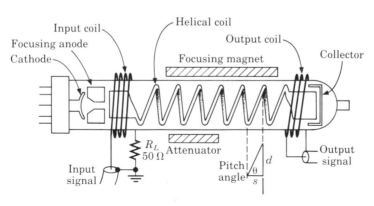

FIGURE 11.36
A traveling wave tube (TWT).

construction is in the form of a ring that sends out a hollow beam with maximum intensity close to the helix to intensify oscillations. The collector voltage may be changed to accommodate an enormous range of frequencies, from 1 GHz to 100 GHz. This oscillator is commonly used in military countermeasures to jam the radar systems of opposing forces.

Summarizing the use of microwave tubes, the multicavity klystron functions as an amplifier, and the single-cavity klystron functions as an oscillator only and is called a reflex klystron. As an amplifier, klystron gains over 65 dB are uncommon for continuous power levels of about 20 kW for frequencies to 25 GHz, but the output power may reach 2 MW at 10 GHz under pulsed power conditions. The reflex klystron is a low-noise device and finds use in 25 W systems below 25 GHz.

The magnetron can be operated as an oscillator only, but is capable of the highest combination of power and frequency. Continuous power levels of 25 kW at 100 GHz are common, and pulsed power limits can reach 10 MW, although not at the same high frequencies.

The traveling wave tube is best known for its low noise features and offers gains to 50 dB while covering the frequency range from 1 GHz to 100 GHz. Its power range may not exceed 10 kW even under pulsed conditions. The TWT finds favorable usage as an oscillator, called the backward wave oscillator because of its wide frequency range.

11.8   PARAMETRIC AMPLIFIERS

Parametric amplifiers use a reactor (only varactor diodes currently) as the active circuit element. The diode capacitance value is changed rapidly, in a manner that results in apparent amplification of the output signal. Because the circuits have low resistance-to-reactance ratios, the accompanying noise figures at microwave frequencies range from 0.2 dB to 6 dB. The basic circuit of a parametric amplifier is shown in Figure 11.37 along with simulated voltage waveshapes.

The pump amplifier, a reflex power klystron in this case, supplies a square wave voltage to the varactor diode at twice the frequency of the signal to be amplified. The pump voltage appears in parallel with the signal across coil L_s, where, at the peak of the signal voltage wave, capacitor C_v rapidly changes value due to the rise time of the pump square wave voltage. From basic electronics, it may be recalled that $V = q/C$, which means that when the charge on the capacitor (q) reaches a value, and then the value of capacitance (C) is greatly reduced, the voltage across the capacitor (V) will increase. This increase in voltage will add to the signal voltage and come from the power furnished by the pump amplifier. The voltage of the pump amplifier returns to its low state at a time when the signal voltage is passing through zero, and no change occurs at this time. The progression of signal voltage and pump voltage will increase with each successive cycle until a level is reached that is the sum

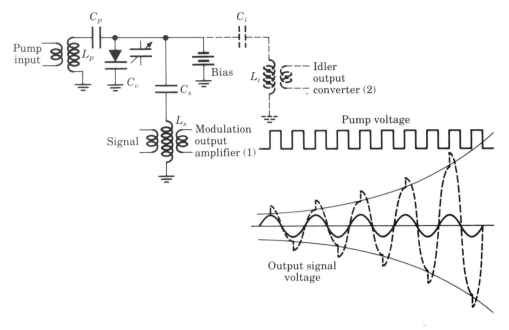

FIGURE 11.37
The circuit and waveshapes of a parametric amplifier.

of the peak-to-peak signal voltage and the maximum pump voltage as a function of its power. The rate of change using a varactor diode as a switch will allow operation well into the microwave frequencies.

Allowing the pump frequency f_p to go higher than the signal frequency f_s, with no special relationship, will cause a third frequency to be generated, called the *idler* signal frequency f_i, which exhibits substantial gain. The idler signal is akin to frequency conversion with gain provided the pump signal is a square wave or a sine wave. Therefore, in terms of straightforward amplification, where only f_p and f_s are present (no idler circuit), the power gain is a ratio of the frequencies:

$$\text{Power gain} = \frac{f_p + f_s}{f_s} \tag{11.40}$$

The noise factor (F) becomes

$$F = 1 + \left(\frac{2T_d}{279}\right)\left(\frac{1}{Q} + \frac{1}{Q^2}\right) \tag{11.41}$$

and the bandwidth is

$$\text{BW} = \left(\frac{2V_p}{V_s}\right)\left(\frac{f_p + f_s}{f_s}\right) \tag{11.42}$$

When this circuit is used as a lower-sideband up-converter, where $f_i = f_p + f_s$, the gain, noise factor, and bandwidth are comparable to the straight amplifier. However, when it is used as a lower-sideband down-converter, where $f_i = f_p - f_s$, the noise factor and bandwidth go unchanged, but the gain results in a loss.

Typical performance characteristics are:

Gain	10 dB to 40 dB
Frequency range	1 GHz to 60 GHz (pump to 100 GHz)
Noise figure	0.2 dB to 6 dB
Bandwidth	about 10% center frequency

It has been proposed that multiple reactive elements in parallel would have a traveling wave effect and widen the bandwidth to about 50% of center frequency. Therefore, single-diode parametric amplifiers are considered narrowband devices.

11.9 CIRCULATORS, MAGIC TEES, AND HYBRID RINGS

The **circulator,** the **magic tee,** and the **hybrid ring** pass a signal to some output ports while preventing signal to other ports within the same device. They control the signal by phase addition or cancellation due to path length or position. The circulator controls selection by the position of an external magnetic field.

Examine Figure 11.38 to see that the magic tee can have an input at port 1 with an output at ports 2 and 3 but not at terminal 4. This happens because the signal entering port 3 is changing phase, and the phase difference appears across the two walls of output 4 to cancel. By the same reasoning, an input to port 2 will output to ports 3 and 1 but not 4. An input to port 3 will exit at ports 1 and 2 but not at 4. An input to port 4 will exit at ports 1 and 2 but not 3. The circulator in Figure 11.38 uses these same input-output patterns for the same port numbers. The difference in operation is that the circulator uses magnetic

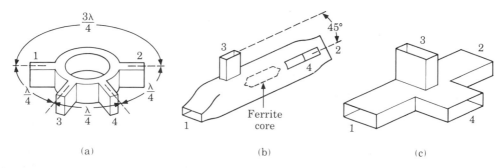

(a)   (b)   (c)

FIGURE 11.38
Wave-steering devices. (a) Hybrid ring. (b) Circulator. (c) Magic tee.

field rotation to cause phase cancellation. The hybrid ring uses path length to achieve cancellation. An input signal to port 1 will exit at ports 2 and 3 but not port 4. An input to port 2 will exit at ports 1 and 4 but not at 3. Input to port 3 will exit at ports 1 and 4 but not at 2. Finally, an input to port 4 will find an outlet at ports 2 and 3 but not at 1. When the path length differs by one-half wavelength, the signals cancel. In all other cases, the path length is either even or a full wavelength out of phase and will add. These devices are used as directional couplers: one antenna is connected to the transmitter output and the receiver input.

11.10   GAS DISCHARGE TUBES

Switches are a necessary part of many electronic circuits. They have been characterized in the past by either manual switches (hand-operated), electromechanical switches (solenoid-operated), or, more recently, electronic switches such as transistors. Switches can be self-sensing and self-activating, as is seen in the **gas discharge tube.**

The gas discharge tube in Figure 11.39 is made of a short section of waveguide that is enclosed at both ends by clear glass windows. The glass is almost totally transparent to the microwave signal but seals the chamber, which contains a gas mixture such as argon, ammonia, or hydrogen. At the center of the chamber is a set of electrodes that form a spark gap to be used at the time of gas ionization. The closely spaced electrodes have a tendency to increase the electric field at this point. Gas discharge tubes are used in the system of Figure 11.1 as the TR tube (7) and the ATR tube (9).

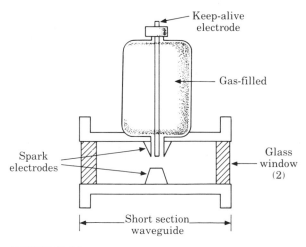

FIGURE 11.39
Gas discharge tube, used as an ATR (with keep-alive voltage) and TR (without keep-alive voltage).

When a low-level signal, such as the signal from an antenna, reaches the gas tube, it passes through the windows (and tube) with a loss of less than 0.5 dB. This is the case for the TR tube, where the signal continues on to the mixer circuit. The same low-level signal goes past the receiver input terminals to the ATR tube. It does not ignite the gas in this tube. The ATR tube is an open circuit and reflects the signal back to the receiver input to be in phase with the next cycle of the input signal, adding to the receiver input signal strength.

When the transmitter sends out a high-power pulse, the gas inside the ATR tube ionizes and looks like a short circuit across the waveguide. For the ATR tube, this is a series connection that permits the transmitted signal to pass through the guide to the antenna. For the TR tube, the short-circuited gas tube prevents the high-power transmitted signal from entering the receiver and damaging the mixer circuit. A low-level DC voltage, called a *keep-alive* voltage, is applied to an electrode of the TR tube and allows a trickle current to flow in the tube (too small to ionize the gas). This holds the tube in *near* conduction, so that when a high-power pulse is transmitted, the TR tube will ionize rapidly. The signal will be attenuated to a level of -80 dB, one-ten millionth of the transmitted signal power.

11.11   MICROWAVE DIODES

The role of the diode in microwave systems has, until recent years, been one of detecting. That is, it has been used to rectify the carrier signal to recover the modulation and as a power sensor. Advances in technology have elevated the diode to a powerful tool that can function as a mixer, a modulator, or a microwave switch.

A diode makes an excellent switch because its resistance may be changed rapidly from very high to very low by a fast-acting bias reversal. Second, diodes now have the ability to handle high power levels at these fast switching rates, and it is easy to place several diodes in parallel when the required power level exceeds that of the single diode. The advantage to the diode switch is its smaller size and its predicted life expectancy. The disadvantage is its slower switching time under high power levels (as compared to tubes). In addition, the natural nonlinearity of the diode curve makes diodes attractive candidates for microwave mixers and detectors. Particular types of diodes include the *pin* diode and the *avalanche* diode. Because these two diodes have far-reaching application in fiber optics, they will be covered in depth in Chapter 14. Other common diodes are the *Gunn* diode, the *tunnel* diode, and the *IMPATT* diode, which have a negative resistance characteristic. Resistance is linear with the applied voltage; we expect an increase in current for an increase in voltage. However, the current conduction of these diodes is such that over a portion of the curve (see Figure 11.40) the resistance *decreases* as the applied voltage is increased. This provides an easy feedback path that is conducive to oscillation at or near 100 GHz. The same negative resistance feature lends itself well to frequency

FIGURE 11.40
Negative resistance curve of a tunnel
diode.

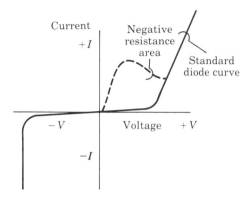

multiplication. The efficiency of such circuits is low (8% to 20%), while the multiplication factors range as high as 12 times the input frequency per device with
output frequencies up to 10 or 11 GHz. The power levels of negative resistance
devices is in the milliwatt range, except for a few isolated cases.

11.12   MICROWAVE FIELD EFFECT TRANSISTORS

Interelectrode capacitances and lead inductances plague the transistor products just as they did the tube amplifiers. The junction capacitance in transistors
is related to the width of the depletion layer, which in turn is controlled by the
bias voltage. Therefore, bias voltage affects the input capacitance of the circuit.
The transit time for electron flow is about the same for the transistor and tube,
but the multilayer solid state devices cause a phase delay that distorts the
output waveshape. Equations used to evaluate transistor amplifiers use the S
(scatter) parameters, which in themselves are not complicated, but because
they become overly involved for microwave frequencies, they are not included
in this coverage. The maximum obtainable signal power of 1 or 2 W (per transistor) at these frequencies further confine the solid state microwave products
to the low-power sections of the system. Paralleling transistors will produce an
all–solid state amplifier of up to 5 W that will operate at 8 GHz.

   In addition to the advantages of higher efficiency from low power drain
and small size, the transistor's life expectancy of 10^9 hours (compared to 10^3
hours for tubes) makes it a valued device in underwater systems and satellites.
Some functions that are not practical for tube circuits are becoming commonplace for transistors. The leading systems seem to favor dual-gate Mesa Field
Effect transistors (MESFET) using gallium arsenide (GaAs) material in an
epitaxial package for discrete components. Monolithic microwave integrated
circuits (MMICs) are used for some sophisticated systems.

   The circuit for an 8.75 GHz two-stage amplifier shown in Figure 11.41 has
a gain of 10 dB and a bandwidth of 200 MHz. Two things become immediately
apparent as this circuit is viewed. First, discrete components are almost com-

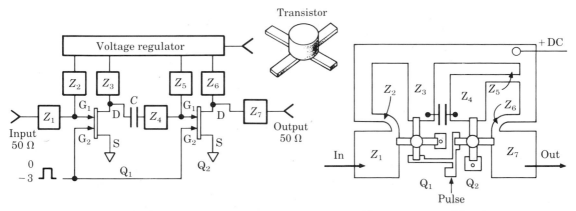

FIGURE 11.41
An 8.75 GHz microwave amplifier circuit and associated circuit board. The size of the copper foil pad is a measure of the impedance.

pletely eliminated and have been replaced by microstrip components. Second, lead lengths are held to an absolute minimum to reduce the circuit inductive and capacitive reactances. Not apparent from the diagram is the minimal presence of solder; excessive soldering would change the values of the microstrip components. Grounding techniques are critical at these frequencies, but this is learned from experience and cannot be taught in a textbook. The physical circuit shown in Figure 11.41 is a microstrip copper foil arrangement based on transmission line theory. Developments in stripline circuits came about from the need for very small reactance values at high frequencies.

Such a circuit is first designed using discrete component technology, and some critical measurements are then taken to verify the theoretical conditions. The transmission line counterparts are then calculated, constructed, and tested. It is not unusual to experience several rounds of modification from the original prototype to the finished product.

11.13   STRIPLINE AND MICROSTRIP CIRCUITS

Circuit functions that commonly lend themselves to micro circuitry include impedance-matching transformers, resonant circuits, and filters. One consideration is the DC current need of each component. Should the component be able to conduct DC current, or should it block DC current?

Each of the two formats shown in Figure 11.42 has advantages and disadvantages. The **stripline,** in Figure 11.42b, has the advantage of behaving like a shielded coaxial cable. Circuit signals do not radiate to cause interference or oscillations, nor do external signals get in to disrupt circuit

FIGURE 11.42
Microwave transmission line circuit parameters. (a) Microstrip. (b) Stripline.

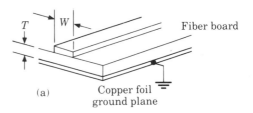

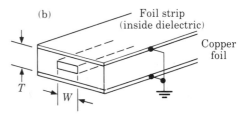

performance. However, the initial cost is much higher and once constructed, the circuit cannot be modified without replacing the entire board. The **microstrip** package, in Figure 11.42a, has the advantage that the foil width may be trimmed during the test phase to alter the circuit component values. The disadvantage is that some of the signal energy is conducted through the copper foil and some is conducted in a field immediately outside of the foil. This external field conduction permits noise interference and signal radiation to parts of the circuit where it is not wanted.

Before an example is presented, an examination of the circuit relationships is in order. The characteristic impedance is determined by

$$Z_0 = \frac{377T}{W\sqrt{\epsilon}} \tag{11.43}$$

where T = board thickness (inches)
W = foil width (inches)
ϵ = dielectric constant of the board material
(4.8 for glass epoxy, 2.5 for teflon epoxy)

Experimenting with Equation 11.43 will show that the characteristic impedance Z_0 could range from 1 Ω to about 5 kΩ. The impedance will increase in value when the board thickness (T) goes up, when the board dielectric (ϵ) goes down, or when the foil width (W) decreases. Impedances higher than 5 kΩ are obtainable, but care should be exercised when entering this zone. Give some forethought to frequency and board thickness, recalling from transmission line theory that when the conductors (foil and ground plane) are separated by 0.2

wavelengths or more, they act as antennas and radiate excessively. The wavelength inside of the foil λ_g is

$$\lambda_g = \frac{\lambda_0}{\sqrt{\epsilon}} \qquad (11.44)$$

where λ_0 is the free-air wavelength. The size of the input impedance pad Z_1 in Figures 11.41 and 11.43 is determined after several choices are made and some basic data are confirmed.

EXAMPLE 11.6

The carrier frequency of 8.75 GHz is known. This amplifier is one of several within a single package, so the input and output impedances are chosen at 50 Ω for good impedance matching. This decision was influenced by the general use of generators with 50 Ω output impedance and 50 Ω coaxial cable used for testing.

   The choice of circuit boards at these frequencies is limited to the fiberglass family. Here a teflon-glass board is chosen, with a dielectric constant of 2.5 and a thickness of 1/8 in.

   The FET input impedance, from earlier measurements, was found to be $R_L - jX = 82 - j30$ Ω. The line impedance is then found:

$$Z_0 = \sqrt{R_s R_L} \sqrt{1 - \left(\frac{x^2}{R_L(R_s - R_L)}\right)} \qquad (11.45)$$

$$= \sqrt{(50)(82)} \sqrt{1 - \left(\frac{(30)^2}{(82)(50 - 82)}\right)} = 74.17 \ \Omega$$

Rearranging Equation 11.43 to solve for W leaves us with

$$W = \frac{377T}{Z_0\sqrt{\epsilon}} = \frac{(377)(0.125)}{(74.17)(1.5811)} = 0.402 \text{ in}$$

The length of the pad (L) in Figure 11.43 with complex numbered impedances calls for the use of a Smith chart. To plot the location of points on the chart, both the input impedance and the load impedance need to be normalized:

$$Z_{\text{in}}(\text{normalized}) = \frac{R_s}{Z_0} = \frac{50}{74.17} = 0.674 \ \Omega$$

$$Z_{\text{out}}(\text{normalized}) = \frac{Z_L}{Z_0} = \frac{82 - j30}{74.17} = 1.106 - j0.405 \ \Omega$$

Figure 11.43 shows these two points plotted on the Smith chart and reveals that the distance between them is 0.16 wavelengths. (Note that the normalized values of Z_{in} and Z_{out} fall on the same circle.) SWR = 1.5.

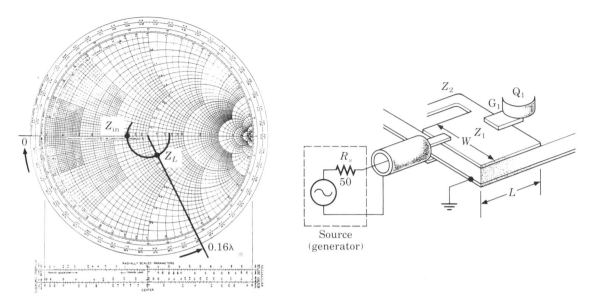

FIGURE 11.43

The input pad for the amplifier of Figure 11.41 and the Smith chart used to determine length L.

From Equation 11.43, the wavelength in the foil strip is found to equal 0.85351 in. The pad length of 0.16 times the strip wavelength (0.13 in) is too small for practical use, so the length was made one-half wavelength longer. (Z repeats every $\lambda/2$.) The line length now equals 0.56×0.85351 in $= 0.56$ in (see Figure 11.43, where $W = 0.4$ in and $L = 0.56$ in).

QUESTIONS

1. What single condition must be met for a signal to be classified as a microwave?
2. What practical range of frequencies is considered microwave?
3. Calculate one wavelength (in centimeters) at 1428.6 MHz.
4. What is the distance to the radar target (in miles) when the return pulse arrives home 42.93 μs after launch time?
5. A radar receiving antenna is _____ the transmitting antenna. (a) placed 25 wavelengths from, (b) elevated higher than, (c) mounted below, (d) also, (e) faced in the opposite direction as.
6. Radar systems transmit (a) a continuous wave, (b) bursts of energy, (c) an AM signal, (d) an FM signal.
7. The TR and ATR tubes are (a) oscillators, (b) amplifiers, (c) rectifiers, (d) mixers, (e) gas-filled tubes.

8. The bandwidth of the radar receiver IF amplifiers is (a) wideband, (b) narrowband, (c) not a factor.

9. Waveguide is similar to coaxial cable with the center wire removed. (T) (F)

10. Waveguide has a characteristic that resembles a (a) high-pass filter, (b) low-pass filter.

11. State the relationship between the a inside dimension of the guide and wavelength at the cutoff frequency.

12. What is the cutoff frequency of the guide when $a = 1.773$ in?

13. What is the cutoff frequency for a waveguide that measures 0.738 in by 1.476 in on the inside?

14. What is the cutoff frequency when the inside of the waveguide measures 5 cm by 2 cm?

15. What would be a good frequency at which to operate the K band waveguide for low losses?

16. What is the attenuation, in dB/ft, of copper K band waveguide at 18.74 GHz?

17. How many JAN waveguide bands could be operated at 16 GHz?

18. Which of these bands is best suited for 16 GHz?

19. "Probe" type coupling is similar to (a) capacitive coupling, (b) conductive coupling, (c) resistive coupling, (d) inductive coupling, (e) none of the above.

20. "Loop" type coupling is similar to (a) capacitive coupling, (b) conductive coupling, (c) resistive coupling, (d) inductive coupling, (e) none of the above.

21. A probe should not be placed in which wall? (a) a wall, (b) b wall, (c) back closed wall.

22. A loop should not be placed in which wall? (a) a wall, (b) b wall, (c) back closed wall.

23. Explain the meaning of the subscripts that follow the waveguide mode designation (such as $TE_{1,0}$).

24. What is the propagation constant for air-filled JAN WR159 waveguide at 4.95 GHz?

25. What is the propagation constant for foam-filled JAN WR159 waveguide at 4.95 GHz?

26. What is the guide wavelength (λ_g) for the two conditions in Questions 24 and 25?

27. Assuming $TE_{1,0}$ mode of propagation, what is the wave impedance for the two waves in Questions 24, 25, and 26?

28. Air-filled TE mode waveguide will always have a wave impedance equal to or greater than 377 Ω. (T) (F)

29. The input impedance of waveguide is a function of (a) frequency, (b) probe or loop position, (c) depth of probe or loop insertion, (d) mode of propagation, (e) all of the above.

30. Will air-filled WR112 waveguide be able to carry 0.25 MW of pulsed power without fear of failure?

31. JAN WR137 is listed as operational over the frequency range of 5.85 GHz to 8.2 GHz. At which of these frequencies will the guide handle the greatest power?

32. How does altitude affect waveguide power ratings?

33. What is done to compensate for the effects of altitude on the waveguide power rating?

34. How is the minimum bend radius defined for E- and H-plane bends?

35. Why is a minimum bend radius specified?

36. What is the minimum length (in wavelengths) for an E-plane twist to effect a 180° phase change?

37. With a signal into port 1 of Figure 11.10 of an E-plane tee, which output port (2 or 3) will be in phase with the input signal of port 1?

38. With a signal into port 1 of an H-plane tee in Figure 11.10, what will be the polarity of the signals out of ports 2 and 3 compared to the input signal?

39. Directional couplers are essentially power taps. (T) (F)

40. The power at the auxiliary output port of a directional coupler is a matter of choice selection (rated in dB). (T) (F)

41. Why would a one-hole directional coupler *not* work as well as a two-hole coupler?

42. What purpose is served by placing a metallic obstruction across one-third of the E plane, inside of the guide?

43. What purpose is served by placing a metallic obstruction across one-third of the H plane, inside of the guide?

44. Obstructions across *both* the E and H planes so as to reduce the cross-sectional area inside the guide to one-third of its original area will serve what purpose?

45. What purpose is served by placing a ferrite strip in the center of the E plane along the length of the guide?

46. What is the minimum mechanical flare angle of a pyramidal horn antenna (in degrees)?

47. What is the maximum mechanical flare angle of a pyramidal horn antenna (in degrees)?

48. Define the terms *minimum length, optimum length,* and *maximum length* for pyramidal horn antennas.

49. What is the minimum length (in centimeters) of a horn with a mouth 13×18 cm and a throat size of 2.8×4.8 cm *(a > b)*?

50. What is the optimum length of the same horn antenna (in centimeters)?

51. If the optimum length of a horn antenna were 24 cm, then the second half of Equation 11.21 would cancel. What would be the horn gain at 4.165 GHz for the throat and mouth dimensions of Questions 49 and 50?

52. Find the gain of a 12.5 in long horn antenna at 2.98 GHz with a mouth size of 9×5.25 in connected to S band waveguide.

53. What is the −3 dB beamwidth for the horn antenna in the preceding question in both the *A* plane and *B* plane?

54. What is the focal length of a 36 in diameter parabolic reflector that measures 6 in deep?

55. What is the dB gain of the reflector in the preceding question at 4.1 GHz?

56. What is the −3 dB beamwidth of the reflector in the preceding question?

57. Name some advantages of the Cassegrain feed reflector compared to a simple parabolic reflector.

58. Name some disadvantages of the Cassegrain feed reflector as compared to the simple parabolic reflector.

59. What is meant by the *shadow effect?*

60. How can the shadow effect be minimized?

61. What is the major advantage of the Cutler feed antenna?

62. What is the dB gain and beamwidth angle of a dielectric rod antenna with a 20:1 length-to-wavelength ratio?

63. Comparing the three basic microwave tubes, the klystron, the magnetron, and the traveling wave tube, establish which will operate as oscillators and which will operate as amplifiers.

64. Which of the three microwave tubes rely on electron bunching for operation?

65. How does the repeller current figure into the operation of the reflex klystron?

66. Which voltage of a reflex klystron is applied first, the cathode voltage, the anode voltage, or the repeller voltage?

CHAPTER TWELVE

WAVE PROPAGATION

12.1   INTRODUCTION

In each of the preceding chapters the idea of transmitting a signal through the air has been mentioned but not explained. Great pains have been taken to save this topic until now, where it can be covered as a comprehensive unit. The only assumption to be made here is that the transmitting and receiving antennas are polarized in the same plane.

It was stated earlier that the transmitted wave contains both a magnetic field and an electric field. The antenna is polarized in the plane of the electric field, and the wave is propagated in a direction that is perpendicular to both the electric and magnetic fields. A three-dimensional image must be visualized to fully understand this principle. Imagine a flat sheet of paper on which the electric field is drawn from left to right and the magnetic field is drawn from top to bottom, then imagine the electric field radiating into and out of the page. This is what you must see in your mind before you can master the theory of propagation.

Now that satellites are a major part of the communication system, we need to rethink the titles we have assigned to each category of wave propagation. The overall groups can be divided into waves that leave our atmosphere and those that do not. If they leave our atmosphere, they will be called **space waves.** If they stay within our atmosphere, they will take one of two forms, called **sky waves** or **ground waves.** We will study these waves in order of simplicity.

12.2   GROUND WAVES

Ground waves are the form of electric field radiation that follows the curvature of the earth. The wave is vertically polarized and will transmit most efficiently at a frequency less than 2 MHz. (This is not a limit, however. Citizens band radio, for example, is propagated by vertically polarized ground waves at 27 MHz.) The ground wave takes its name from the fact that the antenna current, in part, is conducted through the ground by a principle called *eddy currents*. These currents decrease as the distance and frequency are increased. The amount of moisture in the soil is the secret. Salt water is a better conductor than most other materials, and, therefore, ground waves travel best over salt water. Marshy lands are next best, followed by open plains, then mountains, and, finally, dry desert lands. A greater amount of moisture in the soil is the key to better conductivity.

In Figure 12.1, the wave is seen leaving the antenna and traveling in all directions with a characteristic that gives the impression of a runner who cannot move his legs as fast as his body is moving and leans too far forward, eventually falling on his face. This is the actual outcome of the ground wave. As it moves further away from the transmitter, it leans forward to eventually short-circuit against the ground.

In general, the ground wave antenna is shorter than a half wavelength, and emphasis is placed on the *wavelength* distance between the transmitter and receiver. The controlling factors are the distance in meters (D), the power to be transmitted (P), the frequency (f), and the effective cross-sectional area of the antenna, which for a dipole shorter than one-half wavelength is considered to be 0.119 square wavelengths. The relationship for determining receiver power (P_r) takes the form

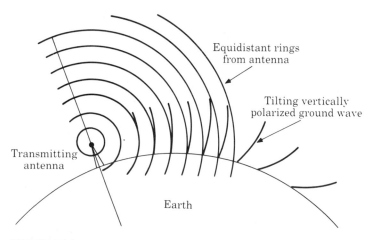

FIGURE 12.1
The ground wave follows the curvature of the earth and leans forward as it moves further away from the antenna.

$$P_r = P_t \left(\frac{A_{\text{eff}} V_c}{D f} \right)^2 \quad \text{(in watts)} \tag{12.1}$$

where $A_{\text{eff}} = 0.119$
V_c = velocity of light, in meters
D = distance, in meters
f = frequency, in hertz
P_t = transmitter power, in watts

EXAMPLE 12.1 ═══

A 600 kHz transmitter, using a short vertical antenna, sends a 1000 W signal to a receiver 20 km (12.46 mi) away. The power at the receiving antenna will be 8.85 mW. The same system, using an antenna and transmitter adjusted for 1.785 MHz, would deliver only 1 mW to the receiving antenna. The reason for the lower received power is that there are more 1.785 MHz wavelengths in 20 km than there are 600 kHz wavelengths.

12.3   SKY WAVES

The sky wave may arrive at its destination by any or all of three major routes and may be propagated from a horizontally or vertically polarized antenna. However, at frequencies higher than 2 MHz horizontal polarization is preferred; for this reason sky waves are highly dependent on line-of-sight communications. That is, the transmitting antenna must be able to "see" the receiving antenna.

In Figure 12.2, the strongest signal follows the line-of-sight path, labeled the *direct* path. However, the signal may also arrive at the receiving antenna

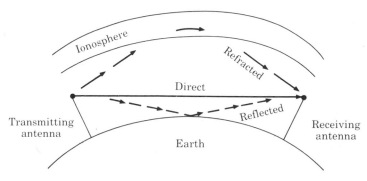

FIGURE 12.2
The three routes of the sky wave.

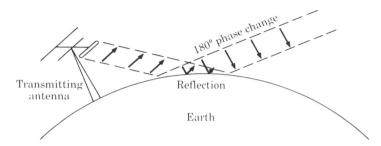

FIGURE 12.3
The resulting phase change of the sky wave due to reflection.

by a *reflected* wave, which strikes the ground and bounces back up to the receiving antenna. This reflected wave is seriously attenuated as a result of striking the ground, but this is a bonus condition, because the wave also changes phase by 180°, as shown in Figure 12.3. The reflected wave cancels some of the direct wave energy.

The third route that the sky wave may travel is called the *refracted* wave. Here a skyward-bound wave is bent by the thinning atmosphere and rerouted to the receiver. We will also refer to the refracted wave as the *skip* wave. Sky waves are those RF waves that *stay within* our atmosphere.

The **refracted** wave differs from the **reflected** wave in that the refraction process simply bends the wave due to the difference in the density of the air. The refracted wave does *not* change phase and therefore will add to the wave that arrives by the direct path. There will be a small phase delay due to the lengths of the paths taken (the refracted wave path is longer), but generally this is relatively small. More in-depth discussion will be provided later in this chapter on how the shell of air mass surrounding the earth affects these waves.

12.3.1 The Radio Horizon

The distance to the optical horizon (in miles) is specified in physics manuals as

$$D = \sqrt{\frac{3h}{2}} \quad \text{(in miles)} \tag{12.2}$$

where h is the antenna height, in feet. Thus, a person 6 ft tall standing on the shoreline would see the horizon at a distance of 3 mi off shore; hence the 3 mi limit that borders every country's shoreline. If you can see it, it belongs to you.

The **radio horizon** differs slightly in that radio waves have a slight bending and fill-in effect behind tall, obstructing objects. That is, a receiving antenna immediately behind a tall hill may receive no signal from a station, but if

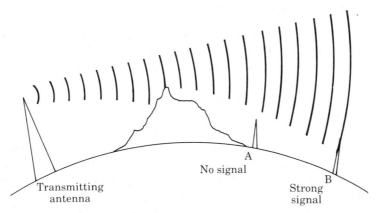

FIGURE 12.4
The shadow effect and how the signal can be improved.

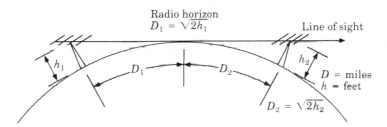

FIGURE 12.5
Radio horizon compared to antenna height.

the same antenna is moved further from the station, the signal strength increases, as seen in Figure 12.4. This void condition is called the *shadow effect*. Figure 12.5 shows the relationship of antenna height to horizon distance, where the distance from the transmitting antenna to the *radio* horizon is found as

$$D = \sqrt{2h} \quad \text{(in miles)} \tag{12.3}$$

where height h is again in feet.

EXAMPLE 12.2

A horizontally polarized antenna is placed on top of a 612.5 ft tower. The distance to the radio horizon is 35 mi, as derived from Equation 12.3. A receiving antenna 53 mi from the transmitter would need to be raised to an elevation of 162 ft to "see" the direct path from the transmitting antenna. Use the same equation for the second antenna, but solve for height instead of distance, using the distance of 18 mi: $(53 - 35)^2/2 = 162$ ft.

12.3.2 The Ionospheric Layers

Our planet has a diameter of approximately 7720 mi and is surrounded by a gaseous substance called the atmosphere that extends 250 mi from the surface. Only 2 mi of this atmosphere contains oxygen in sufficient quantity to support life as we know it. From this level on out, the oxygen content lessens and the composition of ionized molecules increases. The ultraviolet radiation from the sun and the cosmic ray radiation from space cause this ionized layer to be thinner at distances farther from the earth's surface. Figure 12.6 shows the atmospheric layers and their approximate distances from earth. There is no sharp division between the layers, as the diagram implies, but rather a smooth measured changes in the ion layers, which have been cataloged and named.

The oxygen layer, the troposphere, the stratosphere, and the ozone layer are generally fixed in altitude and composition. The ozone layer is a protective shell that shields us from the ultraviolet and cosmic rays of outer space. The stratosphere is also called the *isothermal layer* because of its constant temperature. The oxygen layer, the troposphere, and the stratosphere are not shaded in Figure 12.6.

The ionosphere, on the other hand, changes in both density and thickness from daytime to nighttime and also undergoes seasonal changes. The changes

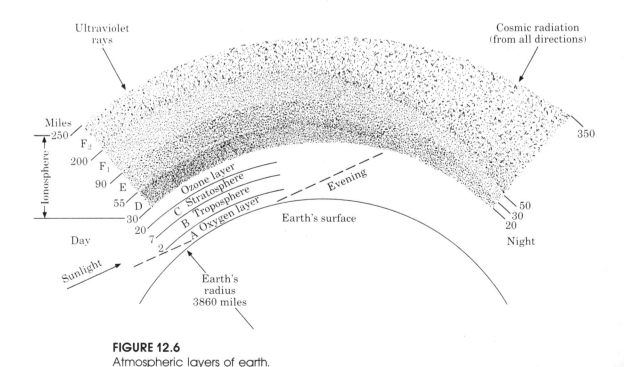

FIGURE 12.6
Atmospheric layers of earth.

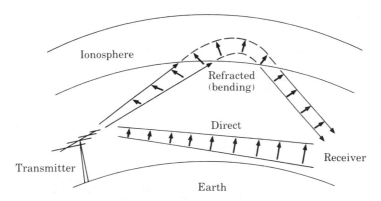

FIGURE 12.7
Refracted sky waves.

occur from the sun's heating the molecules in the layers, which releases positive and negative ions. Bombardment of the molecules by cosmic and ultraviolet radiation intensify ionization. The effects are shown in Figure 12.6 (drawn out of proportion). The D layer disappears at night, the E layer almost completely disappears at night, and the F_1 and F_2 layers combine into a unified F layer. The density of the D layer prevents low-frequency daytime skip wave propagation, but at night these low-frequency waves may propagate great distances. Under these conditions, an offending station can change its antenna radiation pattern to eliminate the interference, reduce its transmission power, or go off of the air.

As the sky waves penetrate the layers of the ionosphere, they are curved around to be directed back to earth. This comes about when the waves enter the thinner layers of atmosphere, where their velocity is changed slightly. Entering at an angle of a less-than-critical value causes one part of the wave to accelerate relative to the total wave, causing a slow curvature of the wave path (see Figure 12.7). The change in direction is the result of the gradual change in density, called the *index of refraction* (see Chapter 14).

12.3.3 Power Density

Sky waves as a whole (direct, reflected, and refracted) react to the power density principles developed by Maxwell. The wave behaves as a point source of power in free space at a distance D from a possible reception point. A point source of energy radiating in every direction would cover the total surface area of a sphere with radius R, as portrayed in Figure 12.8. One square meter of surface area is designated as the receiver area; therefore, the power density P_d at the receiver is the power transmitted divided by the area of a sphere, which equals $4\pi R^2$. In our example, the radius is the distance D, so the power density is

FIGURE 12.8
Power density of a point source
radiator.

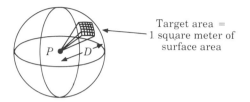

Target area =
1 square meter of
surface area

$$P_d = \frac{P_t}{4\pi D^2} \quad \text{(in watts/square meter)} \tag{12.4}$$

where P_t = transmitter power, in watts
D = distance, in meters

Note that the distance factor is a squared function in the denominator of the
equation. Thus, the universal inverse square law is applicable and establishes
that the power will diminish as a function of the inverse of the square of the
distance between the transmitter and the receiver. Doubling the distance will
reduce the power density at the receiver by a factor of $4(0.5^2 = 0.25 P_d)$.

Equation 12.4 uses the point source as a radiator, but this is only a theoretical antenna, so the final value must be multiplied by the directivity gain of
the transmitting antenna as compared to the point source antenna. Realisti-
cally, Equation 12.4 changes to

$$P_d = \frac{P_t G_t}{4\pi D^2} \tag{12.5}$$

A 5000 W station with an antenna gain of 3.45 would realize a power density at
a point 25 mi distant (40.225 km) of 0.848 μW/m^2.

12.3.4 Electric Field Strength

The voltage at the receiver location can be expressed by the standard power
equation

$$E = \sqrt{PZ} \quad \text{(in volts)}$$

where Z is the impedance of free space, which is 377 Ω = 120 π, and the power
in question is the power density at the receiver. The voltage equation can then
be converted to a form that includes the antenna gain. Thus, the equation for
field strength E is

$$\begin{aligned}
E &= \sqrt{\frac{P_t}{4\pi D^2} (G_t)(120\pi)} \\
&= \sqrt{\frac{(120\pi)(P_t)(G_t)}{4\pi D^2}} \\
&= \frac{\sqrt{30 P_t G_t}}{D} \quad \text{(in volts per meter)}
\end{aligned} \tag{12.6}$$

The antenna gain is the gain of the practical antenna used for this transmission as compared to an isotropic source radiator. The field strength, then, describes the voltage in the changing field at the receiving location that will induce a voltage across a wire parallel to the electric field lines of force. The magnitude of the voltage induced into the wire will be affected by the length of the wire. Therefore, a standard length of 1 meter is specified to compare all such measurements in terms of volts per meter.

The power at the receiver compared to the power at the transmitter can be expressed in decibels by looking at the total system gain and the total system losses. The transmitter antenna gain (G_t) in decibels is the calculated or measured gain minus the losses in the transmission line that connect the antenna to the transmitter. The receiver antenna gain (G_r) in decibels takes into account the same factors as the transmitter antenna gain.

The system losses for sky waves are termed the *free wave path losses,* and as for ground waves, the distance between the transmitter and receiver is specified in wavelengths. For distance and wavelengths in feet, the free wave path loss becomes

$$\text{Free wave path loss} = 20 \log \cos \text{BW} \ \frac{4\pi D}{\lambda} \quad \text{(in decibels)}$$

where D and λ are in feet and BW is the beamwidth, in degrees. Therefore, the ratio of receiver power to transmitter power is

$$\frac{P_r}{P_t} = G_t + G_r - 20 \log \cos \text{BW} \ \frac{(4\pi D)}{\lambda} \qquad \text{(12.8)}$$

EXAMPLE 12.3

A transmitter antenna gain is 18.8 dB minus 4.5 dB waveguide loss, so $G_t = 18.8 - 4.5 = 14.3$ dB and BW = 28°. The receiver antenna gain is 24 dB minus 5 dB waveguide loss, so $G_r = 19$ dB. At 3.2 GHz and 60 mi distance, the receiver-to-transmitter power ratio is

$$\frac{P_r}{P_t} = 14.3 \text{ dB} + 19 \text{ dB} - 20 \log (0.883) \left[\frac{(12.566371) \ (316{,}800)}{0.3075} \right]$$

$$= 33.3 \text{ dB} - 141.1622 \text{ dB} = -107.8662 \text{ dB}$$

The transmitter power will be 107.8622 dB greater than the signal power arriving at the receiver. When 0.5 μW are required at the receiver input terminals, the transmitter must provide

$$P_t = \frac{(0.5 \times 10^{-6})}{\text{antilog} \left[\dfrac{-107.8622}{10} \right]} = 30{,}563 \text{ W}$$

minus the 33.3 dB antenna gain for a total of 14.3 W from the transmitter.

12.3.5 The Skip Wave

The refracted wave is also called the skip wave because it strikes the iono-sphere, refracts back to earth, and reflects off of the surface of the earth back into the atmosphere, where it may again refract back to earth. Under ideal conditions, this skip process could continue for many cycles and cover great distances. The majority of long-distance communications used this phenome-non as the highway for global communications before the introduction of sat-ellites.

The transmission of skip waves is only marginally predictable, depending on the constantly changing ionospheric conditions, the distance to the target location, the number of required skips, and the sending frequency. The receiver power is generally not calculated. Approximate power requirements relative to distance, based on years of experience in skip wave transmission, are a matter of record. Predicted changes in the ionosphere over a three-month period are available from the federal government and aid in forecasting the success of skip wave transmission to any distant location. The skip wave is also called *ducted* transmission because the pocket of atmosphere between the earth and the ion-osphere acts like a waveguide or duct.

When very long distances (many skips) are involved, the signal strength is subject to **fading.** That is, the reception will change dramatically from a weak to a strong signal and will also experience varying degrees of distortion. Figure 12.9 shows only two skips of wave A and one for wave B, but imagine a transmission from Phoenix, Arizona to London, England, which is one-third the distance around the world. Depending on frequency and conditions, there could be 15 to 18 skips over this distance. There could be several different paths, each changing according to its ionospheric encounters and the day/night conditions along the way.

Diversity Reception   A well-designed receiver automatic volume control (AVC) circuit will have very good effects on a severely fading signal, but may not completely eliminate the problem. It has been found that using two anten-

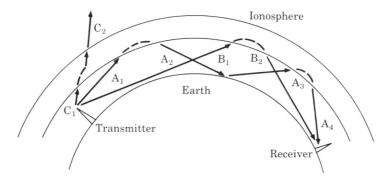

FIGURE 12.9
Sky wave skip patterns.

nas a few hundred feet apart will have a remarkable effect toward controlling the problem of fading. Thirty wavelengths is the distance generally recommended. The two antennas are connected to separate RF amplifiers, mixers, and IF amplifiers but may share a local oscillator, second detector, AVC system, and audio amplifiers. The antenna with the strongest signal will develop the strongest AVC voltage, which will bias off the other IF amplifier. Signals that would normally experience severe fading problems with a single antenna may have *no* noticeable fading with this two-antenna procedure, known as *space diversity*. The common local oscillator ensures that both receivers are tuned to the same transmission.

Further improvement can be achieved by modulating the message signal onto two separate carrier frequencies for transmission. The two separate receiving antennas are attached to two separate RF amplifiers, mixers, and local oscillators but share an IF amplifier, detector, AVC, and audio amplifiers. This procedure is called *frequency diversity*. Space diversity and frequency diversity circuits are shown in Figure 12.10. The bandwidth of the signal is also

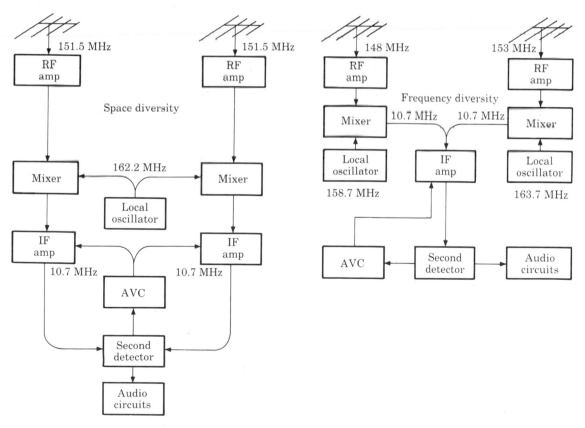

FIGURE 12.10
Space diversity and frequency diversity receiver setup.

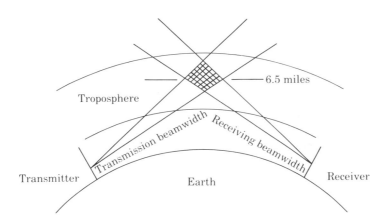

FIGURE 12.11
Tropospheric scatter reception.

a contributing factor in that a wideband signal such as FM is more susceptible to fading than a SSB signal with a very narrow bandwidth.

Tropospheric Scatter   **Tropospheric scattering** is a system of transmission that falls in the same category as magnetism, gravity, and light energy. We can explain what happens in its presence. We can predict and control its behavior to make it work for us, but no one really knows what it is.

Every form of modulated signal is present in a crowded urban area, and there is little interference or cross-modulation. Yet when the receiving antenna and transmitting antenna beams both occupy the same space at an altitude of about 6.5 mi above the earth in the frequency band between 350 MHz and 10 GHz, they interact in a manner similar to ducting. As shown in Figure 12.11, a receiving antenna aimed at a point that would intercept a transmitted signal in mid-air will pick up a wave having the same effects as a refracted signal. This is true only when the conditions listed above are in effect. Although there is still much to be learned about reception through tropospheric scattering, it has been used successfully in a number of instances.

12.4   SPACE WAVES

By definition in this text, space waves are those waves that leave our atmosphere. The refracted wave was found to strike the ionosphere and return to earth, but only if a critical angle above the horizon was not exceeded. When this critical angle *is* exceeded, the RF wave will leave our atmosphere and venture into deep space.

The difference between sky waves and space waves hinges on the frequency of the carrier, the angle of attack above the horizon, and the density of the ionosphere at the time of transmission. The minimum usable frequency (MUF) that will leave the atmosphere is stated as

$$\text{MUF} = \frac{f_c}{\sin \theta} \tag{12.9}$$

where θ is the angle of elevation above the horizon and f_c is the critical frequency, which is a function of the electron density of the ionospheric layer. Typical critical frequencies for the various layers are

Layer	Critical Frequency
D	300 kHz
E	4 MHz
F_1	5 MHz
F_2	6 MHz (night)
	8 MHz (day)

Equation 12.9 is given here for illustration only, not for calculating the minimum frequency. It establishes that for very low angles of elevation, the transmission frequency must be quite high, and as the angle above the horizon increases, the minimum usable frequency decreases.

The power density determination for space wave transmission uses all of the same data required for the sky wave, but seeks a solution in terms of power at the receiver input terminal, in decibels. The equation is developed as follows:

A. From Equation 12.5

$$P_d = \frac{P_t G_t}{4\pi D^2}$$

B. The receiver power is

$$P_r = P_d A_{\text{eff}}$$

C. The effective area is

$$A_{\text{eff}} = \frac{\lambda^2 G_t}{4\pi}$$

D. Thus, receiver power becomes

$$P_r = \frac{P_t G_t}{4\pi D^2} \times \frac{\lambda^2 G_r}{4\pi}$$

E. Simplifying statement D and comparing the received power to the transmitted power results in a ratio of

$$\frac{P_r}{P_t} = G_t G_r\left(\frac{\lambda^2}{4^2\pi^2 D^2}\right) \quad \text{(not in dB)}$$

F. For $\lambda = V_c/f$,

$$\frac{P_r}{P_t} = G_t G_r\left(\frac{V_c}{4\pi f D}\right)^2$$

G. Considering that 4, π, and V_c are constants, and that frequency can be stated in megahertz and distance in kilometers, all of the constants can be combined:

$$\text{Constants} = \left(\frac{3 \times 10^8}{4\pi \times 10^6 \times 10^3}\right)^2 = 0.57 \times 10^{-3}$$

H. Replacing the value of the constants into statement F,

$$\frac{P_r}{P_t} = G_t G_r\left(\frac{0.57 \times 10^{-3}}{D^2 f^2}\right)$$

I. A quick review of logarithms indicates:

$$\log AB = \log A + \log B$$

$$\log \frac{A}{B} = \log A - \log B$$

$$10 \log A^2 = 2 \times 10 \log A$$

so that the statement H can be expressed in decibels as

$$\frac{P_r}{P_t} = G_t + G_r + 10 \log 57 \times 10^{-3} - 10 \log D^2 - 10 \log f^2$$

J. Equation 12.10 can now be written in its final form:

$$\frac{P_r}{P_t} = G_t + G_r - (32.44 + 20 \log D + 20 \log f) \qquad \text{(12.10)}$$

where power and gain are in decibels, D is in kilometers, and f is in megahertz.

EXAMPLE 12.4

When it is known that the gain of the transmitting antenna is 22 dB, the gain of the receiving antenna is 18 dB, the distance is 22,400 mi (14,543.2 km), and the frequency is 5.904 GHz, then the receiver-to-transmitter power ratio is expressed as

$$\frac{P_r}{P_t} = 22 \text{ dB} + 18 \text{ dB} - (32.44 \text{ dB} + 20 \log 14{,}543.2 + 20 \log 5904)$$

$$= 40 \text{ dB} - 191.116 \text{ dB}$$

$$= -151.116 \text{ dB}$$

When the transmitting power is 25 W, the power at the receiver input terminals is

$$P_r = 25\left(\text{antilog} \frac{-151.116}{10}\right)$$

$$= 1.9335 \times 10^{-14} \text{ W}$$

$$= 19.335 \text{ fW}$$

The receiver power equation (12.10) will be used again in Chapter 16 on satellites, so it is advantageous to understand its operation now. Note the similarity between the equations for sky waves and for space waves.

QUESTIONS

1. An antenna is polarized in the plane of the (a) magnetic field, (b) electric field, (c) field perpendicular to the electric and magnetic fields, (d) path parallel to both fields.

2. The transmitting and receiving antennas should be polarized (a) vertically, (b) horizontally, (c) either vertically or horizontally, but both in the same plane, (d) in opposite planes from each other.

3. Before satellites, major global communications were carried on by (a) ground waves, (b) sky waves, (c) space waves.

4. Ground waves propagate best when polarized (a) vertically, (b) horizontally, (c) either, (d) neither.

5. Ground waves propagate best in what frequency range? (a) below 5 MHz, (b) between 5 MHz and 25 MHz, (c) between 25 MHz and 250 MHz, (d) above 250 MHz.

6. Name the three groupings of wave propagation.

7. Which of the wave groupings is most likely to follow the curvature of the earth?

8. How much power would be received by ground wave propagation for a receiver located 42 mi from a 5 kW transmitter at 1.6 MHz?

9. Under what conditions are ground waves effective?

10. How many transmission paths are typical for sky wave propagation?

11. How do the reflected and the refracted waves differ?

12. Which sky wave path has the greatest signal strength?

13. Why is the optical horizon different from the radio horizon?

14. What is the distance to the radio horizon for an antenna elevated 242 ft above ground?

15. How far above ground would the receiving antenna need to be elevated when the transmitting antenna is 61 mi away and on top of a 1250 ft building?
16. Two identical ships at sea have their antennas mounted on the mast, 338 ft above the water line. How far apart can they separate and still maintain line-of-sight communications? (a) 13 mi, (b) 26 mi, (c) 39 mi, (d) 52 mi, (e) 65 mi.
17. Which atmospheric layers refract radio waves? (a) troposphere, (b) stratosphere, (c) ozone, (d) ionosphere, (e) all of the above.
18. Which atmospheric layers change density? (a) troposphere, (b) stratosphere, (c) ozone, (d) ionosphere, (e) all of the above.
19. What is meant by the inverse square law? Give an example.
20. In what units is power density measured?
21. In what units is electric field strength measured?
22. Describe the receiving antenna used to measure electric field strength.
23. What is the relationship between the antenna and the polarization of the transmitted wave?
24. How are earthbound long-range radio communications transmitted?
25. Do sky waves travel farther during the daytime or at night?
26. Which element of the sky wave is responsible for the strongest signal to the receiver?
27. Are sky waves horizontal or vertical?
28. Are ground waves horizontal or vertical?
29. What factors affect the critical frequency for skip waves?
30. What factors affect the critical angle for space waves?
31. Name two types of diversity reception.
32. Describe how the two types of diversity reception are similar.
33. Describe how the two types of diversity reception are different.
34. What is meant by tropospheric scatter?
35. What is the power density 30 mi from a 2.5 kW transmitter with a horizontally polarized antenna having a gain of 7.4?
36. What is the electric field strength at the receiving antenna from the preceding question?
37. How far is it to the radio horizon for a line-of-sight antenna at 578 ft above the average terrain?
38. What is the receiver-to-transmitter power ratio for a satellite with an antenna gain of 32 dB at 23,400 mi altitude on an uplink at 12.5 GHz into a receiving antenna with 49 dB gain?
39. What is the ground wave power at a receiver 250 mi from a 1 kW transmitter that drives a Marconi antenna at 920 kHz?
40. What is the receiver input power for a horizontally polarized microwave link at 3.2 GHz when the antennas are 47 mi apart? The transmitting dish (with 20° beamwidth) has a gain of 22 dB, less 3.5 dB waveguide loss, and the receiving antenna is identical but has 1.9 dB waveguide loss.
41. What is the signal attenuation from a 25 W satellite at 22,400 mi altitude, transmitting at 3.2 GHz from an antenna with 27 dB gain into a receiving antenna with 60 dB gain and 2.8 dB waveguide loss?

CHAPTER THIRTEEN

DIGITAL MODULATION

13.1 INTRODUCTION

Digital modulation came into being in the early 1800s with the invention of the telegraph by Samuel Finley Breese Morse (1791–1872) and its use of a semaphore code of dots and dashes. Telegraphy was limited to hard-wire, point-to-point transmission until the invention of the radio in 1909 by Guglielmo Marconi (1874–1937), when its mode of transmission was converted to an electromagnetic wave by keying a carrier on and off (A1 emission) using the same semaphore code.

13.1.1 Codes

The Morse code, although adopted as the international code, lacks several important features. The symbol length varies from one dot, for the letter "E," to five dots and two dashes, for the dollar sign ($). The rate and rhythm of the code, although standardized, varies with the skill of the sender. A dash is three times the length of a dot, and the space between digits is the same length as a dot. The dollar sign symbol takes 17 times more transmission space than the letter "E," and this is an inconsistency that the automatic equipment could not tolerate. The human receiver was found to make fewer errors than the automated equipment and was often able to identify the sender by the rhythmic beat of the signal.

Because of these shortcomings, other codes began to emerge that were better suited to automated equipment. The Baudot code, invented by a French postal worker, had 52 characters each containing five units of measure. The American Standard Code for Information Interchange (*ASCII*) has 78 characters with seven units of measure per character and offers more versatility than the Baudot code.

465

The ARQ code (automatic request for repetition) is another variation of the Baudot code and represents all characters using three 1s and four 0s. The seven units, called **bits,** can be arranged to make up 128 different characters, but because the ARQ code contains only 32 characters, there is a good deal of redundancy. *Redundancy* describes the fraction of the character that can be eliminated without removing meaning from the code. Only 5 bits are needed to make up 32 characters; therefore, 2 bits out of 7 (28%) are redundant.

The terms *digital communications* and *data communications* were used interchangeably until it was reasoned that *digital* represents the formatting of the signal and *data* represents the material to be formatted in a digital code. Many messages are generated in code and thus constitute data to be stored or transported in the digital transmission, but some messages start out as a time-varying (analog) message and need to be converted to code (data) before being transmitted in the digital format. This chapter will deal only with **digital modulation** techniques and the transmission systems used to transport messages to their destinations. The codes themselves are left to another discipline and will only be discussed as a support tool.

13.2   FOUNDATIONS

A *bit* is the smallest unit of information that can be identified. The word is a conjunction of *binary digit*. The **binary** codes (i.e., made up of two parts) specify a 1 and a 0 as the only possible states. In a circuit, a 1 is a condition of *high* voltage, typically only +5 V, and a 0 is a condition of *low* voltage, usually ground or 0 V. In telephony, a high is called a **mark** and a low is called a **space.** The terms *on* and *off* are reserved for describing the circuit conditions and should be avoided when referring to the signal condition.

The number of bits of data that can be transmitted in one second is called the **bit rate** and is stated in bits per second (BPS). Figure 13.1a shows a digital signal in which each bit is alternately a 1 and a 0. A sine wave of equal timing is shown in Figure 13.1b, and these two signals are compared to the signal in Figure 13.1c, which contains all 1s. It can be seen that a bit of data occupies the same time interval as *one-half cycle* of the sine wave. Thus, the bit rate is twice the analog frequency, that is, 10K BPS = 5K Hz.

Note also that for the first 3 bits of the signal of Figure 13.1c, there is a decay to zero after each bit, while for the last 3 bits there is not. A bit message can be conditioned in either form, but needs to correspond properly to the RZ (return to zero) system or the NRZ (nonreturn to zero) system.

The term **baud** is used to express the rate at which changes occur and is determined by a very strict equation. However, it boils down to be the number of character symbols per second. For instance, a code that uses only 1 bit to represent a symbol can have two conditions, a 1 and a 0. In this case, the bit rate is the same as the baud rate. A code that uses 2 digits (2 bits) to represent

FIGURE 13.1
Bit rate compared to frequency with return-to-zero (RZ) and nonreturn-to-zero (NRZ) output.

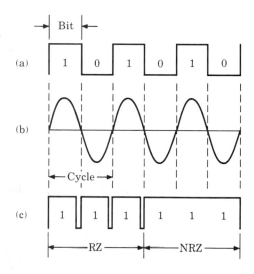

each character can have four conditions: 00, 01, 10, and 11. The bit rate here is twice the baud rate. The equation 400 BPS = 200 baud, for instance, means that each character contains 2 bits. A 3-bit code, 000, 001, 010, 011, etc., would have a bit rate three times higher than the baud rate because 3 bits make up one character. Bits can be compared directly to frequency, but baud cannot unless more information is known.

One of the major advantages of digital signaling is its high level of immunity to noise. The receiver looks for a level change through the use of a zero crossing comparator, for instance, and the noise in the signal would need to be one-half the signal level for the detector to recognize it.

13.3   ANALOG TO DIGITAL CONVERSION

A portion of the systems to be analyzed here originated as far back as the 1930s and were known then as *pulse modulation* or *pulse code modulation* (PCM). These systems coexisted with the early digital systems, but the two system types were rarely used together, and they should not be confused with modern-day A/D converters. They were totally analog systems and were the industry's effort to conserve frequency space or improve signal-to-noise ratios. However, they found limited application because of their bulk and low efficiencies at that time. Over the years, better circuit design has yielded higher efficiencies and better technology has reduced the bulk, making these systems more attractive to industry. Following these improvements, the two families of digital data and analog data converted to digital began to be used in the same systems. The systems to be discussed in the following sections are

1. PAM (pulse amplitude modulation)
2. PWM (pulse width modulation)
3. PPM (pulse position modulation)
4. FSK (frequency shift keying)

13.3.1 Pulse Amplitude Modulation (PAM)

One of the simplest ways to digitize an analog signal is seen when the sine wave message signal is mixed nonlinearly with a low-duty-cycle square wave. The circuit of a class D audio amplifier, shown in Figure 13.2, is a typical **pulse amplitude modulator** with a 20% duty cycle.

The amplifier is held in a cutoff state by the DC base bias circuit between $+V_{cc}$ and $-V_{ee}$ even when a strong sine wave signal appears at the emitter. The only time the amplifier is allowed to switch to the "on" condition is when the base is driven in a positive direction by the large amplitude square wave pulse at the input. However, the amplifier is never driven hard enough to reach collector current saturation. The "on" condition caused by the base pulse combined with the instantaneous level of the emitter signal will produce a pulsed voltage at the collector having a pulse amplitude that is proportional to the level of the message signal on the emitter at that instant.

In this class D power amplifier, the peak power output can be five times greater than the power rating of the transistor, because the amplifier is on only 20% of the time and is cooling for 80% of the time. The *average* power dissipated in the transistor over the cycle is equal to the duty cycle multiplied by the peak power. In this case, the power dissipated (P_d) is 0.2 times the peak power and will be equal to the power rating of the transistor.

The pulsed square wave signal must have two qualities with respect to the audio message signal:

1. The frequency (or repetition rate) of the square wave is generally selected to be five times higher than the highest frequency in the audio signal. In the case where the highest audio frequency is 15 kHz, the square wave will be $5 \times 15\,kHz = 75\,kHz$. There is no magic in the selection of the square wave signal frequency. It should be reasonably higher than the highest audio frequency, and when it is an odd harmonic of this frequency, the chances of successive pulses occurring at the audio signal zero crossing are greatly reduced. The fidelity is held intact, and no information is likely to be lost. Digital systems have identified the *Nyquist rate* (two times the highest analog frequency) as the absolute minimum sampling rate.
2. The peak-to-peak amplitude of the square wave at the collector should be 10% greater than the anticipated peak-to-peak audio signal at the collector. This will allow modulation from 5% to 95% of maximum signal and avoid compression, which causes overmodulation.

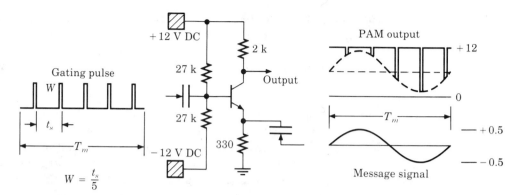

FIGURE 13.2
Pulse amplitude modulation (PAM).

When pulse amplitude modulation is used in audio amplifiers, the PAM signal is applied directly to the speakers with no loss of intelligibility and with no required demodulation of the signal. The speaker will filter out the pulsed frequencies.

Digital circuits perform more efficiently with a flat-top square wave signal than with a rounded top. Figure 13.3 shows a *sample-and-hold* circuit that will modify a PAM signal taken directly from an analog input and convert it to a flat-top PAM signal. However, a square wave pulse will have a wider bandwidth than the original analog signal because of the high harmonic content of the square wave. The input to the driver amplifier is the analog signal directly

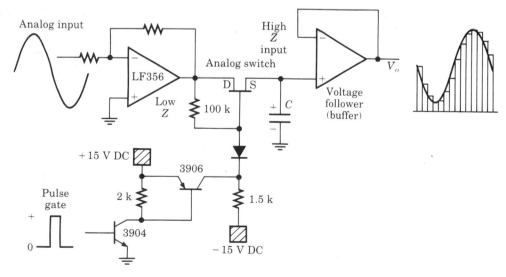

FIGURE 13.3
Sample-and-hold circuit.

from the source. However, less quantizing error (degradation of the higher analog frequencies) will be present when the narrow-pulse-width PAM signal is used.

The electronic switch is turned on by the gating pulse, allowing the holding capacitor C to charge instantaneously (almost) to the level of the analog signal present at that time. The switch is turned off at the termination of the gating pulse, and capacitor C will retain its charge until the next gating pulse. The capacitor's charging time constant is made extremely fast to accommodate an almost instantaneous level change each time the capacitor is exposed to the analog signal. (It should be noted here that the capacitor charging time denotes either an increase or decrease of voltage charge on the capacitor.) The output of the sample-and-hold circuit is a stairstep approximation of the analog signal, which is defined by the curve connecting the start of each sampled voltage, as shown in Figure 13.3.

A more accurate PAM signal is needed for encoded data than for analog signals. A second switching transistor may be connected in the circuit following the hold capacitor such that the input to the voltage follower is disconnected or returned to zero at the termination of each gating pulse, as shown in Figure 13.3. The PAM signal is often used as the input to a pulse-code-modulated A/D converter that will change analog signals to binary code, as shown in Section 13.8.

Recovery of the PAM Signal   When the PAM signal is transported over telephone lines, a simple low-pass filter at the receiving end will bypass the pulse rate frequency and fill in the areas between pulses sufficiently to restore the fidelity of the message signal. When the PAM signal is used to directly modulate a higher carrier frequency for radio transmission, the AM detector at the receiver will act as the low-pass filter to remove the pulse frequency. Again, no fidelity is lost. The only precaution to be observed in the recovery process is to ensure that the low-pass filter has a flat frequency response over the entire baseband frequency range and provides sufficient attenuation at the pulse rate frequency. (You may want to review the sections of Chapter 3 on filter attenuation and required circuit Q when faced with this requirement in practice.)

13.3.2 Time-Division Multiplexing (TDM)

The original purpose for PAM was to conserve power dissipation in low-frequency amplifiers. However, it has also proven to be the perfect vehicle for frequency conservation.

The use of a low-duty-cycle pulse at the introduction of PAM left the amplifier (and system) idle for 80% of the time. It is desirable for the amplifier to work at the 20% duty rate, but also desirable to allow the system to perform other functions for the remaining 80% of the time. This application is called **time sharing** or **time-division multiplexing** (TDM). Another system used for transmitting large blocks of data, called time-division multiple access (TDMA), is not the same as TDM. TDMA is covered later in this chapter.

The multiplexing system of Figure 13.4 permits four PAM signals to be transported over a single pair of wires at the same time without interference and without increasing the baseband frequency response beyond that of any of the individual messages. Each of the four message signals in Figure 13.4 contains all frequencies between 300 Hz and 3 kHz; as such, the figure can be taken to represent four telephone voice conversations. The clock pulse generator output is a 20% duty cycle square wave at 8 kHz and is used to gate on Q_1, a circuit similar to the class D amplifier in Figure 13.2. The same gating pulse is delayed by 90° (one-quarter cycle) and is used to gate on amplifier Q_2. It is then delayed another 90° (180° total) to turn on amplifier Q_3, and amplifier Q_4 is gated on after another quarter-cycle delay. Note that each amplifier is on for 20% of the time, for 80% total time. There is a 5% time space between the turn-off time of one amplifier and the turn-on time of the next amplifier. This quiet period is necessary to prevent two amplifiers being on at the same time, even for an instant.

The four PAM signals are added linearly in the summing amplifier of Figure 13.4, which has an output waveshape corresponding to the time-shared

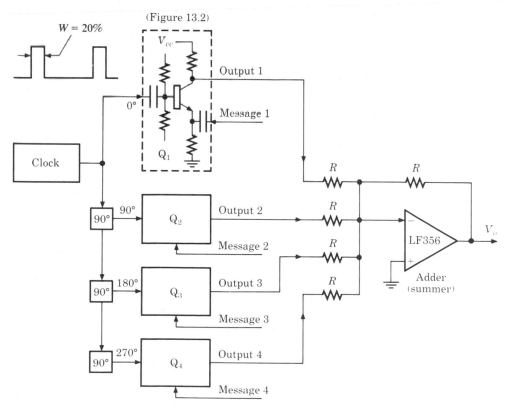

FIGURE 13.4
Time-division multiplexing of four PAM signals.

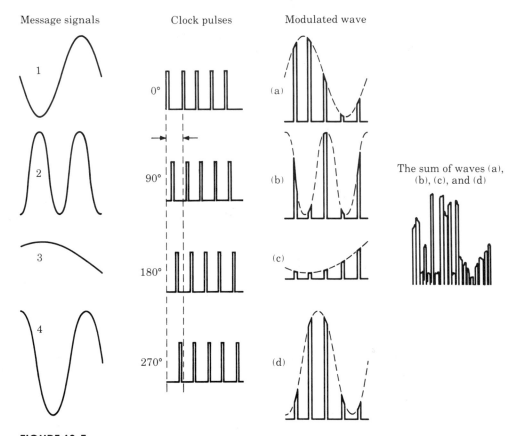

FIGURE 13.5
The waveshapes of Figure 13.4 showing the message inputs, the clock pulses, the modulated pulse's output, and the sum of the four waves.

sum of the voltages in Figure 13.5. The output baseband frequency range contains any and all frequencies between 300 Hz and 3 kHz plus the gating signal frequency of 8 kHz. The 8 kHz signal is one octave above the baseband frequency and may be filtered out if desired.

An eight-message TDM system would need to work with a 10% duty cycle and a 2.5% quiet time between pulses. There is a limit to the fraction of the duty cycle that should be used, defined in terms of high-frequency fidelity both for analog messages and binary coded signals.

Recovery of a TDM Transmission   When a TDM signal is transported over a wire network, a framing or synchronizing pulse is added to the transmission. In early systems, when the commutation of message signal multiplexing was accomplished by rotary equipment, one voice channel was sacrificed in order to carry the sync pulse. Without this pulse, there was no assurance that a channel 1 transmission would be delivered to the channel 1 receiver.

In more recent systems, all inputs are coded in some form, and the synchronizing signal is embedded in the data. The sync pulses are given some unique feature, such as an oversized amplitude or exaggerated pulse width, which the sync oscillator at the receiver could recognize as a timing pulse. A zero digit in the code transmission is held to a nonzero level to prevent missing some pulses and to preserve a constant pulse rate. The nonzero condition is a requirement used for synchronization at the receiver when TDM is used. Systems that still use a true PAM-TDM arrangement use a receiver like that of Figure 13.6.

The receiver has a synchronized gating oscillator with an output that is delayed in the same fashion as the one used at the modulator. In our example, a four-channel PAM trunk had each channel's gating pulse delayed by a factor of 90°. Channel 1 would have a circuit like that of Figure 13.2 and would be turned on at the time the channel 1 message signal entered the system. Channels 2, 3, and 4 would be turned on at their respective times to allow passage of their message signals. Once the four signals have been separated into individual circuits, a low-pass filter (integrator) will reconstruct the pulsed information back into an analog signal.

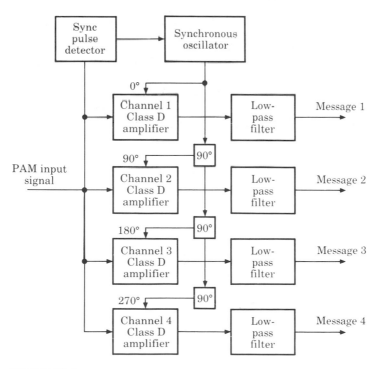

FIGURE 13.6
Separation and recovery of the four message signals from the TDM and PAM transmission.

13.3.3 Pulse Width Modulation (PWM)

Another simple form of modulation is **pulse width modulation.** It converts an amplitude-varying message signal into a square wave with constant amplitude and frequency, but which changes duty cycle to correspond to the strength of the message signal. The signal that changes amplitude will be referred to as the audio to simplify the description. The frequency of the audio includes all frequencies between 50 Hz and 15 kHz, such as the quality audio used in FM broadcasting.

The audio signal is mixed with a triangular wave in a linear adder, as shown in Figure 13.7. The triangular wave must have the same two qualities with respect to the audio as stated in Section 13.3.1 for the PAM signal, in terms of frequency and amplitude. The triangular wave may originate from any function generator, such as a 555 timer. The timer output wave is passed through an integrator (low-pass filter), which, if properly selected, will change the square wave into a triangular wave.

In the linear mixer (adder), neither of the two input signals will change the basic shape or frequency of the other. The result of two signals added algebraically is that the higher-frequency signal (triangular wave in this case) will have a new reference line in the shape and size of the lower-frequency signal (audio). See wave C in Figure 13.8.

The combined triangular and audio waves are applied to a zero crossing comparator, which will have an output that varies between $+V_{cc}$ and $-V_{ee}$, depending on the exact instantaneous level of the input signal. The output of the comparator will have a constant amplitude and a fixed frequency (75 kHz), but one that has a changing duty cycle (the "on" time is compared to the time of one cycle). The PWM signal may now be transported to its destination by telephone pairs, or it can be used to amplitude- or frequency-modulate another, higher-frequency carrier signal.

The idea of double modulation is not new, as you may recall from the discussion of stereo transmission. Moreover, double modulation is one of the main principles behind color television.

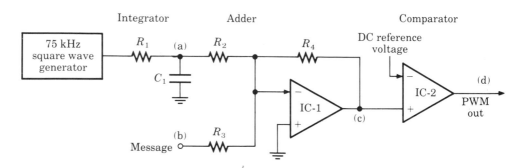

FIGURE 13.7
Pulse width modulator, consisting of a triangular wave generator, an adder, and a comparator.

The Harris Corporation, a manufacturer of quality broadcast equipment, uses PWM as a premodulation format of the audio signal before applying the message signal to the modulated power amplifier of a standard AM transmitter. The transmitter output power amplifier is thus operated as a class D amplifier with greater efficiencies and less distortion.

PWM Recovery   When the PWM signal arrives at its destination by way of telephone lines, the recovery circuit used to decode the original signal is a simple integrator (low-pass filter). The charge on the filter capacitor will be the average of the voltage in any cycle of the PWM wave. When the pulse width is wide, say, 95% of the time of one cycle, the voltage charge on the capacitor will be approximately 95% of the peak carrier voltage. When the pulse width is narrow, say, 5% of the time of one cycle, the voltage charge on the capacitor will

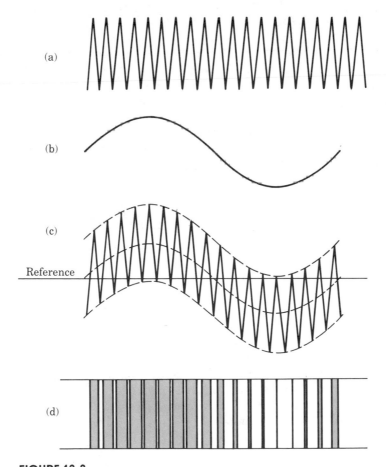

FIGURE 13.8
Waveforms for the pulse width modulator of Figure 13.7 (a) Triangular wave at 75 kHz. (b) Message signal. (c) Sum of (a) and (b). (d) PWM output signal.

be approximately 5% of the peak carrier voltage. The recovered output voltage will change in amplitude corresponding to the width of the pulses in the PWM wave.

When the PWM wave arrives at its destination by way of an antenna-radiated signal, such as described for the Harris system, it is a double-sideband full-carrier A3 signal that must first be detected by a diode rectifier and then passed through an integrator (low-pass filter). Looking back to Chapter 4, it is seen that the detector in every AM radio receiver is just as we have described. This means that the detector in a normal AM radio receiver will decode the PWM (double-modulated) AM transmitted signal without further circuit modifications.

13.3.4 Pulse Position Modulation (PPM)

The analog signal is changed to a PWM signal first, and then the PWM signal is converted to a **pulse position modulated** signal. This double modulation at the transmitter may seem redundant, but the improvement in noise immunity is well worth the added effort. The reason the original PWM signal is not used becomes apparent when we compare the presence of errors at the receiver under high-noise conditions for the three modulation forms (PAM, PWM, and PPM). The PPM transmission is far superior over the other two systems in rejecting noise that will introduce errors. This is its major advantage. The major disadvantages are a more complex circuit and higher costs.

The PWM signal from Figure 13.8 will be used as the input signal to the modulator in Figure 13.9. The goal here is to develop a square wave pulse of short but fixed duration and then change the pulse position relative to the timing of the cycle. How the signal is treated between the PPM input terminals and the input of the square wave generator will depend on which circuit you use for the generator. The 555 timer of Figure 13.9a requires a negative-going trigger pulse. The Schmitt trigger of Figure 13.9b requires a positive-going pulse for a trigger input. Therefore, a phase inverter (common-emitter amplifier) is used in one circuit and not in the other. However, both circuits will require a differentiator (high-pass filter). There will be no change in the end result with regard to which function is performed first (differentiating or inverting).

The 555 timer circuit is used in this explanation because it is a simpler circuit (no inverter). The PWM input signal of Figure 13.9c is first differentiated, changing the square wave pulses to spiked pulses, shown in Figure 13.9d. Note that the negative spikes of Figure 13.9d are now time-related to the trailing edge of the PWM input signal. These pulses will fire the 555 timer, which will generate a 10 µs output pulse (see Figure 13.9e, which shows the resulting PPM signal).

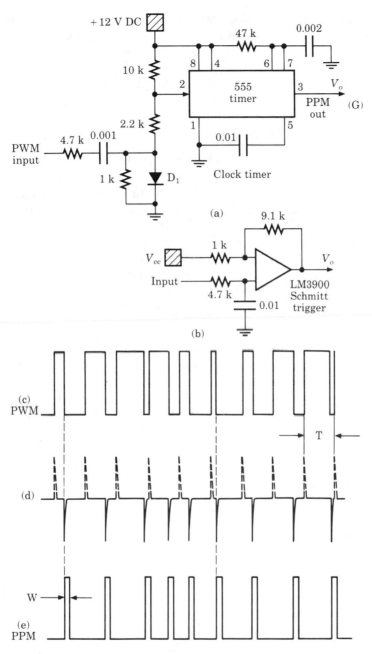

FIGURE 13.9
Converting a PWM signal to pulse position modulation.

Recovery of the PPM Transmission   At the receiver, a fixed-frequency reference pulse is generated from the PPM input signal to activate a flip-flop (bistable multivibrator). The PPM signal is also applied to the RESET terminal of the flip-flop for shutdown. This recreates the PWM signal, which in turn may be demodulated by a simple low-pass filter (voltage averager).

13.4   MODES OF OPERATION

Before we get in over our heads, it would be a good idea to look at some overall systems. Transmitters and receivers are tied together in a number of ways that describe the kind of system rather than the kind of transmitting format used. We will look at these systems in terms of how they are tied together, which is independent of the modulation technique used or what the system is to carry. Refer to Figure 13.10 in the course of the following discussion.

One of the first questions to come to mind is, "In how many directions will the data be transmitted?" This can be best answered by defining the classifications of systems we use today.

1.   *Simplex* (SPX): Information can be sent in only one direction. Commercial radio and television use this mode. They transmit, and you receive. You can never interact by the same communications link, regardless of how the signal reaches your receiver (RF transmission or cable).

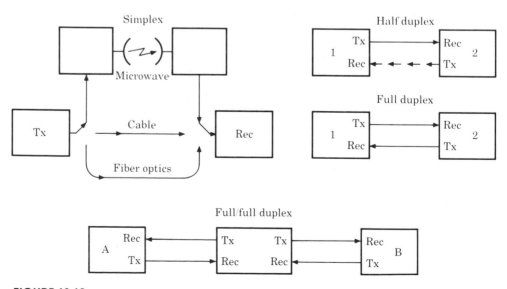

FIGURE 13.10
Modes of transmission: simplex, half duplex, full duplex, and full/full duplex.

2. *Half duplex* (HDX): Information can be sent in both directions, but not at the same time. Citizens band or ham radio are prime examples. The operator must press a control to transmit and release the control to receive. In the transmit mode, the receiver is disabled, and in the receive mode, the transmitter is disabled.

3. *Full duplex* (FDX): Information can be sent in both directions at the same time without interference. The telephone is probably the best example. Both parties may talk at the same time and each is able to hear the other. The human brain has a hard time separating (and understanding) what the ear is receiving at the same time it is trying to control what the vocal cords are transmitting. Machinery does not have this problem.

4. *Full/full duplex* (F/FDX): This is the latest mode to be developed. It describes a unit that is able to receive a message from one remote station while it is transmitting a different message to a third station. Machines are able to carry on two or more conversations at the same time.

Although RF signals are used in these examples, most of the world's land communications are handled by the interconnecting network of telephone companies. The same definitions hold regardless of which route is taken. In RF transmission, when the system has one carrier frequency, the land line has one twisted wire pair. When the RF system uses two different frequencies, the land line has two twisted wire pairs as would be required for FDX.

13.4.1 Station Interconnections

The majority of digital traffic is processed through the telephone networks (see Figure 13.11). An outgoing message is carried through twisted wire pairs to the local branch exchange, then to the receiver's local branch exchange, and finally to the subscriber's receiver. Long-distance interconnections go from the transmitter's local branch exchange to a toll exchange, from there to the destination toll exchange, then to the subscriber's branch exchange, and then to the subscriber's receiver. Depending on the distance between the stations, the message may undergo any number of modulation and demodulation processes, including satellite links and fiber optic cabling, without the knowledge of the transmitter or receiver. Since the telephone company systems are almost all automatic, the equipment will pick the route of the interconnection that is open at the time of call origination.

Such interconnection systems are used for what the phone company calls "dial up" lines. These are the same lines that are used for the home phone. The sender uses a standard telephone to dial the number of the receiver and then places the handset of the phone into an acoustical coupler attached to the transmitting digital terminal. In this mode, a telephone company is able to handle a data speed of about 9600 BPS maximum, and the equipment is compensated for

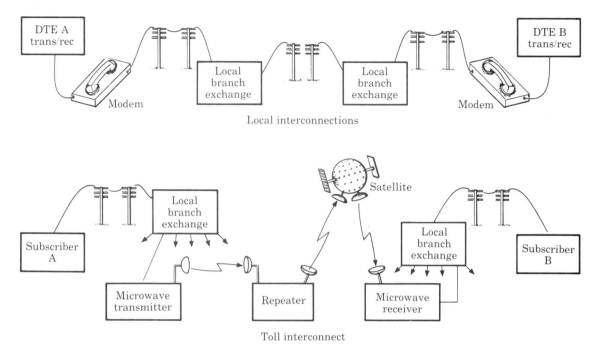

FIGURE 13.11

Interconnections between subscribers: local and toll.

this data rate. Although the engineering has been completed to allow data rates in the megabits, it may take a number of years and a lot of dollars to upgrade the worldwide network of systems to handle this higher capacity.

When high data rates are a requirement, arrangements can be made for special routing, which the subscriber pays for on a leased basis. These lines are called "private" or "dedicated" lines. The subscriber may request a hard copper connection, which means that there will be no switches or transformers in the line between the transmitter and the receiver.

The above description applies to serial code transmission, using a two-wire or four-wire interconnection, but could also apply to parallel code transmission, where eight wires would be required for an eight-digit transmission. The tradeoffs between series and parallel transmission lie in the distance between stations and daily leasing costs.

Because of the growing number of computer manufacturers, it became necessary to establish a fixed arrangement of interconnecting computers to match the connections of any unit with the connections of any other unit. The solution became known as the RS232C interconnecting cable. Table 13.1 shows the assignments to each pin of a 25-pin connector as specified by the Electronics Industries Association (EIA). The full specifications are available from the EIA in Washington, D.C. The RS232C cables are specified to have less than 50 pF capacitance/ft, not to exceed 2500 pF total, which limits the cable usage to about 50 ft maximum length.

The serial data stream from the host computer, called the **data terminal equipment** (DTE), is often applied to a modem, called the **data communication equipment** (DCE). (See Section 13.6 on modems.) Because the DTE and DCE can be from different manufacturers, the RS232 standards are recommended for the interconnecting harness between such devices. In addition to a standard 25-pin connector with assigned pin functions, the RS232 also recommends that a binary 0 and a control line "on" have a voltage level greater than +3 V, but not to exceed +15 V. A binary 1 and a control line "off" must have a voltage level between −3 V and −15 V. These voltage levels are termed **negative logic** because they are the opposite of the standard definitions for a binary 1 and 0.

TABLE 13.1

Pin	EIA Designation	Direction DTE	Direction DCE	Common Usage
1	Protective ground			
2	Transmit data	⟶		Send data (TD, SD)
3	Receive data	⟵		(RD)
4	Request to send	⟶		(RS, RTS)
5	Clear to send	⟵		(CS, CTS)
6	Data set ready	⟵		Modem ready (MR)
7	Signal ground			
8	Receive line signal detect	⟵		Carrier on detect (COD)
9	Data set testing			+V (+10 V DC)
10	Data set testing			−V (−10 V DC)
11	Unassigned			
12	Secondary (see 8)	⟵		Local mode (LM)
13	Secondary (see 5)	⟵		
14	Secondary (see 2)	⟶		New sync (NC)
15	Transmission timing	⟵		Serial clock transmit (SCT)
16	Secondary (see 3)	⟵		Divided clock transmit (DCT)
17	Receiver timing	⟵		Serial clock receive (SCR)
18	Unassigned			
19	Secondary (see 4)	⟶		
20	Data terminal ready	⟶		(DTR)
21	Signal quality detector	⟵		(SQ)
22	Ring indicator	⟵		(RI)
23	Data rate select	⟶		(SS)
24	Transmit signal timing	⟶		Serial clock transmit external (DSTE)
25	Unassigned			

13.4.2 Data Networks

The stations involved in a system are defined as the **host** and the **secondary.** The central processing unit (CPU) is considered the host, and the work terminals are the secondary station. The CPU may have a DTE as a control board and still be the host. The relationships between host and secondary are:

1. Both the host and secondary units may transmit and receive, but only the host may originate a call.
2. A secondary is generally connected to only one host.
3. A host is generally connected to all secondaries.
4. A secondary conducts activities only within its own complex.
5. A host is active over the sum of all the complexes.
6. A station may be a secondary within a very large complex and be a host in a smaller complex.

The **star** complex is generally the smallest system configuration, an example of which is found in a personal computer. The CPU has separate lines to the monitor, the disk drives, and the printer. The advantage of the star complex is that all secondaries are always at the ready. The disadvantage is the high cost of leased lines, where the secondaries are some distance from the host. In other disciplines, the star configuration is called a *home-run line system.*

The **ring** complex can be a system of all host units or one host and any number of secondaries. The advantage of the ring complex is that each unit in

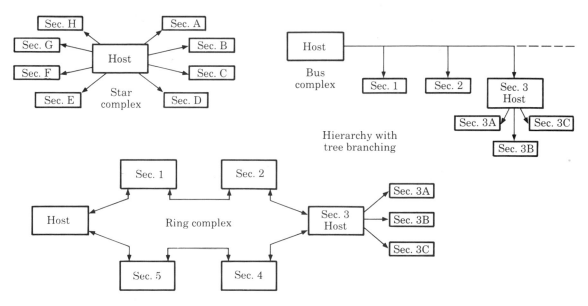

FIGURE 13.12
Forms of system interconnecting complexes.

the system can extract or insert information wherever there is space. The disadvantage is some redundancy in cabling.

The **bus** or **multidrop** complex is the most common because it has the advantages of the other systems and few of the disadvantages. It has good control, easy access, and minimal wiring. The host CPU controls when the secondaries may transmit or receive. Any of the three systems shown in Figure 13.12 may have two-wire, four-wire, or RS232 harness cabling, depending on the system. The bus-configured networks transmit data in bytes of 8 bits over eight parallel lines (parallel data transmission).

13.5   FREQUENCY SHIFT KEYING (FSK)

Frequency shift keying (FSK) is a form of FM that, with minor exceptions, resembles processes covered in Chapter 5. The modulating signal is a digital voltage (coded 1s and 0s) that will *key* the carrier to *shift* between only two distinct, fixed *frequencies*. In this case, the low carrier frequency represents a digital 1 (mark), and the higher carrier frequency is a 0 (space). The second variation from standard is that the carrier is not much higher in frequency than the frequency of the encoded message signal. An example is the Bell model 202 modulator, which transmits a 1200 Hz signal for a mark and a 2200 Hz signal for a space and processes a data signal of 1200 BPS (600 Hz). The model 202 is an HDX low-speed system using negative logic. Digital modulation has an inherently low noise characteristic that, coupled with the noise immunity feature of FM, constitutes a system that is hard to beat.

In keeping with high tech trends, a specially designed VCO leads the field as the generator used in this form of modulation. A VCO is also a part of the PLL used as the detector at the receiving end of the FSK communications system. The digital input of the XR2206 transmitter, pin 9 in Figure 13.13, is the control terminal of an electronic switch (see insert) that will select the resistance at pin 8 (R_2) and the capacitance, C_o, so that the VCO will oscillate at a frequency of $f = 1/(R_2 C_o)$ (in hertz) when the pin 9 voltage is zero. At this time, transistor Q_1 is turned on, causing Q_2 to turn off. ("Turn-on" versus "turn-off" here is used to describe current flow for the "on" condition versus no current flow for the "off" condition.) When Q_2 is turned off, Q_3 is turned on so that the low drop in voltage across Q_3 places the reference voltage V_1 at the emitters of A_1 and A_2. V_1 biases A_1 and A_2 to the "off" condition, and resistor R_1 at pin 7 is ineffective. Also, when Q_2 is turned off, Q_4 is turned on and shuts down Q_5. With Q_5 effectively an open circuit, V_1 is removed from the emitters of B_1 and B_2, allowing them to turn on using R_2 at pin 8 as the emitter resistor of B_1-B_2 (VCO), to generate the space frequency. The VCO charging capacitor C_o is between the collectors of the VCO differential amplifiers A_1-A_2 or B_1-B_2 as each set is alternately turned on or off.

When pin 9 goes more positive than 1.35 V, Q_1 is turned off, reversing the status of all of the transistors in the switch. The net result is that Q_3 becomes

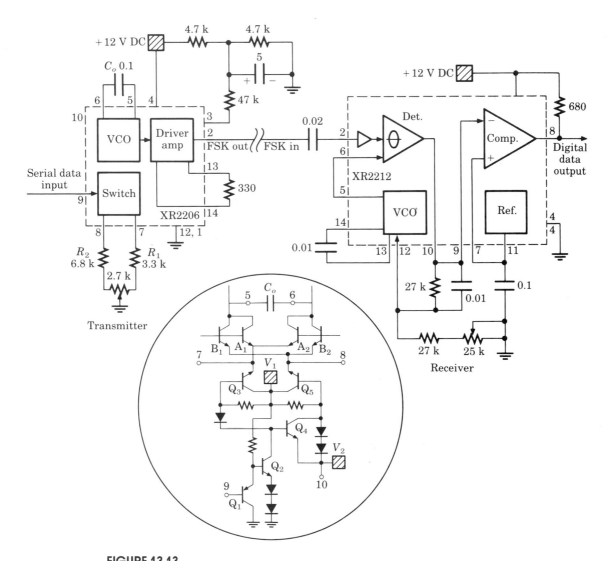

FIGURE 13.13
Frequency shift keying with the XR2206 transmitter (electronic switch insert) and the XR2212 receiver.

an open circuit, placing resistor R_1 at pin 7 as the emitter resistor for the A_1-A_2 differential amplifier of the VCO, to generate the mark frequency. Transistor Q_5 is almost a short circuit that applies V_1 to the emitters of B_1 and B_2. Transistors B_1 and B_2 turn off, making the resistance R_2 at pin 8 ineffective. The switching time is measured in microseconds. Voltage V_1 is a fixed internal reference voltage typically equal to $V_{cc}/2$, and V_2 is equal to $V_{cc}/4$.

13.5.1 FSK Demodulation

The FSK receiver, also shown in Figure 13.13, is a combination PLL and comparator. The frequency-changing signal at the input to the XR2212 PLL drives the phase detector to result in rapid change in the error voltage, which is applied to the input of the comparator. At the space frequency, the error voltage out of the phase detector is below the comparison voltage of the comparator. The comparator is a noninverting circuit, so its output level is also low. As the phase detector input frequency shifts low (to the mark frequency), the error voltage steps to a high level, passing through the comparison level, causing the comparator output voltage to go high. This error voltage change will snap the comparator output voltage between its two output levels in a manner that duplicates the data signal input to the XR2206 modulator.

The free-running frequency of the PLL (no input signal) is set midway between the mark and space frequencies. A space at 2200 Hz and a mark at 1200 Hz will have a free-running VCO frequency of 1700 Hz. The example cited earlier had a data rate of 1200 BPS (600 Hz), which can be used to find the modulation index: ± 500 Hz/600 Hz = 0.83. Remember from Chapter 5 that a modulation index less than 1 is narrowband FM and results in only one significant upper sideband and lower sideband. From Table 5.4, a modulation index of 0.8 places 85% of the voltage at the carrier center frequency (1700 Hz in this case). The first set of sideband pairs has 36% of the voltage, and the second set of sideband pairs has 8% of the voltage, (<1% power), each separated by 600 Hz. The total bandwidth becomes $4 \times 600 = 2400$ Hz, within the baseband frequency range of 300 Hz to 3 kHz. The output of the demodulator comparator should be filtered by a low-pass filter to ensure the removal of the unwanted parts of the carrier transmission.

13.6   MODEMS

A device (or circuit) that combines a *mod*ulator and a *dem*odulator is called a **modem.** An example is the XR2206 modulator and the XR2212 demodulator combined into a single package, shown in Figure 13.14. The modem has a data input terminal and a modulated analog output terminal for the transmit mode, plus a separate set of modulated analog input terminals and data output terminals for the receive mode. Therefore, modems are basically defined as digital-to-analog converters in the transmit mode and analog-to-digital converters in the receive mode. Normally under these conditions, the connection from such a unit to a distant companion modem is a four-wire (two twisted pairs) interconnection for this modem to be operated in full duplex.

When a two-wire interconnection is to be used, a hybrid transformer or hybrid amplifier, also shown in Figure 13.14, is used to tie the transmitter modulated output of one unit to the modulation input terminals of the receiver

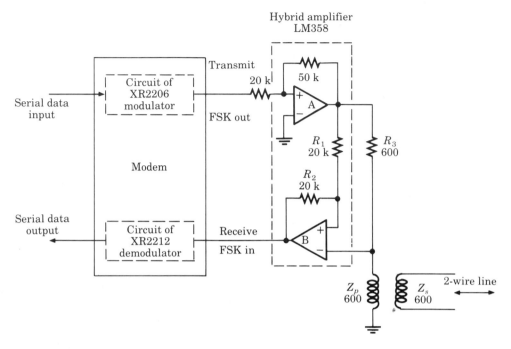

FIGURE 13.14

Modem with hybrid amplifier circuit. Converts four-wire system to a two-wire system for half duplex operation.

to form a common two-wire connection. The hybrid amplifier of Figure 13.14 takes the transmitted output signal from amplifier A, through two half-voltage voltage dividers to both the inverting and noninverting inputs of amplifier B. Taking advantage of the common-mode rejection (equal inverting and noninverting signals) will cancel the transmitter signal output from the XR2206 so that *no* signal arrives at the input terminals of the XR2212 receiver (through amplifier B). The same transmitted signal will divide across the 600 Ω resistor and the 600 Ω impedance of the transformer primary to deliver a −6 dB signal to the line and the receiving unit.

A signal that arrives from the distant unit will appear across the inverting input of amplifier B and ground. The same signal will be greatly attenuated when it passes through the 600 Ω (R_3) and the 20 kΩ (R_1) resistors in series to the noninverting input of amplifier B. Although still in the common-mode configuration, the two signals at the input of amplifier B will be vastly different in level and will *not* cancel, and so the signal will be passed on to the XR2212 demodulator. The number of interconnecting cables is now reduced to one-half, and when translated to monthly lease charges, this represents a sizable cost reduction.

The Bell system model 103 is a full-duplex FSK modem for low data rates of 300 BPS that uses an alternate scheme to deliver full duplex onto a single-

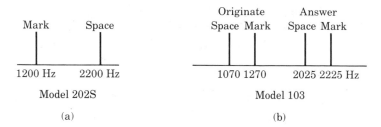

FIGURE 13.15
The mark and space frequencies for (a) the model 2025 modem and (b) the model 103 modem.

wire-pair cable. Remember that for full duplex, either separate cable pairs or two different frequency bands are required. Model 103 uses two different frequency sets, as seen in Figure 13.15, which shows positive logic. The host CPU is considered to be the originator and uses the lower frequency set of 1070 Hz (space) and 1270 Hz (mark) when transmitting. The remote may answer immediately on the higher set of frequencies, using 2025 Hz for a space and 2225 Hz for a mark frequency. The bandwidth of either message is:

$$\text{BW} = 2(\Delta f_c + f_s) = 2(100 + 150) = 500 \text{ Hz}$$

Two of these messages fit easily into the 300 Hz–3 kHz band-pass of the telephone company cable frequency response. Again, observe the negative logic for model 202 in Figure 13.15 and the positive logic for model 103.

13.7   FREQUENCY DIVISION MULTIPLEXING (FDM)

All of the standard broadcast systems discussed so far are subject to division of the frequency spectrum. The commercial AM broadcast band has channel 1 at 540 kHz, channel 2 at 550 kHz, and so on up the band. Possibly the biggest difference between the systems in earlier chapters and the system of **frequency division multiplexing** is that the former all originated from separate transmitters not synchronized with each other. Here we will develop a system where all of the messages are brought to a central location, either singularly or in packets, before being combined for transmission down a single coaxial line. The differences are twofold.

1. The information to be sent may be raw data from its original source without modulation (which cannot be done in an RF system), or it could be data that are shifted up in frequency (heterodyned upward) in groups.
2. The interconnecting link between the transmitter and receiver is most generally coaxial cable.

In this world of cable, you are the only one on the system. You may use any frequency that is compatible with your equipment without regard to user interference. This places all of the limits on the equipment and not on the transmission media. The standard commercial frequencies are used here for purposes of explanation.

Consider many voice frequency signals occupying the baseband frequency range of 300 Hz to 3 kHz, each to be amplitude-modulated onto a different carrier frequency starting at 64 kHz and extending upward in steps of 4 kHz. Suppress the carrier frequency and attenuate the upper sidebands of each signal. When 12 such signals are combined, you would have 12 SSSC (single-sideband suppressed-carrier) AM signals spaced 4 kHz apart occupying the frequency band of 60 kHz to 108 kHz, called a **group,** as shown in Figure 13.16a.

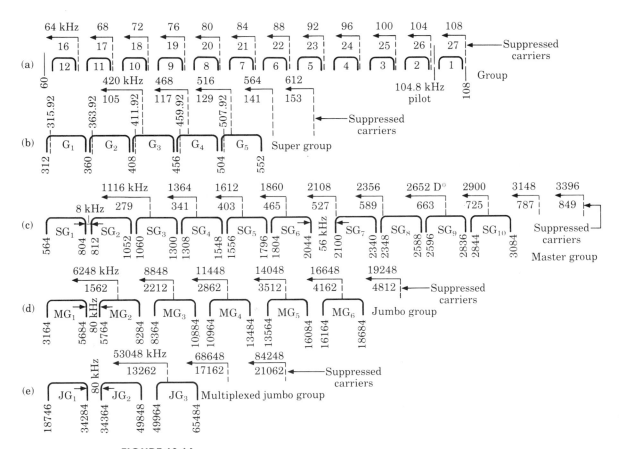

FIGURE 13.16
The frequency spectrum of a group, super group, master group, jumbo group, and multiplexed jumbo group.

Channel 12 is shown as the lower sideband of a 64 kHz carrier, channel 11 as the lower sideband of a 68 kHz carrier, and so on for the 12 channels of $A3_b$ emission, ending with channel 1 as the lower sideband of the 108 kHz carrier to form the group. Each channel has a different voice message that can be recovered at the receiver. Because each voice signal is contained between the frequencies of 300 Hz and 3 kHz, and each is allocated a 4 kHz bandwidth, this provides a natural 1.3 kHz guardband between each message in the group. A pilot signal is transmitted at 104.8 kHz so that synchronization and signal separation are simplified at the receiver.

Five such groups are now combined (by heterodyning each group to a higher frequency band) to make up a **super group,** still with a 1.3 kHz guardband between each message. Ten supergroups are combined to make a **master group,** with an 8 kHz guardband between each master group (except between groups 6 and 7, where a 56 kHz guardband exists). Next six master groups are combined to make up a **jumbo group.** A jumbo group contains 3600 separate voice conversations, each 3 kHz wide: 12 messages $\times$ 5 groups $\times$ 10 super groups $\times$ 6 master groups = 3600 messages = 1 jumbo group.

Occasionally, three jumbo groups are combined (with 80 kHz guardbands) to form a **multiplexed jumbo group,** which will transmit 10,800 messages on a single coaxial cable, as shown in Figure 13.16e. A multiplexed jumbo group will occupy the frequency spectrum from about 18.764 MHz to about 65.5 MHz. A special combination of three master groups (1800 voice messages) is a common transmission by the telephone company, called a *U600 group.*

The carrier frequency signals used to balance-modulate each of the message signals is some multiple of a 4 kHz oscillator signal. For example, the carrier for the first group in the super group signal has a frequency of 420 kHz, the 105th harmonic of 4 kHz. By the time this signal is multiplied up to the last carrier of the multiplexed jumbo group, it would be harmonic 21,062 of 4 kHz. A one-cycle change of the 4 kHz oscillator signal frequency would result in a 21,062-cycle change in the last carrier signal frequency. From Chapter 3, we found that a crystal will operate best for a fundamental frequency between 100 kHz and 10 MHz. Therefore, a master oscillator at 4.096 MHz can easily be divided down to 4 kHz by ten divide-by-2 networks, then converted to any of the other frequencies as required. This holds true for all carriers up to the seventh carrier for the master group cluster, noted by a D^* in the figure to indicate that the carriers from here on up in frequency are derived from a second oscillator source. This also accounts for the 56 kHz separation between supergroups 6 and 7 in the master group.

Figure 13.17 is the block diagram of the network. A block diagram of a balanced modulator is included in the figure as a reminder of the circuit configuration. One important thing to understand is that each voice message is modulated with a carrier by a balanced modulator, each group is modulated with a carrier by a balanced modulator, and each super group has a carrier and a balanced modulator, as does each master group. To form a multiplexed jumbo

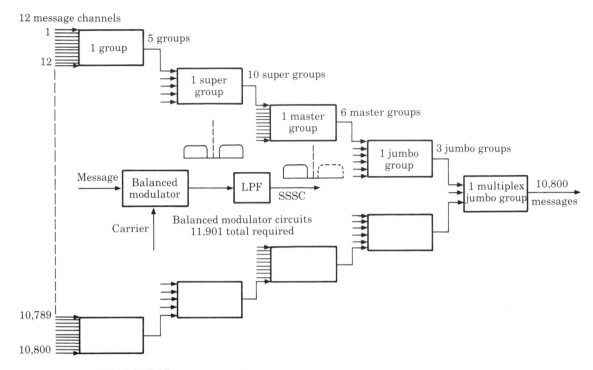

FIGURE 13.17
Frequency division multiplexing (FDM). This system can modulate as many as 10,800 messages on one carrier.

group, the system will incorporate 11,901 balanced modulators (and associated filtering systems) and use 36 different carrier frequencies.

Figure 13.18 is the outline of a circuit used to multiplex a jumbo group, in two steps, to the uplink frequency of 6.175 GHz to be transmitted via satellite. The block of frequencies of the jumbo group is first frequency-modulated up to a 70 MHz intermediate frequency with a carrier deviation of ±8.3 MHz for 100% modulation. Then the 70 MHz IF signal is amplitude-modulated onto the 6.175 GHz satellite carrier uplink frequency.

13.7.1 Decoding the FDM Signals

Retracing the FDM procedure covered above in reverse order will indicate the procedure required to decode the 10,800 voice messages signals back to the original subscriber level. First pick up the satellite downlink signal, heterodyne the carrier down to the 70 MHz IF signal frequency, and then FM detect the multiplexed super group.

High-Q, sharp cutoff band-pass filters are used to separate the multiplexed jumbo group into three jumbo groups. Each jumbo group is filtered into six master groups. Each master group is further filtered into the ten super

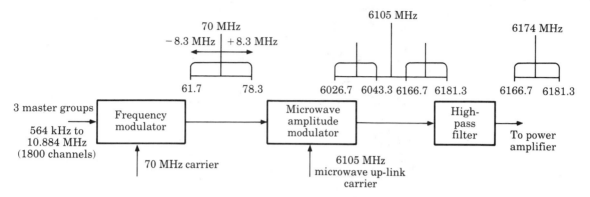

FIGURE 13.18
Satellite uplink transmission system. A master group cluster is converted to an FM intermediate frequency, then to an amplitude-modulated satellite carrier.

groups, then into five groups, and finally into the 12 message signals. The messages can now be demodulated in a product detector (balanced modulator) back to the original subscriber signals. A total of 11,901 balanced modulators are required to recover the 10,800 signals. The circuitry is elaborate, but the end results are well worth the effort.

13.8   PULSE CODE MODULATION (PCM)

We live in an analog world. All of our sensory systems, whether mechanical, electronic, or human, measure their inputs or generate an output on an analog scale. However, we can accomplish some tasks better or faster when the impulse data are in digital form. **Pulse code modulation** is the name of the process used to convert the analog signal into a digital signal and, after processing, to convert the digital signal back into an analog message. The modulator is called an *analog-to-digital converter* (A/D converter, or ADC). The code detector is a *digital-to-analog converter* (D/A converter, or DAC). Unlike analog systems, where the modulator and detector are generally a single circuit, the digital modulator and detector are each made up of a group of circuits that, when combined, form an ADC or DAC. The PCM system converts an analog signal to a true binary signal—in this example, a 4-bit code. In the following discussion, the opposite approach will be taken, that is, the recovery system will be described before the modulator is discussed.

Assume for a moment that the input signal to the circuit of Figure 13.19 is a 4-bit binary 9 (1001) in a serial stream with the least significant bit (LSB) arriving first. The data are stored momentarily in a 4-bit shift register, one register for each bit. A clock pulse is set so that when all four registers are loaded, they are dumped simultaneously to provide a parallel output signal. The up/down counter is a set of four flip-flop circuits (driven by the clock signal)

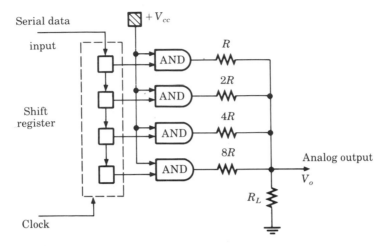

FIGURE 13.19
PCM decoder.

that have AND gates attached to the parallel output terminals. There must be a high clock pulse *and* a high data bit (digital 1) in a register for there to be an output level from that register to the resistive ladder.

The resistive ladder network is generally referred to as the D/A converter because of the ratio of the resistances. The relative resistor values have the same weight as the exponent term to the base 2 in the binary number system.

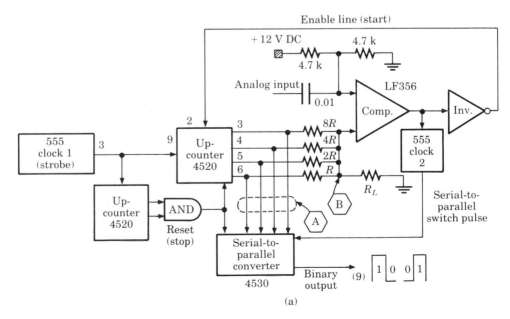

FIGURE 13.20
A/D converter for PCM transmission.

That is, exponents of 0, 1, 2, 4, 8, etc. correspond to resistor values of R, $2R$, $4R$, $8R$, $16R$, etc., with R_L having a value greater than any of the resistances in the ladder. The summing of the voltage drops across R_L resulting from the output of the counters that divide across R and R_L, $2R$ and R_L, etc. will reproduce an analog voltage in the form of a stairstep waveshape to the step 9 level, representing the digital 9 (1001) sent from the transmitter. Each binary number between 0 and 15 will represent a separate voltage level of the recovered analog signal.

You can see from Figure 13.20 that the PCM transmitter has many of the same circuits as the receiver of the previous diagram. The transmitter converts an analog signal into a 4-bit serial binary code output signal. The building blocks include a D/A converter (the resistive ladder), and you might have serious doubts about the reliability of an ADC circuit containing a D/A converter.

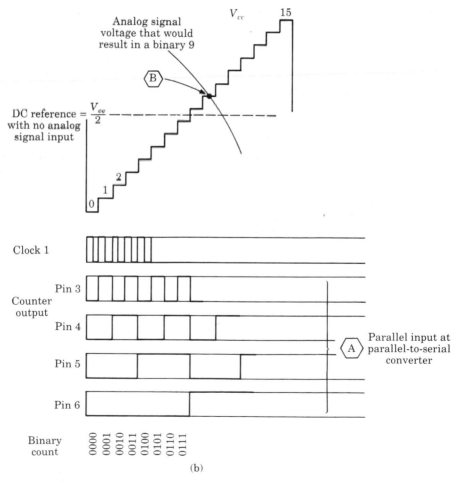

FIGURE 13.20 *(continued)*
A/D converter for PCM transmission.

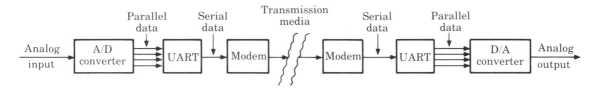

FIGURE 13.21
Pulse code modulation system.

However, the DAC is used in the transmitter to develop a stairstep voltage waveform to be compared to the input analog signal in order to measure at what point the converter will stop counting and convert that step level to a digitally coded number. The stairstep voltage is applied to the noninverting input of the comparator, and the analog signal is applied to the inverting input of the comparator. In this circuit, a DC level is also applied to the inverting input to establish a zero AC signal level. The system of Figure 13.20 uses a 16-step waveshape (0–15), so the analog signal can be sampled to only 16 points on the waveshape. This system is good for large analog input signals that change the equivalent of eight steps higher and eight steps lower than the DC reference level, but is poor for analog signals that change only a small amount from the DC reference level, such as ±0.05 V. The nonlinear stairstep waveshape would be more practical for both high- and low-level analog input signals but introduces distortion that causes a poor signal-to-noise ratio. The degrees of step voltage change are called **quanta,** which means "energy broken up into small parts"; the distortion resulting from the difference between the true analog signal and the approximated digital signal is called *quantum noise* or *quantization noise.*

Note that the digital output from the up-converter that feeds the ladder network (DAC) is also the digital information that is fed to the parallel-to-serial signal converter. The counter output waveshapes are also shown in Figure 13.20b for the sake of comparison. Although this form of D/A and A/D conversion is probably less efficient than some current systems, it is the easiest circuit to explain and is the most comprehensible to the novice. Later circuit descriptions will simply show the block diagram (see Figure 13.21), labeled "a converter" with the implication that you already know how the system is made up and how it works.

13.8.1 UARTs and USARTs

The terms **UART** and **USART** are acronyms for "universal asynchronous receiver and transmitter" and "universal synchronous/asynchronous receiver and transmitter." They are basically serial-to-parallel converters and parallel-to-serial converters similar to the converters in Figures 13.20 and 13.21, but they represent a sending and receiving unit in a single package, like the modems. The message is *not* modulated onto a carrier when using UARTs or USARTs.

13.9   ## DELTA MODULATION

The major advantage of **delta modulation** is the simplicity of the circuit. However, the tradeoff is that it is not easily multiplexed. Figure 13.22 shows the delta transmitter and the delta receiver.

The binary output from the transmitter is solely dependent on the *change* in the analog voltage, that is, whether the signal has increased or decreased since the last sample was taken. When an analog increase takes place, the transmitter outputs a binary 1; an analog decrease will output a binary 0. When the analog signal is a steady level (a rare case), the output alternates between 1 and 0. When the analog signal is on a rise, that is, the current sample is more positive than the previous sample, the counter will output a rising stairstep voltage. A decreasing analog voltage will cause a downward stairstep voltage. When the analog voltage changes too fast for the stairstep voltage to follow it, diagonal clipping similar to that in an AM detector will result. Couple this distortion with a poorer signal-to-quantum-noise ratio (compared to PCM), and you trade off performance for low cost.

13.10   ## QUADRATURE PHASE MODULATION (QPSK)

A quick review of single-sideband modulation in Chapter 3, in particular Figure 3.8, will help you to understand **quadrature phase modulation.** The output of the balanced modulator is a signal at both side frequencies, with *no* signal at the carrier frequency.

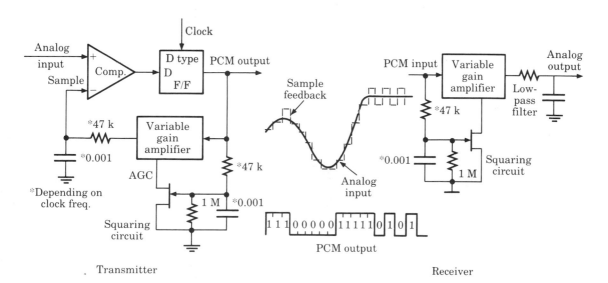

FIGURE 13.22
Variable-step-size delta modulator and demodulator.

The modulator with no carrier phase shift is called the I modulator because it is "In" phase with the carrier oscillator. The modulator with the 90° phase-shifted carrier is a quarter cycle out of phase from the carrier oscillator, so it is called the Q (for "quadrature") modulator. This circuit will shift phase with modulation in the same way that the FSK circuit shifted frequency with modulation, so this process is often abbreviated QPSK for "quadrature phase shift keying."

The input serial data stream is sampled as an entire character by a 2-bit shift register and converted to a parallel output signal by a clock pulse. The circuit is called a **bit splitter.** The output of the most significant bit (MSB) register of the bit splitter is channeled to the I balanced modulator, while the output from the least significant bit (LSB) register of the bit splitter is directed to the Q balanced modulator. The output level of the bit splitter (to either or both modulators) is a positive voltage for a binary 0 and a negative voltage for a binary 1.

Writing a truth table that substitutes a minus phase for a binary 1 and a plus phase for a binary 0, and then letting I be the most significant bit and Q the least significant bit, will result in the QPSK constellation shown in Figure 13.23. A decimal 0 input (binary 00) will dispense a $+V$ to both the I balanced modulator and the Q balanced modulator. The output of the I balanced modulator is a vector that is in phase with the carrier oscillator, as shown at the 0° reference in the constellation of Figure 13.23. The $+V$ to the Q balanced modulator will output a vector that is delayed by 90° from the I signal, as also seen in the constellation. The vector sum of these two output voltages will be a vector at $+45°$ to represent a binary 00. A binary 01 will generate a $+I$ and a

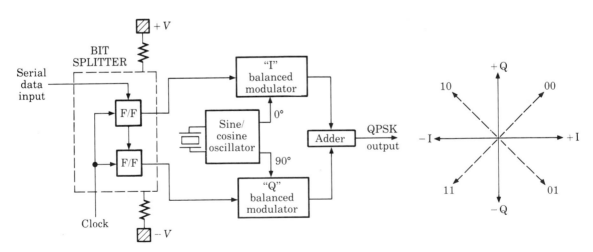

FIGURE 13.23
Quadrature phase shift keying modulator (QPSK).

−Q in the constellation, a binary 10 will be a −I and +Q, and a binary 11 will be a −I and a −Q, to complete the constellation.

The QPSK system described here is capable of 1200 BPS data rate, but there are expansions of this system beyond a single digit. These systems use the expression *M*-ary, where *M* stands for the change from the binary (2-bit) digit. Level converters are used that increase the number of levels to 4, 8, and 16. The *M*-ary level thus becomes 4-ary, 8-ary and 16-ary, and the input data rates are 2400, 4800, and 9600 BPS respectively. Two data bits at the input represent one phase at the output. Therefore, Figure 13.23 will have 4800 BPS input for 1200 BPS output.

13.10.1 Recovery of the QPSK Signals

For single-sideband systems, it was stated that a suppressed-carrier transmission required carrier reinsertion at the receiver to effect signal detection. For analog systems the phase is not a concern, and a single product detector is sufficient to recover the message signal. In QPSK, a 45° phase signal could appear as a +45° or a 225° signal, making the recovery ambiguous.

The circuit of Figure 13.24 identifies a carrier recovery circuit that senses the carrier phase based on the sum of the information received and applies this to two phase detectors (one shifted by 90°) as a means of reference. Recall that the phase detector requires two input signals, as does the product detector. The I phase detector receives the data and an in-phase carrier signal. The Q phase detector receives the data and a quadrature (90°) phase-shifted carrier. The outputs from the two phase detectors are first filtered through a low-pass filter, and then each is applied to its own voltage comparator before they are added together. The output of the linear adder is a serial binary data stream, the same as was applied to the modulator of Figure 13.23.

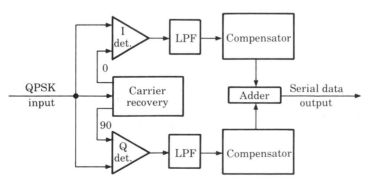

FIGURE 13.24
Carrier recovery loop for QPSK demodulation.

13.11   8-PSK

The 8-PSK modulator accepts data in groups of 3 bits, regardless of the coding process. Each group of 3 bits is applied to the serial-to-parallel bit splitter and output simultaneously, as shown in Figure 13.25. The first and second bits (a and b) are applied directly to separate two-to-four-level converter circuits. When bit a or $b = 0$, the polarity from the level converter is negative. When a or $b = 1$, the polarity from the level converter is positive. When the c bit is 1, the output from the two-to-four-level converter is equal to 1. Where c is 0, the output is ($\sqrt{2} - 1$). The I modulator has a 0° reference carrier and represents the "sine" output. The Q modulator has a 90° carrier and represents the "cosine" output. From this information, the output constellation can be constructed. The selection of levels and polarities used here are most common, but it can be seen that other choices are available so long as they are compatible with your own system. The output levels should always result in a vector that is ±22.5° from each major axis, that is, $\theta = \tan^{-1}(\sqrt{2} - 1)/1 = 22.5°$.

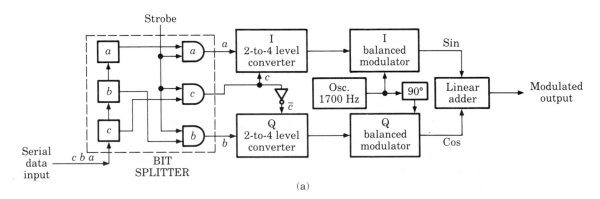

(a)

c	b	a	Sin I	Cos Q
0	0	0	− 0.414	− 1
0	0	1	+ 0.414	− 1
0	1	0	− 0.414	+ 1
0	1	1	+ 0.414	+ 1
1	0	0	− 1	− 0.414
1	0	1	+ 1	− 0.414
1	1	0	− 1	+ 0.414
1	1	1	+ 1	+ 0.414

(b)

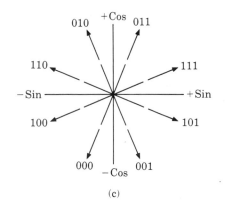

(c)

FIGURE 13.25
(a) 8-PSK modulator. (b) Truth table. (c) Constellation.

Assume for the moment that the first 3 bits of a serial data stream are all zeros (000). The bits to the I two-to-four-level converter are $a = 0$ and $c = 0$. When $a = 0$ the polarity is negative, and when $c = 0$, the level is $(\sqrt{2} - 1) = 0.4142$. Therefore, the sine output of the constellations will be -0.4142.

The bits to the Q two-to-four-level converter are $b = 0$ and $c = 0$, but c is inverted to $\bar{c}$, or 1. The b bit sets a negative polarity, and $\bar{c}$ sets the level to 1 so that the cosine output equals -1. The results for sin $= -0.4142$ and cos $= -1$ are shown in the complete constellation of Figure 13.25. Try any 3-bit data sample to see how your results match those of the constellation.

13.12   LOCAL AREA NETWORKS (LANs)

Local area networking is a technology devised to interconnect any number of electronic tools (from a variety of manufacturers) into a system that will allow sharing of resources and equipment. The intent is to expand a system's capabilities and flexibilities while avoiding a ballooning cost factor. Although the possible size of a network and the number of units in the system are almost without bounds, it will soon become apparent that the complexity increases exponentially with the number of units and the distance between units in the system. The general topic of LANs can be treated exhaustively only in a microprocesser textbook, where more emphasis can be given to protocol and to individual pieces of equipment. However, because it is an electronic communications system and because it may aid the reader in other areas, it is covered here as a development in the field of communications.

The following terms need to be defined before they are encountered in the text:

ASCII	American standard code for information interchange
CRC	Cyclic redundancy check
CSMA/CA	Carrier sense, multiple access with collision avoidance
CSMA/CD	Carrier sense, multiple access with collision detection
EBCDIC	Extended binary-coded decimal interchange code
GPIB	General-purpose interface bus (IEEE 488 bus)
HDLC	High-level data link control
HPIB	Hewlett-Packard interface bus
ISO	International Standards Organization
LAPB	Link access procedure balanced
OSI	Open systems interconnect
SDLC	Synchronous data link control

Simple systems such as a personal computer use the star topology and are designed to work with standard companion pieces such as monitors, disk drives, and printers. Manufacturers of different brands of computers have incorporated setup procedures to work with any brand of peripheral equipment, but they have avoided compatibility measures allowing one brand of computer to work with another brand of computer, claiming that their particular system offered the greatest advantages over the entire range of operating features. Networking represents the industry's effort toward compatibility. In the *star* configuration, the computer is the controller, and any piece of peripheral equipment must communicate to any other piece of equipment (if it has the ability) through the controller, much the same as a person talking through an interpreter. Because of this rigid, single-unit control, star systems are usually relatively slow and are therefore practical for only the smallest of networks (rarely larger than ten units). Figure 13.26 depicts such a controlled system.

In Figure 13.26 the mainframe may be a permanent record storage file for accounting, sales, purchasing, inventory, shipping, or any manufacturing record keeping. These files are generally much too bulky, vulnerable, or classified to be stored or handled at the department-level work station and so never leave the mainframe. A readout of the file may be delivered to the work station for information retrieval or updating, but a privileged code is required to effect changes to the mainframe files. This results in minimal handling of the permanent records so that they are less subject to damage or loss. In the star configuration, if the controller is disabled, the network becomes inoperative, although the work stations may still function as individual microprocessors separate from the network.

The *ring* or *loop* system uses a circulating line that connects all of the units in series to form a continuous circular path. One of the units in the loop may be designated as the controller, in which case the system works in much the same manner as the star system. That is, a unit wishing to send a message must request permission from the controller, identify itself as the talker, and include the address of the listener. The controller checks the transmit bus

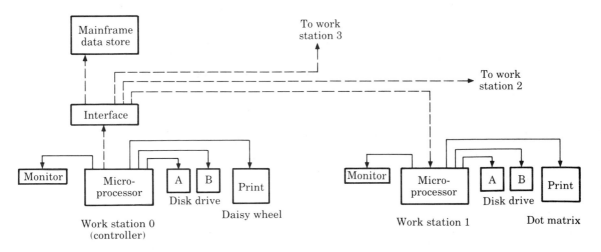

FIGURE 13.26
A star network showing one control station and three work stations.

status for a "busy" or "ready" condition and then specifies which work station is to be the talker (sender) and which work stations will be listeners (receivers). "Talkers" include tape readers, data storage files, digital voltmeters, frequency counters, and any other measuring devices. Listeners may be printers, display devices, programmable power supplies, or programmable signal generators or disk drives. The controller decides who talks and who listens. The controller or the CPU of the computer processor may be a talker or a listener.

In an *ideal loop,* there is no controller. Any of the talkers may transmit whenever they are ready, provided the transmit bus is free. The most popular method used to avoid chaos in such a system is a *token-passing* arrangement. Figure 13.27 shows the input-output port arrangement of a unit in a loop token-passing system, called an **interface.** Because the input and output ports are isolated from each other, the signal can go around the loop only in one direction. One of the units in the system may have a special circuit allowing it to act as a *guardian* (*not* a controller) of the system. This guardian generates a *token* signal (such as two 01111111 start/stop flags) and feeds it into the circulating line. When the token propagates around the full loop and returns to the guardian, it is discarded and a fresh token is issued to the loop. In this way, a token will not be degraded over a period of time in an idle network. The guardian is often equipped with a timer so that, if a token is not returned within a specified time period, the guardian will assume that it is lost, in which case the guardian will purge the data bus and issue a new token.

The interface between the loop and the processor, in essence, operates as follows. A normally closed (NC) switch, electronic or relay, ties the input and

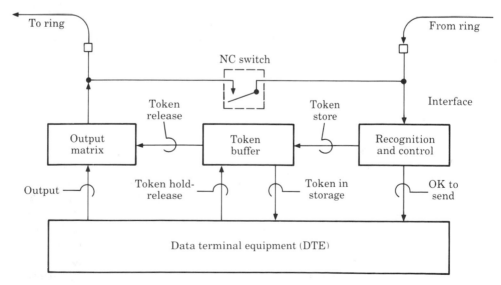

FIGURE 13.27
A typical interface for a station in a token-passing loop.

output lines together when no power is applied to this processor. This maintains the integrity of the loop whether the station is on-line or off-line. When the unit is powered up, the switch is opened and the next token around the loop is channeled into a recognition-and-control circuit. The token is routed to a storage buffer, and a "Your turn to transmit" signal is issued to the processor. The processor checks the buffer to make sure that the token is in storage and directs the buffer to hold the token until further notice. The processor releases the start flag, then sends its transmission through the output matrix onto the loop. When the processor has completed sending its message, it directs the buffer to release the end flag token, which is then fed through the output matrix back onto the bus line of the loop. The token continues to circulate around the loop, giving each processor a turn to examine the message before coming back to "our" processor for another turn. The unit that originated the message will examine it on the return trip, and if no errors are detected, it will remove its message and release the token for others to use. If an error has occurred, it will retransmit the same message.

It is easy to see now why the star topology is reserved for the smallest networks. The processing time is slow, only one unit may transmit at a time, and the cabling between units is the most extravagant. The ring or loop network is an improvement over the star topology because the interconnecting cables are fewer and shorter. The token concept ensures that the messages will not overlap or collide, but it also means that the processing time will increase when the number of units on the system increases. Therefore, the number of units in the loop topology rarely exceeds a few tens of processors.

The principle of local area networks was around long before the microprocessor became the industry tool with the bright future that we see today. It did not become a separate area of concern until we began asking machines with synthetic intelligence to communicate with each other in different languages, an effort further hampered by the demand for high-speed transmissions in very large, complex systems. The term *local area network* was first used to refer to the bus configuration but its meaning has since been broadened to include the lower-echelon systems.

The *bus* configuration realizes the highest data transfer speed and offers a bonus in that the speed will not diminish greatly even when the system grows large. A 1000-unit system is within practical limits, and the wiring requirements are often less than other networks of equal unit size. Work has already been started on a MAN system (metropolitan area network).

A system in which information is transferred as raw (unmodulated) data is called a *baseband* system. The baseband signal may be at any data rate (generally not in excess of 20K BPS). However, this means that only one signal can be on the line at a time without interference from other signals occupying the same frequency spectrum. Standard hookup wire is used in single strands, in bundles, or in ribbon cable form (for example, RS232C cable).

Data modulated onto a carrier constitutes the definition of a broadband system. The data may modulate the carrier as an FSK or QPSK modulated signal, which may then be spread over the band by frequency or time-division multiplexing (many signals on a single bus at the same time). These higher-frequency formats common to loop and bus systems have capabilities to 10M BPS, but also require coaxial cables in most instances to reduce interference, crosstalk, and attenuation. This will increase the cable costs.

The bus topology uses a single cable bundle that extends to all units on the network and requires that the coaxial cable (50 Ω bus) be terminated at both ends, consistent with good transmission line installation techniques. Systems management on a bus-oriented network is typified in Figure 13.28, which includes the systems interface management cable set (five lines), the data byte transfer cable set (three lines), and the data bus (eight lines). The interconnecting cable is a 25-wire set with both a male and female connector on each end to facilitate series-stringing the units together.

The five management lines include an "interface clear" (IFC) line, which will reset all units to their starting status. The "attention" (ATN) line will signal all stations to listen on the data bus line. A unit that is ready to transfer data notifies the controller by activating the "service request" (SRQ) line low, similar in effect to an interrupt signal. The controller then polls all stations to establish the origin of the request and activates the "remote enable" (REN) line to assume control of the transfer. The transmission is terminated by the sender when the "end or identify" (EOI) line is activated, which signals the successful completion of the transfer of a data block.

The "data valid" (DAV) line, the "not ready for data" (NRFD) line, and the "not data accepted" (NDAC) line are all handshaking control lines used to coordinate the transfer of data bytes on the data bus.

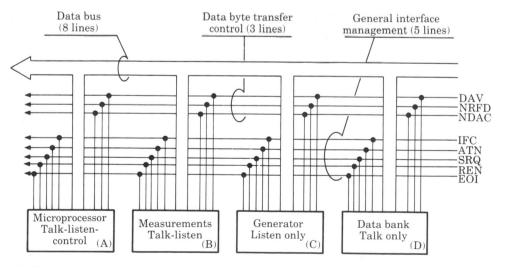

FIGURE 13.28

A bus network using carrier sense, multiple access with collision detection (CSMA/CD).

In a high-level data link control (HDLC) system, the message is sent in frames having a maximum length of 20,048 bits, as shown in Figure 13.29. The data bus has eight parallel lines; therefore, one bit on each line is called a *field* (or a byte or octet). All succeeding fields (one bit on each line) follow in immediate order until the message is concluded. The frame is referred to as having a *parallel-bit, serial-byte* format.

The **frame** is a series of fields that conform to a specific ordered information set. There are three basic frame types, the *information* frame (I-frame), the *supervisory* frame (S-frame), and the command/response frame, which is unnumbered (U-frame). Since this treatment on LANs is just an introduction, only the information frame will be covered here.

Figure 13.29 details the I-frame in accordance with the standards set up by the International Standards Organization (ISO). Its format is designated as high-level data link control. IBM developed a synchronous data link control (SDLC) format, but because the two formats are so much alike, they will both be treated here as HDLC except as noted. The frame begins with a starting flag of one field in an octet (8-bit character), which designates a "busy line" condition signaling all senders to hold their transmissions. The second field is an octet identifying the destination link address. The third field is the link control octet and is divided into bits as follows. Bit 8 is a "don't care." Bits 7 and 6 are the command code and have one of the following meanings:

00	Universal command
01	Unlisten command
10	Untalk command
11	Ignore

Bits 5 through 1 represent the address of the unit sending the command. The message packet may hold any number of octets up to 2500 (20,000 bits) and is followed by the frame check field. The frame check is a 2-octet section for parity

Start flag 01111110	Destination link address	Link control	Message any number of octets	Frame check	End flag 01111110
8 bits 1 field 1 octet	8	8	8–20,000 bits 2500 octets	16	8

FIGURE 13.29
An information frame in an HDLC system.

and CRC verification. The ending flag octet is the same as the starting flag and acts to bracket the frame.

In the bus network, a station must take control of the data bus in order to transmit. The two most widely used arrangements are token passing and "carrier sense, multiple access with collision detection" (CSMA/CD). A station wishing to transmit in the CSMA/CD mode will first look at the data bus to see if any other station is transmitting a message. If so, the inquiring station will wait a length of time before looking at the data bus again. It will continue to do this until it senses no message on the data bus, at which time it will begin transmission. At this time, other stations that need to transmit will sense our carrier on the bus and hold off transmission until the bus is free. This arrangement satisfies all conditions for the majority of the time. Remember that we are talking about fairly large networks that may be separated by considerable distances. Measured in propagation time, the distances could equal many milliseconds. It is possible for two stations to examine the data bus at the same time and, both finding it free, begin sending out messages. This will result in a *collision* of the two message packets. Both stations will sense a collision and immediately terminate the transmissions. Each station will then go back to random checking of the data bus for carrier presence until they can successfully complete the transmissions. Under these conditions, there is very little difference between the CSMA/CD and the CSMA/CA systems.

There are many data management systems in operation in local area networks. Some are listed below and have publicized information readily available.

EITHERNET   A baseband HDLC system that uses CSMA/CD at a 10M BPS rate. It was developed by the Xerox Corporation, which was later joined by Digital Equipment Corporation (DEC) and Intel Corporation in refining and implementing the system.

DOMAIN   Developed by the Apollo Computer Corporation as a broadband system uses a token-passing ring structure.

LOCALNET   A broadband system that uses the CSMA/CD format for a bus distribution network. It was developed by the Systek Corporation.

WANGNET   A broadband system that uses CSMA/CD for bus distribution, developed by the Wang Computer Corporation.

Other ring-structured management systems are OMNI-LINK, developed by Northern Telecom, and PRO-NET, developed by Proteon Associates. Other bus distribution systems are NET/ONE, OMNINET, PRIMENET, and Z-NET.

As early as 1977, the International Standards Organization (ISO) recognized the need to formalize an architecture for defining standards to enable data handling/processing units from different vendors to communicate with each other in a compatible network. They adopted the "open systems interconnect" (OSI) model, structured in a seven-layer hierarchy such that the system could be subdivided into smaller pieces for standardization, yet remain faithful to the overall plan. Remember that the OSI model is not a standard, but simply a framework around which standards can be developed. The layers are set up as follows.

The Physical Layer   The physical layer is the lowest, most fundamental layer. It specifies the *mechanical, electrical, functional,* and *procedural* standards for accessing a network. The RS232C cable is one of the standards developed for this layer. A 25-pin plug and jack is the mechanical standard. The voltage levels of +5 to +15 V for a space and −5 to −15 V for a mark define the electrical standard. The assignment of pin numbers constitutes the function standard. The procedural standard is set by declaring, for example, that pin 4 be designated "request to send" (CA), followed by "clear to send" (CB) on pin 5, and data are transmitted on pin 2, "transmitted data (BA)."

Data Link Layer   Level 2 of the ISO protocol format is responsible for assembly and formatting of the message packet. It ensures reliable transfer of data over the data link and standardizes error detection and retransmittal. It establishes, by bit location, when the message frame is a command or a response. It designates the source service access point address (SSAP), the destination service access point address (DSAP), and the bit order of the frame.

The Network Layer   The prime function of the network layer is to establish flow control between systems across a communications network. It must determine the most appropriate network configuration for the function provided by the system (e.g., dial-up, leased line, or packet). Dialogue between the user's DTE and DCE is carried on at this level, as is communication between the DCE of one terminal and the DCE of another terminal, which is also talking with its own DTE. Virtual circuits (dial-up and leased lines) and datagrams (unacknowledged small message packets) are examples for this layer.

Layers 1 through 3 are considered to be the fundamental layers, and layers 4 through 7 are referred to as the higher-order layers. There are no standards for some of these higher layers as of this printing, but that does not eliminate the definition for the intended use of the layers. For this reason, the descriptions for layers 4 through 7 are brief. Table 13.2 establishes where (and what) standards have been established. The abbreviations used in the table are identified as follows:

OSI	Open systems interconnect
CCITT	Consultative Committee on International Telegraphy and Telephony
ANSI	American National Standards Institute
IEEE	Institute of Electrical and Electronics Engineers
NBS	National Bureau of Standards
DOD	Department of Defense

TABLE 13.2
The ISO protocol layer framework and the current corresponding standards

ISO Layers	CCITT	IEEE 802	ANSI X3T9.5	NBS	DOD
Applications					
Presentation					
Sessions				Session	TCP
Transport				Transport (TP)	
Network	X.25			IP	IP
Data link	LAP-B	Logical link control	Data link		
Physical	X.21	Medium access control	Physical		
		Physical			

Transport Layer   The transport layer is responsible for the type of service, the grade of service, and the connection management. It deals heavily with the exchange of data between processes in different systems. Briefly defined, the transport layer is responsible for the end-to-end integrity of the transmission.

Sessions Layer   The sessions layer determines when a message may be transmitted, the quantity of data to be accumulated before transmission may begin, who is authorized to use the transmission, and the availability of the network link. Message priority, processor capabilities, and buffer storage access are also the responsibility of this layer.

Representation Layer   Layer 6 defines the assembly protocol of the parts and elements (syntax) of the data message, which include the coding, synchronization, interrupt, and termination of the message dialogue. Translations from EBCDIC to ASCII fall into this layer.

Applications Layer   Applications is the highest layer of protocol and specifies the general management of the network services providing access to common data buses and common peripherals. The applications layer is the layer that communicates directly with the user's applications program or process. It controls the sequence of events between the computer applications and either

FIGURE 13.30
The IEEE 802 standards format.

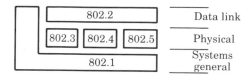

the user or another application. It is responsible for the transfer of information between applications processes and end-to-end encryption.

The Institute of Electrical and Electronics Engineers (IEEE) formed a group in 1980, called the 802 Committee, to establish a set of standards for local area networks that would conform to the ISO framework. The committee was composed of nearly 150 experts from industry (both IEEE members and nonmembers). These writings, formatted in Figure 13.30 became known as the "802 standards" and are summarized as follows:

802.1	Describes the relationships among the standards and their relationships, as a whole, to the ISO OSI reference model.
802.2	Standard logical link control protocol to be used in conjunction with the medium access standards.
802.3	A bus-oriented system using CSMA/CD as an access method.
802.4	A bus-oriented system using token passing as an access method.
802.5	A ring-oriented system using token passing as an access method.

The peer protocol procedures described in the ISO layer structure utilize some of the concepts and principles, as well as commands and responses, of the "asynchronous balanced mode" (ABM), as defined in ISO 7809-1984 and ANSI X3.66-1979. The ABM procedures provide the basis upon which the CCITT recommendations X.25 level 2 LAPB procedures are defined. The frame structure defined in data link layer procedures is also covered in these references to the various medium access control (MAC) procedures. The combination of a MAC sublayer address and a logical link control (LLC) sublayer address is unique to each data link layer service access point in the local area network.

13.13   CELLULAR COMMUNICATIONS

When an existing system undergoes a unique improvement, it is wise to give that system a new name. This is what happened with the mobile two-way telephone communications system. Under the original plan, a single medium-power transmitter was placed at the center of population in an urban community having a service area about 50 mi in diameter. The central transceiver had a capacity of between 100 and 500 channels; however, because mobile telephones are full-duplex, the system was limited to about 250 conversations at a time. Although the costs were high and subscriber lists were small, the system served the community very well. As the community and the subscriber lists grew, users experienced long waiting periods before they could get into the system. Most of the systems were owned and operated by the local telephone companies, which were capable of supplying mobile-to-mobile as well as mobile-to-home or mobile-to-business telephone interconnections.

The geographical shape of the service area was controlled by the radiation pattern of the transmitting antenna, but problems resulted when the community grew or changed shape. Figure 13.31 shows the city limits of Tucson, Arizona, in heavy dashed lines with a ring indicating a 500 μV field strength pattern from a 10 W transmitter in the business district. This power level would service the bulk of the population well, but provide less-than-ideal service to the fringe areas of the city. A 500 μV ring from a 40 W transmitter is partially seen to the right in the same figure and provides good signal to the entire city; however, it also transmits to the west over empty desert lands. The system designer or installer will generally elect to use the high-power

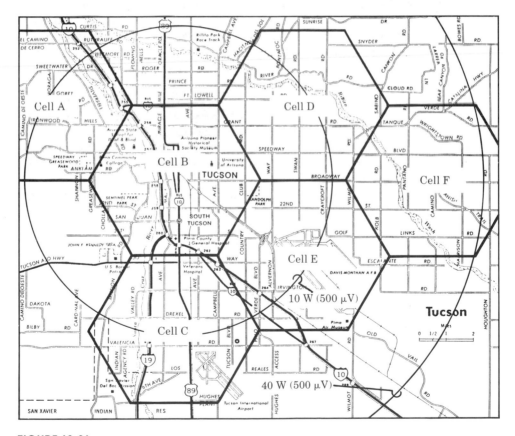

FIGURE 13.31
Tucson, Arizona, showing a system of six cells in a network mobile interconnecting link.

transmitter in favor of good signal strength within the service area, despite the waste of power in the sparsely populated areas.

The changes to this system had to be planned and implemented over a period of time. The first step was to use new, system-ready mobile units in new installations, which offered capabilities far beyond those of the old system. Once the mobile units were in place, the second step was to convert the one large service area into several smaller service areas, called **cells.** The cells took on the hexagonal shape of a honeycomb. The one high-power transmitter was replaced by six separate 5 W transceivers tied together by land line interconnections through the branch exchanges of the telephone companies. A mobile unit in cell A would transmit to cell A's transceiver, then through the telephone system to cell C, where the message would be retransmitted to a mobile unit in cell C's area. These two steps introduced several advantages. The cell transmitter and mobile transmitter each operate at a lower power. Second, a mobile unit in cell A, one in cell C, and a third in cell F could use the same frequencies without cochannel interference. The reuse of frequencies will increase the system's capacity by as many cells as can use the same frequencies. The only restriction is that adjacent cells may *not* use the same frequencies. Third, the system is easily expanded because adding a new cell will not affect any of the established cells. (Chicago had 91 cells at last count.) Fourth, cells provide better management of the total service area. In Tucson, there is a mountain range to the northeast, so the city will probably not grow in that direction. Using cell building blocks, the system can be expanded in other directions without affecting this service area.

These are the most obvious differences. Each cell's transceiver is connected through the telephone system to a central *mobile telephone switching office* (MTSO) for better overall system control. The cells are not independent of each other or of the total system. Suppose a mobile unit in cell A is using channel 24, and another unit in cell C is also using channel 24. So far, no problem since the same frequencies may be used in widely separated cells. However, suppose unit A, being mobile, now moves into cell B. The MTSO keeps track of which frequencies are in use in each cell, knowing that adjacent cells should not use the same frequency. As soon as cell A gives up control of unit A to cell B, the MTSO will change the transmit/receive frequencies of either mobile unit depending on the in-use channels of each cell. The communications link will thus continue on without cochannel interference. The switchover takes about 250 ms and will generally go unnoticed by either mobile unit operator.

Other intersystems equipment conversations are going on at the same time. Now that you have a general idea of what happens, a step-by-step trip through a complete transaction will clarify other differences from the original mobile telephone systems.

To begin, each cell transceiver sends out an identification signal of equal strength for all cells. When a mobile unit operator picks up the handset, a scanning system in the unit measures the signal strength of all cells' identification code signals. The mobile unit then sets up contact with the cell having the strongest signal. The data channel in each cell transceiver, called a *setup channel,* operates at a 300 baud rate and uses the standard ASCII 7-bit code. The mobile unit also sends the cell its identification code, which is then passed on to the MTSO by telephone line for recognition, frequency assignment, and future billing. The mobile operator then gets a dial tone and initiates a call in the usual manner. The call transaction is the same as any standard telephone call.

During the time of the call, the cell transceiver monitors all of the active channels in its area as well as in the six surrounding cell areas. It maintains a constant conversation with its six adjacent cells and the MTSO. Together, they control which mobile unit is on what channel, which cell is the controlling cell (by comparing the strength of the mobile signal in each area), the location of a mobile unit in each cell, and what action is to be taken when a mobile moves from one cell to another. The controlling cell uses a form of reverse AGC; that is, it can order the mobile unit to alter its output power so that cell interference is minimized. The mobile output power is maximum at 3 W, but it can be reduced in 1 dB steps (seven steps total) to as little as 0.6 W. All of these control functions are carried on without the phone operator's knowledge.

Consistent with most mobile communications systems, the antennas are vertically polarized to guarantee uniform reception and transmission in all directions, regardless of the direction in which the mobile vehicle is traveling. The mobile antenna is a half-wave vertical whip, usually mounted at the top center of the rear window of the mobile vehicle, and is used for receiving and transmitting.

The cell transmitting antenna, again, is a single, driven, vertical, omni-directional element mounted on top of the transmitting tower (see Figure 13.32). The cell receiving antenna system is shown one-third of the way down the transmitting tower in the figure and consists of six half-wave vertical

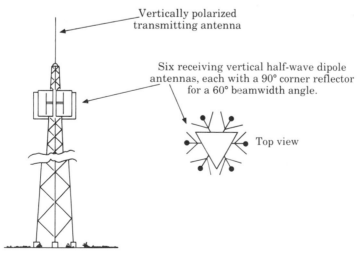

FIGURE 13.32
A cell receiving and transmitting antenna tower.

dipole antennas, each with a 90° corner reflector. The position of the dipole antenna with respect to the corner reflector gives each assembly a 60° beamwidth radiation pattern ($6 \times 60 = 360$) with about 17 dB gain. The array of the six directional receiving antennas is used to locate the mobile unit within the cell area and to compare the signal strengths being received by several cell transceivers to determine which cell will maintain control of the transmission. In this manner, the cells can detect the location of the mobile within a 60° arc and can also determine in which cell the mobile unit is located.

These voice channels are frequency modulated to ± 12 kHz, equal to 100% modulation. The voice message is limited to the frequencies between 300 Hz and 3 kHz, as are most voice channel transmissions. The transmission bandwidth (from Carson's rule) is therefore $2(\Delta f_c + f_s) = 2(12$ kHz $+ 3$ kHz$) = 30$ kHz. However, 30 kHz is also the separation between channels, which means that there are no guardbands between channels. The transmitting and receiving channels are separated by 45 MHz. The mobile transmitting frequency band (and cell receiving band) is fixed at 825 MHz to 845 MHz (665 channels), while the mobile receiving frequency band (and cell transmitting band) is fixed in the range of 870 MHz to 890 MHz (665 channels). As an example, 825.015 MHz would be the mobile transmitting frequency and 870.015 MHz the mobile receiving frequency for channel 1.

It should be noted that these are the frequencies for channels 73 to 76 and 80 to 83 in the retired UHF television band. This frequency band may be extended in the future to include 845 to 851 MHz for mobile transmitting and 890 to 896 MHz for mobile receiving, which will add another 200 channels to each cell, for a total of 865 channels. This will permit higher-density traffic in congested metropolitan areas such as Chicago and New York City. Some of the initial start-up systems are still using the older business band frequencies of 150 to 170 MHz and the 470 MHz bands, but these frequencies are being phased out in favor of the new, higher-frequency bands.

A typical set of specifications for a mobile unit follows.

General:
    Battery voltage   9.0 to 16.0 V DC (neg. ground)
    Receive current   1.1 A (max)
    Transmit current   3.0 A (max)
Receiver:
    Frequency range   870–890 MHz
    Channel spacing   30 kHz
    Frequency stability ± 2.5 ppm
    Sensitivity   1 µV for 12 dB sinad
    Selectivity:
        Adjacent channel   Better than 50 dB
        All other channels   Better than 65 dB
    Audio response   300 Hz to 3 kHz ± 1 dB
    Harmonic distortion $<5\%$
    Intermodulation -65 dB

Transmitter:

Frequency range	825–845 MHz
Channel spacing	30 kHz
Carrier stability	±2.5 ppm
Load impedance	50 Ω
Output power	3 W (nominal max level)
Power steps	Seven 1 dB steps
100% Deviation	±12 kHz peak
FM hum and noise (66% mod., 1 kHz tone)	<−40 dB (C message weighting)
Distortion	<5%
Tx attach/inhibit time	<2 ms
Carrier power inhibit	−60 dBm

Other specifications include the mechanical dimensions and accessories that are available with the unit; however, these are determined only on a personal basis.

QUESTIONS

1. What is the smallest unit of binary code information, and where does the name come from?
2. What is the relationship between the data rate and the equivalent analog frequency?
3. Define the terms RZ and NRZ.
4. Define baud rate.
5. A 3-bit binary code transmitted at 1200 BPS would have what baud rate?
6. What is the major advantage of digital signaling over analog signaling?
7. What is the Nyquist rate?
8. What is a possible difference between a class D audio amplifier and a pulse amplitude modulator?
9. Name the two major components of a sample-and-hold circuit.
10. What type of circuit is required to recover a low-frequency PAM transmission?
11. What is the duty cycle of a PAM transmission that has a high output voltage for 63 ms and a zero output for 252 ms?
12. What is the total transmission bandwidth of four voice signals from 300 Hz to 3000 Hz each that are sent down a twisted pair cable using TDM modulation?
13. Why is a zero bit in a TDM signal held to a nonzero level?
14. After an individual message has been separated from the TDM transmission, what type of circuit will restore the original message?
15. Is it necessary to synchronize a TDM transmission at the receiver?
16. What function is served by returning a zero digit to a nonzero level in a TDM transmission?

17. How wide is the pulse duration ("on" time) for a 20% duty cycle of a 500 Hz square wave carrier?

18. What is the carrier frequency when the "on" time is 2.5 μs for a wave with a 20% duty cycle?

19. What is the change in frequency (in hertz) for a PWM carrier at 5 kHz when the duty cycle changes by 20%?

20. From the following list, state which characteristics of a PWM change with modulation:
    a. frequency
    b. amplitude
    c. phase
    d. duty cycle
    e. time per cycle

21. Which of the following circuit types would recover the message of a PWM transmission? (a) diode rectifier, (b) phase-locked loop, (c) ratio detector, (d) high-pass filter, (e) low-pass filter.

22. What advantage does PPM have over PAM or PWM?

23. The pulse position modulation system requires a detector that (a) is a diode, (b) changes modulation form before filtering, (c) needs a carrier reinserted, (d) is a low-pass filter, (e) is a high-pass filter.

24. Define *simplex, half duplex, full duplex,* and *full/full duplex.*

25. Standard telephone lines are capable of handling what data rate on a consistent basis? (a) 2400 BPS, (b) 9600 BPS, (c) 39K BPS, (d) 150K BPS, (e) 600K BPS.

26. An RS232C is a (a) system concept, (b) form of modulation, (c) grade of coaxial cable, (d) standard connector pin out.

27. What is the maximum length of standard RS232C interconnecting cable? (a) 50 ft, (b) 150 ft, (c) 500 ft, (d) 1000 ft, (e) 5000 ft.

28. What operation does a host computer (or terminal) have that a secondary (or DTE) does not have?

29. Draw circuits for the star, ring, and bus system configurations.

30. Which of the three systems (star, ring, bus) is best suited for short cable runs?

31. Which of the three systems can handle the greatest traffic?

32. In FSK, the carrier rest frequency is (a) a binary 1, (b) a binary 0, (c) halfway between a 1 and a 0, (e) no output.

33. Describe the relationship that a mark and space have to a binary 1 and 0 in an FSK system.

34. What is the modulation factor when an 800 BPS data signal modulates an FSK system to a 3600 Hz space and a 2400 Hz mark?

35. Is FSK considered to be narrowband or wideband FM? Is it true frequency modulation?

36. A voltage-controlled oscillator is a (a) frequency up-converter, (b) frequency down-converter, (c) frequency-to-voltage converter, (d) voltage-to-frequency converter, (e) voltage-to-amplitude converter.

37. What value of resistance would be used with a 0.01 μF capacitor to resonate an FSK transmitter having a space frequency of 4 kHz and a mark frequency of 3.4 kHz?

38. What kind of a circuit will demodulate the FSK transmission?

39. Describe the operation of a modem.

40. What is the major contribution of the hybrid circuit?

41. A modem may operate in (a) simplex only, (b) half duplex, (c) full duplex, (d) full/full duplex.

42. In an FDM system, what is the first change to be performed on a newly arrived message signal?

43. At the lowest group carrier frequency, what filter Q is needed to attenuate the upper sideband by a ratio of 100:1 for a 300 Hz to 3 kHz message signal? (Review Equation 3.1)

44. In an FDM system, how many groups make up one super group and how many voice messages are in the super group?

45. How many voice messages are in a U600 group?

46. How many forms of modulation are used in a satellite transmission between the voice frequency input to the final satellite signal?

47. How wide is the guardband between each message in a group?

48. What is the pilot carrier frequency in a group?

49. Does each group have a pilot carrier?

50. What is the function of the pilot carrier signal?

51. Does any other higher-order group (super group or above) have a pilot carrier? If so, why? If not, why not?

52. How wide is the guardband in a master group?

53. How wide is the guardband in a multiplexed jumbo group?

54. How many balanced modulators and sideband filters are needed to form a multiplexed jumbo group?

55. How many balanced modulators are needed to recover the 10,800 voice messages in a multiplexed jumbo group?

56. In the PCM circuit referred to as the D/A converter, where do the resistors of the ladder network get their ratio of values?

57. What function does the D/A converter perform in the circuit of an A/D converter?

58. To what primary function of the A/D converter does the number of steps in the stairstep wave contribute?

59. What function is served by the clock frequency in the A/D converter of the PCM modulator?

60. How may a PCM A/D converter system be tailored to accommodate both high-level and low-level analog input signals?

61. Define *quantization noise*.

CHAPTER FOURTEEN

FIBER OPTICS

INTRODUCTION

One glance at the simplified system of Figure 14.1 is enough to provide a basic understanding of fiber optics. For a working knowledge of fiber optic systems, however, it is necessary to look at each component part separately. Fiber optics can be described in one sentence as a transmission system employing a light-emitting source, turned on and off very rapidly by electrical impulses, whose emissions are sent through a glass pipe to a light-sensitive receiver to convert the changing light intensities back into electrical impulses.

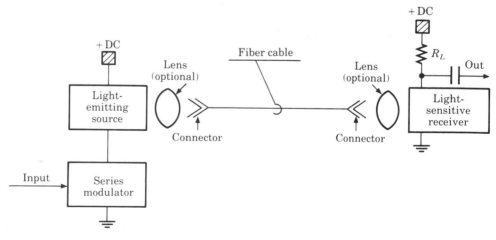

FIGURE 14.1
Basic fiber optics system.

In this chapter we will investigate the light sources, some of the modulators, the coupling into and out of the light guide, the light receivers, and, of course, the fiber cables themselves. This chapter will emphasize the budgets and other major factors needed to plan a system and will close with some design considerations.

14.2   MODULATION TECHNIQUES

Give a moment's thought to the headlights on your automobile. How many different ways can you change the light output? The direction is already fixed by the reflector, constructed to concentrate the greatest quantity of light in a specific direction.

1. You can turn the lights on and off.
2. If you have a light dimmer, you can change the brightness between two levels.
3. You can change the length of time that the lights are on or off.

That's about it. This boils down to three basic forms of modulation. "On" and "off" switching represents digital signaling ("on" for 1 and "off" for 0), which can switch very quickly indeed. Changing the light intensity over a very small range is a counterpart to linear amplitude modulation. Changing the on time compared to off time represents pulse width (or pulse duration) modulation.

14.3   FREQUENCIES

Care is advised in using the word *frequency* in association with fiber optics. We will encounter two different applications of this word, and one must be careful not to confuse the two. The most direct link to frequency as we know it in fiber optic systems is in reference to the quantum vibration of the energy emitted from the light source. The frequency of these vibrations is then converted to free-air wavelength, and *frequency* is used in this sense for all future notations. The light source frequency is sometimes called the *carrier* frequency, although it is not a true carrier in the sense of modulation as we have discussed it so far. The second use of *frequency* refers to the modulating (or message) signal frequency, which is regularly specified in bits per second. One bit can occur for each half cycle of the message frequency, so the bit rate is twice the highest analog signal frequency.

Figure 14.2 shows a situation that everyone can relate to. It depicts the electron beam of a color TV picture tube and shows the electron beam striking the red phosphor dot. The rapid exchange of energy from the beam to the dot excites the phosphor into radiating **photons** of energy, which agitate (wiggle) at 4.2857×10^{14} times per second. The wavelength of light given off by this phosphor is $V/f = 0.7 \times 10^{-6}$ m (0.7 μm). Through misalignment, the same electron beam could be made to collide with the *green* phosphor dot (the phos-

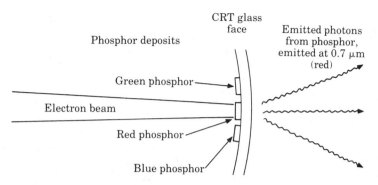

FIGURE 14.2
Color dots from a color cathode ray tube.

phor above the red dot); the photon energy given off by the green phosphor would agitate at 5.263×10^{14} times per second, resulting in a wavelength of 0.57 μm. The electron has no control over the color that is emitted by the phosphor; the chemical composition of the individual phosphors determines the color. The velocity of the electron beam and the number of electrons in the beam will change the *intensity* of the light emitted (from $E = mc^2$) but will not change the color.

The range of frequencies and the corresponding wavelengths that make up the infrared, visible, and ultraviolet light bands are cataloged in Figure 14.3. In electronics, it is customary to chart increasing frequency toward the

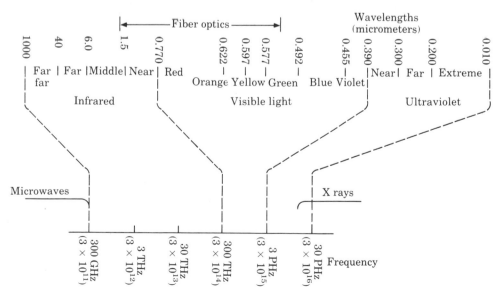

FIGURE 14.3
Light frequencies and wavelengths.

right of the diagram, which results in wavelength decreasing toward the right. This format will continue to be followed in this chapter and is *different from many other texts* on fiber optics. Be careful to observe this variation when comparing data from other sources.

14.4   FIBER OPTIC CABLES

The "core" of the fiber optic cable is a very thin strand of highly refined cylindrical glass. One ton of dry silica sand will yield about 10 lb of 99.999% pure glass. However, when you consider that about 20 mi of 0.010 in diameter core (254 μm) can be drawn from a 4.5 in cube of glass, the yield is not too bad. The glass core of the cable may have a diameter as small as 4.5 μm (2/10,000 in) or as large as 400 μm (16/1000 in). The core is surrounded by a second layer of glass called the **clad.** The clad is fused directly to the core so that the naked eye cannot see the boundary between the two materials. The clad has a different *optical density* from the core material. Both the core and the clad may be made of glass, the core may be glass while the clad is plastic, or both may be made from plastic materials of different densities. Fiber optics is founded on the theory of reflection that results at the interface between two materials of different densities. In metallic waveguide, the energy is reflected along the guide when one-half wavelength of energy is shorter than the size of the waveguide. In fiber optics, the energy will reflect down the glass waveguide when the **angle of reflection** remains smaller than a critical angle determined by the ratio of the densities of the core and clad materials.

The cross section of Figure 14.4 illustrates the construction of a cable having a glass core 50 μm in diameter with an index of refraction (see the next section) of 1.45. The cladding around the core has an outer diameter of 100 μm and an index of 1.3. The clad always has a lower index of refraction than the core material. Reflections with *zero loss* will take place at the interface surface

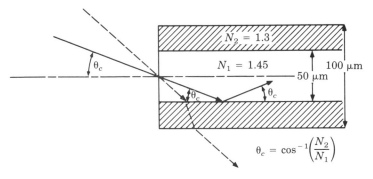

FIGURE 14.4
Light reflection inside the guide from N_1 and N_2.

of the core and clad materials, provided the light energy approaches the interface at an angle that is *less* than the critical angle θ_c.

In very early systems, highly polished mirror surfaces at the interface resulted in losses of about 0.25% per reflection. A 50 μm diameter core with a critical angle of 26.3° will result in a reflection about every 100 μm. Within about 2 in (0.05 m) the light energy would be totally dissipated. These losses do not occur when the two materials have a different index of refraction. Cable lengths of 30 mi (50 km) are successfully used in systems now, and lengths are increasing on a daily basis.

14.4.1 Refraction

Light energy traveling through a vacuum will move at a velocity of 3×10^8 m/s (984×10^6 ft/s or 186,400 mi/s). It is considered to have the same velocity in our atmosphere. As light enters any transparent medium, it slows down slightly depending on the optical density of the new material. When comparing the velocity of light in free air to the velocity of light in the given medium, the ratio is a unitless number called the **index of refraction** N:

$$N = \frac{V_c}{V_m} \qquad (14.1)$$

where V_c = velocity of light in air
V_m = velocity of light in the new medium

The index of refraction is a number larger than 1, which means that light in any transparent material moves slower than it does in air. Indices for various materials are given in Table 14.1.

TABLE 14.1
Index of refraction (N)

Material	Index
Vacuum	1.0000 (reference)
Air	1.0003
Water	1.33
Ethyl Alcohol	1.36
Magnesium fluoride	1.38 (antireflect)
Fused quartz	1.46
Glass	1.38–1.9
Zinc sulfide	2.29
Diamond	2.42
Silicon	3.4
Gallium arsenide	3.6

EXAMPLE 14.1 ══

For water, $N = 1.33$. From Equation 14.1, the velocity of light in water is $1/1.33 = 0.75$, or 75% of the velocity of light in air.

──

The term *refraction* identifies a directional change to a ray of light, as well as a velocity change when light crosses between two materials of different refractive indices. Refraction is also dependent on the angle of penetration. This principle can be demonstrated easily. Figure 14.5 illustrates a person looking over the side of a boat into a clear pond. The images we see are light rays that are reflected off of the subject and detected by the eye, so we can use the lines of sight as representing rays of light.

When the person looks straight down into the water, there is a change in velocity (which we are unaware of) but no change in direction; the seaweed appears directly below the boat. As the viewer looks toward the shore, he thinks he sees a fish at A. The angle of penetration is θ_1, so the light rays refract (bend) as well as reflect, which means that the fish is actually at B. When the viewer looks at the rock, the penetration angle θ_2 is smaller than θ_1, and the illusion is greater than when the fish was viewed. The rock appears to be at location A but is really at location B. As the viewer gazes closer to the shore line, he finds that he can no longer see into the pond but rather sees the reflection of the tree on the shore. This is because the angle of entry has become smaller than the *critical* angle θ_c.

All of these conditions depend on the refractive index of the two media (air and water). The **angle of refraction** is found by

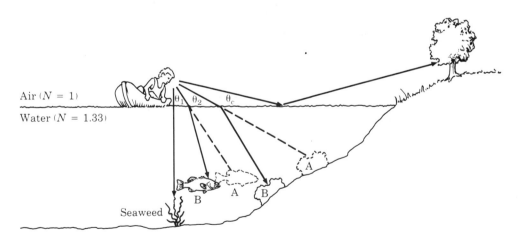

FIGURE 14.5
Refractions due to a change in index.

$$\theta_r = \cos^{-1}\left(\frac{N_0 \cos \theta_i}{N_1}\right) \qquad (14.2)$$

where N_0 = index of source (air in this case, $N = 1$)
N_1 = index to enter (water in this case, $N = 1.33$)
θ_i = angle of entry

In the case of the fish, where the entry angle is 75°, the refractive angle is

$$\theta_r = \cos^{-1}\left(\frac{(1) \cos 75}{1.33}\right) = 78.8°$$

This represents a 3.8° change in angle. For the case of the rock, where the entry angle is 45°, the refractive angle is

$$\theta_r = \cos^{-1}\left(\frac{(1) \cos 45}{1.33}\right) = 57.88°$$

representing a 12.88° change in angle.

To find the *critical angle* θ_c where reflection will begin, all we need to know are the indices of the two materials:

$$\theta_c = \theta_i = \cos^{-1}\left(\frac{\text{smaller } N}{\text{larger } N}\right) \qquad (14.3)$$

The critical angle at the fish pond is

$$\theta_c = \cos^{-1}\left(\frac{1}{1.33}\right) = 41.4°$$

We can use this information to understand how a beam of light is reflected inside a glass conductor by comparing the fiber optic cable of Figure 14.6 to the conditions at the fish pond.

Ideally, we would like all of the light from the source to enter the core of the cable on a path that is parallel to the center line of the core. This is often difficult to achieve because the light source may send out energy at every possible angle. Therefore, we need a statement that defines an angular limit for

FIGURE 14.6
Light refraction and reflection.

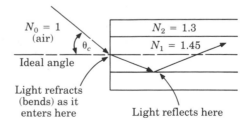

$N_0 = 1$
(air)
θ_c
Ideal angle

Light refracts (bends) as it enters here

$N_2 = 1.3$
$N_1 = 1.45$

Light reflects here

which energy will enter and propagate down the guide and that includes the critical angle of reflection. Second, at the glass core interface, our concern is the angle of penetration rather than the angle of reflection, as is the case inside of the cable, so the cosine function changes to its complement, the sine function. Third, there are now two angles to consider, one as the light moves from air to glass and another at the core-clad interface. These steps could be done separately, but in the interest of saving time in future calculations, they are combined into one equation. The only assumption is that the light is in air ($N = 1$) and enters the core.

Now the source critical angle, referenced from the center line, becomes

$$\theta_c = \sin^{-1}\sqrt{N_1^2 - N_2^2} \tag{14.4}$$

For Figure 14.4, where $N_1 = 1.45$ and $N_2 = 1.3$, the critical angle is

$$\theta_c = \sin^{-1}\sqrt{1.45^2 - 1.3^2} = 39.96° \ (40°)$$

This says that any light approaching the cable from any direction at an angle of 40° (or less) from the center line of the cable will propagate down the core of the cable. A boundary 40° away from the center line in all directions will form a cone, called the **cone of acceptance.** All light arriving at the core within the cone of acceptance will enter the core and propagate through the light guide, as shown in Figure 14.7.

14.4.2 Numerical Aperture

The **numerical aperture** is the cable's ability to collect light. It is derived from the index of the cable for a given system. It is usually specified by the cable manufacturer and is always a number less than 1. The numerical aperture (NA) may be calculated as the sine of the critical angle:

$$NA = \sin \theta_c \tag{14.5}$$

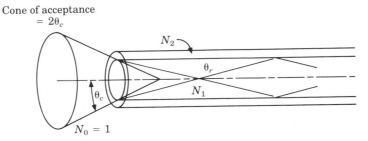

FIGURE 14.7
Cone of acceptance.

For the critical angle of 40° used in the preceding example, the numerical aperture is

$$NA = \sin 40° = 0.64226$$

When only the indices of the cable are known, the numerical aperture is found by

$$NA = \sqrt{N_1^2 - N_2^2} \qquad (14.6)$$

In the example, N_1 (core) = 1.45 and N_2 (clad) = 1.3.

$$NA = \sqrt{1.45^2 - 1.3^2} = 0.64226 \text{ (rounded to 0.64)}$$

14.4.3 Graded Index Cables

The fiber optic cables discussed so far have a sharp index change at the clad-core interface and are called *step index* cables. The cross section of a *graded index* cable is shown in Figure 14.8, where each layer of clad has a smaller index number relative to its distance from the core. Light entering the cable from a wide angle (away from center axis) will propagate in the light guide as a result of several *refractions* rather than a single reflection, as for the case of the step index cable. In Figure 14.8, θ_3 represents an angle of reflection, and all other angles are angles of *refraction*. Equations 14.2 and 14.3 can be applied to determine the individual angles resulting in a cone of acceptance: $2\theta_c = 2 \times 56.25° = 112.5°$. The graded index cable has greater light gathering ability (larger numerical aperture) than the step index cable. It also has a smaller number of modes by a factor of 2, which reduces the pulse dispersion (see the following sections) and allows for *either* higher data-rate-handling ability *or* longer cable length at the present data rate. The tradeoffs are higher manufacturing costs and slightly larger cable diameters.

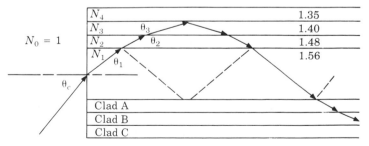

FIGURE 14.8
Graded index cable.

14.4.4 Single-Mode versus Multimode

In general, a low-order mode refers to a small-diameter cable, and a high-order mode to a large-diameter cable. More factors are taken into account and other parameters are affected by the transmission mode, but cable size is a good place to start.

When all light rays enter the optic guide parallel to the center line of the core, the light can travel the full length of the cable with no reflections. The light would pass from one end of the cable to the other end in a *single mode*. Laser light emission offers the most nearly collimated (parallel) rays. Other light sources radiate omnidirectional light, which may enter the core at some angle off center (but still within the cone of acceptance) and go through many reflections as it travels down the guide. These rays are said to be in a *high-order mode*. The light from a single source can be shown to project down the optic cable in more than one mode, and large cable size increases the probability of a number of higher-order modes, called *multimodes*. The number of modes that a given cable can support depends on the core diameter, numerical aperture (core-clad index), and the wavelength of light energy. The number of modes M differs, by design, for step index and graded index cables:

$$M = \left(\frac{\text{NA}2.22d}{\lambda}\right)^2 \quad \text{(step index)} \tag{14.7}$$

$$= \left(\frac{\text{NA}1.57d}{\lambda}\right)^2 \quad \text{(graded index)} \tag{14.8}$$

EXAMPLE 14.2

Given a core diameter of 50 μm, a numerical aperture of 0.23, and a wavelength of input light energy of 0.82 μm, the number of modes in a step index cable is

$$M = \left(\frac{0.23 \times 2.22 \times 50 \times 10^{-6}}{0.82 \times 10^{-6}}\right)^2 = 969$$

The number of modes in a graded index cable is

$$M = \left(\frac{0.23 \times 1.57 \times 50 \times 10^{-6}}{0.82 \times 10^{-6}}\right)^2 = 485$$

14.4.5 Pulse Dispersion

It will become obvious after a casual examination of a multimode transmission system that a light ray traveling down the center axis of the light guide will arrive at the output end of the optic cable before a light ray that has to reflect many times in its travel down the cable. If the reflections are numerous enough and the cable long enough, the time delay may be sufficient for one pulse to

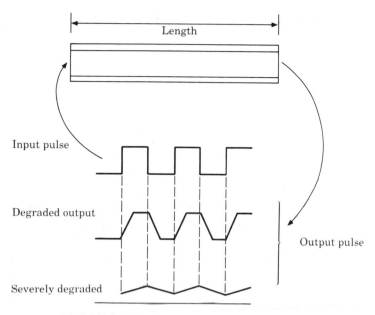

FIGURE 14.9
Pulse dispersion of input pulses in multimode cable.

completely cancel another pulse. In Figure 14.9 all of the light enters the cable at the start of the input pulse. Some of the light rays enter the cable on center axis and pass straight through the guide with no reflections (theoretically) to arrive at the output. This output energy is only a fraction of the total light input, so the output energy is low.

Other light rays entering the cable at exactly the same time, but on a path away from the center axis, could reflect thousands of times down the cable and will arrive at the output later than the on-axis light rays. The net result is that (1) the output pulse is no longer the true square wave that was put into the cable, and (2) the pulse *width* is broadened. When the pulse width is a measure of the time occupied between 50% of the rise time to 50% of the decay time, then the output pulse is observed as being wider than the input pulse. **Pulse dispersion** can degrade the output pulses to a point where the output detector can no longer tell where one pulse ends and the next pulse begins.

The second condition related to pulse dispersion in multimode cables is controlled by the number of modes, the core-clad index, the entry angle, and the cable length. It can be shown experimentally that for certain critical angles of light entry and for certain cable lengths (or multiples of those lengths), that *destructive* or *constructive* interference may occur. This interference may be viewed in the same terms as time delay of analog signals. When the pulse of a multimode input is delayed by a time period equal to the width of the input pulse, destructive interference is encountered. When the time delay is equal to two times the width of one input pulse, constructive interference results. The time delay is determined as

$$\text{TD} = \frac{N_1 L}{V_0}\left(\frac{1}{\cos\theta} - 1\right) \quad \text{(in seconds)} \tag{14.9}$$

where N_1 = core index
L = cable length
V_0 = velocity of light in free space (in the same units as the cable length)
θ = angle of light entry (worst-case $\theta = \theta_c$)

EXAMPLE 14.3

$N_1 = 1.3$, $N_2 = 1.28$, and $L = 5000$ m. From Equation 14.3, we find the critical angle to be 10°. Using these values in Equation 14.9 yields a time delay of 0.3385 µs. You would then need to compare the delay time to the time of one input pulse and make a decision from there. The most favorable conditions are found for a small angle and a low core index.

14.4.6 Cable Construction

Fiber optic cables are manufactured in three phases.

1. A highly refined silica glass tube is exposed to dopants and heat to form layers with different indices on the inside of the tube. This process is called the *modified chemical vapor deposition* (MCVD) system. Other processes in limited use are the *vapor phase axial deposition* system and the *outside vapor phase deposition* system. The MCVD system is used worldwide and is the only system discussed here.
2. The tube is heated to collapse it into a solid rod called a **preform,** which completely eliminates the center opening.
3. The preform is then heated and stretched. As the length increases, the diameter decreases. The thin-diameter cable is then jacketed for protection of the finished product.

Step 1 A highly refined pure silica glass tube, which will become the outer clad layer, is given a mild acid bath and then flushed with de-ionized water to remove surface contaminants. Filtered air drying completes the preparation. The cleansed silica tube is about 1 m long and 5 cm in diameter (40 in × 2 in). The tube is placed in a glass-working lathe, as shown in Figure 14.10. The tube must be meticulously aligned with the center line of the lathe chuck so that when the tube is rotated, the applied torch heat is evenly distributed and no sagging or bending will result from the application of heat.

Vaporized dopants called *reactants* are fed into the center of the glass tube while the tube is rotated and heated to 1300°C. The heating torch is slowly

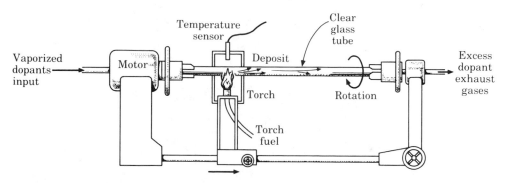

FIGURE 14.10
Glass-working lathe.

moved down the length of the glass tube. The reactants pass through the heated area, are softened, and are deposited on the inside surface of the glass tube. The excess reactants are exhausted out the far end of the tube. Several lateral passes are made by the heating torch depending on the thickness of the layer to be deposited. The rotating velocity of the tube, the torch temperature and travel velocity, and the reactant quality and quantity are all controllable factors to ensure the exact composition and thickness of each layer. Silicon tetrachloride ($SiCl_4$) is the basic dopant. When it is combined with either germanium oxides (GeO_2) or phosphorus oxides (P_2O_5), the index of refraction will increase. When boron oxides (B_2O_3) are added, the index decreases.

Step 2   When all of the layers have been deposited, the glass tube is heated to 1900°C and the atmospheric pressure causes the tube to collapse, completely eliminating the center hole. This solid rod is called a preform. The finished preform may have a cross-sectional area that is 10,000 times the area of the finished cable. The extreme care taken in the fabrication of the preform will be improved by a factor of 10,000 when the preform is altered to the diameter of the finished cable.

Step 3   Drawing the light guide fiber from the finished preform requires as much care as the fabrication of the preform. Figure 14.11 shows a typical fiber-drawing apparatus. In most cases, the fiber is drawn down in a vertical posture to avoid the distortions associated with horizontal sag and stretch, which could defeat the accuracy of the system.

The preform is mounted in a holding chuck that is capable of up-and-down movement. The preform is slowly fed down into a heating furnace that must provide high-temperature, clean heating and must not contaminate the fiber nor disturb the surface of the newly drawn fiber. The zirconia induction furnace seems to fill this requirement better than the graphite resistance furnace, the CO_2 laser furnace, or the oxyhydrogen torch furnace.

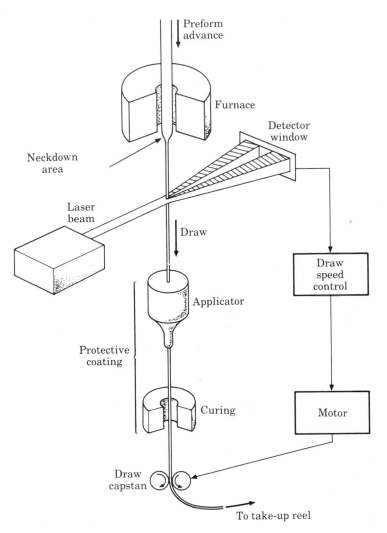

FIGURE 14.11
Core drawing process.

The cable diameter is monitored by a laser beam striking the draw and casting a shadow across the detector window. When the draw is too thick, the shadow is wide and the detector current is low, which causes the draw velocity motors to increase the speed of the draw, making the diameter thinner. The opposite effect takes place when the draw diameter is too thin. The draw diameter is maintained to ±1% in this fashion. The drawing procedure controls the mechanical strength of the cable and can contribute heavily to the cable losses both inside the cable and at the cable ends where connections are to be made. The pristine fiber leaving the neckdown area is processed through a vat containing liquid protective coating. The coating is cured before the cable reaches the fiber-drawing capstans.

14.4.7 Cable Losses

So far we have discussed the ideal case of a lossless cable, knowing full well that this condition exists in theory only. In practice, only a part of the light captured in the core will appear at the cable output. The loss is expressed in decibels as

$$\text{Loss (dB)} = 10 \log \frac{P_{in}}{P_{out}} \tag{14.10}$$

where P_{in} and P_{out} are the light input and output power, respectively. Losses, in general, are caused by impurities and irregularities in the glass core/clad preform and are carried over into the finished product. Losses are categorized as

1.   absorption losses
2.   scattering losses
3.   impurity losses
4.   mechanical variation losses

These losses are a function of the wavelength.

A graph plotting losses versus wavelength is given in Figure 14.12. The wavelength region between 0.6 µm and 1.6 µm is the most commonly used section of the light spectrum for light guide communications because it is the region of lowest overall losses. The dB loss scale in Figure 14.12 is arbitrary and depends on the cable being measured. The dB scale can change by a factor of 10 to 1 (0 to 60 on one graph, 0 to 6 on another graph). The shape of the plot, however, does not change.

Absorption losses result when the light photons in the guide have enough energy to excite the electrons in the glass core material. The oxygen ions in pure silica have very tightly bonded electrons, and only the ultraviolet light photons have enough energy to be absorbed. However, in silica light guide, the dopants and transitional metal impurities have electrons that can be excited in the visible and near-infrared light regions. These absorption trails are minimal between 1.2 and 1.3 µm wavelength. The peaks in the graph due to *hydroxyl ions* (OH−) are commonly referred to as "water peaks" and are caused by the dopant additives to the preform. The hydrogen from one dopant combines with the oxygen from another dopant to form water. The fundamental vibrations of the OH− ion occur at 2.7 µm wavelength (outside the band of interest) with overtones at 1.38, 1.24, and 0.95 µm wavelengths, which extend into the window region of optical transmission. Extreme care is taken to reduce the moisture content of the material, and to control the area where the cables are made so as to minimize the water content in the preform.

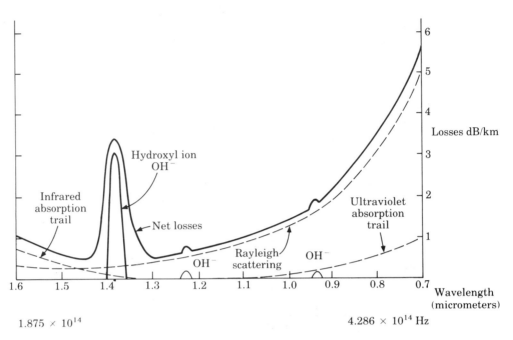

FIGURE 14.12
Wavelength vs. losses in typical fiber optic cables.

Rayleigh scattering (pronounced *ray-lay*) is a fundamental condition that results from inconceivably small variations in the density and compositional structure of the fiber preform. The scale of these impurities is smaller than the wavelength of light that they affect. They are basic and cannot be eliminated, and they set the lower limit of the fiber optic cable losses. Rayleigh scattering is identified in Figure 14.12 and again in Figure 14.13, and is proportional to the reciprocal of the fourth power of the wavelength.

Any number of *mechanical variations* can occur during the preform and draw process. The most common are microbending, core size change, bubble penetration, and microcracks. Microbending, shown in Figure 14.13a, is movement of the core off the cable center axis for a short length, with no change in core diameter. Changes in core diameter could cause the same effects. Although extreme care is taken in the preform stage, microbubbles can appear in either the core or the clad material. When they appear near the surface of the core-clad interface, they will cause scattering of the light beam and increase light losses, as shown in Figure 14.13b.

Cable Loss Summary

1. Minimum absorption losses of 0.05 dB occur at 1.38 μm wavelength, and losses increase in both directions from this wavelength. Losses at higher wavelengths are due to infrared electron absorption, while at lower wavelengths they are due to ultraviolet electron absorption.

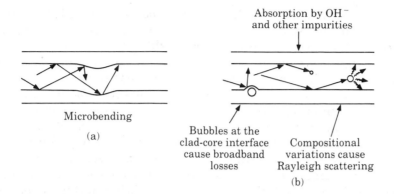

FIGURE 14.13
Cable losses resulting from variations in the construction process.

2. Hydroxyl ion contamination causes water peaks at 1.38 μm in the light guide band and at 2.4 μm outside the band. If care is not taken, hydroxyl ion losses could reach 50 dB.

3. Rayleigh scattering is due to impurities in the core material and increases with the fourth power of frequency.

4. Minimum overall losses occur at 1.28 μm and again at 1.45 μm. It is probable that most future light guide communications will be at these two wavelengths.

5. The numerical aperture will decrease as influenced by microbending and will cause an increase in the optical attenuation and pulse spreading.

6. A temperature change from 0° C to −60° could add as much as 5 dB to the cable losses. Stress (strain and tension) could add another 10 dB.

7. The glass fiber could break when the bend radius is less than 150 times the diameter of the core and clad together (150 × 125 μm = 1.875 cm or 0.75 in). Attenuation will begin to be measurable when the bend radius drops below 1000 times the diameter.

Cable losses are always stated by the manufacturer. You must be sure that the light source and the light sensor are at the same wavelength as the cable. Differences in source or detector wavelength from that of the cable could add as much as 30 dB/km of loss to your system.

14.5 LIGHT SOURCES

Remember that not just any light can be aimed successfully through the light guide. The wavelength attenuation graph of Figure 14.12 shows that when the light source is selected without foresight of the cable losses, the results could be extremely harmful. Materials that will produce light at favorable wavelengths must be considered along with source size, power output, life expectancy, and cost. The two sources that meet these requirements best are the **light-emitting diode** (LED) and the **injection laser diode.**

The light-emitting diodes for fiber optics use the same principles as the LEDs in pocket calculators. They are pn junctions that, when forward-biased, will cause minority carriers to be injected across the junction. Once across the junction, the minority carriers will recombine with the majority carriers and (in certain materials) will give up their energy in the form of light photons. The energy given up is approximately equal to the energy gap for the material. LEDs used as fiber optic light sources are treated with greater care than the simple light sources for other uses. Still, the LED is easy to construct compared to lasers, so its cost is relatively low. Epitaxial LEDs emit light peaked at about 0.94 µm (but are wide bandwidth) and are capable of about 3 mW light output at 100 mA of forward bias. LEDs have relatively slow response time, with a turn-on–turn-off time of about 150 ns, and, except for their life expectancy of 10^6 hours (114 years), are most practical for low-bit-rate, short-distance fiber optic systems.

A planar diffused-surface LED is shown in Figure 14.14a. It is formed by a controlled diffusion of zinc into a tellurium-doped gallium arsenide n wafer. A typical diode has an output power of about 500 µW at a wavelength peaked at 0.9 µm. Turn-on and turn-off times range from about 15 to 20 ns, making them better than the epitaxial LEDs, and they are useful to about 30 MHz (60M BPS). Both the epitaxial and planar diffused LEDs have a Lambertian radiation pattern similar to a directional antenna (see Figure 14.14b) and, with their large emitting surface areas, have low efficiencies for coupling into small-diameter fiber cables.

The planar heterojunction LED is constructed from six layers of semiconductor materials, as shown in Figure 14.15a. This process concentrates the

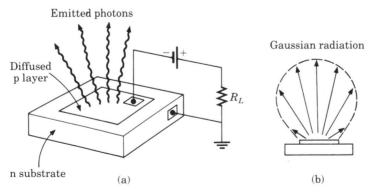

FIGURE 14.14
Planar diffused-surface LED.

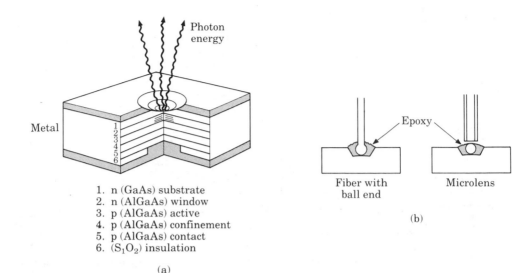

1. n (GaAs) substrate
2. n (AlGaAs) window
3. p (AlGaAs) active
4. p (AlGaAs) confinement
5. p (AlGaAs) contact
6. (S$_1$O$_2$) insulation

(a)

FIGURE 14.15
Planar heterojunction LED.

light energy to a small area at the center of the diode's diffused surface and
keeps the beam narrow enough to couple the light into small-diameter fiber
core. Concentrating the light beam also concentrates the heat energy. A *well* is
etched into the planar surface to bring the heat-generating element closer to
the surface so that the heat can be easily dissipated. The Burrus (etched-well)
LED is shown in Figure 14.15a, but the figure cannot show the precision of the
well, which is 25 μm deep in a wafer that is only 100 μm thick. Although the
heterojunction diode still has the Lambertian radiation pattern, it can be easily
coupled into a 50 μm diameter core using microlenses, as shown in Figure
14.15b (two cases).

Edge-emitting LEDs called *semiconductor injection laser diodes* were
developed to reduce the size of the light-emitting surface without sacrificing
power output. These diodes evolved in several stages, but only the most current
families of injection lasers need be discussed to provide a full understanding of
the theory behind their operation (Figure 14.16).

Injection lasers are made of an n-type gallium arsenide (GaAs) wafer into
which a p-type dopant is diffused. In the cutting process from wafer to chip
form, the two narrow ends are sliced to produce very smooth mirrorlike surfaces called *facets,* while the two larger sides are deliberately roughened in the
cutting process to discourage light emission. When this heterojunction diode is
forward-biased with a DC voltage, both ends of the laser chip will emit radiant
energy having a wavelength of 0.85 μm. When one polished end surface is
gold-plated, only the other end will emit light.

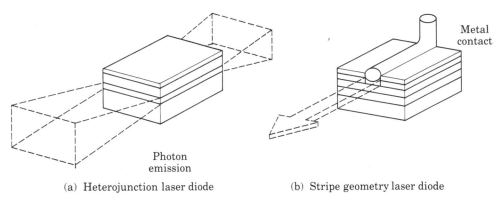

(a) Heterojunction laser diode   (b) Stripe geometry laser diode

FIGURE 14.16
Injection laser diodes.

Injection lasers are basically light-emitting diodes and behave as LEDs until the threshold bias voltage is reached, at which time they begin to lase. Above the threshold bias, the laser power output increases at an exponential rate for a continuing increase in bias voltage, as shown in Figure 14.17a. The confinement of light occurs when the refractive index of the p-type semiconductor material is higher than the refractive index of the n-type semiconductor material that borders it. The two heterojunctions act as waveguide to support a resonant frequency and several whole multiples of the fundamental frequency, which is related to the size of the cavity. When lasing takes place, the spectrum of emitted light narrows (by a ratio of 10 to 1) compared to the radiant energy spectrum before lasing. This reduction in wavelength emission is shown in Figure 14.17b.

Analyzing the wavelength chart of Figure 14.17b, note that the bandwidth of the LED (at the half-power points) covers a wavelength spread from about 1.1 μm to about 0.7 μm. If this span were converted to frequency, the

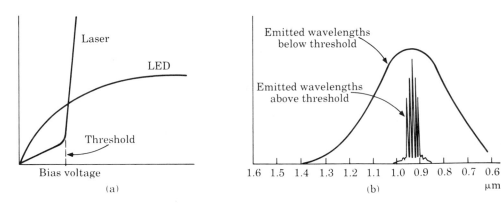

FIGURE 14.17
Spectral response of the injection laser diode.

half-power bandwidth would be about 155.8 *million* MHz. The narrower bandwidth of the laser at the half-power points would be only a mere 14.8 million
MHz. These bandwidths are light energy only, with no modulation. It may
seem that if you are the only one sending a message down your own private
cable, it makes no difference how much of the band you use; you won't be
interfering with any other message. However, the bandwidth of the emitted
light spectrum shows up as a time dispersion loss through the cable, since each
wavelength of light has a different propagation velocity in the light guide. The
greater the number of light wavelengths entering the guide, the greater the
dispersion losses. This means that the data rate will have to be reduced, or the
distance shortened, or both. Ideally, what is needed is a single-frequency light
source, called a *monochromatic* light source.

The cleaved-coupled-cavity (C^3) semiconductor laser, developed by Bell
Labs, comes close to being a true monochromatic light source. The wavelength
spectrum of the C^3 laser is better than that of the injection laser by a ratio of
400:1, as shown by the comparison plot in Figure 14.18, and better than the
LED by a ratio of 4000:1. The cleaved-coupled-cavity shown in Figure 14.18a is

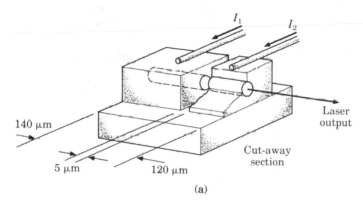

(a)

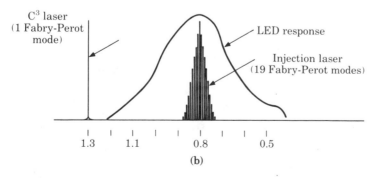

(b)

FIGURE 14.18
Cleaved-coupled-cavity laser and spectral response (monochromatic).

a gallium arsenide indium phosphide (GaAsInP) semiconductor of the injection laser family that has been sliced into two unequal lengths. The laser is constructed on a single-substrate film and then sliced so that the active channel of each section remains in perfect alignment with the channel of the other section. The gap is about 5 μm wide, and the two sections are about 120 μm and 140 μm in length. The exact size of each laser section is not critical since its electrical properties can be adjusted during operation. Removal of the size restriction makes mass production much simpler. Each section of the C^3 laser has a separate contact so that the current through each diode may be controlled individually, thus fine-tuning the output bandwidth. Each section by itself would behave as an injection laser with a half-dozen or so Fabry-Perot modes, as shown in Figure 14.18b, each mode about 20 Å apart (1 Å = 1 angstrom = 10^{-10} m). With both diodes active, the coupling between the sections can support only those modes that would be resonant in *both* sections at the same time. The lasing modes in one diode that do not match the wavelength of the lasing modes in the other diode will be canceled out. Independent control of the injection currents will alter the effective index of refraction so that the wavelength of the stimulated emission shifts the Fabry-Perot modes to cancel all but one mode. The wavelength and the injection currents are inversely proportional to each other.

The optic fibers are at their clearest at a wavelength of 1.55 μm (a range of 1.3 to 1.6 μm). When a monochromatic light source is adjusted for this wavelength and then coupled to a graded index, small-diameter cable, both the attenuation factor and the pulse dispersion will be minimized.

The industrial measure of a system's performance is called the "bit rate-distance product." It specifies the highest bit rate that can be transmitted over a 1 km cable. A system that can send 60M BPS/km can also send 30M BPS/2 km or 15M BPS/4 km. Bell Labs has announced an operational system at 2.46×10^{11} bits/km, which translates into 1×10^6 bits over a 246 km (75 mi) cable.

14.6   LIGHT DETECTORS

Of all the things you learn regarding light receivers, there are two points you should keep in mind at all times:

1. All light receivers are reverse-biased diodes.
2. The current induced by the photodiode is proportional to the photon energy that causes it.

The types of diodes used in light sensors to be examined here are

A. photosensitive transistors
B. Darlington phototransistors
C. pin diodes
D. avalanche diodes

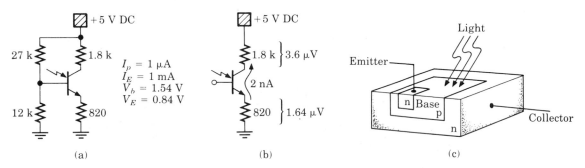

FIGURE 14.19
Photosensitive transistor.

It may seem strange that phototransistors are listed as a category of diodes, but it must be remembered that the two junctions of a transistor are diode junctions. The photosensitive area of a phototransistor is the base material of the base-collector diode. When light photons strike the base, the energy of the photons is absorbed by the electrons in the valence band. These electrons are given enough energy to break free of the valence band and enter the conduction band. The result is a flow of free electrons (current flow). When the junction is forward-biased, the net increase in current will be barely noticeable. However, when the junction is reverse-biased, the photoinduced current will be very noticeable compared to the reverse-bias current, called *dark current*.

A phototransistor is shown in Figure 14.19a, forward-biased for class A operation and having a collector current of about 1 mA. Suppose that the photosensitive base-collector junction adds 1 μA of current for 10 μW of optic input energy. The total collector current will change to 1.001 mA, a barely noticeable difference. The base circuit has been removed in Figure 14.19b, which has placed the transistor in cutoff. Base voltage is zero, emitter voltage is zero, and the collector approaches V_{cc}. The only current that will flow is the leakage current through the emitter-base diode and the base-collector diode. The leakage current (dark current) may be as low as a few nanoamperes. Now when 1 μA of photoinduced current is introduced, the change in collector current could be hundreds of times larger than the dark current. This change in collector current develops a signal voltage across the collector load resistor as the output signal voltage.

The simple circuit of Figure 14.20 has a discrete phototransistor receiver circuit for a low data rate of 25K BPS and an optical input sensitivity of about 5 μW. The phototransistor, Q_1, acts as the emitter resistance of Q_2. Changes of current in Q_1 match the changes of current in Q_2. The output voltage of Q_2 is the collector current of Q_2 through the 100 kΩ variable collector load resistance. Q_3 and Q_4 are direct-coupled voltage amplifiers. The response time of this circuit is 2.5 μs rise and fall time, with a responsivity of 18 μA/μW. The dark current is 25 μA.

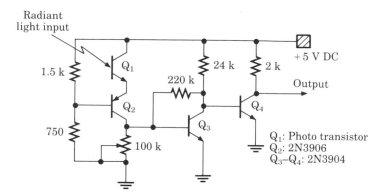

FIGURE 14.20
Discrete phototransistor optic receiver circuit.

There is little to add to the description when the phototransistor is a Darlington pair, as shown in Figure 14.21. The Darlington pair dark current is 100 μA and the responsivity is 500 μA/μW, an increase by a factor equal to β over the single transistor. Less amplification is needed for an output equal to the single-phototransistor output. The rise and fall time of this circuit is about 50 μs. Both the phototransistor and the Darlington phototransistor are practical for receivers in short-distance fiber optic systems with a data rate below 100K BPS.

14.6.1 Pin Diodes

The depletion region in a reverse-biased pn junction diode starts at the border of the two materials and extends into the n material by a width dependent on the DC bias. The absorption region begins at the crest of the depletion event and penetrates into the n material until all of the minority carriers are recombined with majority carriers. The response time of the pn diode is the time it takes for the carriers to totally recombine and is a function of the width of the depletion region compared to the width *and location* of the absorption region

FIGURE 14.21
Discrete Darlington phototransistor receiving circuit.

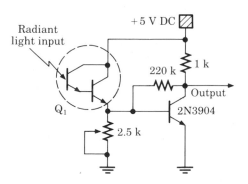

(see Figure 14.22a). To increase the response time (high-speed devices) of the diode, it is necessary that the depletion and absorption regions occupy the same space within the n material. This can be done by increasing the DC bias voltage or by decreasing the doping of the n material. The bias is limited by the peak inverse voltage rating of the diode; therefore, the n material doping level is reduced until the n material is *intrinsic* (almost no longer an n-type semiconductor). A high ohmic contact to the external circuit is required, so another large n-type layer is added to dissipate heat. This assembly develops into a *p-type, intrinsic-type, n-type* diode, or **pin diode.**

When normally biased for DC, the pin diode will perform exactly as any other diode. That is, it will rectify the AC component of the wave (below light

FIGURE 14.22
Pin diode compared to a pn diode.

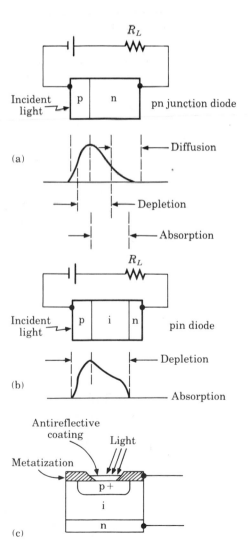

FIGURE 14.23
Pin diode with low-noise, high-gain,
current-to-voltage comparing amplifier.

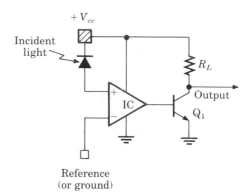

Reference
(or ground)

frequencies in this case). When the wavelength of light energy striking the
photosensitive element is shorter than the distance between the p and n layers,
the intrinsic layer begins to store a charge. The intrinsic layer now acts like a
variable resistor, easily controlled by the DC bias voltage. High bias makes the
diode look like a low resistance (as low as 1 Ω), whereas low bias makes the
diode look like a high resistance (about 1 kΩ) to the high signal frequencies.
Note here the two different uses of the word *frequencies:* (1) The wavelength of
the light frequencies, 5×10^{14} Hz, is equal to 0.6 μm and sets the length of the I
layer. (2) The signal frequencies, to the gigahertz range, represent the modulating signal and are affected by the resistance value of the intrinsic layer. At
wavelengths of 0.6 μm, silicon diodes can be made to have a response time of
0.5 ns and efficiencies approaching 90%. More typical response values range
from 2 ns at 20 V to 6 ns at 5 V DC bias.

Several words (with one meaning) are used interchangeably in reference
to the sensitivity of photodiodes. They are *collector light current, radiation sensitivity,* and *responsivity.* All are stated in amperes of diode current for watts of
light energy. A typical sensitivity value for pin diodes is 0.5 μA/μW, which is
high compared to photodiodes, with values of 0.2 to 0.3 μA/μW. A typical circuit using a pin diode with a low-noise, high-gain current-to-voltage comparing
converter and amplifier is illustrated in Figure 14.23.

14.6.2   Avalanche Diodes

Avalanche diodes are the same devices referred to in microwave systems as
IMPATT diodes and are similar to varactor and zener diodes but have a different thermal design. Avalanche diodes are made from materials that respond to
light photons and, like photomultiplier tubes, are capable of generating a signal current *many* times greater than the current produced by the incident light
photons. The avalanche diode is a four-layer device consisting of an n layer, a p
layer, a high-resistivity n-intrinsic layer, and a p$^+$ (highly doped p-type) low-
resistivity layer. The disadvantage of the diode is the high reverse bias of 70 V
to 300 V required for avalanche.

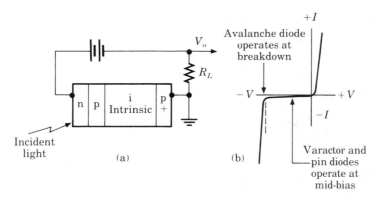

FIGURE 14.24
Avalanche diode.

The avalanche diode is reverse-biased at just below the zener knee of the curve, as shown in Figure 14.24b. High-velocity photons enter the diode at the n layer, penetrate the p layer, and are absorbed in the intrinsic region. In their flight, the photons release hole-electron pairs in the p layer. The p layer is loosely bonded so that each evacuated electron will collide with a second electron. Each of these electrons will collide with still another pair of electrons and set them free. The process continues on, resulting in an *avalanche* of electrons (more correctly, hole-electron pairs). The holes migrate toward the p^+ layer, and the electrons move toward the positive potential at the n layer, thus creating current flow. When the amount of current flow under dark-current conditions is compared to the ratio of photon-to-electron current flow, the effect resembles a high current gain. Dark current and surface leakage current are typically around 10 nA at 200 V DC bias, and projected life expectancies exceed 1000 years.

The major point to observe here is that even a small change in bias voltage will result in a magnified change in current flow. Because the diode is extremely sensitive, this small change in voltage could be mistaken as a signal pulse across the output load resistor. To control this false pulse condition, the avalanche diode bias voltage must be rigidly regulated, as shown in Figure 14.25. With no signal input to the circuit, the output from the AGC amplifier is low, as is the output from the peak detector, which results in no control voltage to the AGC control system. The high-voltage supply is high, the AGC amplifier is at maximum gain, and the output signal is the system noise only. When an optical signal activates the avalanche photodiode (APD), the low-noise amplifier applies signal to the AGC amplifier and to the peak detector. The AGC control now has an output that lowers both the high-voltage supply and the gain of the AGC amplifier. The overall effect is a constant output signal and a better signal-to-noise ratio at the signal output (V_o). Varying the input level by a factor of 5 to 1 and the AGC gain by a factor of 1000 to 1, a 37 dB optical (74 dB

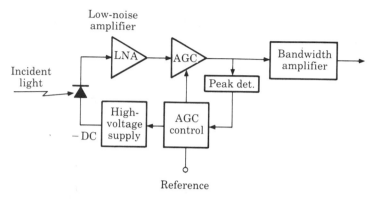

FIGURE 14.25
Practical avalanche photodiode receiver.

electrical) dynamic signal range can be accommodated without system over-load.

Good sensitivity, high current gain, fast response time, and an output peaked at one of the better fiber wavelengths makes the avalanche diode the favorite detector for use in low-level, high-speed fiber transmission systems, in spite of the expensive control circuitry needed to tame the DC bias drifting. Typical specifications are

Responsivity	10 $\mu A/\mu W$
Current gain	30–100
Response time	2.5×10^{-10} s
Wavelength	1.3 μm

14.7   CONNECTORS

There are two basic types of connectors used in fiber optics, the single-fiber and the multifiber connector. Use of one or the other depends on the power require-ments of the intended system and, therefore, the size or shape of the source exit port. Very high power sources may have a large exit port to couple the energy efficiently into the fiber system. In such cases, a bundle of optic fibers may be required rather than just a single fiber. Systems with very high data rates require practically a single-mode cable. This suggests a small-diameter fiber and a single-conductor cable connector.

The major difference between the metallic connectors used for coaxial cables and the connectors used in fiber systems is that the core of the optic system cable must be butted against the core of the light-emitting source, end to end. In Figure 14.26a, we see a *ferruled* light source housing package, which is shown fitted into the chassis-mounted threaded holder in Figure 14.26b. A cable connector is attached to the fiber end and then to the opposite end of the

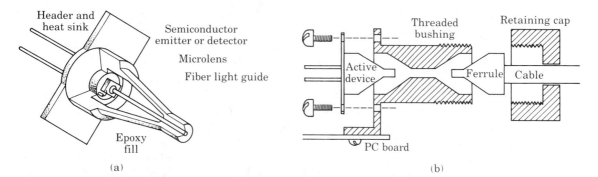

FIGURE 14.26
(a) Ferruled source housing. (b) Threaded ferrule holder on PC board.

threaded holder to butt the two cable ends together. It is on the mating of these two cable ends that the assembler should focus the most attention. Remember that we are butting together the ends of two cores with a diameter smaller than that of a human hair (50 μm = 0.002 in).

Figure 14.27a, b, and c identifies some of the conditions that the industry technology can control. Figure 14.27a shows the core and clad perfectly round and centered, the ideal condition. Figure 14.27b shows both the core and clad as round, but the core is off center (eccentric) within the clad. Figure 14.27c shows the core centered within the clad, but out of round (elliptical). These sketches

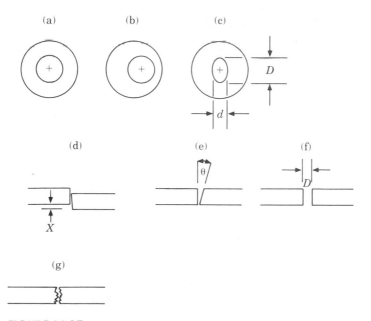

FIGURE 14.27
(a), (b), (c) Some cable variations. (d), (e), (f), (g) Installer-controlled variations.

reemphasize how control of the preform and draw processes are so critical at the manufacturing level.

Figure 14.27d, e, f, and g illustrates connector features over which a great deal of control can be maintained. Figure 14.27a shows two cores off center by a factor of X. Using the example of a 50 μm core diameter, the normalized value of offset is X/r, where r is the core radius. When the cores are offset by 1 μm, the loss factor is $1/25 = 0.04$. This translates into a loss of 2 dB, as indicated on the graph of Figure 14.28 (leftmost plot).

Figure 14.27e shows the effect when the adjoining ends of the core are not squared; one core is squared, and the other is angled. The loss factor resulting from angular misalignment is determined as the sine of the angle. Again using a 25 μm core radius with an angular misalignment of 10°, the normalized loss factor is 0.17365. Translating this factor to the graph of Figure 14.28 (center plot) indicates a loss of about 0.8 dB.

When the core is honed too harshly, so that it is short when fitted into the socket connector, there will be a separation between the ends of the two cores. The loss factor due to separation is represented by the distance D divided by the radius r. A separation of 10 μm corresponds to a normalized loss factor of $10/25 = 0.4$, a loss (from Figure 14.28, rightmost plot) of about 1 dB.

The elliptical loss depicted in Figure 14.27c is the ratio of the narrow dimension (d) to the wide dimension (D) of the ellipsoid. For a 50 μm core, if the ellipsoid measures 37.5 μm × 62.5 μm, then the normalized loss factor will be $37.5/62.5 = 0.6$. When this factor is carried over to Figure 14.28 (rightmost plot), the loss is found to be about 1.8 dB.

The rough surface depicted in Figure 14.27g is not measurable, but is included as a category of less-than-ideal conditions for optimum transmission of light from one fiber to another.

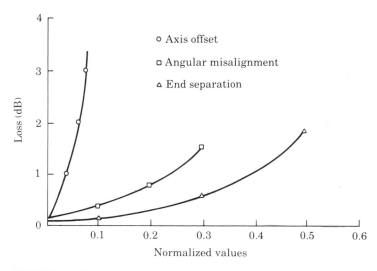

FIGURE 14.28
Normalized values of connector end conditions.

For a single-fiber connection, the individual losses are summed to constitute the total loss per connection. For a connection having all the flaws shown in Figure 14.27, the losses would add up to 5.6 dB, indicating a very poor connection. Typical losses are between 0.5 and 3.0 dB per connection. Metal connectors have greater precision than molded plastic connectors and therefore have lower losses, but they are much more expensive.

The connector assembly procedure is presented in picture form in Figure 14.29. Figure 14.29a is an exploded view of the connector subassembly. After stripping the cable, as per Figure 14.29b, slide the crimping ring over the cable first, and then slide on the retainer cap. Apply a light coat of permanent glue to the clad and the cable jacket before sliding the cable end into the ferrule. Allow

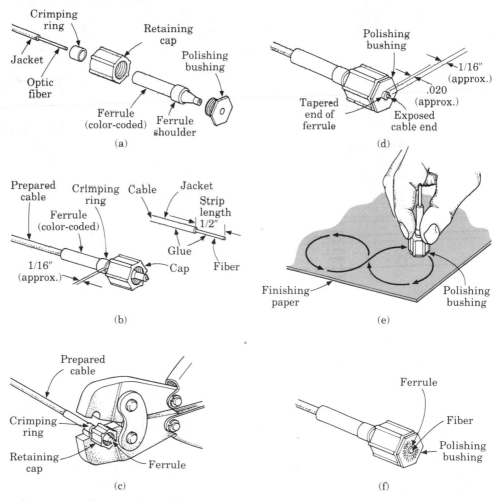

FIGURE 14.29
Step-by-step cable connector assembly.

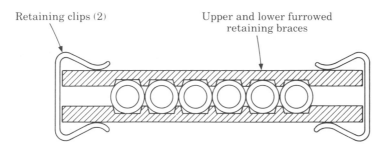

Retaining clips (2)

Upper and lower furrowed
retaining braces

FIGURE 14.30
Six-fiber conductor multicable retaining assembly.

a few minutes for the glue to dry. Move the crimping ring forward to a position over the ferrule and allow free movement of the retainer cap. Lock the crimping ring in place. Attach the threaded polishing bushing, making sure that the core extends out past the polishing surface. Holding the cable in a vertical position, polish the cable end by using a figure-8 motion across a piece of finishing paper (grade 000 sandpaper). Apply only light pressure. Replace the finishing paper with a soft polishing cloth and buff the cable end. Remove and discard the polishing bushing, and the cable is ready for installation.

A six-strand bundle of fibers is shown in Figure 14.30, where each strand has a 125 μm (clad) diameter. The multiconnector is butted against a light source with an output port of 750 μm $\times$ 50 μm cross-sectional area. The efficiency of coupling is the ratio of the total core area to output port area. When the core size is greater than the output port size, the efficiency is typically around 80% (-1 dB).

14.8   ADVANTAGES/DISADVANTAGES

Now that the basic facts about fiber optic systems are known, it will be easier to discuss the relative features of the systems. The following list is in no order of importance and does not take into account any differences that may exist in the external circuitry.

Advantages

1. *Resource conservation:* The main ingredient in glass is sand, and there is an almost unlimited supply of sand in the world compared to the supply of copper or aluminum.
2. *Safety:* Photons of light rather than an electrical current move through the optic fiber. Therefore, there is no chance of a spark flash, which could be dangerous.
3. *RFI:* Since the fiber system carries no electrical current, the energy transmitted through the fiber cannot radiate RF interference, nor can it be contaminated by any external noise or RF fields.

4. *Security (privacy):* It is nearly impossible to eavesdrop on fiber optic systems without being easily detected. Because of the absence of current flow through the fiber, criminal intrusion into the system is also prevented. Confidential information cannot be routed to unwanted receivers, nor can false information be fed into the data stream.

5. *Low losses:* Examination of 46 fiber types revealed a range of losses from 2 dB/km to 385 db/km, with an average loss of 27 dB/km. If we discount the two high-loss types, the average loss drops to 12.5 dB/km. Metallic cables at equivalent information rates (but not light frequencies) show an average loss of 49 dB/km. The 30 types of coaxial cable examined showed a minimum loss of 27 dB/km and a maximum loss of 146 dB/km.

6. *Bandwidth:* Within a totally closed system, the number of signals that can be modulated on a fiber optic light beam exceeds the number that can be modulated on a very high frequency RF carrier by a factor of about 1000. Considerations include equal bandwidth per message signal and the highest ratio of carrier to message frequency in both cases.

7. *Deterioration:* Glass is immune to corrosive and oxide degradation and will stand up well in harsh environments. Moisture, toxic vapors, and acids will not degrade the glass fibers. *Do not* use plastic fibers in toxic environments.

8. *Size and weight:* The size of the core and clad of a single fiber conductor is much smaller than the diameter of a common copper wire conductor; however, when the insulation is included, the sizes are similar. Bundles of fiber cables are smaller by a factor of 10 and weigh less by a factor of 14 than an equal number of copper wire conductors. More signals may be transmitted on a single fiber than can be carried on a single copper wire. When weight is a major factor, make sure that the savings in cable weight are not offset by the added weight of the extra electronics.

9. *Temperature:* Excluding the protective insulation, the melting point of glass is much higher than copper, and that of copper is much higher than the plastics used for cores and clads. Melting points are: glass, 1900°C; copper, 1083°C; and plastics, 100–400°C.

10. *Long life:* Based on the history of glass and due to its immunity to harsh environments, the life expectancy of glass fibers is predicted to exceed 100 years. Predictions for plastic fibers have yet to be determined.

Disadvantages

1. *Limited application:* All fiber optic systems are limited to *fixed* point-to-point *ground* installations. They cannot leave the ground nor be associated with a mobile communication station.

2. *Low power:* Popular light-emitting sources are restricted to very low power devices. There are higher-power devices available, but they carry a higher price tag than the repeater amplifiers they would replace.

3. *Distance:* Because of the low-power sources, the distance between repeater amplifiers must be relatively short for the high data rates demanded in some systems.

4. *Modulation:* The ways in which the light source can be modulated are limited, although advanced forms of modulation are applicable to the subcarrier signals prior to the fiber optic cable systems.

5. *Nuclear radiation:* It is a known fact that glass will darken when exposed to neutron bombardment. The harder the glass, the more quickly it will discolor. The half-life recovery time of the glass is a function of its hardness and is measured in terms of decades.

6. *Fragility:* Although vibration tests have been made on glass fibers to predict their longevity, insufficient time has passed since their development to ensure freedom from hazing (microscopic cracking) due to age and vibrations.

Products and systems are improving on a daily basis, and predictions are that the majority of point-to-point fixed installations will see their metallic conductors replaced by fiber optic systems.

14.9   BIT ERROR RATE

It has been adopted as an industry standard to tolerate no more than one error for every one *billion* bits of information (1×10^{-9}). A higher error rate would degrade the system performance, while a lower rate would put too great a demand on the devices. The quantity of 1×10^{-9} errors is called the *bit error rate* (BER) and represents the amplitude to which noise peaks will rise only once in 10^9 attempts. Mathematically and experimentally, it has been found that a peak signal voltage of 12.3 times the rms noise voltage is necessary to establish a 10^{-9} BER. These values calculate to a signal-to-noise ratio of 21.8 dB.

14.9.1 Minimum Input Signal

In planning a system, one practical place to start is at the detector. Here you want to know the minimum input signal that would not exceed the acceptable BER. To make this decision, select the photodetector with the highest SNR. A second specification of the detector is its *responsivity*, which is a measure of the photodetector's response, in amperes (or volts), to the photon light energy input, in watts.

EXAMPLE 14.4 ==

A detector consisting of a pin diode and preamplifier has these specifications: responsivity = 4.5 mV/μW, measured at V_{cc} = 5 V DC, λ = 0.82 μm,
and P_{in} = 10 μW. SNR = 24 dB at 2 μW peak input power.

The procedural steps for converting responsivity and SNR into the
minimum input signal power that would exceed 10^{-9} BER is as follows:

1. The SNR is stated at 2 μW peak input power, and the responsivity
   is stated in mV/μW. Therefore, at 2 μW the output must be 2 $\times$
   4.5 mV = 9 mV$_{pk}$.
2. Convert 9 mV$_{pk}$ to mV$_{rms}$: $9 \times 10^{-3} \times 0.7071 = 6.634$ mV$_{rms}$.
3. Find the rms noise voltage at 24 dB lower than the 6.634 mV$_{rms}$
   signal:

$$24 \text{ dB} = 20 \log \frac{6.634 \text{ mV}}{e_{\text{noise(rms)}}} = 0.4 \text{ mV}$$

4. Find the peak message signal voltage needed to override the noise
   and establish the 10^{-9} BER: 0.4 mV $\times$ 12.3 = 4.92 mV.
5. Using the responsivity value, set up a ratio to compare the known
   facts to the unknown facts, that is, 4.5 mV is to 1 μW as 4.92 mV
   is to the unknown input power:

$$\frac{4.5 \text{ mV}}{1.0 \text{ } \mu\text{W}} = \frac{4.92 \text{ mV}}{P_{in}}$$

Therefore, P_{in} = 1.094 μW.
6. Convert P_{in} to dBm:

$$\text{dBm} = 10 \log \frac{1.094 \text{ } \mu\text{W}}{1 \text{ mW}} = -29.6 \text{ dBm}$$

14.10   FLUX BUDGET

Planning a fiber optic communications link may be started by knowing either
of two major limiting factors: the highest data bit rate or the transmission path
length. When the system combines high data rates with a long transmission
path, then repeater amplifiers may be required at appropriate intervals along
the way. In such cases, then each link is treated as a separate system. It is
reasonable to assume that all component parts of the system (except the light
source) will add losses to the system. Calculating the overall loss is called
planning a *flux budget*. Cost is always a factor and should be included in each
decision that is made in the planning stage. However, cost will take a back seat
to technical excellence in the following listing of planning steps. Two separate
approaches are described here in example form:

1. *Plan A*. Know the requirements at the receiver, calculate the losses for a specific length, determine the source power, find the system rise time budget, and then modify as needed.
2. *Plan B*. Know the requirements at the receiver, know the power and bit rate at the source, determine the maximum cable length and rise time budget, and then modify as needed.

Plan A

EXAMPLE 14.5

The system is to operate at 20M BPS over a 1500 m cable. Using the detector from Example 14.4, we have
Detector:

$$\text{minimum input power} = -29.6 \text{ dBm for } 10^{-9} \text{ BER}$$
$$\text{numerical aperture} = 0.7$$
$$\text{core diameter} = 200 \text{ } \mu\text{m}$$
$$\text{wavelength} = 0.82 \text{ } \mu\text{m}$$
$$\text{rise-fall time} = 10 \text{ ns}$$

Cable:

$$\text{wavelength peak} = 0.82 \text{ } \mu\text{m}$$
$$\text{diameter} = 100 \text{ } \mu\text{m}$$
$$\text{loss at } 0.82 \text{ } \mu\text{m} = 7 \text{ dB/km}$$
$$\text{numerical aperture} = 0.3$$
$$\text{response} = 20 \text{ MHz at } 0.9 \text{ } \mu\text{m}$$

Source:

$$\text{power output } (I \text{ DC} = 100 \text{ mA}) = 700 \text{ } \mu\text{W at } 0.82 \text{ } \mu\text{m}$$
$$\text{diameter} = 200 \text{ } \mu\text{m}$$
$$\text{numerical aperture} = 0.5$$
$$\text{rise-fall time} = 12 \text{ ns}$$

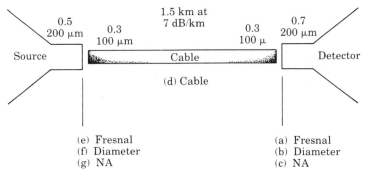

FIGURE 14.31
Simple system flux budget diagram.

Using these specifications and Figure 14.31, the following losses are tabulated.

Losses	Net
	Detector minimum input = −29.6 dBm
A. Fresnel = 2 × 0.2 = −0.4 dB	−29.2 dBm
B. Diameter = 20 log 100 μm/200 μm = −6.02 dB	−23.18 dBm
C. NA = 20 log 0.7/0.3 = +7 dB (omit)	
D. 1.5 × 7 dB = −10.5 dB	−12.68 dBm
E. Fresnel = 2 × 0.2 = −0.4 dB	−12.28 dBm
F. Diameter = 20 log 200 μm/100 μm = +6 dB (omit)	
G. NA = 20 log 0.3/0.7 = −4.437 dB	−7.843 dBm
H. Long-term degradation = −3 dB (total losses)	−4.843 dBm
I. Assume an additional 1 dB per connector for axial and separation losses (−2 dB)	−2.843 dBm

Convert −2.843 dBm to microwatts:

$$\text{Input power } (P) = (0.001)\,\text{antilog}\,\frac{-2.843\ \text{dBm}}{10} = 520\ \mu\text{W}$$

The interdevice losses are defined as follows. **Fresnel losses** are another name for surface reflection losses and occur as light leaves one medium to enter another; 100% exchange of light energy is not achieved. In the case of glass to air or air to glass (which is the only case of concern), there is about 4.5% light loss. This amounts to about 95.5% efficiency. In decibels,

$$\text{Loss (dB)} = 10 \log 0.955 = -0.2 \text{ dB power loss per surface}$$

Diameter loss occurs when the diameter of the core from which light is leaving is larger than the diameter of the core into which the light is entering:

$$\text{Loss (dB)} = 20 \log \frac{\text{entry core diameter}}{\text{exit core diameter}} \quad \text{(Omit if dB value is positive)}$$

The **numerical aperture loss** occurs when the NA of the sending device is larger than the NA of the accepting device:

$$\text{Loss (dB)} = 20 \log \frac{\text{NA in}}{\text{NA out}} \quad \text{(Omit if dB value is positive)}$$

A typical light-emitting source circuit is shown in Figure 14.32. The data (from whatever source) are applied to add to or subtract from the diode voltage altering the diode current. The drive control adjusts the bias to set the steady-state diode current.

Next the *rise time budget* must be found. Because of the pulse spreading effects in multimode fiber optic cables, the total system should be analyzed to ensure proper handling of the intended data signal. The system rise time is determined as

FIGURE 14.32
Typical injection laser drive circuit.

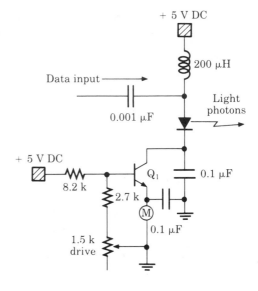

$$T_s = \sqrt{T_c{}^2 + T_d{}^2 + T_t{}^2} \qquad\qquad (14.11)$$

where T_s = system rise time
T_c = cable rise time
T_d = detector rise time
T_t = transmitter rise time

EXAMPLE 14.6

The cable specifications state the bandwidth-distance product to be 20 MHz/km, so that at 1.5 km the bandwidth is 13.333 MHz. The cable rise time is:

$$T_c = \frac{0.35}{f} = \frac{0.35}{13.33 \text{ MHz}} = 26.25 \text{ ns} \qquad\qquad (14.12)$$

Adding this quantity to the detector rise time of 10 ns and the light source rise time of 12 ns gives the system rise time:

$$T_s = \sqrt{26.25^2 + 10^2 + 12^2} = 30.545 \text{ ns}$$

Converting the system rise time back to frequency,

$$f = \frac{0.35}{T_s} = \frac{0.35}{30.545 \text{ ns}} = 11.458 \text{ MHz}$$

Since the frequency is one-half the bit rate, the bit rate is 2×11.458 MHz = 22.916M BPS (greater than the system data rate of 20M BPS). This system is operational at 520 μW, which is 75% of capacity and allows for freedom of adjustment.

Plan B

EXAMPLE 14.7

The source is selected as a gallium aluminum arsenide double-heterojunction injection laser diode with the following specifications:

$$\text{power output} = 6 \text{ mW at } 0.82 \text{ } \mu\text{m}$$
$$\text{diameter} = 200 \text{ } \mu\text{m}$$
$$\text{numerical aperture} = 0.4$$
$$\text{rise-fall time} = 8 \text{ ns}$$

We will run the laser at 80% of capacity at signal peaks of 4.8 mW (+6.8125 dBm).

The detector is selected as an avalanche photodiode (APD) for very high sensitivity, with at least these specifications:

$$\text{core diameter} = 250 \text{ } \mu\text{m at peak response } 0.82 \text{ } \mu\text{m}$$
$$\text{numerical aperture} = 0.5$$
$$\text{responsivity} = 80 \text{ } \mu\text{A}/\mu\text{W}$$
$$\text{dark current} = 0.05 \text{ } \mu\text{A}$$
$$\text{breakdown voltage} = 375 \text{ V DC}$$
$$\text{gain} = 120$$
$$\text{rise-fall time} = 2 \text{ ns}$$
$$\text{capacitance} = 2 \text{ pF}$$
$$\text{quantum efficiency} = 85\%$$

The BER calculations for pin diodes are extremely rigorous (even more so for avalanche diodes with gain) and somewhat variable. Rather than presenting long calculations, the graph in Figure 14.33 plots bit rate versus sensitivity directly in dBm for 10^{-9} BER for a whole family of pin diodes and avalanche diodes. This graph has been proven accurate to within 1 dB, both mathematically and experimentally, in laboratory measurements.

The graph is used here by picking a data rate (the same as was used for plan A) and then checking the graph for the 10^{-9} BER sensitivity. At 20M BPS, the sensitivity is −62 dBm. The difference between the source output power (+6.8125 dBm) and the detector input level (−62 dBm) is 68.8 dB. This difference will include the two connectors and the cable losses. If we use the same cable as in plan A as a start, we can seek improvements after completing a trial flux budget.

At the detector:

$$\text{Fresnel } = 2(2 \times 0.2/200 \text{ } \mu\text{m}) = -0.4 \text{ dB}$$
$$\text{Diameter} = 20 \log 100 \text{ } \mu\text{m} = +6 \text{ dB (omit)}$$
$$\text{Numerical aperture} = 20 \log 0.5/0.3 = +4.437 \text{ dB (omit)}$$

At the source:

$$Fresnel = 2(2 \times 0.2/100) = -0.4 \text{ dB}$$
$$Diameter = 20 \log 250 = -7.959 \text{ dB}$$
$$Numerical \ aperture = 20 \log 0.3/0.4 = -2.5 \text{ dB}$$
$$Axial \ and \ separation \ losses = -2 \text{ dB}$$
$$3 \text{ dB for aging} = -3 \text{ dB}$$

The losses total -16.259 dB. When these losses are subtracted from the remaining budget balance, the cable to be selected can have only 52.54 dB total loss. At 7 dB per kilometer, the cable length may not be longer than 7.5 km.

Obstacle 1: The fiber cables are supplied in 1 km lengths. Therefore, seven splices would be required. Estimating 2.5 dB loss per splice adds 17.5 dB more losses than the total calculated, so the cable cannot be a full 7.5 km long. Using five splices adds 12.5 dB of loss. Thus, the original allowable cable loss of 52.54 dB less 12.5 dB leaves us with only 40.04 dB, divided by 7 dB/km = 5.72 km.

Obstacle 2: Our selected fiber cable has a bandwidth-distance product of 20 MHz/km. At a data rate of 20M BPS (10 MHz), the cable length should not exceed 2 km.

The only alternative is to seek a different cable. Examining the cable spec sheets, an all-glass step index cable is found having a bandwidth-dis-

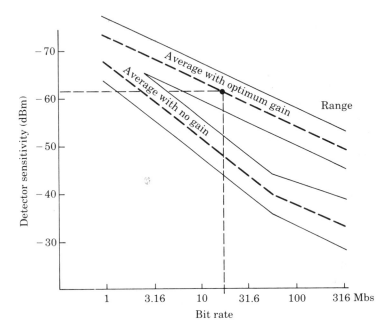

FIGURE 14.33
Typical receiver sensitivity curves (minimum input for 10^{-9} BER).

tance product of 50 MHz/km and 8 dB/km. It has a diameter of 62.5 μm and
an NA of 0.16, which means that the diameter and NA losses will have to
be recalculated, adding on losses for four in-line splices at 2.5 dB per splice.
It is left to the reader to rework the steps of plan B, using the new cable
with four splices, and then to recalculate the rise time budget. The results
are:

Cable length = 4.3675 km (2.7144 mi)
Total rise time = 3.1665×10^{-8} for 11.053 MHz (22.1M BPS)

This system is said to be cable-loss-limited, as opposed to rise-time-limited.

If we took the best fiber cable on the list, at 2 dB/km and 400 MHz/km,
with a diameter of 4.5 μm and an NA of 0.1, and then developed a 10 W
light source, the maximum distance would be 25 km (15.5 mi) at a maximum frequency of 7.85 MHz (15.7M BPS).

This example may help you to understand the problems of achieving a high-
quality fiber system and should promote respect for a system that has a one
billion BPS signal transmitted over a one-piece 75 mi (246 km) cable without
additional amplification.

14.11   LASERS

Although it is not the intent of this chapter to delve heavily into the theory of
lasers, it would be remiss to omit them entirely since lasers are surely a part of
fiber optics. The classifications of lasers are

1. semiconductor lasers
2. gas lasers
3. solid lasers
4. liquid lasers

All lasers have an active material to convert the input power to radiant
light in a fashion that resembles oscillations and amplification, a feedback
optic system to reflect the beam back and forth through the active material,
and an output mechanism. Laser light is monochromatic (one color) and is
theoretically described as a single frequency of concentrated, collimated, in-
phase rays of light, referred to as *coherent light*. The collimated light will hold
the rays in a narrow beam for great distances.

The semiconductor laser was covered in some detail under light sources. It
is referred to as an injection laser diode. The excitation mechanism is the DC
current through the diode to control the amount of light output. The DC current can be easily modulated.

Gas lasers use helium or neon as the active medium and operate on the
principle of discharging a current into a gas for stimulation. The gas is

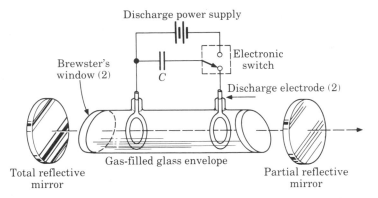

FIGURE 14.34
Typical gas laser.

contained in a cylindrical tube that has highly reflective mirrors at both ends outside of the tube. One mirror is 100% reflective, and the other mirror permits passage of a small amount of light as an output coupling system. The output beam is continuous in nature.

When external mirrors are used as the feedback device, the ends of the glass tube are angled, as shown in Figure 14.34. This introduces *zero loss* through the glass end surfaces. This zero loss angle is called *Brewster's angle* θ_B after its discoverer:

$$\theta_B = \tan^{-1}\left(\frac{N_2}{N_1}\right) \qquad (14.13)$$

Recalling the discussion of Fresnel losses, we found that these losses are due to surface interface reflections. The amount of reflectance is determined as follows (see Figure 14.35a):

$$\text{Reflectance} = \left(\frac{N_1 - N_2}{N_1 + N_2}\right)^2$$

EXAMPLE 14.8

When $N_1 = 1$ (air) and $N_2 = 1.5$ (glass), the reflectance is 0.04, or 4% loss. Converting this to decibels,

$$10 \log (1 - 0.04) = 0.1774 \quad \text{(approximately 0.2 dB)}$$

Brewster's angle is

$$\theta_B = \tan^{-1}\left(\frac{N_2}{N_1}\right) = 56.3°$$

Tilting the glass face at 56.3° will avoid reflections.

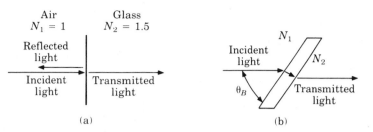

FIGURE 14.35
Brewster's angle for 100% light transmission.

The solid laser is most often a ruby aluminum oxide rod lightly doped with chromium. The ends of the rod are *perfectly* parallel and coated with a reflective material such as gold (when end mirrors are not used). One end coating is less dense than the other and acts as an output port. The atoms within the ruby rod change energy levels when triggered into excitement. In doing so, they emit photons of energy. These photons collide with other atoms that emit more photons, and a chain reaction results, called *stimulated emission*. The ruby rod must be repeatedly pulsed to sustain emission, and this is accomplished by a xenon flash tube, as shown in Figure 14.36. The current through the xenon flash tube is furnished by a capacitor discharge switching system.

Liquid lasers (see Figure 14.37) use an organic dye such as rhodamine 6G in methanol as an active medium. It is pumped through an enclosed glass tube and will luminesce when exposed to pulses of high-intensity light. A flash lamp is used to supply the light energy that sets the rhodamine 6G into a magnetic

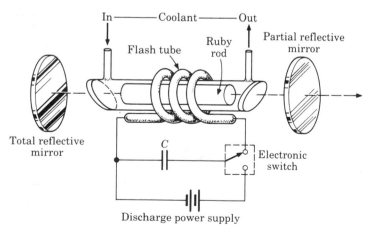

FIGURE 14.36
Solid laser. Ruby rod pulsed by flash tube.

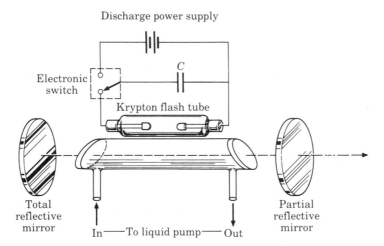

FIGURE 14.37
Liquid laser and flash tube.

spin orbit of excited electrons. The liquid laser can be tuned over a wide range of the visible light wavelength spectrum. Rhodamine 6G emits red light between 0.605 and 0.635 μm, but a dozen or so other gases may be selected to cover the visible light spectrum between 0.450 to 0.635 μm. See Table 14.2 for a list of materials used in lasers, and their wavelengths. Note that a quarter-inch diameter Nd:YAG rod is capable of emitting several hundreds of watts output.

TABLE 14.2

Elements	Wavelengths (μm)	
Hydrogen cyanide (HCN)	773.0	
Water vapor (H_2O)	118.6	
Carbon dioxide (CO_2)	10.6	(infrared)
Hydrogen fluoride (HF)	3.1	
Yttrium aluminum garnet (Nd:YAG)	1.064	
Glass (Nd:glass)	1.060	
Gallium arsenide (GaAs)	0.9100	
Krypton (K)	0.7525	
Ruby (AL203)	0.6943	(visible)
Helium-neon (HeNe)	0.6328	
Rhodamine (6G)	0.5900	
Argon (Ar)	0.5145	
Helium-cadmium (HeCd)	0.4416	
Molecular xenon (Xe_2)	0.1700	(ultraviolet)
Hydrogen (H_2)	0.1600	

14.12   SYSTEM COMPONENTS

A quick overview of a 64-channel data system will help put the fiber optic transmission path into perspective. Section A of Figure 14.38 shows 64 subcarrier generators, each 100 kHz apart at the mark frequency and shifting 2 kHz to the space frequency. There is a wide guardband between channels. The modulators are VCOs that are frequency-shift-keyed between the space and mark frequencies at a rate of 48K BPS. Each subcarrier modulator is followed by a band-pass filter to preserve its integrity.

Section B is the summing section where 64 different frequency signals are added together. Each *line summer* adds four channels together. Each *group summer* adds four lines together, and each *trunk adder* adds four groups together. The output from the trunk adder will be the full frequency range of 100 kHz to 6.4 MHz containing the 64 channels of data. The output from the trunk adder

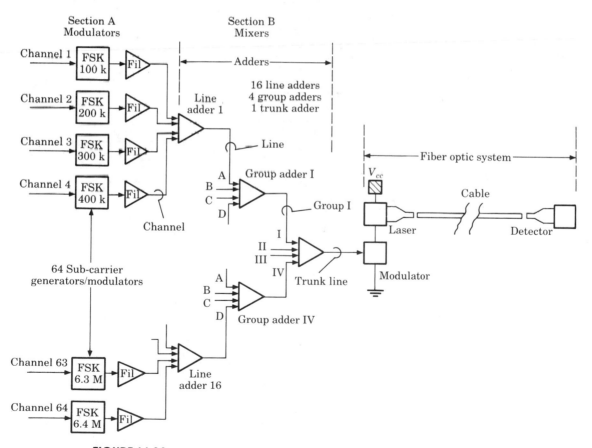

FIGURE 14.38
A 64-channel multiplexing section that makes up the input signal to the fiber optic laser modulator.

is then applied to an injection laser transmitter (see Figure 14.39), where the total signal is pulse-modulated onto the light beam. Each channel has 48K BPS, so all 64 channels use about a 3 MHz data rate. The optic system must have low losses over the distance traveled and have a favorable pulse dispersion (rise time budget) to handle 3 MHz.

At the receiving end, a light detector such as the one shown in Figure 14.25 (avalanche photodiode) will convert the light signal into an electrical signal. At this time, the 64 channels are separated through a series of active band-pass filters, shown in Figure 14.40. The trunk signal is separated into four group signals, each group is broken down into four line signals, and each line is broken down into the four channels. Obviously, the bulk of the electronics comes before and after the fiber systems, and the methods used are the same as those studied in previous chapters. The major difference is the *path* between the transmitter and the receiver, which, for the case of fiber optics, is able to handle many more signals over a single conductor at a much faster rate.

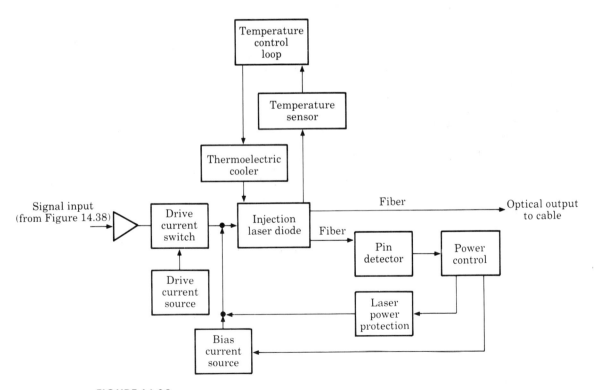

FIGURE 14.39
Injection laser diode modulator (part of Figure 14.38).

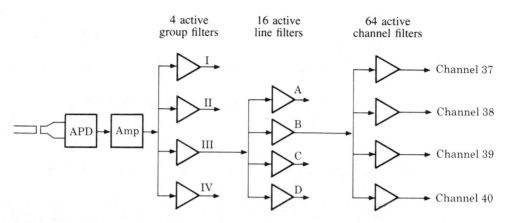

FIGURE 14.40
Receiving end of the fiber optic system carrying 64 channels of data. Channel separation requires 84 active filters.

QUESTIONS

1. A bit rate of 20M BPS is equivalent to what analog frequency?
2. A photon is radiant light energy. (T) (F)
3. The wavelength is a ratio of frequency to _____.
4. A frequency of 3.409×10^{14} has what wavelength?
5. As frequency goes up, wavelength gets (longer) (shorter).
6. The clad of a fiber optic cable has a refractive index that is (larger) (smaller) than the core index.
7. Light reflections at the boundary between two different indices of glass will be (lossy) (lossless).
8. Light reflections are more likely to occur when the incident light approaches the core-clad interface at a (small) (large) angle away from the center axis.
9. To refract means to (a) bounce off of, (b) penetrate, (c) bend, (d) disregard, (e) stop.
10. The index of refraction compares the _____ to the _____.
11. The reference index is the index of _____.
12. Which has the higher index, water or glass?
13. What is the angle of refraction when one is looking into a pond of water at 60° from the pond's surface?
14. The example used with Equation 14.3 finds the critical angle at the pond to be 41.4°. Will the critical angle for an air-to-water interface *always* be 41.4°?
15. When light enters the core from the end of the cable, the cone of acceptance takes into account the angle of _____ at the air-glass interface and the angle of _____ at the core-clad interface.

16. What is the numerical aperture of a cable whose critical angle is 26.1°?

17. What is the numerical aperture of a cable with a clad index of 1.378 and a core index of 1.546?

18. The numerical aperture is a figure of merit to express the cable's ability to _____.

19. A large numerical aperture is better than a small numerical aperture. (T) (F)

20. When the numerical aperture increases, the critical angle (increases) (decreases).

21. For highest efficiency, a (large) (small) NA is better.

22. Fiber optic cables (can) (cannot) be used in areas where explosive gases are known to exist.

23. The crosstalk and electrical interference problems in fiber optic systems are (poorer than) (the same as) (better than) the conditions found with hard-wire systems.

24. The index of refraction is related to (a) the density of the glass, (b) the velocity of light in the core, (c) the length of the cable, (d) the diameter of the core, (e) none of the above.

25. The reflective index of the core material is a larger number than the clad material. (T) (F)

26. The critical angle is the arc cosine of N_2/N_1. When light enters the guide at an angle away from the center line that is (larger) (smaller) than the critical angle, it will dissipate out the walls of the guide and be lost.

27. The closer the index of refraction of the clad is to the index of the core, the _____ the angle of reflection will be inside the core. (a) smaller, (b) less effective, (c) more important, (d) larger, (e) less important.

28. The numerical aperture of the cable is a measure of _____.

29. More light will enter the core when the numerical aperture gets (a) larger, (b) smaller, (c) NA has no effect on the amount of light entering the core.

30. The core has $N = 1.6$, and the clad has $N = 1.3$. What is the value of the critical angle?

31. The core has $N = 1.6$, and the clad has $N = 1.3$. What is the angle of the cone of acceptance?

32. The NA will be (larger) (smaller) when the clad index is close in value to the core index.

33. The critical angle is an angle referenced to (a) Brewster's angle, (b) the angle of reflection, (c) the center line of the core, (d) a perpendicular to the core center line, (e) the core diameter.

34. What is the numerical aperture of an optic cable that has a critical angle of 41°? (a) 0.411, (b) 0.500, (c) 0.656, (d) 0.723, (e) 0.886.

35. What is the angle of the cone of acceptance for a cable with a 28° critical angle? (a) 7°, (b) 14°, (c) 28°, (d) 56°, (e) 84°.

36. The light frequencies are in the range of (a) 10^{11} Hz, (b) 10^{12} Hz, (c) 10^{13} Hz, (d) 10^{14} Hz, (e) 10^{15} Hz.

37. Light at a frequency of 5×10^{14} Hz has a wavelength of (a) 0.6 μm, (b) 0.7 μm, (c) 0.8 μm, (d) 0.9 μm, (e) 1.0 μm.

38. Collimated light rays are (a) perpendicular to each other, (b) at large angles to each other, (c) at small angles to each other, (d) parallel to each other, (e) none of the above.

39. Which of the following light sources give off the narrowest spectral radiation pattern? (a) LEDs, (b) diode lasers, (c) cleaved coupled devices.

40. Which of the following devices would best serve as a single-mode light source? (a) LEDs, (b) diode lasers, (c) cleaved coupled devices.

41. As the diameter of the core increases, the number of modes of light propagation will (a) increase, (b) decrease.

42. In optical waveguide, the energy group velocity is (a) the same as the speed of light in air, (b) less than the speed of light in air, (c) greater than the speed of light in air.

43. In optical waveguide, the lowest-order mode to propagate in the guide is the axial mode. (T) (F)

44. In calculating the system losses, the term *BER* means (a) before entering receiver, (b) below energy received, (c) better energy ratio, (d) beyond errors retrieved, (e) bit error rate.

45. The acceptable level of BER is stated as one error in (a) 1000 bits, (b) 1,000,000 bits, (c) 1,000,000,000 bits, (d) 1,000,000,000,000 bits.

46. The BER should be stated in −dBm. (T) (F)

47. The BER is a quality associated with the (a) detector, (b) cable, (c) light source.

48. The system losses are added together to find the (a) current of the detector, (b) minimum cable length, (c) required supply voltage, (d) minimum source output to exceed the BER.

49. A flux budget, in the context of fiber optics, is a (a) measure of electrical interference, (b) pattern of light output, (c) cost analysis, (d) treatment of the glass core, (e) count of the system losses.

50. The flux budget is independent of cable length. (T) (F)

51. Fresnel losses are losses due to (a) diameter, (b) numerical aperture, (c) a change in medium, (d) cable length, (e) BER.

CHAPTER FIFTEEN

TELEVISION

15.1 INTRODUCTION

Emission code A5 is reserved for commercial television in the North American continent. This is a unique transmission of an amplitude-modulated video signal with a full carrier and upper sideband, but only one-third of the lower sideband. The assigned channel bandwidth is 6 MHz in this region, with the visual carrier placed 1.25 MHz above the lowest frequency in the channel. Europe and Asia uses a channel bandwidth of 5 MHz to 14 MHz along with the same vestigial sideband modulation principles used by the Western block of nations. All but one country use an FM sound carrier placed 4.5 MHz (North America) or 6 MHz (Europe) above the picture carrier. France uses an AM sound carrier.

Figure 15.1 shows a condensed commercial television station transmitter. Two separate transmitters are used because two different forms of modulation are used and because the aural transmitter carrier power need be only 15% of the visual carrier power. This will permit the station to announce by audio when the visual transmission is in difficulty or to show visually that the sound transmitter is in trouble.

15.2 THE SCANNING PRINCIPLES

It will be easy to see how the various frequencies fit into the transmission plan after you understand how the picture is displayed. Figure 15.2 shows that the television picture tube (hereafter called the cathode ray tube or CRT) is operated in the same fashion as an oscilloscope. That is, the assembly in the neck of the CRT, called an *electron gun,* produces a narrow beam of electrons that are electrostatically attracted to the phosphor screen on the face of the CRT.

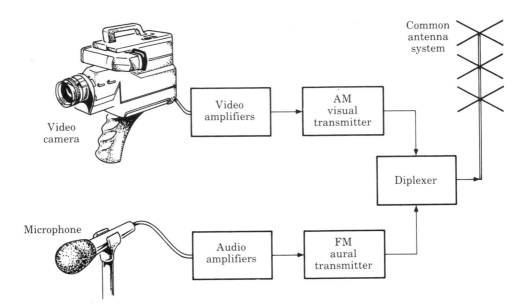

FIGURE 15.1
A condensed version of a television transmitter. The visual transmitter is totally separate from the aural transmitter.

With no further action, the electron beam would energize the phosphor at the center of the screen to produce a single dot of light. Like the oscilloscope, when this dot is moved rapidly from side to side, it gives the illusion of a line of light. This illusion is maintained by two factors: the persistency of the phosphor (that is, it glows after the energy source is removed) and the persistency of vision of the human eye, which is about one-sixteenth of a second. Based on this persistency principle, television (and the motion picture industry from which it originated) is able to display 30 pictures every second to make viewers imagine that they see a moving object.

Position this horizontally scanned line of light at the top of the CRT and move it down the screen slowly as it moves rapidly from side to side until 262.5 lines are traced out (see Figure 15.3, where only 11 lines are shown for simplicity). This pattern of 262.5 lines is called one **field** and ends at the bottom center of the CRT. Next move the beam back up to the top center of the CRT, and trace out a second field of 262.5 lines. The second set of scan lines are shown in Figure 15.3 as dashed lines to indicate that they are placed midway between the first field's scan lines. The first field traces out all of the odd-numbered lines (1, 3, 5, etc.) and the second field fills in all of the even-numbered lines (2, 4, 6, etc.). This results in two fields of 262.5 lines each, or 525 lines total, which is the same as one frame or one complete picture, repeat-

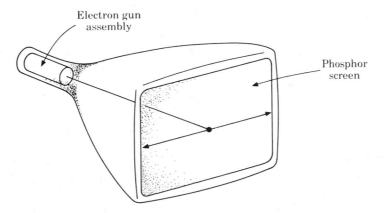

FIGURE 15.2
A cathode ray tube with the electron beam energizing the phosphor screen to form a single dot.

ed 30 times every second. This alternating line display technique (carried over into most video display devices) is called *interlaced scanning*. The presence of the scan lines alone on the face of the CRT is called a **raster**. No picture need be present on the face of the CRT for a raster to be observed.

When 525 lines are constructed 30 times every second, the time per line is the reciprocal of the frequency: $30 \times 525 = 15{,}750$ Hz and the time per scan is 63.5 μs. The scanning action is that of a linear sawtooth wave taking 52 μs to trace and 11.5 μs to retrace. The CRT is blanked out (turned off) for the duration of the 11.5 μs retrace time.

FIGURE 15.3
Interlaced scanning of the CRT screen reduces flicker rate.

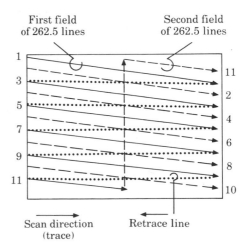

15.3   THE DEFLECTION SYSTEMS

There are two systems in use today for moving the electron beam around on the face of the CRT. They are the **electrostatic** and the **electromagnetic** deflection systems. The type of deflection will depend on the ultimate use of the CRT display. Electrostatic deflection is accomplished by placing four metal plates inside the neck of the CRT with connecting wires to the outside for voltage application. Two of the plates control the vertical movement of the beam, and two plates control the horizontal movement.

The two vertical plates are placed above and below the path of the electron beam, as shown in Figure 15.4, while the two horizontal plates are on either side of the electron beam path. When one plate is grounded and a positive AC voltage is applied to the other plate, the path of the electron beam will bend toward the positive plate. The beam is moving too fast to strike the deflection plate, so only its direction is changed. The advantage of electrostatic deflection is that it performs equally well at all frequencies, from DC to the limits of the CRT's ability to produce a trace. The disadvantage is the small deflection angles of about $\pm 15°$. A 21 in picture tube would be about 48 in long.

Electromagnetic deflection uses two sets of coils in place of the deflection plates. A sawtooth current is required through the coils in order to achieve linear trace and retrace. The coils, called the **deflection yoke,** fit around the outside of the neck of the CRT and have an inductive component as well as the wire resistance considered to be in series with the inductance. A sawtooth voltage applied to the resistive element will produce a sawtooth current through the resistance. However, a square wave voltage must be applied to the inductive portion of the yoke to produce a sawtooth current through the inductance. The sum of the square wave voltage and the sawtooth voltage is called a **trapezoidal** voltage waveshape. The advantage of the electromagnetic deflection system is the wide deflection angles of about $\pm 60°$. The disadvantage is

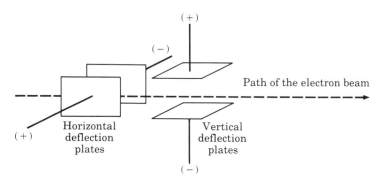

FIGURE 15.4
Electrostatic deflection plates.

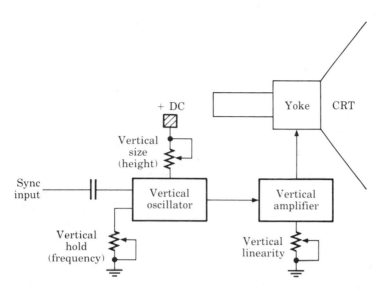

FIGURE 15.5
Vertical deflection block diagram.

that coils work best at only one frequency. However, this is not really a disadvantage because in television, each half of the yoke is required to work at only
one frequency—60 Hz for the vertical and 15,750 Hz for the horizontal.

The vertical circuit consists of an oscillator that will operate near 60
cycles to generate the voltage wave needed for deflection. The vertical frequency adjustment (called the *vertical hold* control) is part of the oscillator circuit.
This is a free-running oscillator that will be synchronized by the station signal
at the exact frequency of that station. The vertical size control (or height control) adjusts the DC voltage to the oscillator and controls the strength of the
output signal, which determines the size of the vertical scan of the CRT (see
Figure 15.5).

The oscillator signal is amplified by the vertical output power amplifier to
convert the voltage signal to a current signal to drive the deflection yoke. The
vertical **linearity** control is generally a part of this circuit. The linearity control is a gain control that varies the operating Q point of the amplifier and
distorts the waveshape by compressing the top or bottom of the wave as
required to achieve the best distribution of the vertical scan from top to bottom
of the CRT screen. The size control and the linearity controls interact for best
picture results.

Vertical yoke currents of 1 A are common. The vertical oscillator and the
vertical amplifier with the deflection yoke comprise the entire vertical circuit.
A typical vertical circuit is shown in Figure 15.6.

The horizontal deflection section has the same functions as the vertical
system but is developed in a slightly different manner. The horizontal

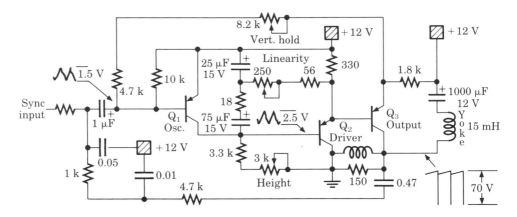

FIGURE 15.6
Vertical oscillator and output amplifier circuit.

oscillator used for television was the first practical use of the phase-locked loop. The oscillator has a free-running frequency of 15,750 cycles adjustable by a control setting (*horizontal hold* control), as shown in Figure 15.7. That is, with no input signal, the circuit components are selected to control the frequency. The oscillator is a voltage-controlled circuit that depends on the DC voltage at the input to establish the frequency of operation.

A feedback signal from the horizontal circuit is applied to a phase detector, which compares the frequency of an incoming pulse from the station (when

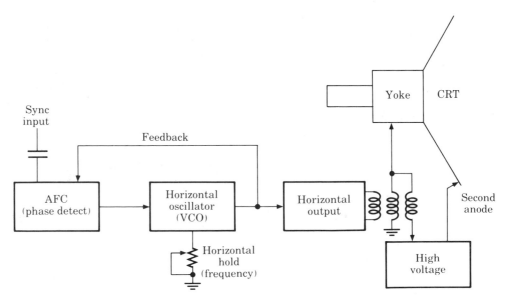

FIGURE 15.7
Horizontal deflection and high-voltage block diagram.

a station is selected) and develops the DC voltage to control the oscillator. The DC voltage is the error voltage of the phase detector. The oscillator output is amplified by a power amplifier to develop the current signal used to drive the horizontal section of the deflection yoke. This exchange from voltage to current is accomplished through a horizontal output transformer. Because of the higher horizontal frequency (compared to the vertical), the horizontal yoke current need not be as great as the current in the vertical deflection system.

15.3.1 The High Voltage

An additional secondary winding on the horizontal output transformer is used to step up the horizontal signal so that it can be rectified and filtered to generate the high voltage (see Figure 15.8). The process of high voltage generation is much the same as that used in most automobile ignition systems. That is, a pulsed DC voltage slowly builds up a magnetic field around a coil and then allows it to collapse rapidly. The rapidly collapsing field induces a large voltage amplitude into the additional secondary, which is rectified for the high DC operating anode voltage of the CRT. This high voltage, sometimes called the *ulter* voltage, ranges from 10–12 kV, for monochrome CRTs, to 25–30 kV, for color tubes. The collapsing magnetic field in the horizontal output transformer is called a *flyback* action because it appears to stretch the voltage out slowly and then lets it snap back to a low level. Thus, the horizontal output transformer, the high-voltage transformer, and the flyback transformer are all names for the same device.

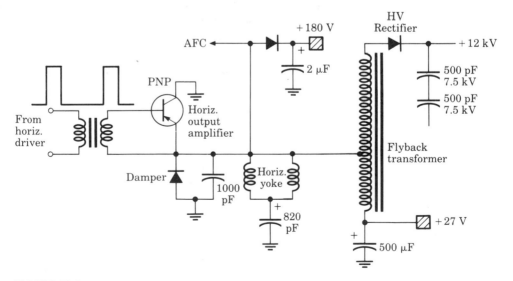

FIGURE 15.8
Horizontal output amplifier and high-voltage transformer circuit.

It is important to realize that when the horizontal oscillator or the horizontal amplifier fail for any reason, there will be no signal to the high-voltage section and no high voltage. As a result, there will be no raster displayed on the face of the CRT. Finally, it should be noted that a raster (brightness on the face of the CRT caused by the scan lines) will be observed when the CRT is correctly biased, the low-voltage power supply is operating, the vertical deflection system is working, and the horizontal deflection and high-voltage sections are working. These five sections produce the brightness on the face of the CRT in most video display systems.

15.4   THE VIDEO PICTURE SIGNAL

As the camera scans one line, the camera tube senses changing brightness levels along that line corresponding to the different brightness levels in the scene being viewed. The camera tube changes these variations in light level into voltage changes. When these impulses are recovered at the receiver, they are applied to the CRT in a fashion that will cause the CRT to increase or decrease the light intensity of the beam at that instant in order to reconstruct the original scene on the CRT face. The simultaneous movement of the scanning beam as it changes intensity is the action that displays the picture.

With a horizontal scan time of 52 μs per line (**trace time**), how many on/off cycles can the CRT display along one scan line? The answer lies in the limits placed on the video frequencies, which, for commercial television, is 4.2 MHz. The CRT would display approximately 220 on/off functions of a 4.2 MHz signal in 52 μs. These would appear as vertical bars on the face of the CRT if all 525 scan lines had the same information. This display is referred to as **lines of resolution** and is a measure of how many individually recognizable vertical lines can be resolved (displayed) on the CRT screen. The number of lines of resolution equals the trace time per line divided by the time for one cycle of the highest frequency of the signal to be viewed; this formula is simplified to scan time multiplied by frequency (52 μs × 4.2 MHz = 218.4 lines). This represents the response of a square wave modulating the beam and in this case would result in 218 black vertical bars separated by 218 white vertical bars. For sine wave rather than square wave conditions (not quite as sharp, but still recognizable), multiply the calculated number by the square root of 2: 52 μs × 4.2 MHz × $\sqrt{2}$ = 309 lines of resolution. This is a typical value for commercial television receivers.

Two points should be noted here. First, the vertical lines of resolution should not be mistaken for horizontal scan lines. Second, the resolution quality of the CRT far exceeds the quality of the applied video signal of 4.2 MHz. Closed-circuit video systems generally display 1000 lines of resolution because the video amplifiers and signal have a frequency response to almost 15 MHz. In addition, most personal computer display monitors use 15 MHz of video bandwidth to produce the 1000 lines of resolution needed to give a clear, crisp image for 80-column word processing.

15.4.1 Blanking and Synchronizing Pulses

The wave in Figure 15.9a is the wave of changing voltage that would produce a particular pattern of information on the CRT. Notice that a blanking pulse has been added in the "black" direction indicating that the CRT is turned off for the retrace time. A synchronizing pulse is also added to tell the receiver when to go to the next scan line in order to stay matched with the transmitter. The camera at the television station is doing exactly the same thing that your receiver is doing and at the same time. In fact, it is the camera control system that tells your receiver when to blank out and when to start each new trace. It is this practice of requiring the station to add a blanking pulse and a synchronizing pulse that, in part, makes this modulation technique so different from all of the other systems.

The sync pulses instruct the receiver to end the trace time and start the retrace so that the next scan line can be correctly synchronized. When a wave such as this is viewed on an oscilloscope, each successive video cycle overlays all of the cycles that preceded it, as shown in Figure 15.9b. The picture information is changing constantly and is different for each scan line; however, the blanking and synchronizing pulses are the same for each line, and the oscilloscope will lock onto these repeated waveshapes.

The relationship of the picture *intensity* to the blanking and sync pulses is also shown in Figure 15.9b. A *white reference* level is established at 12.5% of the signal peak, a *black reference* level is set at 70% of signal peak, the blanking level (called the *pedestal level*) is at 75% of signal peak, and the remaining 25% of signal is the *synchronizing* signal. This signal will be applied to the CRT base to modulate the electron beam; this process is called *Z axis* modulation. This term is taken from geometry for a three-dimensional graphic display, where the X axis is left and right (horizontal deflection), the Y axis is up and down (vertical deflection), and the Z axis is in and out (on/off cycles for the CRT).

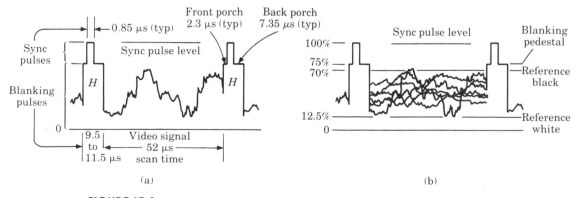

FIGURE 15.9
The horizontal scanning waveshape. (a) One line of information with horizontal blanking and sync pulses. (b) Several overlapping horizontal scan lines.

Two fields are scanned vertically 30 times every second. This sets the vertical frequency at 60 Hz; therefore, a vertical sync and blanking pulse appears between every set of 262.5 horizontal pulses. Figure 15.9 shows the horizontal blanking and sync pulse on an oscilloscope. If you adjust the scope sweep frequency control lower you will see many horizontal pulses on the scope display. Continue this action until 262.5 horizontal pulses are seen at one time, very close together. Framing these 262.5 horizontal pulses will be one vertical blanking and sync pulse. This sketch of the wave is shown in Figure 15.10.

During each vertical blanking and sync pulse (1.333 ms) the horizontal system of the receiver still needs to be synchronized. Figure 15.10 shows the horizontal sync pulses superimposed on the vertical blanking and sync signals as dotted lines. Dotted lines are used here to make the pulses stand out from the balance of the wave; in reality, they would be true voltage changes occurring at the times indicated.

15.4.2 Positive/Negative Picture Phase

When a sine wave (or voice signal) is amplitude-modulated onto a carrier, as covered in earlier chapters, the polarity of the message signal has no influence on the end product of the modulated wave. When the message signal is recovered at the receiver and applied to a speaker, the effects on the eardrum are the same whether the speaker cone is moving out or moving in; the ear senses changes in air pressure but not direction. However, the video signal has polarity that is meaningful to the CRT and that changes what the eye will see.

The electron gun in the neck of the CRT is basically a pentode tube. A negative pulse applied to the control grid of the CRT (bipolar base) will drive the gun toward a cutoff condition. A positive pulse applied to the cathode of the CRT (bipolar emitter) will have the same effect. These two signal conditions

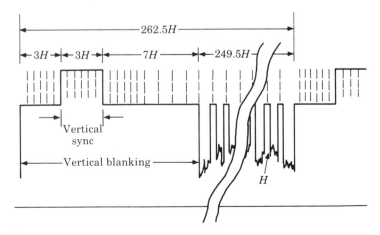

FIGURE 15.10
One vertical scan, shown as one field with relative times for vertical blanking and sync pulses.

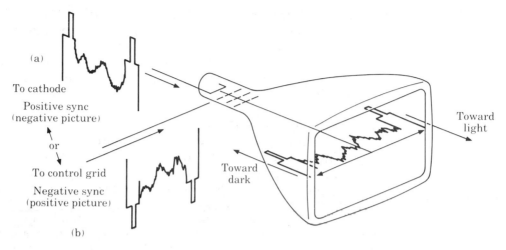

FIGURE 15.11
The effects of applying a negative picture phase signal (a) to the cathode and a positive picture phase signal (b) to the control grid of the CRT.

are shown in Figure 15.11. In Figure 15.11a, the sync and blanking pulses are the more positive portion of the signal voltage, and the white peaks of information are toward the negative direction in the wave. This polarity signal is called *negative picture phase* because the white level (considered to contain the majority of picture information) is negative compared to the blanking and sync levels. The signal in Figure 15.11b is called *positive picture phase* for similar reasons.

However, there is one aspect of the picture phase that is more important than which element of the CRT receives the video signal. That is the way in which the video signal is polarized to amplitude-modulate the picture carrier. Figure 15.12a shows a modulated carrier where the sync pulses cause the

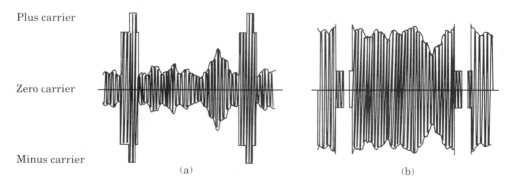

FIGURE 15.12
A video signal amplitude-modulating a carrier signal. (a) A negative picture phase modulated carrier. (b) A positive picture phase modulated carrier.

greatest carrier amplitude; this is called *negative picture phase modulation* (the white levels are toward zero). Figure 15.12b shows a modulated carrier where the sync pulses cause the smallest amount of carrier amplitude, called *positive picture phase modulation* (the white levels cause the largest carrier change). Most national systems favor negative picture phase modulation for the following reasons.

1. The average voltage in the wave is smallest for bright scenes. Since the transmitted scenes will be bright more often than they are dark, the required power in the transmission will be smallest for the maximum message signal.
2. Overmodulation will not cause compression of the sync pulses.

Positive picture phase modulation has the opposite conditions. More average power is required for the same brightness and overmodulation *will* compress the sync pulses, causing loss of picture synchronization.

15.5   VESTIGIAL SIDEBAND TRANSMISSION

Amplitude-modulating a carrier signal with a 4.2 MHz picture signal will generate sidebands that extend above and below the carrier frequency by an amount equal to the message signal frequency. In this case, an 8.4 MHz bandwidth modulated signal will result.

In the North American continent, 6 MHz per channel is the bandwidth limit for a commerical telecast. By placing the carrier frequency 1.25 MHz above the lower limit of the assigned channel frequency band and attenuating *part* of the lower sideband, the 4.2 MHz upper sideband can remain intact. Figure 15.13 shows this condition. The lower sideband amplitude is equal to the carrier amplitude at a few cycles below the carrier frequency. At 0.75 MHz below the picture carrier frequency, the lower sideband is at a reduced level of -3 dB, and at 1.25 MHz below the picture carrier frequency, the lower sideband amplitude is at a reduced level of -60 dB.

The upper sideband is maintained at the carrier level out to several megahertz and is allowed to roll off to -3 dB below the carrier amplitude at 4.2 MHz above the carrier frequency. However, it must drop to -60 dB below the carrier level at 4.475 MHz above the carrier frequency. This leaves a space of 0.275 MHz in the allotted band of 6 MHz for the frequency-modulated sound signal.

15.5.1 The Sound Carrier

As stated several times earlier in this chapter, the aural portion of the television transmission is a frequency-modulated carrier wave. The carrier rest fre-

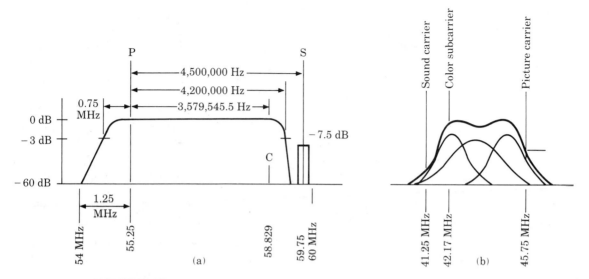

FIGURE 15.13

Vestigial sideband modulation. (a) The band-pass response curve of a vestigial sideband transmission. (b) The IF amplifier response curve to compensate for the loss of part of one sideband.

quency is specified at 4.5 MHz above the accompanying picture carrier (in North America). The aural signal is limited in frequency to the range of 50 Hz to 15 kHz and will modulate the carrier to 100% for a carrier deviation of ±25 kHz. The deviation ratio is, then,

$$m_{f(\max)} = \frac{f_{c(\max)}}{f_{m(\max)}} = \frac{25 \text{ kHz}}{15 \text{ kHz}} = 1.67 \text{ rad}$$

From the Bessel function chart (Table 5.4) it is seen the there are two sideband pairs. However, this deviation ratio, by definition, places the television sound transmission very close to narrowband FM where the ratio is $\pi/2 = 1.57$ rad.

15.5.2 Channel Assignments

Table 15.1 shows the channel assignments for television transmission. Channel 1 was an experimental channel for receiver manufacturers in the days before the abundance of daytime programming. Decreasing use of the channel caused the FCC to reassign the frequencies. Channels 70 through 83 will be lost to commercial television before this book is in print, for similar reasons. The higher UHF channels are being reassigned to mobile land, mobile satellite, and cellular communications.

TABLE 15.1
Television channel assignments

Channel Number	Band (MHz)	Visual (MHz)	Aural (MHz)	Channel Number	Band (MHz)	Visual (MHz)	Aural (MHz)
2	54–60	55.25	59.75	36	602–608	603.25	607.75
3	60–66	61.25	65.75	37	608–614	609.25	613.75
4	66–72	67.25	71.75	38	614–620	615.25	619.75
5	76–82	77.25	81.75	39	620–626	621.25	625.75
6	82–88	83.25	87.75	40	626–632	627.25	631.75
7	174–180	175.25	179.75	41	632–638	633.25	637.75
8	180–186	181.25	185.75	42	638–644	639.25	643.75
9	186–192	187.25	191.75	43	644–650	645.25	649.75
10	192–198	193.25	197.75	44	650–656	651.25	655.75
11	198–204	199.25	203.75	45	656–662	657.25	661.75
12	204–210	205.25	209.75	46	662–668	663.25	667.75
13	210–216	211.25	215.75	47	668–674	669.25	673.75
14	470–476	471.25	475.75	48	674–680	675.25	679.75
15	476–482	477.25	481.75	49	680–686	681.25	685.75
16	482–488	483.25	487.75	50	686–692	687.25	691.75
17	488–494	489.25	493.75	51	692–698	693.25	697.75
18	494–500	495.25	499.75	52	698–704	699.25	703.75
19	500–506	501.25	505.75	53	704–710	705.25	709.75
20	506–512	507.25	511.75	54	710–716	711.25	715.75
21	512–518	513.25	517.75	55	716–722	717.25	721.75
22	518–524	519.25	523.75	56	722–728	723.25	727.75
23	524–530	525.25	529.75	57	728–734	729.25	733.75
24	530–536	531.25	535.75	58	734–740	735.25	739.75
25	536–542	537.25	541.75	59	740–746	741.25	745.75
26	542–548	543.25	547.75	60	746–752	747.25	751.75
27	548–554	549.25	553.75	61	752–758	753.25	757.75
28	554–560	555.25	559.75	62	758–764	759.25	763.75
29	560–566	561.25	565.75	63	764–770	765.25	769.75
30	566–572	567.25	571.75	64	770–776	771.25	775.75
31	572–578	573.25	577.75	65	776–782	777.25	781.75
32	578–584	579.25	583.75	66	782–788	783.25	787.75
33	584–590	585.25	589.75	67	788–794	789.25	793.75
34	590–596	591.25	595.75	68	794–800	795.25	799.75
35	596–602	597.25	601.75	69	800–806	801.25	805.75

15.6   THE RECEIVER BLOCK DIAGRAM

The most noteworthy observation concerning Figure 15.14 is that this is a *superheterodyne* receiver and carries with it all the characteristics that the name implies. Earlier chapters went into great detail on superheterodyne

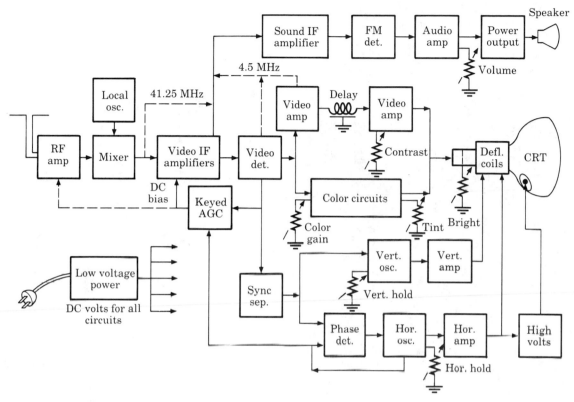

FIGURE 15.14

A typical block diagram for a television receiver.

receivers; therefore, this section will cover only the unusual features that relate to television.

15.6.1 The Tuner

The tuner portion of the receiver has an RF amplifier, mixer, and local oscillator. It is sometimes called the "front end" of the receiver based on its job of selecting the carrier frequency and reducing it to the intermediate frequency. The tuner in a television receiver is almost always on a separate, shielded subchassis away from the main receiver chassis. Because of this, the tuner from one receiver will work equally well in any other receiver provided that the DC operating voltages are compatible.

Digital systems have replaced mechanical tuning systems. A counter circuit sets a DC voltage level, which is applied to a varactor diode in a tuned circuit, which in turn controls the resonant frequency of the RF amplifier, mixer, and local oscillator (see Figure 15.15).

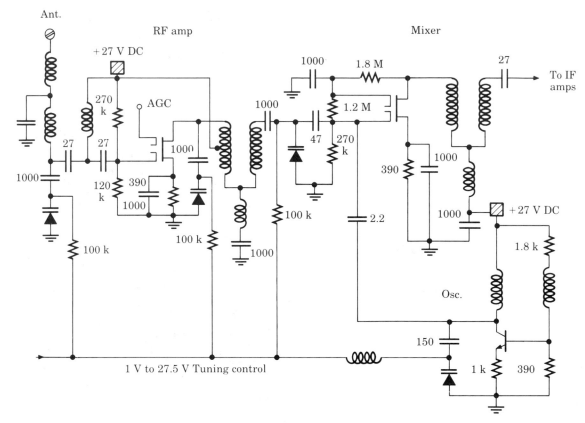

FIGURE 15.15
Varactor diode tuning in a television front end.

A condition exists in television tuners that separates them from all other receiver front ends. Because there are two carriers transmitted 4.5 MHz apart, the mixing action causes an exchange in the carrier frequency position. Channel 2, for example, assigns the station picture carrier at 55.25 MHz and the sound carrier at 59.75 MHz. The local oscillator frequency for this channel is set at 101 MHz. The difference between the oscillator frequency of 101 MHz and the picture carrier is 45.75 MHz, but with two carriers being received, the sound carrier also mixes with the oscillator frequency, for a difference frequency of 41.25 MHz. Note that the sound carrier was above the picture carrier in the transmitted signal, but in the heterodyne process, the sound carrier has been placed below the picture carrier at the output of the tuner. The difference in frequency has remained unchanged (45.75 MHz − 41.25 MHz = 4.5 MHz), so no difference in performance will be seen in the operation of the receiver.

15.6.2 The IF Amplifiers

Stagger-tuned circuits broaden the frequency response of the IF amplifiers and greatly improve the selectivity of the system. Stagger-tuning is a variation of the response curve of Figure 7.7 in that each tuned transformer is set for a slightly different frequency to spread the gain over a wider range. Remember that in the transmitted signal, the lower sideband extends 0.75 MHz away from the carrier frequency. This means that there will be twice as much voltage in the signal for the first 0.75 MHz of bandwidth than there will be for all higher frequencies. To correct this condition, the picture carrier is set halfway down the slope at the high-frequency end of the IF response curve.

Many video IF amplifiers are now constructed in microchip circuits, and the video detector is usually included in the last of the chip circuits. However, this should not deter the investigator from recognizing where the detector is located by reading the frequency ranges of the circuit.

15.6.3 Automatic Gain Control

The AGC circuit also has a facet not found in other systems. Consider first the function of the AGC, that is, to examine the strength of the signal at the output of the detector and convert this to a DC voltage, which is used to adjust the bias on the IF amplifiers controlling gain in a manner that will maintain a constant signal level to the low-frequency amplifiers. Figure 15.14 indicates that the average DC voltage in the wave is a direct result of *scene brightness* as well as signal strength. The only part of the wave that truly represents the signal strength is the amplitude of the blanking and sync pulses. Therefore, a pulse from the horizontal deflection section is applied to the AGC network so as to turn on the AGC amplifier *only* during the horizontal blanking time. This will make the AGC independent of scene brightness. Such a system is called *gated AGC* or *keyed AGC*.

15.6.4 The Video Amplifiers

The amplifiers that increase the signal strength from about 2 V p-p at the output of the video detector to about 100 V p-p to drive the CRT are low-gain (per stage), wide-bandwidth amplifiers. The high-frequency response of an amplifier is directly related to and inversely proportional to the stray capacitance of the system. The stray capacitance includes the interelectrode capacitance of the active device and the wiring capacitance, which together form a capacitive reactance in parallel with the signal path. As the frequency goes higher, the reactance gets smaller to reduce the gain of the amplifier at the higher frequencies.

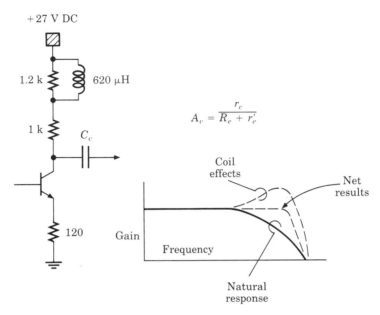

FIGURE 15.16
Peaking coil in a video amplifier to counteract the effects of circuit capacitances.

In Figure 15.16, note the collector load impedance as a 1 kΩ resistor in series with the parallel *RL* circuit. This circuit is called a *peaking coil* circuit and improves the high-frequency response of the amplifier. Because gain equals collector impedance divided by the emitter impedance, the gain is compared at two different frequencies. At 100 kHz, $X_L = 390$ Ω and the parallel combination with the 1200 Ω resistor totals about 300 Ω. The collector impedance is thus 300 Ω + 1000 Ω = 1300 Ω. Assume an emitter impedance of 130 Ω and a gain of 10. At 4 MHz the coil reactance is 16 kΩ in parallel with 1200 Ω for a value of about 1114 Ω, plus 1000 Ω, for a total collector impedance of 2114 Ω. Substituting this value into the gain equation gives a gain of 16.25. The gain at 4 MHz is 1.625 times greater than the gain at 100 kHz, which will compensate for the loss in gain due to the capacitive reactance of the circuit, resulting in a flat frequency response over the frequency range of interest. The coil reactance is selected for the frequency to be peaked, and the parallel resistor is selected to control the *Q* of the circuit (bandspread). When the coil *Q* is low, a very high resistance is selected as a coil form.

15.6.5 The Sound Carrier

The receiver separates the sound carrier from the composite signal at some point between the last video IF amplifier and the contrast control. Possible takeoff points are shown in Figure 15.14. In the video IF circuits, the sound carrier takeoff frequency is the difference between the two carrier frequencies, or 4.5 MHz. After the video detector, the sound carrier is reduced to 4.5 MHz, and this can be used as the sound carrier frequency directly. In either case, the sound system is identical to any other FM receiver circuit with the exception that the center frequency is 4.5 MHz.

Very early receivers were called *split-sound* because the sound and video were separated immediately after the tuner and each had its own set of IF amplifiers. An economy drive in the mid 1960s sent both IF signals through the same set of amplifiers, called *intercarrier receivers*. The only real disadvantage is the presence of raspy noise when lettering overlays the picture during commercials. Correct AGC adjustment can minimize this noise. With the increased use of chip circuits, and especially with the release of stereo sound for television, look for the return of the split-sound receivers.

15.6.6 The Synchronizing Circuit

A simple class C amplifier with signal bias is used for synchronization. The purpose of the sync separator is to pass all of the sync signals (both vertical and horizontal) while bypassing the video signal to ground. The bias (which sets the clipping level) must be a function of signal strength so that only the sync signals are separated, regardless of whether there is 1 V or 5 V of signal at the input to the sync separator. A set of parallel branching filters at the output of the separator will divide the sync pulses into a vertical set and a horizontal set to allow separate control of the frequencies of their respective oscillators.

15.7 COLOR

Light and light frequencies were discussed in Chapter 14 on fiber optics and will be reviewed here only as they pertain to color television. The three primary colors found in nature are red, yellow, and blue. All other colors can be developed through mixing these three basic colors. When white light (made up of all wavelengths of visible light) strikes an object, all wavelengths will be absorbed except the dominant wavelength. The dominant wavelength will be reflected, and this is the color that the eye will see. This process is called the *subtractive* method of color mixing; that is, start with all wavelengths, subtract the colors not wanted, and reflect the color that is wanted. In nature, when all colors are added together, the mixture is a dark gray or near black; all light wavelengths are absorbed.

Color television uses red, green, and blue as its primary colors because as a man-made light source, television can produce a wider combination of colors with these primaries than with the natural three primaries. Color television uses the *additive* method of color mixing; that is, as a light source, it adds red light to green light to produce yellow light. When all colors of light are added together, the result is white light. The standard white is present when 30% of the light is supplied as *red,* 59% of the light is supplied as *green,* and 11% of the light is supplied as *blue,* as measured on a light meter.

In the color CRT, there are three identical electron guns that develop three separate beams of electrons (see Figure 15.17). The CRT face is a pattern of dots composed of three different phosphor materials. One will glow red when bombarded with electrons, another will glow blue, and the third will glow green. A *shadow mask* is placed behind the phosphor screen to prevent the electron beams from striking the wrong color dots. Good mechanical alignment is critical here.

The green electron gun is aligned so its beam energizes only the green phosphor dots. The red gun energizes only the red dots, and the blue gun energizes only the blue phosphor dots. To display the color yellow, the blue electron gun is turned off, the green and red electron guns energize their respective phosphors, and the eye will see yellow as a result. This implies that a varying signal voltage is needed to represent each of the primary colors in the chosen spectrum.

15.7.1 Subcarrier Modulation

The transmission channel bandwidth (6 MHz) is completely occupied by the video signal and the sound carrier. However, three new signal voltages must be added to the transmission for color telecasts to become a reality. The black and white video signal (hereafter called the **luminance** signal) contains white, which was determined to be 30% red, 59% green, and 11% blue. If only two colors were transmitted, they could be subtracted from the luminance signal at the receiver to develop the third color:

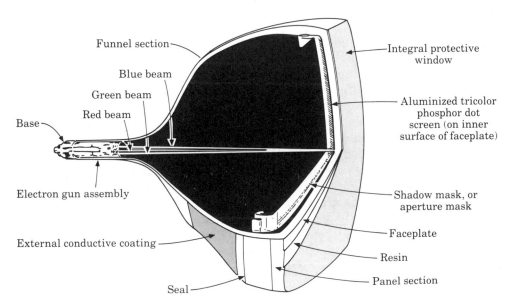

FIGURE 15.17
Cutaway view of a color cathode ray tube.

$$\begin{array}{lccc}
\text{white} = & 30\% \text{ red} + & 59\% \text{ green} + & 11\% \text{ blue} \\
\text{subtract} - & 30\% \text{ red} & & - 11\% \text{ blue} \\
\hline
\text{remainder} = & 0\% & 59\% \text{ green} & 0\%
\end{array}$$

Thus, only two color signal voltages need be added to the transmission.

A very common, recognizable color on the CRT is flesh. Since flesh tones contain a large measure of red (one of the desired colors for transmission), it was decided to amplitude-modulate a double-sideband signal and suppress the carrier using a voltage corresponding to the red color as a modulating signal. The carrier to be used was "in" phase with the color flesh, so this signal became known as the I signal or the I modulated signal, depending on its status. The color message need not contain much detail, so the bandwidth of the I signal is restricted to frequencies below 1.5 MHz.

Two carrier signals that are 90° out of phase (Q signals, for "quadrature") can be summed without intermodulating and can be separated at the receiver without difficulty. The signal representing the third color (blue) is restricted in bandwidth to 500 kHz because blue colors have almost no detail. The voltage representing the color blue is amplitude-modulated as a double-sideband suppressed-carrier signal, whose carrier is 90° out of phase from the carrier of the I modulated signal.

When these two suppressed-carrier signals are added to the luminance signal (a temporary combination), all three signals can be used to amplitude-modulate the station carrier frequency. The only question to be answered is the selection of the subcarrier frequency. A spectrum-analyzed display of the luminance signal (see Figure 15.18) reveals that all of the energy is clustered

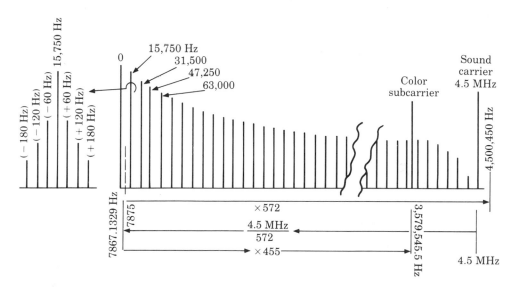

FIGURE 15.18
A spectrum analysis of the video signal and the placement of the color subcarrier within the same frequency range.

around the harmonics of the horizontal frequency, 15,750 Hz. Therefore, the color signals may be modulated on a carrier that is an *odd* harmonic of one-half the horizontal frequency (7875 Hz). When this was first tried, a disturbing beat tone was heard that came from the harmonic signal heterodyning with the sound carrier of 4.5 MHz ($572 \times 7875 = 4,504,500$ Hz). Working backward from the sound carrier of 4.5 MHz and dividing by 572 gave 7867.1329 Hz as one-half the horizontal frequency, for a *new* horizontal frequency of 15,734.266 Hz (within 1% of 15,750 Hz) and a corresponding *new* vertical frequency of 59.94 Hz. Harmonic 455 of one-half the horizontal frequency (455×7867.1329 Hz = 3,579,545.5 Hz) became the new subcarrier frequency onto which the color signals were modulated. In shop talk, 3.58 MHz is substituted for the exact numerical frequency.

15.8   MULTICHANNEL TELEVISION SOUND (MTS): TV STEREO

The principles for stereo in FM radio worked so well that they required little change to work for TV FM sound. The audio frequencies for both systems are limited to the range of 50 Hz to 15 kHz, so only the subcarrier frequencies and the relative voltage amplitudes are open to arbitration.

The problem that existed in early color transmissions, the presence of an audible beat note when the 4.5 MHz sound carrier heterodyned with harmonic 572 of one-half the line frequency, was overcome for stereo TV by incorporating preventions into the system from the start. The pilot carrier signal and the stereo subcarrier were selected as multiples of the color horizontal line frequency (H):

$$1H = 15,734.266 \text{ Hz}$$
$$2H = 31,468.532 \text{ Hz}$$
$$5H = 78,671.330 \text{ Hz}$$
$$6.5H = 102,272.73 \text{ Hz}$$

The main channel sound (monaural) is simply the linear sum of the left and right (L + R) channel sound signals. The main channel sound occupies the frequency range of 50 Hz to 15 kHz and is of such an amplitude as to deviate the station's aural carrier frequency to ±25 kHz (standard TV sound deviation), as shown in Figure 15.19.

The right channel sound is first inverted and then linearly added to the left channel sound to form the left minus right (L − R) sound signal, which occupies the same frequency range of 50 Hz to 15 kHz. The L − R signal is then amplitude-modulated in a balanced modulator to produce a double-sideband suppressed-carrier (DSSC) signal. This AM L − R signal is the stereo signal. A carrier signal centered at $2H$ (31,468.532 Hz) is the second input to the balanced modulator. The modulated stereo signal occupies the frequency band between 16,468 Hz and 46,468 Hz as the modulated lower sideband of the L − R

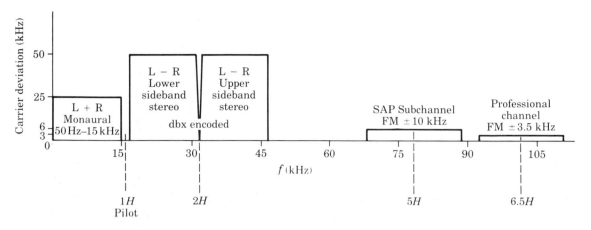

FIGURE 15.19
The message signal to frequency-modulate the TV sound carrier.

stereo signal ($2H - 15$ kHz) and the modulated upper sideband of the $L - R$ stereo signal ($2H + 15$ kHz). The amplitude of the modulated stereo signal is 6 dB greater than that of the monaural signal and will frequency-modulate the station's aural carrier frequency to ± 50 kHz. This changes the deviation ratio of the modulated stereo signal to 50 kHz/15 kHz = 3.33 rad and changes the stereo portion from narrowband FM to wideband FM, having a bandwidth of $2(\Delta f_c + f_s) = 2(50$ kHz $+ 15$ kHz$) = 130$ kHz. The increase in deviation is necessary to overcome the added noise at these higher frequencies.

The pilot signal, to be used for demodulation of the stereo signal at the receiver, is included in this audio range at $1H$, equal to 15,734.266 Hz. It is found in Figure 15.19 between the $L + R$ audio range and the modulated $L - R$ stereo signal. The pilot signal by itself has an amplitude that would deviate the station carrier by ± 3 kHz.

A third message signal, the equivalent of the SCA signal in radio FM stereo, is known as the *second audio program* (SAP) in television stereo. The second audio program has uses for the hearing-impaired and has an upper frequency limit of 10 kHz. This message signal is frequency-modulated onto a $5H$ (78.76 kHz) carrier, where 100% modulation is ± 10 kHz, and its amplitude is such as to frequency-modulate the station carrier to ± 15 kHz. Here again is an example of double frequency modulation. The final deviation ratio is 1.5 rad and establishes the SAP signal as narrowband FM. The stereo program (modulated $L - R$) is increased in amplitude by a factor of 2 to mask the noise at this higher frequency (higher than the monaural signal). Logic indicates that the third message signal should be still larger in amplitude to improve the signal-to-noise ratio. However, the third sound channel has a carrier ($5H$) that was found to cause a disturbance in the picture, so the amplitude was kept low to minimize this effect. This means that the third sound channel is much noisier than the stereo sound channel.

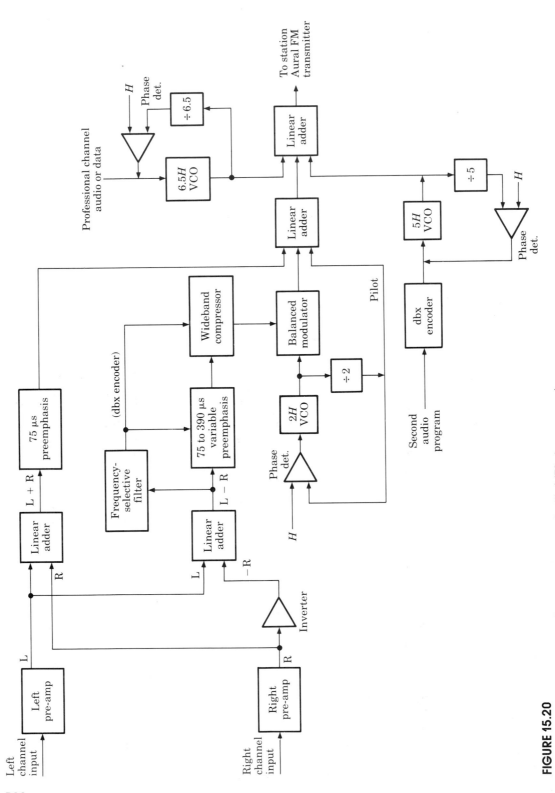

FIGURE 15.20
The studio portion of the audio processor used with MTS stereo sound.

A fourth sound channel, called the "professional channel," is used for communications between the station's facilities. It may carry voice transmissions not to exceed 3.5 kHz or data signals not to exceed 1500 BPS. When used for voice communications, the message signal is frequency-modulated onto a $6.5H$ (102.272 kHz) carrier for a maximum deviation of ± 3.5 kHz. This, again, is narrowband FM with a deviation ratio of 1 rad. When the professional channel is used for data transmission, it is frequency-shift-keyed between 100.77 kHz as a mark frequency and 103.77 kHz as a space frequency. Note the negative logic format for data transmission.

The entire message signal now covers the frequency range of 50 Hz to 103.77 kHz, and it is this signal that will be called the message signal used to frequency-modulate the station carrier.

Figure 15.20 shows that the 75 μs preemphasis treatment is applied to the monaural (L + R) channel sound, but the noise increases at $2H$ (the stereo channel) by 3 dB and at $5H$ (the SAP channel) by 9 dB. The second and third channel signals need very special treatment if they are to meet the standards of FM radio stereo quality. The L − R audio is preemphasized and compressed (in amplitude) in a Dolby dbx Spectral Compressor variable-preemphasis system. The dbx system has a fixed minimum preemphasis of 75 μs and a variable preemphasis to extend the range from 75 μs to 390 μs depending on the level of input signal. The dbx system has further capabilities of adjusting the audio level in accordance with the high-frequency content of the audio. When there is much high-frequency content in the audio signal, the output level to the balanced modulator is reduced. When there is little high-frequency content in the audio signal, the output signal to the balanced modulator is increased. The output ratio of the compressor is 2 to 1. For 100 dBm input, the output can be reduced to 50 dBm. The compressor is level-sensitive such that an input signal level of 0 dBm will output a 0 dBm signal. When the input signal increases by 20 dB (to +20 dBm), the output signal will only increase to +10 dBm; when the input signal decreases to −80 dBm, the output will only decrease to −40 dBm. All preemphasis circuits are basically high-pass filters and, as such, cause some overshooting in the signal; therefore, a high-level limiter is included as part of the overall dbx system. Figure 15.21 shows the change in compression ratio for the dynamic range of the compressor.

FIGURE 15.21
The input level and output level of the 2:1 compression ratio encoder.

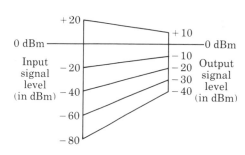

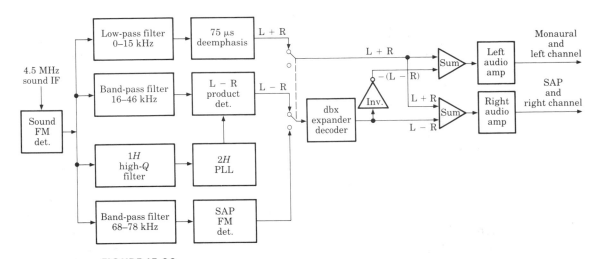

FIGURE 15.22
Typical block diagram for television stereo FM sound and second audio program.

At the receiver, the system to be examined includes only those circuits that follow the FM detector, and only the monaural and stereo signals will be demodulated. A low-pass filter over the frequency range of 50 Hz to 15 kHz will separate the L + R channel (monaural) sound from the total received signal, as shown in Figure 15.22. This signal must now be deemphasized by a 75 μs network.

The band of frequencies between 16.5 kHz and 46.5 kHz is separated from the total signal at the detector output by a band-pass filter, which recovers the modulated L − R stereo signal. The modulated sideband signal is then fed to a balanced modulator (acting as a product detector) to heterodyne with a 2H carrier signal and recover the L − R audio signal. The audio signal is then expanded by the same 2:1 ratio as the compressor, deemphasized, and finally added to the L + R audio signal. The matrixing for TV stereo is identical to the system used for FM radio. A positive phase of the L − R signal is added to the L + R signal to cancel the right channel sound and develop a signal equivalent to two times the left channel sound. An inverted version of the L − R signal, −(L − R), or −L + R, is added to the L + R signal to cancel the left channel sound and produce a signal of two times the right channel sound.

QUESTIONS

1. Vestigial sideband represents what form of emission? (a) A1, (b) A2, (c) A3, (d) A5, (e) A9.
2. The bandwidth of each commercial television channel in North America is (a) 1.25 MHz, (b) 6 MHz, (c) 4.2 MHz, (d) 4.5 MHz, (e) 8 MHz.
3. The aural transmitter power is what percentage of the visual transmitter power? (a) 10%, (b) 15%, (c) 20%, (d) 25%, (e) 30%.

4. What natural phenomenon aids the viewing of television pictures?
5. How many horizontal scan lines make up one field?
6. How many horizontal scan lines make up one frame?
7. How many fields make up one frame?
8. How many fields are reproduced every second?
9. How many frames are reproduced every second?
10. What is the television horizontal scanning frequency?
11. What is the time for one horizontal trace and retrace?
12. What is the time for one horizontal trace only?
13. What is the television vertical scanning frequency?
14. How many lines of resolution could be viewed on a 52 μs scan line at 6 MHz bandwidth?
15. What bandwidth would be required to display 700 lines on a 52 μs scan line?
16. With a trace time of 85% of one cycle, what is the horizontal scanning frequency required to display 480 lines of resolution for a 6 MHz bandwidth signal?
17. What is the maximum bandwidth of the video signal?
18. Where are the blanking and synchronizing signals generated?
19. Make a statement that describes how the blanking and sync pulses are polarized.
20. List the percentage of peak signal for the
    a. reference white level
    b. reference black level
    c. blanking pedestal
    d. sync pulses
21. Define "Z axis modulation."
22. Define "positive picture phase."
23. Define "negative picture phase modulation."
24. Compare positive picture phase modulation to negative picture phase modulation. State the advantages and disadvantages of each.
25. What form of picture phase modulation is used in North America?
26. With respect to the commercial television channel bandwidth, where is the visual carrier placed for vestigial sideband modulation?
27. State, in decibels, the rolloff characteristics of the attenuated lower sideband for vestigial sideband transmission.
28. State, in decibels, the rolloff characteristics of the upper sideband for vestigial sideband modulation as used for television.
29. State the legal location of the sound carrier for commercial television transmission in North America.
30. What is the minimum frequency and the maximum frequency used to modulate the sound carrier in commercial television?
31. State the carrier frequency change for 100% modulation of the aural carrier in commercial television.
32. Is television sound wideband or narrowband FM?
33. What is meant by the "front end" of a television receiver?

34. Does the tuner oscillator operate at a higher or lower frequency than the RF antenna signal?
35. How does the tuner treat a two-carrier input signal in preparation for the IF amplifiers?
36. Where is the picture carrier located on the response curve for the IF amplifiers?
37. What role would scene brightness play in the AGC circuit if special steps were not taken?
38. Name the input and output signals of a gated AGC system.
39. What is the purpose of a peaking coil?
40. Describe the points in the receiver system where the sound may be separated from the video signal.
41. What is meant by an "intercarrier" receiver?
42. What is meant by a "split-sound" receiver?

CHAPTER SIXTEEN

THE COMMUNICATIONS SATELLITE

16.1   INTRODUCTION

The ultimate worldwide communications system will feature a satellite as one of the major components. Although long-range communications took place before the age of satellites, the systems suffered from conditions that required a great deal of effort to overcome. Even today, due to remote location or surrounding terrain, there are isolated communities that are difficult to reach by point-to-point communication systems. The satellite is really nothing more than a radio relay station, but it offers the one advantage that is missing in all other systems; the capability of a direct *line-of-sight* path to 98% of the earth's surface. (The polar caps are still inaccessible to communications satellites.)

The transoceanic cables, which for years had been a major voice link between the continents, had begun to reach saturation. In addition, these cables lacked the ability to handle the growing demands for high-speed data and wideband television signal transmission over their relatively great lengths. These conditions, coupled with continents' growing land communications needs, paved the way for the development of the space relay station.

One may think it would have been cheaper to expand the land systems. However, the technology was advancing, and each new installation was profitable only when allowed to stay in place for a reasonable number of years. A changeover was expensive, time-consuming, and always seemed to be a step behind current developments.

The greatest strides in the development of satellite communications took place between 1965 and 1979, when COMSAT (Communications Satellite Corporation) launched four satellites within 6 years. The first and best known, named Early Bird, was placed in orbit in 1965. This station was the first to

TABLE 16.1
Satellites

Category	Examples	Earth Stations	Functions
Intercontinental satellites	Early Bird Intelsat II Intelsat II Intelsat III Intelsat IV Intelsat IVa Intelsat V Russia's Molniya and Statsionar, Europe's Symphonie	Large, expensive, highly reliable; designed for interconnection to circuits of national telephone administrations	To provide worldwide common-carrier telephone and data circuits Point-to-point TV relays
Traditional common-carrier domestic satellites	Canada's ANIK Western Union Westar RCA's Satcom AT&T and GTE's Comstar Indonesia's Palapa Arabsat	Large, common-carrier earth stations Medium-size private earth stations 10 and 4.5 meter band Receive-only TV earth stations	To enhance common-carrier telephone networks To provide leased long-distance circuits at lower cost Point-to-point TV relays TV broadcasting to local stations and CATV MUSAK music
Domestic multiple-access satellites	SBS (Satellite Business Systems)	5 or 7 meter, 14/12 GHz, installable on corporate premises or on city rooftops	To provide private networks for corporations and government, to relay telephone data and image traffic
Television broadcast satellites	JBS (Japanese Broadcast Satellites) Canada's CTS NASA's ATS-6	Large transmitter stations Inexpensive receive-only antennas small enough for homes and schools	TV broadcasting direct to homes or schools

handle worldwide commercial telephone traffic from a fixed position in space. The next series, Intelsat (International Satellites), was a group of satellites that served 150 stations in 80 countries. Table 16.1 lists some of these satellites and their functions. These systems became a tremendous competitive force among the common carriers when the FCC legislated the "open skies policy" in 1972, which encouraged private industry to enter the domestic communications satellite market. Western Union's *Westar* went up in 1974, followed shortly thereafter by Bell Labs' Telstar.

This chapter will describe how the satellite works and will cover the concepts and parameters involved in establishing earth-space communications.

16.2   THE SATELLITE ORBIT

An orbit is a circular path in space occupied by an object, moving in a direction parallel to the surface of a planet, that has a forward velocity sufficient to create an outward thrust (centrifugal force) equal to the gravitational pull of the planet it orbits. There are three common orbital patterns: the polar orbit, the inclined elliptical orbit, and the equatorial geosynchronous orbit. This chapter will concentrate on the equatorial geosynchronous orbit, called the **Clarke orbit.** The following factors apply equally to all orbits.

1.   The plane of the orbit must pass through the center of the object to be orbited. For instance, a satellite could *not* orbit the earth around a latitude of 42°N because this plane does not pass through the center of the earth.
2.   The time to complete one orbit depends on the mass of the vehicle (as compared to the mass of the earth), the vehicle's velocity (dependent on the initial thrust supplied by the rocket engines and the mass of the payload), and the final orbital altitude.

To place a satellite in a position that appears to be stationary over a selected location on the earth's surface means that the vehicle must move in the same direction as the earth rotates. This final requirement eliminates the polar orbit. An inclined elliptical orbit could be in a direction and at an altitude and velocity that would appear stationary relative to a given longitude, but this orbit shifts its north/south latitudinal position.

The only orbit that meets all of these requirements is one that is *directly over the equator,* moving in a *west-to-east* direction at an altitude of 22,282 mi above sea level (this varies slightly with the size of the satellite), and with a forward velocity of 6874 mph to complete one orbit in 24 hours.

16.3   THE SATELLITE POSITION

By international agreement, satellites placed in the Clarke orbit should be separated by 4 geometric degrees (1833 mi between satellites). This displacement is based on the typical earthbound antenna system with a beamwidth angle of 1.7° and prevents interference between adjacent satellites. This limited the number of satellites in the Clarke orbit to 90. Since that initial agreement, based on logged experience, greater care in selecting equipment, and the rapid growth in the market, the separation was reduced to 2° and legislation is forthcoming that will allow a 1° separation (458 mi) between satellites. By assigning different frequency bands to adjacent satellites and by polarizing the antennas (horizontal, vertical, right-hand circular, left-hand circular), even greater interference reduction may be achieved. Figure 16.1 shows the satellites in the Clarke orbit between 61.5° west longitude and 135° west longitude, in the American orbital parking region. A longitude of 61.5° west at the equator lies directly above the northeast corner of Brazil, and 135° west longitude at the equator lies south of the Hawaiian islands.

The maximum thrust from self-propelled launch vehicles limits the payload for this altitude to about 1300 kg (2900 pounds or 1.45 tons). The development of the space shuttle as a launch platform will increase the possible satellite size to about 5000 kg (11,000 pounds or about 5.5 tons).

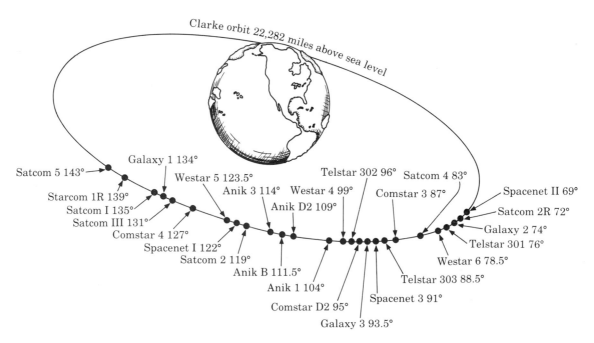

FIGURE 16.1
American parking area in the Clarke orbit.

16.4   LINKAGES

16.4.1 The Up-Link

All of the ground equipment along with the transmission path and the receiving antenna at the satellite are included in the **up-link** system. Basically, this system includes everything before the input terminals of the satellite receiver. To conserve space and weight on board the satellite, the receiving antenna is made much smaller than the satellite transmitting antenna. Although a smaller receiving antenna lowers the gain to the receiver input and widens the beamwidth angle of the reception pattern, these small deficiencies are easily overcome by either increasing the size of the ground transmitting antenna or raising the ground transmitter output power, or both. The ground station up-link antenna should have a narrow-beamwidth angle. The ground transmitter need only supply from 1 to 3 kW of unmodulated carrier power to drive an antenna with a parabolic reflector 10 m (32.5 ft) in diameter. The largest transmission parabolas are 18 m (59 ft) in diameter and have a power gain of 80 dBW effective radiated power.

16.4.2 The Down-Link

The last section did nothing more than introduce the up-link. We now need to define the **down-link** and then take a look inside the satellite before discussing systems. The down-link is described in terms of satellite transmitter output power, down-link antenna gain and beamwidth, and the ground area that the transmitted signal will service (called a **footprint**).

Satellites are solar-powered and provided about 250 W for all operations. The typical transmitter output power is about 7 W per channel for a 24-channel system. By international agreement, each channel has an absolute maximum power of 32 dBW EIRP (equivalent isotropically radiated power) or 29.85 dBW ERP compared to a half-wave dipole antenna, which means an antenna gain of about 140.

Footprint patterns are represented in Figure 16.2, which is not drawn to scale. The shape of the footprint is controlled by the design shape of the parabolic reflector and may differ for each satellite in orbit. The single-beam satellite is the more common in telecommunications networks. The footprint may be very large, as shown here, or it could be made smaller to direct the transmission to a selected area. *Spot beams,* shown directed to the Hawaiian islands and Alaska, may be included in the footprint pattern as well as the larger mainland signal.

The multibeam satellite, also shown in Figure 16.2, is an alternate system that is intended to increase the signal strength to the service area by controlling the beamwidth and signal distribution over the area. However, many more beams are required. Each system has its advantages and disadvantages. The multibeam satellite requires a more elaborate antenna system

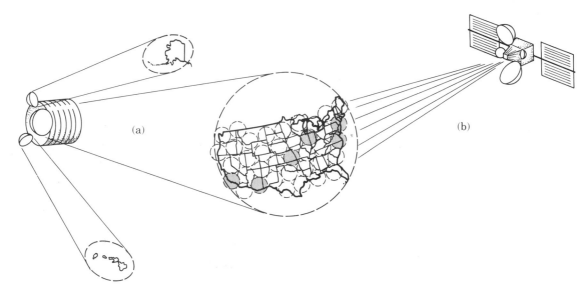

FIGURE 16.2
Footprints of satellite beams. (a) Spot beam. (b) Multibeam.

(higher costs) but is more efficient in terms of power distribution over the entire designated area. The single-beam system transmits to areas where there is no receiving equipment (poorer transmission efficiency) but reduces the complexity of the antenna array. This makes the payload efficiency better for operations conducted once in the lifetime of the satellite as opposed to operations conducted on an hour-by-hour basis.

16.4.3 The Cross-Link

At the altitude of the Clarke orbit, one satellite could command a footprint area of 42.2% of the earth's surface. The beamwidth from the satellite for such coverage would be only 17.174°, as shown in Figure 16.3, but this would not allow total global coverage. A minimum of three satellites, placed 120° apart in the Clarke orbit, would cover all of the earth's surface but the polar caps (98%). This makes it possible for one earth station to transmit to another station on the opposite side of the globe by sending data to its "in view" satellite, from there to its closest neighbor in the Clarke orbit, and from there back to earth. The cross-link distance is 45,450 mi between satellites, making the total distance (earth–satellite A–satellite B–earth) 90,014 mi or 144,832 km. A message would take about one-half second to travel this distance, or almost a full second if a return response is needed. In voice communications, this delay would probably go unnoticed, but in data communications, it represents an appreciable delay.

FIGURE 16.3
Cross-link connections for world
coverage.

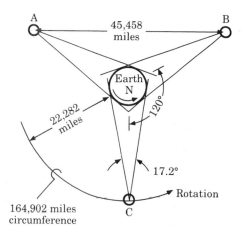

Intelsat was the first to set up such a communications link, with vehicles spaced over the Indian Ocean, the Atlantic Ocean, and the Pacific Ocean at the equator.

16.5   ASSIGNABLE SATELLITE FREQUENCIES

The satellite frequencies that have been assigned through the 1990s are shown in Table 16.2. These frequencies are shared by terrestrial and communications satellites. The term *terrestrial* describes satellites used for weather forecasting, meteorological studies, tracking and data relay systems (TDRS) for launch and low-orbit satellites, business systems, and some military functions. Satellite frequencies are usually classified as (a) military only, (b) multimission for terrestrial and telecommunications, and (c) multiband for telecommunications only, which may include the maritime services at 1.5 GHz to 1.65 GHz for data and communications with ships at sea. Business and entertainment services can share (or reuse) these frequencies provided they use highly directional up-link antenna systems. The introduction of the smaller home receiving dishes (with lower gain and wider beamwidths) are making satellites incompatible in the 12.2 GHz to 12.7 GHz band.

The C band frequencies are expressed as 6/4 GHz, the Ka band as 14/12 GHz, and the Ku band as 30/20 GHz. In these dual frequency specifications, the first frequency is for the up-link and the second frequency refers to the down-link. Reuse of the frequencies in the C and Ku bands is made practical by very careful design of the reflector antenna system and by using dual polarization of the down-link antennas (see Tables 16.3 and 16.4).

TABLE 16.2
Present and potential satellite frequencies

Frequency (MHz)	Bandwidth (MHz)	Direction	U.S. Satellites	Other Satellites
1530–1559	29	Down	Marisat	Inmarsat
1626.5–1660.5	34	Up	Marisat	Inmarsat
3400–3500	100	Down		USSR
3500–3700	200	Down		USSR
3700–4200	500	Down	Comstar Westar Satcom	Intelsat Palapa ANIK, Sakura
5850–5925	75	Up	—	—
5925–6425	500	Up	Same as 3700–4200	
6425–7075	650	Up	Not authorized	
7250–7750	500	Down	DCSC	NATO and
7900–8400	500	Up	DCSC	USSR
10700–11700	1000	Down	None	Intelsat
11700–12100	400	Down	SBS, Gstar SCPC	ANIK B&C
12100–12300	200	Down		
12300–12700	400	Down	BSS	
12700–12750	50	Up	None	
12750–13250	500	Up	None	
14000–14500	500	Up	SBS, Gstar SCPC	ANIK B&C
14500–14800	300	Up	None	
17300–17700	400	Up	None	
17700–18100	400	Up/down	BSS	
18100–20200	2100	Down	None	Sakura
27000–30000	3000	Down	None	Sakura

SBS = Satellite Business Systems
BSS = Broadcast Satellite Systems
DCSC = Data Communications Signal Channel
SCPC = Single Carrier Per Channel (voice-activated)

16.6   INSIDE THE SATELLITE

Figure 16.4, page 605, shows the internal layout of a satellite. There are basically five sections within the satellite, each totally dependent on the other four.

    A.   the electronics section, called the **transponder**
    B.   the antenna systems
    C.   the power package
    D.   the station-keeping section, made up of
        1.   the control and information section
        2.   the rocket thruster section

TABLE 16.3

North American C band satellite frequency band conversion table

Channel Number	Frequency (MHz) Up-Link	Down-Link	Satellite systems A	B	C	D	E	F
1	5945	3720	1 (V)	1V (V)	1A (H)	1 (H)	1D (H)	1 (H)
2	5965	3740	2 (H)	1H (H)	1B (V)	2 (V)	1X (V)	
3	5985	3760	3 (V)	2V (V)	2A (H)	3 (H)	2D (H)	2 (H)
4	6005	3780	4 (H)	2H (H)	2B (V)	4 (V)	2X (V)	
5	6025	3800	5 (V)	3V (V)	3A (H)	5 (H)	3D (H)	3 (H)
6	6045	3820	6 (H)	3H (H)	3B (V)	6 (V)	3X (V)	
7	6065	3840	7 (V)	4V (V)	4A (H)	7 (H)	4D (H)	4 (H)
8	6085	3860	8 (H)	4H (H)	4B (V)	8 (V)	4X (V)	
9	6105	3880	9 (V)	5V (V)	5A (H)	9 (H)	5D (H)	5 (H)
10	6125	3900	10 (H)	5H (H)	5B (V)	10 (V)	5X (V)	
11	6145	3920	11 (V)	6V (V)	6A (H)	11 (H)	6D (H)	6 (H)
12	6165	3940	12 (H)	6H (H)	6B (V)	12 (V)	6X (V)	
13	6185	3960	13 (V)	7V (V)	7A (H)	13 (H)	7D (H)	7 (H)
14	6205	2980	14 (H)	7H (H)	7B (V)	14 (V)	7X (V)	
15	6225	4000	15 (V)	8V (V)	8A (H)	15 (H)	8D (H)	8 (H)
16	6245	4020	16 (H)	8H (H)	8B (V)	16 (V)	8X (V)	
17	6265	4040	17 (V)	9V (V)	9A (H)	17 (H)	9D (H)	9 (H)
18	6285	4060	18 (H)	9H (H)	9B (V)	18 (V)	9X (V)	
19	6305	4080	19 (V)	10V (V)	10A (H)	19 (H)	10D (H)	10 (H)
20	6325	4100	20 (H)	10H (H)	10B (V)	20 (V)	10X (V)	
21	6345	4120	21 (V)	11V (V)	11A (H)	21 (H)	11D (H)	11 (H)
22	6365	4140	22 (H)	11H (H)	11B (V)	22 (V)	11X (V)	
23	6385	4160	23 (V)	12V (V)	12A (H)	23 (H)	12D (H)	12 (H)
24	6405	4180	24 (H)	12H (H)	12B (V)	24 (V)	12X (V)	

Systems

A = Satcom 1R, 2R, 3R, 4, and 5

B = Telstar 301, 302, 303; Comstar 3 and 4

C = ANIK D

D = Galaxy 1, 2, and 3

E = Westar 4 and 5

F = Spacenet 1, Westar 2 and 3, ANIK B

V = vertical polarization

H = horizontal polarization

16.6.1 The Transponder

The transponder is a high-frequency radio receiver, a frequency down-converter, and a power amplifier used to transmit the down-link signal. It receives a modulated signal on one carrier frequency and retransmits the same information on a lower carrier frequency with no demodulation or signal cleanup processing. A simplified block diagram of a typical transponder is shown in Figure 16.5 on page 606.

TABLE 16.4
North American Ku band satellite frequency table

ANIK C2,C3 Transponder	Frequency (MHz) Up-Link	Down-Link	Gstar 1 Transponder	Frequency (MHz) Up-Link	Down-Link
T1 (V)	14017	11717	1 (H)	14030	11730
T2 (V)	14043	11743	2 (H)	14091	11791
T3 (V)	14078	11778	3 (H)	14152	11852
T4 (V)	14104	11804	4 (H)	14213	11913
T5 (V)	14139	11839	5 (H)	14274	11974
T6 (V)	14165	11865	6 (H)	14335	12035
T7 (V)	14200	11900	7 (H)	14396	12096
T8 (V)	14226	11926	8 (H)	14457	12157
T9 (V)	14261	11961	9 (V)	14044	11744
T10 (V)	14287	11987	10 (V)	14105	11805
T11 (V)	14322	12022	11 (V)	14166	11866
T12 (V)	14348	12048	12 (V)	14227	11927
T13 (V)	14383	12083	13 (V)	14288	11988
T14 (V)	14409	12109	14 (V)	14349	12049
T15 (V)	14444	12144	15 (V)	14410	12110
T16 (V)	14470	12170	16 (V)	14471	12171
T17 (H)	14030	11730			
T18 (H)	14056	11756	**SBS 1,2,3,4 Transponder**	**Frequency (MHz) Up-Link**	**Down-Link**
T19 (H)	14091	11791			
T20 (H)	14117	11817	1 (H)	14025	11725
T21 (H)	14152	11852	2 (H)	14074	11774
T22 (H)	14178	11878	3 (H)	14123	11823
T23 (H)	14213	11913	4 (H)	14172	11872
T24 (H)	14239	11939	5 (H)	14221	11921
T25 (H)	14274	11974	6 (H)	14270	11970
T26 (H)	14300	12000	7 (H)	14319	12019
T27 (H)	14335	12035	8 (H)	14368	12068
T28 (H)	14361	12061	9 (H)	11417	12117
T29 (H)	14396	12096	10 (H)	14466	12166
T30 (H)	14422	12122			
T31 (H)	14457	12157	**Spacenet 1,2 Transponder**	**Frequency (MHz) Up-Link**	**Down-Link**
T32 (H)	14483	12183			
			19 (H)	14040	11740
			20 (H)	14120	11820
			21 (H)	14200	11900
			22 (H)	14280	11980
			23 (H)	14360	12060
			24 (H)	14440	12140

V = vertical polarization
H = horizontal polarization

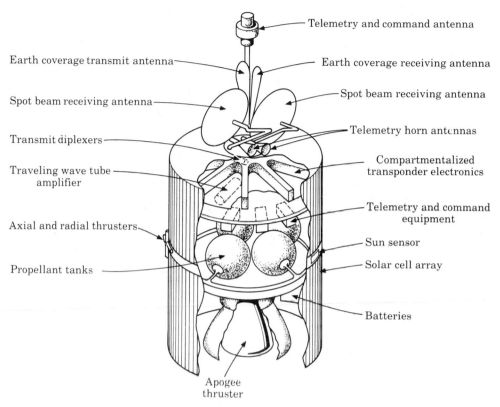

Telemetry and command antenna

Earth coverage transmit antenna

Earth coverage receiving antenna

Spot beam receiving antenna

Spot beam receiving antenna

Transmit diplexers

Telemetry horn antennas

Traveling wave tube amplifier

Compartmentalized transponder electronics

Telemetry and command equipment

Axial and radial thrusters

Sun sensor

Propellant tanks

Solar cell array

Batteries

Apogee thruster

FIGURE 16.4
Exposed view of the systems layout inside a satellite.

The RF signal from the receiving antenna is separated through a band-pass filter and then heterodyned downward to the down-link frequency. It is filtered again to pass only the difference frequency, amplified, in some cases limited, and then fed into a traveling wave tube to be powered for the trip back to earth. In single-transponder satellites in the C band, the local oscillator is operated at a frequency *higher* than the RF antenna signal, and a simple low-pass filter then separates the difference frequency from the higher sum, RF, and oscillator frequencies. Figure 16.6 shows a multitransponder electronic system where a single microwave local oscillator signal is distributed to all channel mixers. Each up-link carrier signal is lowered by the same amount for each channel to its associated down-link carrier frequency. On simple relay channel systems, the multitransponder system consists of 12 or 24 individual receivers that share a common local oscillator at 2225 MHz.

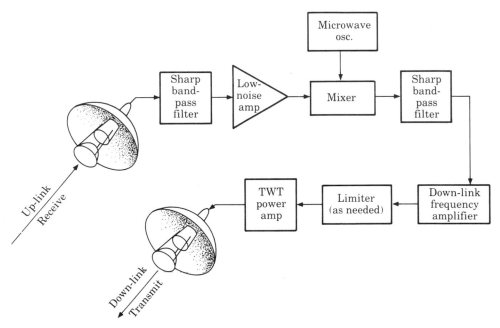

FIGURE 16.5
Typical single-channel transponder.

In more sophisticated systems, either frequency division multiple access (FDMA) or time division multiple access (TDMA) schemes are used to make the on-board systems more efficient and flexible. A more complex system is depicted in Figure 16.6 that includes a routing-controlled channel switching network and address code and frequency synthesis mixer circuits. Address data are mixed in with the message package assembled at the ground station for each message signal group. The address data are extracted before the carrier is reduced in frequency and are used to control a switching system at the input of the mixers. All of the messages from one ground station may not have the same final destination. By time sharing, messages may be routed by TDMA control. By controlling the final output signal, the power to a given area may be modified for the down-link transmission.

16.6.2 The Antenna Systems

The antenna systems, once properly in place, will generally operate trouble-free for the life of the satellite. However, they are complicated mechanical assemblies that are folded into a dense package for launch and must be deployed into a large, accurately aligned assembly during orbit, so their success is always a marvel of accomplishment. When the space shuttle becomes a

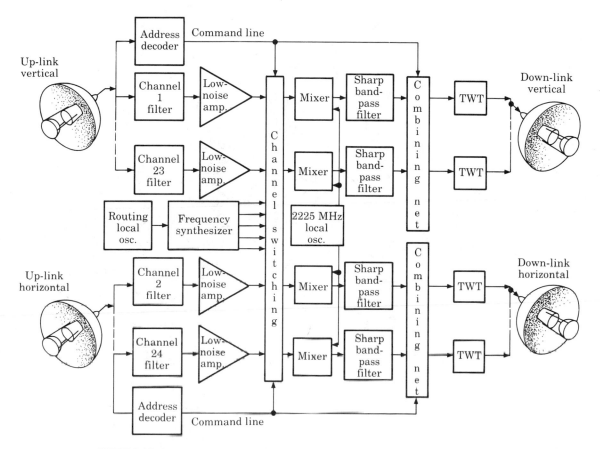

FIGURE 16.6
A multichannel transponder with addressed switching functions.

permanent part of the launch program, the antenna systems will become less of a determinant for the success of a mission.

A satellite may be constructed on the same principle as an electric motor, in which the shaft remains stationary while the housing turns. Figure 16.4 portrays such a system. The rotary shell acts as a flywheel or gyroscope to maintain correct orbital attitude, while the antenna assembly connected to the hub remains stationary. Figure 16.7 is a drawing of a satellite readied for release from a shuttle. The antenna systems are already deployed, and the electronics aboard have been verified.

16.6.3 The Power Package

The satellite must be powered from either a battery or a solar energy system. If a typical vehicle consumes 250 W total power, then a 24 V battery would need

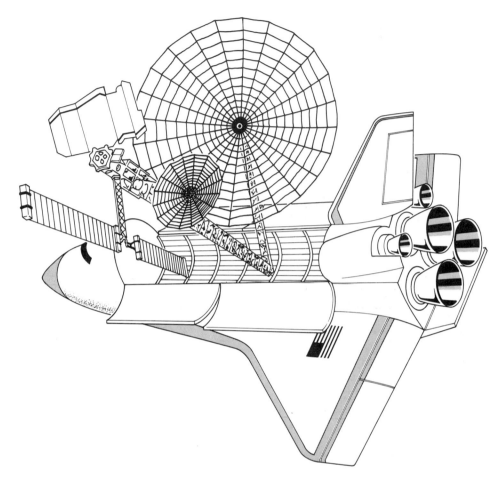

FIGURE 16.7
Artist's rendition of a satellite launch from the space shuttle.

to furnish 10.4 A continuously for the expected life of the satellite. A solar energy system will not fit the bill either because satellites in the Clarke orbit are eclipsed by the earth for about 1.5 hours during each orbit. Therefore, a combination system is used. A solar cell system supplies the power to run the electronics and charge the batteries during the sunlight cycle, and the battery furnishes the energy during the eclipse.

Full sunlight furnishes about 100 W of energy per square meter (10.75 square feet) over most of the earth's surface. Solar cells are about 15% efficient, so from these two statistics the area of the solar energy system is estimated. The typical life expectancy of a solar cell is between 10 and 15 years, and this nearly matches the predicted life of a battery that is charged and discharged at the rate of the satellite's requirements. These two factors are the main limitations to the useful lifespan of a satellite.

16.6.4 Station Keeping

Simply stated, *station keeping* means keeping the satellite in the correct orbit with the antennas pointed in the exact direction desired. In addition to the information channel signals, there is also a constant metering of the satellite's altitude, velocity, roll, yaw, and pitch attitudes. These status reports form a steady stream of data information between the satellite and the ground control station. Minor changes in the gravitational pull of the earth, encounters with meteor storms, and microscopic variations in data reading can all add up to a slow drift of the satellite from its intended path. The telemetering conversation needed to prevent this is a constant two-way chatter between the devices aboard the satellite and the ground computer terminals. These transmissions are sandwiched between all of the communications signaling and represent the "control and information" section of the vehicle.

There are several small rocket thrusters aboard the satellite that modify the forward and retro velocity of the spacecraft. There are also six other small thrusters that control yaw, pitch, and roll. These thrusters (and their fuel) along with the control and information section constitute the two remaining sections of the satellite.

16.7 FORMS OF MODULATION

Nearly every form of modulation covered in this text is being used in present-day satellite communications networks. The transponders aboard satellites are locked to a preset carrier frequency to perform frequency recognition and translation only. There is no sensory device on board the satellite to identify the form of modulation in use. Therefore, all modulation and detection schemes can be controlled by the ground equipment.

The newer, better satellites (late 1980s) have a bandwidth capability of 500 MHz, which is divided equally into 12 channels of 40 MHz each. Commercial television and teleconferencing video signals are frequency-modulated directly onto the satellite carrier frequency. The raw, composite, color-plexed video signal with frequencies to 4.2 MHz is diplexed with a frequency-modulated sound signal riding on a 6.8 MHz carrier. The voltage of the video signal is 84% of the voltage needed for 100% modulation of the carrier to ± 12.5 MHz. The remaining 16% of the voltage comes from the multiplexed FM sound carrier signal.

Two additional bits of information can be gleaned from these specifications.

1. The bandwidth of the modulated video signal, from Carson's rule, is

$$BW = 2(\Delta f_c + f_m) = 2(12.5 \text{ MHz} + 6.8 \text{ MHz}) = 38.6 \text{ MHz}$$

Therefore, one transponder is saturated and totally occupied for each channel of television to be transmitted.

2. The deviation ratio of the system is the maximum deviation divided by the highest modulating signal frequency:

$$m_f = \frac{12.5 \text{ MHz}}{6.8 \text{ MHz}} = 1.838 \text{ rad}$$

This means that satellite television broadcasting is narrowband FM.

Telephone message modulation via satellite was covered in the discussion of digital modulation. The message is an amplitude-modulated, single-side-band suppressed-carrier signal that is frequency-division-multiplexed before being frequency-modulated onto a 70 MHz intermediate frequency carrier (to ± 8.3 MHz). It is then amplitude-modulated onto the satellite transponder carrier frequency signal. This totals four modulation processes in the satellite up-link alone. The label given to this modulation package is FDM/FM, or frequency division multiplexing/frequency modulation.

Many forms of digital modulation have found their way into satellite carrier systems, with QAM (quadrature amplitude modulation) and QPSK (quadrature phase shift keying) as the leading contenders. Systems manufacturers report 300 to 1300 voice channels per transponder, depending on the modulation format. The emphasis is on making maximum use of the satellites already in orbit. Considerable effort is put into FDMA (frequency division multiple access) and TDMA (time division multiple access), where many users may access the satellite at the same time to conserve both time and equipment. For the down-link, a 750 μs sequence using TDMA is favored for data and voice frequency messages. In the cross-link transmissions, however, FDMA is favored because of the longer time delay of response messages.

16.8   FREE-PATH SPACE LOSSES

The principle of losses due to distance and frequency was introduced in Chapter 12. Equation 12.10, for the receiver-to-transmitter power ratio, expressed distance in kilometers and frequency in megahertz, along with the constants, in the following relation:

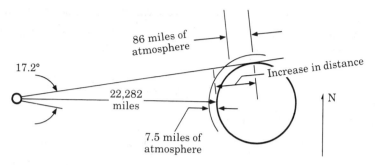

FIGURE 16.8
Propagation distance change as a function of latitude to a satellite in the
Clarke orbit.

$$\frac{P_r}{P_t} = G_t + G_r - (32.44 + 20 \log D + 20 \log f) \quad \text{(in decibels)}$$

In satellite communications, it is sometimes easier to work with miles rather
than kilometers. When this is the case, only the value of the constant
changes:

$$\frac{P_r}{P_t} = G_t + G_r - (36.57 + 20 \log D + 20 \log f) \tag{16.1}$$

Remember that losses become greater as frequency goes higher (more wavelengths between the transmitter and receiver) and as distance increases (the
inverse square law).

Examining the distance factor first, we see that the altitude of the Clarke
orbit at the equator is 22,282 mi as indicated in Figure 16.8. At latitudes away
from the equator, the distance increases until the transmitted signal passes the
horizon at about 85° north or south latitude. The distance increases to a maximum of about 25,900 mi when the satellite and receiving stations are at the
same longitude. When the satellite and receiving station are at different lon-
gitudes, Equation 16.2 is used once for the difference in latitude and again for
the difference in longitude. The results are added to the orbit's distance from
the equator.

$$\Delta D = r(1 - \cos \theta) \tag{16.2}$$

where r = radius of the earth = 3963 mi
$\quad \Delta D$ = distance increase, in the dimensions of r
$\quad \theta$ = difference in latitude *or* longitude (calculated separately)

EXAMPLE 16.1 ══════════════════════════════════

For communications from Phoenix, Arizona, at 34° north latitude and 112.5° west longitude, to Satcom I, at 135° west longitude over the equator, the increase due to latitude is

$$\Delta D = 3963 \ (1 - \cos 34) = 677.5 \text{ mi}$$

and the increase due to longitude is

$$\Delta D = 3963 \ (1 - \cos [135 - 112.5]) = 301.6 \text{ mi}$$

The total distance to the satellite is

$$22{,}282 + 677.5 + 301.6 = 23{,}261 \text{ mi}$$

More importantly, as the distance increases, the approach angle to the earth station causes the signal to pass through a thicker layer of air (more miles of atmosphere), also shown in Figure 16.8, and this increases the atmospheric absorption and adds noise contamination. At the equator, the atmosphere is about 7.5 mi thick, while at 85° latitude, the signal passes through nearly 85 mi of atmosphere. This atmospheric contamination produces an additional loss, expressed as

$$\text{dB loss} = \left[\left(\frac{7.5}{\cos L^\circ} \right) - 7.5 \right]^{1/3} \tag{16.3}$$

At 34° north latitude, the noise losses are

$$\left[\left(\frac{7.5}{\cos 34^\circ} \right) - 7.5 \right]^{1/3} = 1.156 \text{ dB}$$

This only serves to identify that a receiving station at the equator, directly under the satellite, will receive a better signal than a station at any other latitude and longitude by a calculable amount. The increased losses due to distance and noise should not cause a mass migration to the equator for better reception, because they are not stand-alone calculations. They simply add to the losses of the overall system and should reach a maximum of about 4.5 dB.

16.9   THE GROUND STATION

The two most critical areas of the ground receiving station are at the head end of the system. They are the receiving antenna and the input RF amplifier. The low-noise amplifier has aquired the acronym LNA.

It is known that the gain of the parabola, as a function of the diameter-to-wavelength ratio, will increase as the frequency increases. It is likewise

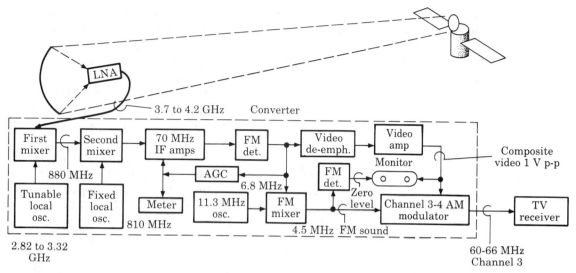

FIGURE 16.9
A receive-only television (ROTV) converter.

known that the free-space path loss will increase as the frequency increases. However, one condition can exactly offset the other for all frequencies, so that the gain of the dish and the free-path space loss need only be calculated once to be effective for all frequencies in the band. Parabola size need be the only consideration when antenna gain is to be changed.

The procedure generally used to determine the input signal level required at the receiver is as follows. First find the distance from your station to the farthest satellite in your system. Second, calculate the power to your station considering the signal level from the satellite and the geometric center frequency of the band in which you plan to operate. At this time, a −3 dB aging factor is usually added to avoid marginal signal reception. The next step is to examine the receiver/converter system and, from these results, to determine the receiving antenna and LNA gain requirements.

The receive-only television converter (ROTV) in Figure 16.9 is designed to receive a signal as low as 250 fW and convert it to channel 3 or 4 of the VHF band, at a level of 500 μV or better, for use with a commercial television receiver. A signal of 1000 μV at the receiver antenna terminals is the industry standard for excellent picture quality. The ROTV converter uses the standard superheterodyne receiver format with double conversion for good image rejection. The first converter accepts signals in the frequency range of 3700 to 4200 MHz and down-converts them to the first intermediate frequency of 880 MHz. There is generally only a crystal or ceramic band-pass filter at this point, with no amplification. The 880 MHz signal is then down-converted a second time to the 70 MHz intermediate frequency where the receiver gain is achieved. The video signal in this service is frequency-modulated in the up-link, so a video

FM detector is used here to recover the composite video signal. The signal is then deemphasized (returned to a flat frequency response) and amplified before being routed to a VHF amplitude modulator and monitor jack.

The sound signal sent to the satellite in this system was frequency-modulated onto a 6.8 MHz carrier and must be down-converted to the standard 4.5 MHz carrier frequency, as shown in Figure 16.9. When a monitor jack is included, the FM sound signal is recovered as an audio signal at zero level (1 mW across 600 Ω). The 6.8 MHz sound carrier is also used as the AGC reference signal level. The AGC voltage is used to control the gain of the 70 MHz IF amplifier and to drive the meter circuit as a signal level indicator.

In some military and industrial applications, the signals from satellites are so weak that they are indistinguishable from the noise at the antenna. A segment of the receiver designed to counteract this condition is shown in Figure 16.10. The receiver down-converts the carrier from the satellite to a 130 MHz IF signal and then converts it down again to 10.7 MHz second IF. The investment here is in the detector circuit, called a *quadrature phase-locked loop* detector, which is capable of recovering extremely weak satellite signals. When the signal is too weak to be detected by phase detector 1, phase detector 2 turns on a quadrature (90° shift) sweep generator, which causes the local oscillator to the second mixer (VCO) to sweep across the 130 MHz frequency range in search of the missing (weakened) carrier signal. Since noise causes an amplitude change in the signal, shifting the VCO phase by 90° causes the signal to stand out as a recognizable difference. Signals as low as 20 to 30 dB below the receiver noise level have been successfully recovered using a quadrature phase detector.

16.9.1 Aligning the Satellite Dish

The degree of difficulty in tracking a satellite depends entirely on the satellite's orbit. In the solar orbit, the receiving antenna will be in motion during the

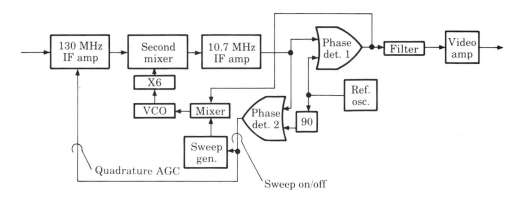

FIGURE 16.10
A typical block diagram of a quadrature phase-locked loop (QPLL) detector for low signal levels.

entire time that the satellite is within view of the receiving station. This could be nearly a 180° antenna swing, but accounts for only about one-quarter of the satellite's orbital time. The first step in selecting an antenna site is to make sure you have a clear view of the satellite's orbit from your receiving location, free of obstruction from hills, buildings, power lines, and sometimes even trees.

Tracking in the Clarke orbit requires a motionless antenna because the satellites are in a fixed relative position. They will be at an angle above the horizon (facing the equator) that is equal to 90° minus the receiver latitude and will follow an arc running from about 60° east to 60° west of your location. Again, this window must be clear of all obstacles.

Although there are several styles of antenna mounting hardware, the polar axis mount is the most popular because it requires only one alteration of the dish position to track any of the satellites in your view of the Clarke orbit (see Figure 16.11). All permanent satellite receiving antenna systems require a solid concrete base. The polar axis system also requires a 3½ in diameter stand pipe about 2 ft longer than the radius of the dish to be mounted. It is critical that this mounting post be perfectly plumb in all directions for the dish to track correctly.

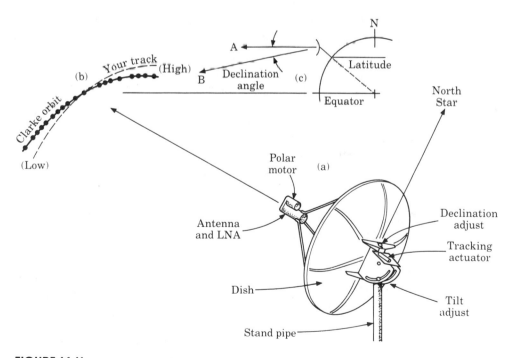

FIGURE 16.11

A ground station receiving antenna system. (a) Receiving antenna and LNA. (b) Alignment with the Clarke orbit. (c) Declination angle.

1. Mount the dish on the post and aim it directly at the equator (due south in the northern hemisphere). Snug the retainer bolts so that the dish will not sway.

2. Tilt the dish skyward until the center line of the horizontal scan hinge points directly at the North Star (in the northern hemisphere; Signa Octans in the southern hemisphere), and tighten the tilt-locking bolts. At this time, your dish will be aimed on a line that is parallel to your latitude and to the equator (see Figure 16.11a and b for clarification).

3. For the dish to point toward the family of satellites in the Clarke orbit, it will need to tilt downward by an angle that is related to your latitude, called the *declination angle*. Use Equation 16.4 to find this angle.

$$\text{Declination angle} = \arctan\left(\frac{r \sin L°}{(A + r) - (r \cos L°)}\right) \qquad \textbf{(16.4)}$$

where r = earth's radius = 3963 mi
A = orbital altitude (Clarke orbit = 22,282 mi)
L = latitude

At a latitude of 34° N, the declination angle is 5.5°. Adjust the turnbuckle to tilt the top of the dish downward by 5.5° more than the center line of the pivot hinge.

Now is the time to connect your converter and receiver to the dish and LNA. From Figure 16.1, find the satellite that is closest to your longitude, aim the dish at this satellite, and make whatever minor adjustments are needed for best picture on the receiver and strongest signal on the converter meter. Make sure the converter is set for the center channel number and that the antenna at the focal point is properly polarized.

Point your antenna west and adjust for the satellite at the end of your field range. Then point the antenna east and repeat this test. You may find that at one end of the range the dish is too low and at the other end it is too high. Loosen the post lock bolts and rotate the entire assembly a small amount toward the direction that is lower than the Clarke orbit. Repeat the test for east and west tracking, moving the mount on the post a small amount each time until the dish tracks correctly.

The receiving antenna system shown in Figure 16.11 includes a parabolic reflector, an LNA, and a polarization shift motor to change from vertical to horizontal phase signals. The LNA is placed as close to the antenna as possible to keep the line losses and noise pickup minimal at the amplifier input. The characteristics that make the amplifier a low-noise circuit are selection of special germanium transistors in a cascode amplifier circuit, edge-wound resistors, tantalum capacitors, and *extreme* care in circuit layout. Gain per stage is the last consideration.

16.10   **SOME FUTURE TRENDS**

A movement toward larger satellites depends on practical economics. Using a shuttle-like transport, a satellite capable of four times the current payload limit is within reach. Each satellite uses only one control system, which leaves room for a much greater electronics package. Units with 3000 channels are already in the design stage, and far-sighted plans call for manned space stations with hundreds of thousands of channels each. The use of larger satellites means more space allowance between satellites to reduce possible crowded conditions in the Clarke orbit and minimize interference between signals.

Larger vehicles also mean greater costs. With the current life expectancy of 10 years, the high cost is unrealistic. Repair satellites are planned that will perform such functions as replacing batteries and solar cells as well as refueling the rocket thrusters. In this way, the life span of a satellite can be extended manifold. The ultimate goal is to reduce the cost per channel per year to only a fraction of what it is today, even though the initial cost of launching the vehicle will be higher.

QUESTIONS

1. What percentage of the earth's surface is visible in terms of direct line of sight for a three-satellite Clarke orbit communications link?
2. What areas of the globe are not reached by the three-satellite Clarke orbit communications link?
3. Name four conditions that a vehicle must meet to remain in orbit.
4. Name the fifth orbital condition, which describes the Clarke orbit.
5. What is the altitude of the Clarke orbit?
6. Originally, what was the maximum number of satellites that the Clarke orbit could contain?
7. Under the 4° separation plan, how far apart (in miles) were satellites spaced in the Clarke Orbit?
8. In the 6/4 GHz frequency allocation, which number represents the up-link frequency?
9. Explain why the up-link (or down-link) frequency was chosen, as compared to the other.
10. What is the realistic maximum effective radiated power (in dBW) permitted for a satellite by international agreement?
11. Define the term *footprint*.
12. What is the maximum earth surface area (in percent) covered by a footprint from a single satellite in the Clarke orbit?
13. What is the minimum distance (in miles and in kilometers) from the earth's surface to the Clarke orbit for a transmission that includes one cross-link path?
14. What is the total delay time for the transmission in the preceding question?
15. What are the minimum and maximum carrier frequencies assigned in the C band satellite frequency range?

16. How are adjacent channels protected against interference in C band satellite transmissions?

17. What is the up-link frequency and polarization of channel 6X for the Westar 4 satellite?

18. How many vertically polarized transponder signals can come from the Spacenet 1 satellite?

19. What electrical operation(s) is(are) performed by the satellite transponder?

20. What kind of a detector is used aboard the Telstar 302 satellite transponder?

21. In a 12-channel multitransponder satellite, what is the frequency of the local oscillator for the lowest C band channel? For the highest C band channel?

22. Where are the address data originated for a TDMA satellite transmission?

23. For how many hours out of each orbit is a Clarke orbit satellite in full view of the sun?

24. How much energy is generated (in watts) by a typical solar panel 4 square meters in size?

25. What are the factors considered to limit the useful life span of pre-1990 satellites?

26. What is the function of the "station keeping" section of the satellite?

27. What is the bandwidth of one satellite teleconference up-link transmission?

28. What form of modulation is used to transmit teleconference signals via satellites?

29. What defines 100% modulation of the satellite carrier by a composite video signal?

30. What percentage of the total modulating voltage is supplied by the video signal used to modulate the satellite carrier?

31. How is the aural signal (associated with a telecast) processed before it is modulated onto the satellite carrier?

32. Video signals to and from the satellites use (a) full-carrier amplitude modulation, (b) wideband frequency modulation, (c) double-sideband suppressed-carrier amplitude modulation, (d) narrowband frequency modulation.

33. The aural signal of a satellite teleconference transmission is double-frequency-modulated. (T) (F)

34. The distance between a Clarke orbit satellite and all earth stations is the same. (T) (F)

35. What is the power at the receiving antenna (in dBW) when the satellite ERP is +25 dBW transmitted at 12 GHz over a distance of 23,500 miles?

36. What is the power at the receiving antenna from the preceding question in dBm?

37. What is the total distance in miles between a station at 55° north latitude to a Clarke orbit satellite when both have the same longitude?

38. What is the additional absorption loss (in dB) for the signal in the preceding question?

39. How much gain must the receiving antenna (and LNA) furnish when the signal to the receiving antenna location is −188 dBW and the receiver sensitivity is 0.25 fW?

APPENDIX

FM VECTOR ANALYSIS

From Equation 5.3, for the FM wave, and the resulting Bessel functions, it can be observed that the vector of the first sideband pair is in phase quadrature (lagging) with respect to the carrier vector. Each succeeding sideband pair is in phase quadrature with the sideband pair preceding it. Therefore, we can use the vector summation process to verify that the modulation index for FM is a true indicator of the radian phase relationship between the modulated signal to be transmitted and the unmodulated carrier signal alone.

Each of the sketches in Figure A.1 represents a carrier, having a value of J_0, and the first sideband pair, with a value of $2(J_1)$, shown as a lagging vector of 90°. The two parts of Figure A.1 are different ways to show the sum of vectors and produce the same resultant, R.

Each additional sideband pair is represented by a progressively lagging sideband vector. Furthermore, a negative value represents a phase alteration opposite to the lagging phasor, that is, a *leading* phase shift of 90°. The sum of the vectors is then compared to the vector of the unmodulated carrier phase at 0.0°.

The scalar diagram in Figure A.2 can be used as a guide if, for this graph only, we disregard amplitude and consider only polarity. The diagram depicts only positive values of the amplitude modulas given in the Bessel function solution chart.

FIGURE A.1
Two ways to draw vector diagram summations.

FIGURE A.2
Vector polarity diagram.

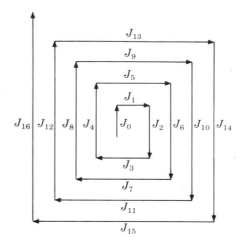

Figure A.3 is a reproduction of Figure 5.18 for $m_f = 1$ rad, where

$$J_0 = 0.7652$$
$$J_1 = 0.4400$$
$$J_2 = 0.1150$$
$$J_3 = 0.0195$$

All J terms are positive values; therefore, all sidebands are as indicated in Figure A.2

Negative values from the amplitude modulas chart can be treated using a modulation index of $10 = 10$ rad $= 573°$ (one full revolution plus 213°). The J terms and graph are shown in Figure A.4, where $J_0 = -0.2454$. Since the positive J_0 term in Figure A.2 points up, a negative J_0 term must point down. J_1, J_2, and J_3 are all positive terms; therefore, they point as indicated in Figure A.2. The J_4 term is another negative value, so since Figure A.2 shows a positive J_4 term pointing up, a negative J_4 term must point down. A positive J_5 points

FIGURE A.3
The resulting carrier vector (57.3°) when the modulation index, m_f, is 1 rad.

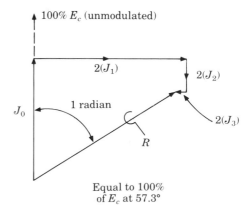

FIGURE A.4

The resultant vector (360° + 213° = 573°) for a modulation index of 10 rad (10 × 57.3° = 573°).

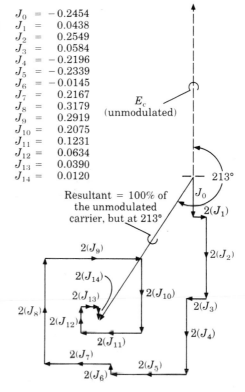

$$J_0 = -0.2454$$
$$J_1 = 0.0438$$
$$J_2 = 0.2549$$
$$J_3 = 0.0584$$
$$J_4 = -0.2196$$
$$J_5 = -0.2339$$
$$J_6 = -0.0145$$
$$J_7 = 0.2167$$
$$J_8 = 0.3179$$
$$J_9 = 0.2919$$
$$J_{10} = 0.2075$$
$$J_{11} = 0.1231$$
$$J_{12} = 0.0634$$
$$J_{13} = 0.0390$$
$$J_{14} = 0.0120$$

to the right; therefore, a negative J_5 term must point to the left. J_6 is also negative and must point in the opposite direction from that indicated in Figure A.2 (that is, negative J_6 points up). The balance of the J terms are all positive values; therefore, they all follow the same directions as in Figure A.2. Remember that each J term (except J_0) must be multiplied by 2 to include the upper and lower sidebands.

The final sideband vector, in this case J_{14}, terminates on the circle representing 100% of E_c and is at an angle of 213° from the unmodulated carrier vector used as the reference. In reality, this modulated carrier has completed one full revolution and has continued on for an additional 213°, for a total of 573° = 10 rad.

INDEX

WE VALUE YOUR OPINION—PLEASE SHARE IT WITH US

Merrill Publishing and our authors are most interested in your reactions to this textbook. Did it serve you well in the course? If it did, what aspects of the text were most helpful? If not, what didn't you like about it? Your comments will help us to write and develop better textbooks. We value your opinions and thank you for your help.

Text Title _____ Edition _____

Author(s) _____

Your Name (optional) _____

Address _____

City _____ State _____ Zip _____

School _____

Course Title _____

Instructor's Name _____

Your Major _____

Your Class Rank _____ Freshman _____ Sophomore _____ Junior _____ Senior

_____ Graduate Student

Were you required to take this course? _____ Required _____ Elective

Length of Course? _____ Quarter _____ Semester

1. Overall, how does this text compare to other texts you've used?

_____ Superior _____ Better Than Most _____ Average _____ Poor

2. Please rate the text in the following areas:

	Superior	Better Than Most	Average	Poor
Author's Writing Style	_____	_____	_____	_____
Readability	_____	_____	_____	_____
Organization	_____	_____	_____	_____
Accuracy	_____	_____	_____	_____
Layout and Design	_____	_____	_____	_____
Illustrations/Photos/Tables	_____	_____	_____	_____
Examples	_____	_____	_____	_____
Problems/Exercises	_____	_____	_____	_____
Topic Selection	_____	_____	_____	_____
Currentness of Coverage	_____	_____	_____	_____
Explanation of Difficult Concepts	_____	_____	_____	_____
Match-up with Course Coverage	_____	_____	_____	_____
Applications to Real Life	_____	_____	_____	_____

3. Circle those chapters you especially liked:
   1   2   3   4   5   6   7   8   9   10   11   12   13   14   15   16   17   18   19   20
   What was your favorite chapter? _____
   Comments:

4. Circle those chapters you liked least:
   1   2   3   4   5   6   7   8   9   10   11   12   13   14   15   16   17   18   19   20
   What was your least favorite chapter? _____
   Comments:

5. List any chapters your instructor did not assign. _____

6. What topics did your instructor discuss that were not covered in the text?_____

   _____

7. Were you required to buy this book? _____ Yes   _____ No

   Did you buy this book new or used? _____ New   _____ Used

   If used, how much did you pay? _____

   Do you plan to keep or sell this book? _____ Keep   _____ Sell

   If you plan to sell the book, how much do you expect to receive? _____

   Should the instructor continue to assign this book? _____ Yes   _____ No

8. Please list any other learning materials you purchased to help you in this course (e.g., study guide, lab manual).

   _____

9. What did you like most about this text? _____

   _____

10. What did you like least about this text? _____

   _____

11. General comments:

   May we quote you in our advertising? _____ Yes   _____ No

   Please mail to: Boyd Lane
                     College Division, Research Department
                     Box 508
                     1300 Alum Creek Drive
                     Columbus, Ohio 43216

   Thank you!